Barcade in Back

Tourism and Sustainability

'If unequal opportunities are large within many countries they are truly staggering on a global scale'. So concludes the World Bank's 2006 *World Development Report*. It is a global unevenness within which the barriers to immigration of Third World migrants to wealthy First World nations go ever higher, while the barriers to travel in the reverse direction are all but extinct. So how exactly can tourism contribute to narrowing this glaring inequality and gap between the rich and the poor? Are ever-expanding tourism markets – and the new, responsible, forms of tourism in particular – a smoke-free, socio-culturally sensitive form of human industrialisation? Is alternative tourism really a credible lever for lifting poverty stricken countries out of the mire of global inequality, setting them on the right track to 'development', and making poverty history?

Tourism and Sustainability critically explores and challenges what have emerged as the most significant universal geopolitical norms of the last half century – development, globalisation and sustainability – and through the lens of new forms of tourism demonstrates how we can better understand and get to grips with the rapidly changing new global order.

This third edition has been extensively updated and includes new material on:

- Poverty reduction, livelihoods and pro-poor tourism
- New forms of tourism in cities
- Continuing growth of the fair trade movement
- Tourism's contribution to climate change
- Volunteer and 'gap' tourism
- The effect of disasters on new tourism

Drawing on a range of examples from across the Third World, *Tourism and Sustainability* illustrates the social, economic and environmental conditions for the growth of new tourism. The book is original in its assessment of tourism through the lens of power – who holds it; how it is used; and who benefits from the exercise of power in the tourism industry. Additionally, the analysis is an interdisciplinary one and the book will therefore be useful to students of Human Geography, Environmental Sciences and Studies, Politics, Development Studies, Anthropology and Business Studies as well as Tourism itself.

Martin Mowforth is a freelance researcher and Research Fellow in Human Geography at the University of Plymouth, where his work focuses on issues of environment, development, sustainability, natural disasters and tourism. He has been and still is an occasional development worker in the region of Central America, this work involving issues of education, environment, the application of sustainability indicators, tourism, disaster recovery, poverty reduction and general community welfare.

Ian Munt is a freelance human settlements specialist and has worked on projects with UN agencies, bilateral donors and non-governmental organisations in Central America, Africa, Asia and the Pacific, and Europe.

On previous editions

'This book should be compulsory reading for all those engaged in tourism research.'
Erlet Cater, *In Focus*, Tourism Concern

'. . . one of the most significant books produced on tourism in the past few years.'
Geoffrey Wall, *Annals of Tourism Research*

'A valuable and overdue contribution to a multi-disciplinary area. This book meets the challenge to say something clear and interesting in a quicksand of ambiguities.'
Professor John Lea, University of Sydney

'Informative, stimulating, and provocative, the book deserves to be read by a wide audience . . . It is absolutely essential reading for all those serious scholars of tourism studies wishing to appreciate "the bigger picture".' Brian Wheeller, *Annals of Tourism Research*

'. . . the book is quite simply one of the most important theoretical contributions to the growing subdiscipline of tourism geography and is likely to be a mainstay for many years to come.' Keith Debbage, *Annals of the Association of American Geographers*

'. . . a far-reaching, timely and quite penetrating critique of some of the forms of tourism that have emerged as a direct response to the clarion call for sustainable tourism development.' Michael Parnwell, *Journal of Development Studies*

Tourism and Sustainability

Development globalisation and new tourism in the Third World

Third edition

Martin Mowforth and Ian Munt

Routledge
Taylor & Francis Group
LONDON AND NEW YORK

First edition published 1998 by Routledge
Second edition published 2003 by Routledge
Third edition published 2009 by Routledge
2 Park Square, Milton Park, Abingdon, Oxon, OX14 4RN

Simultaneously published in the USA and Canada
by Routledge
270 Madison Ave, New York NY 10016

Routledge is an imprint of the Taylor & Francis Group

Transferred to Digital Printing 2009

Typeset in Times and Franklin Gothic by
Keystroke, 28 High Street, Tettenhall, Wolverhampton

British Library Cataloguing in Publication Data
A catalogue record for this book is available from the British Library

Library of Congress Cataloging in Publication Data
Mowforth, Martin.
   Tourism and sustainability: development, globalisation and new tourism
in the third world/Martin Mowforth and Ian Munt. — 3rd ed.
      p. cm.
   Includes bibliographical references and index.
   1. Tourism —Developing countries.   2. Sustainable development—
Developing countries.   I. Munt, Ian.   II. Title.
   G155.D44M69 2008
   338.4′791091734—dc22   2008010913

ISBN 10: 0–415–41402–4 (hbk)
ISBN 10: 0–415–41403–2 (pbk)
ISBN 10: 0–203–89105–8 (ebk)

ISBN 13: 978–0–415–41402–9 (hbk)
ISBN 13: 978–0–415–41403–6 (pbk)
ISBN 13: 978–0–203–89105–6 (ebk)

For Matthew and Joseph

Contents

Foreword to third edition

It is now well over fifteen years since we started talking about and writing the first edition of this book, first published in 1998. New editions provide an opportunity to take a hard and critical look at what we have got wrong and what we believe still holds good. We have been lucky enough to receive the feedback of colleagues and students in many different countries, some supportive, others challenging us where they think we are off target. We originally wrote *Tourism and Sustainability* in part as a critical reaction to what appeared to be at the time a rather uncritical approach to the expansion of tourism (and especially new forms of tourism). On its tenth birthday we have retained much of the original argument for this third edition. Of course, some of our original categorisations may have been offered as a challenge rather than as exhaustively and scientifically researched typologies: hunches designed to provoke critical discussion rather than prescriptive conclusions, and our pessimistic outlook may remain overcooked. And yet, the progress in promoting more locally rooted, more equitable, and environmentally responsive forms of tourism in the past fifteen years has been painfully slow. Understanding why this is so, and overcoming obstacles to change, remains a fundamental challenge.

Some of the issues that we covered, at least briefly, in the first edition but that were relatively 'silent' in the public realm are now right at the top of the global political agenda. Climate change and the need to end poverty, for example, are now daily topics of media discussion and central targets for the United Nations and member governments. The UN's Millennium Development Goals coupled with global campaigns against poverty have imprinted this global challenge in the consciousness of many, and have demanded a response of agencies such as the World Tourism Organisation.

At a time when the 200th anniversary of the abolition of the slave trade has been marked, some have drawn the parallels between these monumental challenges and demanded that it is this generation's duty to step up to the mark and 'Make Poverty History'. Yet in a world where the jewels of new tourism from the Maldives to Kiribati and Tuvalu in the Pacific are in danger of literally disappearing beneath the waves, and where global inequality is rapidly accelerating (not slowing down or reversing), it appears clear that these will be the burning issues for our and subsequent generations. Unless western consumption practices are reined back, and their rapid global spread (through, for example, the breathtaking industrialisation of China and India) is not tempered, the implications will be potentially disastrous. Tourism needs to take its temperature within this broader global context and judge its part in averting or encouraging the onset of crisis.

Much has changed since both the first and second editions of *Tourism and Sustainability*. The growing identification, at least in the western world, of the G8 countries as the single imperial power and the spread of the consumerist lifestyles associated with their political and economic platforms have given rise to a number of counterbalancing movements. Prime among these has been the anti-globalisation movement that makes its presence felt at each meeting of the G8 powers. Also of potentially considerable significance as a counterbalance to the G8's power is the so-called 'Pink Tide' of Latin America – the rise of governments that question and challenge the wisdom and appropriateness of the Washington Consensus – and the alternative G20 powers led by the IBSA axis – India, Brazil and South Africa. Additionally, the so-called war on terror, resulting from the 9/11 attacks on the USA, has brought about major changes in our patterns of behaviour and movement. All of these events, movements and trends have had direct and indirect effects on the development of tourism.

The December 2004 Indian Ocean tsunami in which an estimated 230,000 people tragically lost their lives seemed to stand for both the fragility and resilience of tourism. The rebuilding and rehabilitation that followed the tsunami provided a reminder of the potential advantages of locally generated and led development. Equally, however, the ugly lure of tourist dollars led to land grabbing in some countries (especially the coastline of Thailand) in an attempt to finally deny the poor the right to return to their land. The tsunami brought the battle for land into the open in the words of the Asian Coalition of Housing Rights 'like never before'.

In this edition we have continued to build a more wide-ranging discussion on development, a fundamental requisite for understanding and assessing new forms of Third World tourism. This ranges from a presentation of development theory, to a critical discussion of tourism's fair trade and pro-poor development potential. In particular, the promotion of people-centred approaches to development has continued to find resonance in pro-poor, community-centred, tourism initiatives and as a counterbalance to the top-down and trickle-down approaches to tourism master planning in Third World destinations. We have therefore provided a separate chapter focused on pro-poor tourism and a further chapter looking at the under-researched practice and potential of pro-poor tourism in cities. But it is an approach that remains at the margins, more a form of analysis rather than a form of practice. In the years ahead it will be important to guard against development spin and the liberal use of 'pro-poor' as a development prefix in the same way that 'eco-' became a prefix in the tourism industry.

The core of our argument, however, remains unchanged in that development is an inherently unequal and uneven process, symbolised arguably by the diasporic and increasingly thwarted movements of Third World migrants to the First World, starkly contrasted to the accelerating movements of relatively wealthy western tourists to the Third World and the ideology of freedom of movement that supports this.

The question with which we embarked in 1993 also remains unchanged: can new forms of tourism become a significant force in global development? In the context of increasing global inequality and poverty, the overall size of the tourism industry (a point that tourism advocates invoke to explain why it should be a major force for development) and the advances made, our conclusions remain cautious. Indeed, as we write, the United Nations' annual survey confirms that progress on reducing global poverty has slowed depressingly to 'snail's pace'. Without starting from an assessment of the structures and relationships of power that influence the fate of tourism, it is, perhaps, a little too easy to be seduced by the possibilities inferred from what remain relatively few examples of positive change. To re-emphasise, however, this is not a 'do-nothing' manifesto. Far from it, we hope. Never has practical action been of such importance (regardless of the scale). Rather, it remains a contribution to understanding the wider global context within which tourism operates and from which responses must be forged.

Martin Mowforth and Ian Munt
February 2008

Acknowledgements

We should like to acknowledge the work of and extend our gratitude to all those already acknowledged in the first and second editions of the book for allowing us to draw on their material. For this third edition, we specifically acknowledge permission granted by the following persons and organisations to use extracts from their work: Sage Publications Ltd for Table 2.4; Survival International for material in Boxes 3.1, 6.2, 6.3 and 8.12, and for the data used in Figure 6.2; Oren Ginzburg for use of a cartoon image from his book 'There You Go!' (Figure 8.1); J. & C. Voyageurs for the photograph in Box 3.5; Routledge for the material in Peter Murphy's book given in Table 4.1; Mick Kidd and Chris Garratt of Biff Products for Figures 4.1 and 10.7; the *Guardian* for extracts of an article by Alex Hamilton (1995) (Box 8.9) and Pass Notes 223 (1993) (Box 4.2); the *Independent* for permission to use extracts from an article by S. Boggan and F. Williams in Box 6.4; Tirso Maldonado and the Fundación Neotrópica (San José, Costa Rica) for extracts from their work on carrying capacity calculations (Box 4.6); Paul Fitzgerald (the cartoonist Polyp) for Figure 5.2; the Open University for permission to adapt the Environmental Continuum in Table 6.1; the World Conservation Monitoring Centre for the Table of Protected Area Categories (Table 6.2) and data on the global growth of protected areas (Figure 6.3); Prof. Jules Pretty for his typology of participation (Table 8.1); the Panos Institute for material from its Media Briefing no. 14 (1995) for our Box 5.1; the Overseas Development Institute for Tables 11.3 and 11.4; Tourism Concern for permission to reproduce 'A letter from Belize' (Box 8.6); Earthscan for permission to reproduce Figure 9.1 using data from UN-Habitat; and Reuters for permission to use extracts from an article by Andrew Cawthorne on slum tourism (Box 9.3).

We owe much to the work of the following organisations which have produced reports of special relevance and significance for this field of study: the Catholic Institute for International Relations (CIIR, now called Progressio), London; the International Institute for Environment and Development (IIED), London, especially the work of Jules Pretty on local participation; the Panos Institute, London; Survival International, London, for its reports on tourism and tribal peoples; the *New Internationalist*, especially for its work on BINGOs – Big International NGOs – but also for many other extracts from its articles; and the Overseas Development Institute, especially in the persons of Jonathan Mitchell and Caroline Ashley, for their excellent work on pro-poor tourism. Two organisations deserve special mention: the TIM-Team (Tourism Investigation and Monitoring Team) in Bangkok, especially in the person of Anita Pleumarom, who makes available a constant stream of stimulating material on tourism and development; and Tourism Concern, London, without whose reports, assistance and advice our analysis would be considerably weaker.

The work of the Cartography Unit of the Department of Geographical Sciences at the University of Plymouth deserves special mention for its professionalism and high quality and for the patience with which the Unit's staff, notably Brian Rogers, Tim Absalom and especially Jamie Quinn, accepted the regular changes we required to the maps, graphs, tables and drawings they created for us. We also wish to express our gratitude to three former students from the University of Plymouth: in summer 2007 on our behalf Elizabeth Elliott and Matthew Lane carried out a long and detailed study of the tourism-related activities and policies of a wide range of governmental and supranational agencies, the results of which we have used in Chapter 10, and for which we are extremely grateful to them; Emma Painter was generous with her time in assisting with the final preparation of the manuscript and we are very grateful to her for it.

Many thanks go to Alison Stancliffe, who supplied the material for the case studies of Bali and served as our internal reviewer and extremely helpful critic for the first edition. Thanks also go to the invisible external reviewers of the third edition proposal, whose insightful and encouraging comments made our revisions more purposeful, and to all those who took the time to review the first and second editions and whose praise and criticisms have helped us formulate this third edition. Our editors at Routledge, Jenny Page, Michael P. Jones and Andrew Mould, have constantly urged us onwards with the work for this edition, their encouragement growing ever more urgent as we slipped almost effortlessly past the first two deadlines.

June and Diana deserve more than gratitude for their patience and encouragement.

Figures

⬤ Tables

Boxes

Abbreviations

ACHR	Asian Coalition of Housing Rights
ACPA	Ambergris Caye Planning Authority (Belize)
ADB	Asian Development Bank
AITO	Association of Independent Tour Operators
ANCON	National Association for the Conservation of Nature (Panama)
AONB	Area of outstanding natural beauty
ASEAN	Association of South East Asian Nations
ASTA	American Society of Travel Agents
BA	British Airways
BAG	Burma Action Group
BAH	British Airways Holidays
BASD	Business Action for Sustainable Development
BCES	Belize Centre for Environmental Studies
BCSD	Business Council on Sustainable Development
BEST	Belize Enterprise for Sustained Technology
BINGO	Big International Non-governmental Organisation
BTIA	Belize Tourism Industry Association
CAMPFIRE	Communal Areas Management Programme For Indigenous Resources
CANATUR	Costa Rican Chamber of Tourism
CARICOM	Caribbean Community and Common Market
CARTA	Campaign for Real Travel Agents
CCC	Caribbean Conservation Corporation
CEDLA	Centre for Latin American Research and Documentation (Netherlands)
CERT	Campaign for Environmentally Responsible Tourism
CIIR	Catholic Institute for International Relations (now called Progressio)
COBA	Cost–benefit analysis
CONAP	National Council of Protected Areas (Guatemala)
CSR	Corporate social responsibility
Danida	Danish Agency for Development Assistance
DEFRA	Department for Environment, Food and Rural Affairs (UK)
DFID	Department for International Development
DTI	Department of Trade and Industry
EBRD	European Bank for Reconstruction and Development
ECC	Effective carrying capacity
ECLAC	Economic Commission for Latin America and the Caribbean
ECPAT	End Child Prostitution, Pornography and Trafficking
ECTWT	Ecumenical Coalition on Third World Tourism
EIA	Environmental impact assessment/analysis
EPZ	Export Processing Zone

ESCAP	Economic and Social Commission for Asia and the Pacific
EU	European Union
FAO	(United Nations) Food and Agriculture Organisation
FIDE	Foundation for Investment and Development of Exports (Honduras)
FOE	Friends of the Earth
FOEI	Friends of the Earth International
FTSE	Financial Times Stock Exchange
G8	Group of eight most industrialised nations
G20	Group of 20 middle income nations led by Brazil, South Africa and India and set up as an alternative force to the most developed G8 nations
GAGM	Global Anti-Golf Movement
GATS	General Agreement on Trade in Services
GATT	General Agreement on Tariffs and Trade
GDP	Gross domestic product
GEF	Global Environment Facility
GFI	Green Flag International
GIS	Geographic information system
GMS	Greater Mekong Subregion
GNP	Gross national product
HDI	Human Development Index
HPI	Human Poverty Index
IBLF	International Business Leaders Forum
IDB	Inter-American Development Bank
IFI	International financial institution
IGCP	International Gorilla Conservation Programme
IHEI	International Hotels Environment Initiative
IHT	Instituto Hondureño de Turismo/Honduran Institute of Tourism
IIED	International Institute for Environment and Development
ILO	International Labour Organisation
IMF	International Monetary Fund
INGO	International non-governmental organisation
INGUAT	Guatemalan Institute of Tourism
IPPG	International Porter Protection Group
IUCN	International Union for the Conservation of Nature and Natural Resources (often referred to as the World Conservation Union)
IYE	International Year of Ecotourism
JICA	Japan International Cooperation Agency
KWS	Kenya Wildlife Service
LAC	Limit of acceptable change
MDG	Millennium Development Goal
MGP	Mountain Gorilla Project
MNC	Multinational corporation
MRDF	Methodist Relief and Development Fund
NACLA	North American Congress on Latin America
NAFTA	North America Free Trade Agreement
NCA	Ngorongoro Conservation Area (Tanzania)
NEDA	Netherlands Development Assistance
NGO	Non-governmental organisation
NLD	National League for Democracy (Burma)
ODI	Overseas Development Institute
OECD	Organisation for Economic Cooperation and Development

OPEC	Organisation of Petroleum Exporting Countries
OPIC	Overseas Private Investment Corporation
PAR	Participatory action research
PRA	Participatory rural appraisal
PRG	People's Revolutionary Government (Cuba)
PRM	Participatory research methodology
PRSP	Poverty reduction strategy paper
PUP	People's United Party (Belize)
RA	Rapid appraisal
RAP	Rapid assessment procedures
RAT	Rapid assessment techniques
RCC	Real carrying capacity
REA	Rapid ethnographic assessment
RRA	Rapid rural appraisal
SAP	Structural adjustment programme
SELVA	Somos Ecologistas en Lucha por la Vida y el Ambiente/We are Ecologists struggling for Life and the Environment (Nicaragua)
Sida	Swedish International Development Cooperation Agency
SJAS	San José Audubon Society (Costa Rica)
SLORC	State Law and Order Council (Burma)
SPDC	State Peace and Development Council (Burma)
SSSI	Site of special scientific interest
ST–EP	Sustainable Tourism – Eliminating Poverty
TEA	Toledo Ecotourism Association (Belize)
TEN	Third World Tourism European Network
TIES	The International Ecotourism Society
TIM-Team	Tourism Investigation and Monitoring Team (Thailand)
TNC	Transnational corporation
TOI	Tour Operators Initiative
UDP	United Democratic Party (Belize)
UNCED	United Nations Conference on Environment and Development
UNCHS	United Nations Centre for Human Settlements
UNCTC	United Nations Centre on Transnational Corporations
UNDP	United Nations Development Programme
UNEP	United Nations Environment Programme
UNESCAP	United Nations Economic and Social Commission for Asia and the Pacific
UNESCO	United Nations Educational, Scientific and Cultural Organisation
UNFPA	United Nations Population Fund
UNHCR	United Nations High Commissioner for Human Rights
UNWTO	World Tourism Organisation
USAID	United States Agency for International Development
VSO	Voluntary Service Overseas
WCMC	World Conservation Monitoring Centre
WDM	World Development Movement
WLT	World Land Trust
WTO	World Trade Organisation (Please note: due to possible confusion with the initials WTO, signifying both the World Trade Organisation and the World Tourism Organisation, the latter is now generally referred to as the UNWTO. We follow this practice.)
WSSD	World Summit for Sustainable Development

WTTC	World Travel and Tourism Council
WTTERC	World Travel and Tourism Environment Research Centre
WWC	Ward Wildlife Committee (Zimbabwe)
WWF	World Wide Fund for Nature

1 Introduction

In recent years the image of the Third World in western minds has emerged in part from that of cataclysmic crisis – of famine and starvation, deprivation and war – to represent the opportunity for an exciting 'new style' holiday. Offering the attraction of environmental beauty and ecological and cultural diversity, travel to many Third World countries has been promoted, especially among the middle classes, as an opportunity for adventurous, 'off-the-beaten-track' holidays and as a means of preserving fragile, exotic and threatened landscapes and providing a culturally enhancing encounter. At the same time, some Third World governments have seized upon this new-found interest and have promoted tourism as an opportunity to earn much-needed foreign exchange – another attempt to break from the confines of 'under-development'.

There is something contradictory in viewing the Third World through an analysis of tourism. At one extreme there is the association of the Third World with fundamentalist terrorism, a risk to the security of the west, or overpopulation, poverty and disease. At the other there are also more subtle manipulations by the tourism industry of friendly and culturally diverse Third World peoples, natural and pristine environments and ecological variety. Both extremes, we will argue, form an important element in understanding contemporary Third World tourism. Importantly, and central to the line of argument developed in the following chapters, tourism is a way of representing the world to ourselves and to others. It cannot be understood as just a means of having some enjoyment and a break from the routine of every day, an entirely innocent affair with some unfortunate incidental impacts. Rather, a deeper understanding of tourism is needed to appreciate fully its content and expression as well as its potential impact.

Primarily, this is a book about development and its reflection in tourism; tourism is a metaphorical lens that helps bring aspects of development into sharper focus. It is not just about the role and impact of tourism in Third World development, but also about the roles of First World people and organisations (operators, tourists, non-governmental organisations, etc.) in the manufacture of 'development' as an idea, project or end-state.

Purpose and limits of the book

This book focuses on new and purportedly sustainable forms of tourism to Third World destinations in the context of a world undergoing accelerated processes of globalisation. The focus is relatively tight in two important respects. First, there are now many studies of tourism, especially tourism in the Third World, that catalogue and discuss its growth and impacts. In particular, studies have tended to highlight the economic, environmental and socio-cultural impacts of conventional package tourism.[1] Rather than adding to this body of work, we focus particularly on the under-researched 'new' forms of tourism promoted in the First World and patronised mainly by North Americans, Europeans and

Australasians (and with increasing participation of the middle classes in some Third World and so-called middle income countries). Although proportionately small relative to all forms of Third World tourism, the new forms of tourism are significant in terms of both the claims that are made about them and the rate at which they are growing.

Second, we do not seek to add to the growing number of accounts which attempt to identify sustainable tourism development (in terms of the environment, economy and culture) or prescribe good practice methods and tools for achieving this goal. Many of these have emerged from First World sources and necessarily involve judging the sustainability or otherwise of varying types of Third World tourism development. Instead, we explore the ways in which the claim, or discourse, of sustainability is used and applied to new forms of tourism (for example, the way in which ecotourism – a new form of tourism – is premised upon the notion of sustainability). One way of capturing this important difference in our approach is to state that this is a book about sustainability and Third World tourism, rather than sustainable Third World tourism. The former approach, adopted here, signals the need for a critical analysis of the issues, while the latter implies the need to define and prescribe models of good practice.

Tourism as a multidisciplinary subject

Rather than commencing a study of Third World tourism with the environmental, economic and socio-cultural impacts of tourism (worthy though these are as research considerations in themselves), the starting point here involves seeking to understand how socio-cultural, economic and political processes operate on and through tourism. In other words, it is necessary to take a step back in the analysis of tourism. This stems in part from the weaknesses present in the tourism literature. As Britton (1991) observes:

> Although over-simplifying, we could characterise the 'geography of tourism' as being primarily concerned with: the description of travel flows; microscale spatial structure and land use of tourist places and facilities; economic, social, cultural, and environmental impacts of tourist activity; impacts of tourism in third world countries; geographic patterns of recreation and leisure pastimes; and the planning implications of all these topics . . . These are vital elements of the study of travel and tourism. But these sections are dealt within descriptive and weakly theorised ways.
>
> (Britton 1991: 451)

This problem is of fundamental importance as it has led to an absence of an adequate theoretical critique for understanding the dynamics of tourism and the social activities it involves.

There are, therefore, two identifiable groups of tourism research. The first is concerned primarily with auditing, categorising, listing and grouping the outputs or consequences of tourism; the second approach is concerned primarily with conceptualising the forces which impact on tourism and, through an analysis of these forces, providing a broader context for understanding tourism. The crucial difference in the latter approach is that tourism is seen as a focal lens through which broader considerations can be taken into account, and it confirms the multidisciplinary foundation upon which tourism research is built; and it is in our opinion the only way in which tourism can be comprehended. As a personal activity, tourism is practised by a diverse range of the population; as an industry, it is multisectoral; and as a means of economic and cultural exchange, it has many facets and forms. Any comprehensive analysis of the field must therefore be multidisciplinary; and of necessity a study of tourism must be a net importer of ideas, themes and concepts from the broader social sciences.

Accordingly, our discussion draws on economics, development theory, environmental theory, social theory, politics, geography and international relations, for example. Inevitably, this breadth of consideration will mean that a number of relevant aspects are not examined in depth, and do not necessarily cover the complexity of the matters under discussion. We hope, however, that the discussion will serve as a stimulant to further thinking, discussion, research and study.

At the same time as using the concepts of a range of academic and intellectual fields in order better to understand tourism, the study of tourism helps us to illuminate more general economic, political, social, geographical and environmental processes. We try not to see tourism as a discrete field of study. Rather, it is an activity which helps us to understand the world.

Key themes and key words

We have attempted to draw attention to the principal components of the discussion under the banners of key themes and key words. The relationships between the analysis and the key themes and key words are summarised diagrammatically in Figure 1.1. In reality these ideas are interconnected and dialectic; it is, therefore, principally a means of trying to organise our thoughts and approach. In very broad terms, these ideas respectively are shorthand for the rapid pace of global change and interdependence, an increasing global environmental awareness and a heightened concern with global inequalities and protracted structural poverty. The point we wish to emphasise is that such is the scale, nature and depth of the processes and problems assumed within globalisation, sustainability and

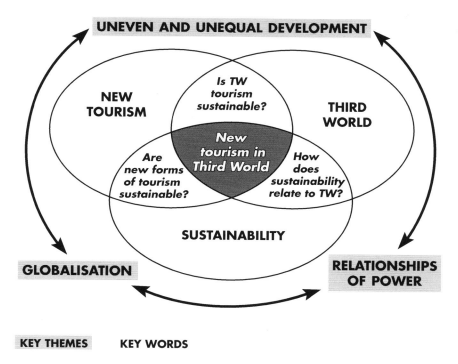

Figure 1.1 Key themes and key words

development, three of the most significant drivers and reflectors of change, that they should therefore hold the centrepiece of our analysis. Chapter 2 is dedicated to analysing these formative ideas. Noteworthy here are the increasing concern and processes of change associated with each of these notions since the 1950s and of their accelerated significance since the 1980s – so much so that it is difficult to consider one without the other: sustainable development, global sustainability, global development, and so on.

Three themes seek to underpin the discussions throughout the book. The first, and most significant, theme is the uneven and unequal development which underlines the relationship especially between the First World and Third World (but also at inter- and intra-regional and intra-national levels), and through which we argue all forms of tourism are best understood. At its most basic level this theme is reflected in the fact that, for the time being (and notwithstanding the rapid growth of tourists generated from China and India in particular), it is people from the First World who make up the significant bulk of international tourists and it is they who have the resources to make relatively expensive journeys for pleasure. Equally, processes of uneven development are reflected through the growing elite and newly wealthy classes in some Third World countries who are now able to participate in tourism. And of course tourism development is also highly uneven geographically. Areas that are in fashion today may fall from grace tomorrow.

Uneven and unequal development seeks to demonstrate why a critical understanding of tourism can be expanded through an analysis that places relationships of power, the second of the key themes, at the heart of the inquiry. These relationships range from the political, economic and military power of First World countries in contrast to Third World countries, the power wielded by international multilateral donors (from the World Bank to the European Union), the power of local elites in contrast to the local populations in tourist destination communities, and the power invested in tourists themselves. Our contention is that an analysis of tourism must start from an understanding and critical assessment of the relationships of power involved.

The final key theme is globalisation, a notion that at its core attempts to capture the idea that we are living in a shrinking world, a world in which places (countries, cities, communities, tourist resorts and so on) are increasingly interdependent.

In addition to these three themes which inform the analysis of Third World tourism, there are a number of key words which will recur throughout the book and which structure our discussion. As Figure 1.1 records, these are sustainability, tourism and the Third World. The idea of a key word is used here because it alerts us to the fact that words are open to debate and differing interpretations; and that, consequently, there are problems in attaching uncontested meanings to words. As Raymond Williams (1988) suggests, the meanings of key words are closely related to the problems they seek to discuss. Take the idea of sustainability, for example. The very many meanings and interpretations of this idea are often a direct reflection of the explicit and implicit connections that people make (Williams 1988: 15). For environmentalists, for example, the problem is one of the degradation of the world's natural resources that activities such as tourism have caused. And the meaning attached to sustainability in this case is, at its core, ecological: the need to preserve and protect the natural environment. To industrialists by contrast, sustainability may represent the opportunity to reduce costs, increase or retain market share, win new customers and increase profit margins. In the former, there is an assuredly ethical dimension associated with the sustenance of the environment for all; in the latter, a primary concern to protect and enhance shareholder interest.

In the context of this book, tourism itself is, of course, the most obvious key word. For the purposes of this discussion, the term 'new tourism' has been adopted. While this is a rather broad term, it helps to indicate that a variety of tourisms have emerged and that in some important respects these seek to distinguish themselves from what is referred to as mainstream or conventional mass tourism – the type of tourism that has attracted the most academic attention and the wrath of many commentators. It will be argued later that the 'new' in tourism also helps us to trace the relationships with new types of consumer (known as the new middle classes), new types of political movements (ranging from new socio-environmental movements to the so-called anti-globalisation movements) and new forms of economic organisation (known as post-Fordism).

The meanings attached to tourism are many and varied, and 'tourism' and 'tourist' have in some quarters become derided and ridiculed. How often are people heard to describe their holidays as being 'well away from tourists and the main tourism areas'? In other quarters, Third World tourism is regarded negatively, and the word symbolises a range of problems: environmental degradation, the distortion of national economies, the corruption of traditional cultures, and, on a trivial level for the individual, unsanitary conditions and food poisoning. The emergence of new forms of tourism (prefixed with sustainable and eco-, for example) is testimony to the identification of the problem and the attempt to signal that these new forms aim to overcome the problem and to be something that plain old 'tourism' is not.

The final key words are the geographical focus of the analysis, the so-called Third World. The history of development studies has thrown up a variety of terms that attempt to represent and categorise countries according to their wealth and social well-being. In particular, the terminology attached to countries lower down the 'human development index' (a widely adopted index ranking countries on a number of criteria) has been keenly disputed – should they be described as 'poorer', 'lower income', 'developing', 'under-developed', 'the South', 'Third World', or indeed 'non-viable economies' (de Rivero 2001) or 'slow economies' (Toffler, quoted in W. Sachs 1999)? All such terms possess their advocates and detractors, reflect political priorities and dispositions, and no one term will suit all audiences. The problem of definition is made more acute when rapidly changing economic factors mean that old categories no longer hold good, and the meaning of development is constantly contested (Nederveen Pieterse 2001).

Principally, in this book, we refer to Third World and First World as shorthand (and a generalisation) for the spatial and socio-economic inequality and unevenness between and within regions, countries and cities. Other terminology is used only where direct quotes from other sources necessitate this or where the meaning conveyed by the alternative term is distinctly different from these two terms (for example, in a differentiation between those relatively wealthy countries with significant new middle classes with a taste for new forms of tourism and those countries where the middle classes are currently demanding conventional forms of tourism, especially those in South Asia and the Pacific Rim). Occasionally, we also refer to the West and the Rest, where the 'West' represents the First World capitalist economies promoting western materialistic lifestyles, and the 'Rest' seeks to emphasise the inherent inequality in global development.

'Third World' also helps to reflect on and convey the way in which we are using this term in relationship to the notion of development. For example, the word 'developing' is avoided, because it implies that there is an end state to the process of development and that all countries will eventually reach a 'developed' state. By contrast, there are strong grounds for arguing that the process of development is one which actually causes under-development elsewhere and at the very least is a state of never-ending flux to the extent that all countries are 'developing'. The term 'Third World', in other words, helps to empha-sise the ways in which power, resources and development are unequally and unevenly shared globally – if anything, the very term 'Third World' requires an acknowledgement that despite sixty years of 'development' activity, profound global inequalities both persist and are increasing. This is not to say, however, that the Third World is easily defined as a neat geographical entity coterminous with nation states. Inequality and poverty are of course not only manifest on a global scale, but also occur within and between countries and in relation to a variety of characteristics, particularly sex, ethnicity and class. As Dodds comments: 'Put simply, there are parts of the Third World in the First World and vice versa' (Dodds 2002: 6; Nederveen Pieterse 2001). Indeed, one of the arguments promoted in the book is that analysis of the economies, societies and the tourism sector at the scale of the 'nation state' (the predominant imagination conjured by the label 'Third World') is inadequate for understanding the spatial expression of contemporary globalisation. In Chapters 5 to 11 we examine a number of examples of inequalities at various different scales.

Although strict geographical divisions along national borders are increasingly meaningless, as a generalisation in talking of the First World we are referring to the power vested in the nation states and institutions of North America, Europe, Australasia and Japan. The Third World refers to those nation states and institutions that make up Latin America, the Caribbean, Africa and parts of Asia, although it is necessary to acknowledge the increasing wealth of some countries in South Asia and the Pacific Rim, and of course, the breath-taking pace of China's industrialisation.

After the 11 September 2001 terrorist attacks on the USA, this simplistic division into two distinctly different worlds was redefined along even more simplistic lines by President George W. Bush's declaration that the world was 'either with us or with the terrorists' – with the 'free' capitalist world's economic fundamentalism or with the fundamentalist jihad. The act itself and President Bush's reaction to it drew more sharply than ever the lines dividing the First World capitalist economies and those which seek a different path of development and between the apparently 'civilised' world and the 'uncivilised' world, as it rapidly became labelled. Bush's declaration and the war that followed it have been referred to as 'the final stage of globalisation' (Montero Mejía 2001), in which a global world with a global army aggressively enforces and imposes the First World's economic agenda on the rest of the world. The events of 11 September 2001 had immediate and drastic consequences for the tourism industry, but despite subsequent related events such as the bombings in Bali and the wars in Iraq and Afghanistan, the effects have not been long-lasting. Nevertheless, any analysis of tourism should pay some attention to the global impacts on the industry of such events, and we briefly investigate the effects again in Chapter 11.

Tourism and geographical imagination

Tourism is one of the principal ways through which our 'world-views' are shaped. This results not only from our holidays but also from the way destinations are represented through travel reviews, travel programmes and documentaries, travel brochures and guides, advertising and the way in which we exchange our holiday experiences.

It is important to note at the outset that we are dealing with the nature of representation. Some geographers have adopted the term 'geographical imagination' as shorthand for these processes: the 'way we understand the geographical world, and the way in which we represent it, to ourselves and to others' (Massey 1995c: 41). It is also shorthand for emphasising that activities, issues, places and so on are subject to competing inter-pretations. There are as many ways to see, explain and interpret as there are positions to view 'reality'. Benidorm, Bangkok and the Himalayas, for example, conjure up very different sorts of representations in our minds. As Harvey (1989a: 1) suggests, the 'eye is never neutral and many a battle is fought over the "proper" way to see'.

It is necessary to acknowledge this and work through the complexities rather than taking things for granted. Much of the argument presented in this book concerns the way in which tourism is a contested activity and how the field of tourism must be as concerned with the nature of representation and interpretation as describing a reality. This involves the way in which we represent both our own activities (how we define ourselves as, for example, tourists, travellers, visitors, and what each of these categorisations entails),[2] and the places in which we holiday (for example, built-up beach resorts or remote regions).

We all have our individual geographical imaginations and these are formed from a variety of factors – sex, age, class, ethnicity, culture, the media and many others. As authors, both white males, writing about Third World tourism from the 'privilege' of our comfortable First World lifestyles, our interpretation of the issues may differ markedly from the interpretation of others who are experiencing very different circumstances. Similarly, tourists interpret and represent their experiences in ways that may be fundamentally opposed to the experience of those being visited; and these interpretations and representations will differ between different types of tourists. Even the World Bank and International Monetary Fund (IMF) have a particular geographical imagination of the Third World. Their representation of tourism and sustainability may also differ sharply from those of local communities in the countries where the policies of these supranational institutions are applied.

These differing geographical imaginations emphasise that representations of the world are socially and politically constructed and that there is an array of factors that contribute to our understanding of the world; Benidorm, Bangkok and the Himalayas are, so to speak, the social products of tourism. Take, for example, the following extract discussing the impact of tourism in the mountainous regions of South Asia. This helps to emphasise that First World views of tropical environments or mountainous regions as those of unrestrained beauty and natural bounty, to be revered and mythologised, may not be held by all. It also suggests that what is seen as worthy of tourism changes over time.

> Mountains for the local mountaineers have obtained an aura of mysticism acquired from the Western ideal ... The typical attitude of the mountaineer toward the mountains once was 'if only they were flat, I could plough them,' has changed through working and guiding Western Alpinists. This Western attitude is reinforced when the elite local porters and climbers are taken to Western countries for climbing workshops and tours. Gradually other notions that Westerners have about mountains such as conservation ... seep into the vocabulary and thoughts of the local mountain populations as interaction with the tourists becomes more frequent and intense.
>
> (Allan 1988: 14)

This example of once divergent imaginations helps to emphasise two other points. First, some individuals, companies, institutions and countries will be better able to diffuse their particular imaginations to others. When we think of Bali, Goa or Hawaii, for example, the images and representations that are called forth are less likely to be of local people struggling to maintain cultural identity in the wake of mass tourism development and more likely to be of palm-fringed beaches and crystal blue waters (often the products of travel brochures, travel reviews and holiday programmes). In short, some imaginations are more powerful than others (Allen and Massey 1995).

Second, and related to this exercise of power, differing and highly divergent geographical imaginations imply that there is a high level of contest as dominant forms attempt to impose themselves on subordinate imaginations. The core of the argument in this book is about these contested views as they are expressed through tourism, ranging from the struggles between multinational tourism corporations and environmentalists, and the divergent understandings and definitions of sustainability, to the social struggle between so-called 'travellers' and tourists.

Finally, geographical imagination also makes a very distinctive contribution to our understanding of globalisation and its impacts (Allen and Massey 1995). There is a sense that we are living in a smaller, more compressed and interconnected world, and tourism is often invoked in this process of globalisation, a process perceived differently by different people in different places. On the one hand, places are drawn into the sphere of global tourism and the feeling of a smaller world encourages consumption of further places. On the other hand, some places deemed unattractive to tourism are marginalised from the processes of global interdependence. The relationship is rather a complex and symbiotic one. Tourism is both cause and consequence within globalisation.

Layout of the book

This book aims to present many of the issues and debates associated with different aspects of new tourism as questions. To some of these questions we give our own interpretations; to others we do not respond, choosing instead to leave the issue open for you to examine further through the references provided.

Given our approach to the study of tourism and the multidisciplinary approach that we feel tourism studies demand, there is clearly a myriad of non-tourism sources that inform the debates and issues within and surrounding tourism. In addition, with a growing interest in and popularity of tourism studies, in recent years there has been a large and growing number of works written on tourism itself. Inevitably, therefore, the references provide but a sample of the breadth of writing on this subject. (It should also be noted that for some authors we have for practical reasons cited a number of works, to enable you to access the range of resources; this does not necessarily indicate that each piece of work develops a discrete and unique argument.)

It should also be noted that we have attempted to introduce useful tools and methods at various points in the book. These techniques are regularly used by development agencies and their partners, consultants and researchers, and help provide practical measures to complement more critical approaches to the field of tourism and development. Examples vary from a listing of the tools of sustainability and how carrying capacity is calculated in Chapter 4, to the application of 'problem tree' analysis (in this case to pro-poor tourism development in the Greater Mekong Subregion) in Chapter 11. More broadly, some boxed readings such as the reporting of slum tourism in Kenya (Chapter 9) also help to demonstrate the importance and effectiveness of presenting arguments and geographical imaginations through journalistic formats.

The book is in two parts. The first, commencing with Chapter 2, attempts to critically discuss the multiple meanings and interpretations of sustainability, globalisation and development, picking up on the relationships between these words and ideas and how in a number of important ways this triumvirate is blended into one interchangeable notion. The discussion also seeks to suggest how and why new forms of tourism have emerged and how they are related to sustainability. Central to this chapter is a consideration of power and the ways in which globalisation, sustainability and development are used to convey the interests of different groups; these include international agencies, national governments, new social movements, tour operators, tourists and the local communities which receive tourists.

Chapter 3 builds upon this notion that power must lie at the heart of tourism analysis. Initially, it considers the tools for analysing and discussing power: ideology, hegemony and discourse. It looks at the existing political economy critiques of tourism and argues that the majority of these are set within the context of mass international tourism. It is suggested that a critique of these forms of tourism is still necessary and a number of characteristics of such a critique are set out.

Chapter 4 looks specifically at the relatively recent emergence of the new forms of tourism and the terminology and definitions associated with them. Principles of 'sustainable' tourism are examined critically along with the tools generally used to define, assess and measure sustainability in tourism. We briefly speculate on how the role of sustainability in tourism will develop in future.

Chapters 2 to 4 therefore provide a conceptual framework within which the remainder of the book is developed. The second part of the book extends the relationships between sustainability, development and new forms of tourism as they apply to the major players and issues: the tourists, the socio-environmental movement, the operators in the industry, the communities at the destination end, tourism developments in urban areas, the national governments and international consortia which make decisions affecting the development of tourism, and finally at pro-poor tourism, one of the most recent forms of new tourism that links poverty, inequality, development and tourism.

Chapter 5 examines the motives, the role and characteristics of those tourists (mostly from the First World) who take part in the new forms of tourism. Building upon the earlier discussion regarding the relationship between the new middle classes and new tourism, we suggest why new forms of tourism are of importance to the new middle classes of the First World and how they protect and enhance this activity.[3]

Many of the 'new' tourists are also represented by new social organisations (or what are referred to in this book as new socio-environmental organisations) which mobilise around issues such as ecology, the environment, human rights and, most recently, anti-globalisation, or more positively, as Susan George argues, 'global justice' (Bygrave 2002). Chapter 6 seeks to explore how new forms of tourism can be positioned in a discussion of environmentalism. A number of forms of environmentalism are presented and the ways in which sustainability and new tourism are related to them are assessed. The relationship between First World and Third World socio-environmental organisations represents a struggle for power and control over problem interpretation and policy. The charge that global environmental organisations often intervene in ways which can be seen as eco-colonialist is also considered.

Chapter 7 analyses the way in which the tourism industry, in all its guises and facets, absorbs and adapts itself to the notion of sustainability. Different players in the industry react in different ways, but they all attempt to define sustainability in a way that reflects their particular interest; and we look at the ways in which this process of self-adjustment and subsumption of the notion of sustainability by different sectors of the industry occurs. This chapter also places the industry within the broader framework of trade, reviews current changes in global trade policy and examines the possibilities for 'fair trade' tourism. It includes a brief investigation into some of the new features associated with the new forms of the tourism industry.

Chapter 8 defines what is implied by the term 'host' populations and examines the importance of local participation (and the notion of 'participatory development') in the activity of tourism. A spectrum of meanings of the word 'participation' exists and a number of case studies are presented and set within this continuum. The techniques for measuring and improving local participation in planning, decision-making and implementing tourism schemes are assessed. Examples of the displacement of local populations by tourism developments are given, and it is clear that area protection measures (such as the designation of national parks) are often implicated in these displacements and subsequent resettlements. The formation of local elites in exploiting the techniques and benefits of both tourism research and tourism activity is examined.

Much research and discussion in new tourism focuses on environmental and eco-logically based tourism predominantly in rural areas. Chapter 9 challenges the bias in this analysis, arguing that the phenomenal growth of cities and the urbanisation of poverty requires that new forms of pro-poor urban tourism need far greater attention. More broadly, the chapter considers why there is such a prevalent anti-urban bias in development theory and practice, and traces this broader lineage in the emergence of new tourism. Two of the more prevalent forms of urban tourism are discussed. The first is the cultural heritage tourism that is found in the Third World's architectural gems, and we consider the degree to which the harnessing of these urban resources and assets can provide benefits but also serious detrimental effects caused by creeping gentrification. The second is so-called slum tourism that exposes new tourists to the 'reality' of informal urban settlements and we assess the degree to which this benefits local communities and raises awareness of such areas or represents further evidence of the pernicious and voyeuristic nature of some forms of tourism.

Chapter 10 looks at the role of governments and supranational institutions in tourism. The importance of political analysis in tourism studies and of analysis of the range of external influences on national governments is noted. With new forms of tourism emerging, many Third World countries have identified the opportunity of developing tourism around notions of ecology and environment, and the adoption of discourses of sustainability by governments is described.

Chapter 11 addresses one of the two most pressing challenges (the other being climate change) of our contemporary world: poverty and deprivation. As consistently argued throughout the book, new tourism is promoted primarily as a means of obviating the negative impacts of conventional mass tourism and as offering a vehicle for development. The emergence of pro-poor tourism initiatives is the most pronounced movement in attempting to demonstrate how new tourism can contribute to the broader pro-poor development agenda. The chapter sets new tourism in the broader context of understanding the 'poor' and 'poverty', and the most significant global framework for addressing poverty, the United Nations Millennium Development Goals. The main part of the chapter then discusses the meaning and content of pro-poor tourism and its applications, and reviews some of the up-to-date analysis of its efficacy in addressing poverty. It is suggested that pro-poor tourism might be best considered as one form of temporary migration that can directly benefit poor communities. The chapter provides a number of detailed case studies of the fragility of tourism and the primary need for human security.

In the final chapter, we draw several general conclusions from the discussion, and indulge in some crystal ball gazing about the future development of new tourisms and their relationships with the economic, social, cultural, political and geographical aspects of development.

2 Globalisation, sustainability, development

There are three essential strands to Chapter 2 which are subsequently followed through in the rest of the book. First, it is necessary to establish the relationship between the Third World and tourism. At first sight this may seem an obvious link, but it is important to establish the dynamic nature of the relationship and frame it within broader, global changes. It is necessary to demonstrate the way in which tourism has expanded in the Third World, and how, so to speak, the frontiers of tourism have increasingly been pushed back. In order to do this we refer to the concept of globalisation (the first of our key words), a notion that attempts to capture the way in which the world has shrunk in relative terms. The analysis starts therefore from changes in the First World and we review the ways in which globalisation is reflected in economics, culture and politics.

We also need to look at the underlying factors which have resulted in some profound changes to tourism and at how new forms of tourism (in contrast to mass or mainstream tourism) have emerged in Third World destinations. This is not to argue that mass tourism to Third World countries has been magically displaced, but that alternatives to such tourism are now well established; subsequent chapters will demonstrate what forms such new tourisms take and discuss the extent to which these are manifest in the Third World. It is the claims made by many of these new tourisms to be sustainable and appropriate alternatives to mass tourism that this book seeks to explore.

The second important strand is the notion of sustainability and sustainable development. We trace the relationship of these ideas to Third World tourism. An important question posed by many observers is whether and in what ways development in Third World countries, including tourism development, is or is not sustainable. This is especially reflected in the debates over Third World tourism where it is often argued that existing forms of mass tourism development are unsustainable in terms of the negative impacts on the environment, the way in which it corrupts and 'bastardises' local cultures and the manner in which any potential economic benefits are frittered away as a result of the First World ownership of much of the tourism industry globally. It is from this negative premise that much new tourism takes its cue, in an attempt to redress the impacts of tourism and establish forms of (new) tourism that are reportedly environmentally, economically and culturally sustainable.

But our interpretation and discussion of sustainability is broader. This chapter takes a step back and examines the ways in which sustainability is reflected through the wider processes in economics, culture and politics in the First World, and it is argued that these changes are closely associated with globalisation. It starts from the important question of what is being sustained, by whom and for whom; do all interest groups have the same intentions or aspirations in terms of sustainability? In other words, who decides what sustainability means and entails, and who dictates how it should be achieved and evaluated?

The final strand, development, looks at the ethical undertow of much new tourism and deals with perhaps the most significant issue of this century – how to address escalating poverty in the Third World and close the widening gap between increasingly rich and poor countries. In the context of tourism, advocates of new forms of tourism are asking if and how the economic benefits of tourism can be shared by Third World communities. We have left the discussion of development until last as it has rapidly become interchangeable with globalisation and sustainability. Our discussion of the evolution and contested nature of development will seek to bring these key strands together.

This chapter (together with Chapters 3 and 4) introduces a range of concepts and debates that are built upon in the second part of the book. They are important concepts in that they provide a way of refocusing the analysis and arguing that the emergence and development of Third World tourism can be understood only within a much broader frame of analysis.

Tourism in a shrinking world

Globalisation is a concept that is increasingly invoked in the analysis of tourism. With the seemingly limitless spread of tourism, the embracing of virtually any form of activity and the general ubiquity of tourists and tourism, the temptation to reference globalisation in discussions of tourism has been irresistible, often through casual and uncritical statements.

So what is globalisation and how can it most usefully be deployed in a critique of tourism? Essentially globalisation is a concept that seeks to encapsulate processes operating on a global scale. It refers to the ever-tightening network of connections which cut across national boundaries, integrating communities in new 'space-time combinations' (S. Hall 1992a: 299) and resulting increasingly in the feeling that the world is a single interconnected and interdependent whole, a shrinking world where local differences are steadily eroded and subsumed in a homogeneous mass or single social order (Allen and Massey 1995). As Giddens (1989: 520) puts it, 'Our lives, in other words, are increasingly influenced by activities and events happening well away from the social context in which we carry on our day-to-day activities'. Clearly then, globalisation is much more than an abstract concept and represents a fact of our everyday lives that the world, in some crucial respects, has shrunk (Bauman 1998).

Globalisation has been an especially appealing concept for geographers because it emphasises the way in which economic, social, cultural and environmental relationships have been stretched and interwoven across the globe. This is not to suggest, however, that globalisation is a new phenomenon. Processes of interconnection have been taking place for hundreds of years as part of an ongoing transition in the development of global capitalism. The qualitative difference today is the pace at which the process of globalisation is happening, with a current extraordinarily intensified phase of global transformation and change. McGrew (1992) captures the richness of the concept and the impact of this acceleration by expanding on two distinct dimensions which he terms scope (or *stretching*) and intensity (or *deepening*):

> On the one hand it defines a process or set of processes which embrace most of the globe or which operate worldwide: the concept therefore has a spatial connotation. Politics and other social activities are becoming 'stretched' across the globe. On the other hand it also implies an intensification in the levels of interaction, interconnectedness, or interdependence between the states and societies which constitute the world community. Accordingly, alongside the

'stretching' goes a 'deepening' of the impact of global processes on national and local communities.

(McGrew 1992: 107)

The intensity or deepening of which McGrew speaks is reflected in the way that Third World countries and communities of once peripheral areas are increasingly drawn into tourism.

Globalisation, then, provides an organising concept through which we are able to explore the extent and impacts of global change in terms of economics, culture and politics (Allen and Massey 1995). Economic globalisation conveys the manner in which economic relationships and flows have been stretched across the globe. In the context of tourism, many point to the phenomenal growth of the industry in a global sense (it is now reputed to be the largest single industry) and the rapidity with which new places are continually drawn into the tourism process. Take, for example, an average travel agent and consider the range of destinations on offer. Not only has the number of holiday destinations increased but also the distances between destinations and markets have increased markedly, and we will be examining how new tourism practices have helped to accelerate this process. This also suggests that globalisation is about capitalising on the revolutions in telecommunications, finance and transport, all of which have been instrumental in the 'globalisation' of tourism. In addition, tourism for an increasing number of Third World countries is big business. It is not just capital and commodities that can be transported and transferred easily across the world, but tourists too. It is necessary therefore to consider how changes in contemporary global capitalism have impacted upon the development of tourism, a point we take up later.

At its crudest, cultural globalisation has been used as shorthand for the emergence of a single global culture reflected in global consumerism, most usually based on US lifestyles. As one of the most prominent tourism commentators, Dean MacCannell (1992), argues, in this 'giant, or global, socio-economic system, a "New World Order" . . . there is a pretence that all the sub-groups and communities are actually phased together in some kind of relational equilibrium' (MacCannell 1992: 169). Burns and Holden (1995) also imply that the new world order has a direct relationship to the emergence of global consumerism, exemplified by 'McDonald's franchises' or what Ritzer (1993) refers to as the 'McDonaldization of society'.

Much commentary on tourism, and particularly new forms of tourism, is dedicated to bemoaning the armies of mass tourists who, it is claimed, voraciously consume places and cultures transforming them into Disney-like extravaganzas where cultural inauthenticity is actively promoted. It is a cultural order, pundits argue, in which sophistication, difference and authenticity are increasingly denied and cultural homogenisation is the norm. The task before us is to try to discern some of the most prominent changes in the cultural sphere and judge whether this supposed homogenisation is a reasonable reflection of reality.

Finally, political globalisation focuses attention on what some argue is the erosion of the sovereignty of nation states with territorial borders dismantled or melting away under the pressures of globalisation and the ascendency of transglobal politics and organisations. As will be argued in the second part of this book, such supranational organisations as the World Bank and IMF have had far-reaching effects on many Third World countries and their development of tourism. Similarly, the globalisation of environmental issues, stressing the way in which our lives are inextricably linked and impact upon one another, has resulted in the emergence of vociferous debates over the environmental sustainability of tourism. Again such debates have had profound effects on the development of Third World tourism, and we shall be exploring them in the following sections.

This process of globalisation locates the growth and expansion of tourism firmly within the 'complex nature of social change' and resonates some of the concepts that are discussed in this chapter. But it also sees the problem largely as one of 'mass tourism' and considers that solutions lie in the development of sustainable alternatives. The following chapters examine the scope of sustainable tourism and argue that the 'humanisation' process which, for instance, the Asian Consultation on Tourism favours, is much harder to achieve in practice, even given alternatives to mass tourism.

Uneven and unequal development

When, as a result of a coup d'état in the West African state of Gambia in 1994, the UK Foreign Office advised British tourists to avoid the country (advice that was subsequently taken up by other European tourist-sending countries), the Gambian economy and tourism industry virtually collapsed. This is a clear reflection of the way in which places have been drawn into one another and have become interdependent, but equally it is a stark illustration of the highly uneven and unequal nature of globalisation, which is highlighted in Figure 2.1. Clearly, Gambia as a holiday destination is far more adversely affected by the cessation of tourism than are First World tourists. The latter need only to consider alternative holiday destinations while Gambia has little if no opportunity to consider alternative tourism markets.

Tables 2.1–2.3 express the unequal nature of global tourism development in quantitative terms, although we would refer the reader to the note of caution about the reliability of the World Tourism Organisation's (UNWTO) statistics.[1] These tables demonstrate the simple points that it is First World countries that receive the bulk of international tourists, that generate the greatest number of tourists and that receive the greatest quantity of income from them. Some evidence exists to show that domestic tourism is now becoming a feature

Figure 2.1 Globalisation: uneven and unequal development

of tourism in some Third World countries (Ghimire 2001), but this has not yet reached a level which could refute the general global trends just remarked. The selected First World countries given in Table 2.1 are generally also able to generate more income per international tourist received than are the Third World countries given in Table 2.2. The list of countries given in Table 2.3 reveals that the greatest proportion of tourist income has its origins in First World countries, although it is noteworthy that India and Mexico are now listed in the top 25 tourism spenders and that China's rank position has gone from 9 to 7 since the beginning of the twenty-first century.

Similarly, the idea of globalisation also fails to acknowledge which places and peoples are included in this process and which are excluded. When Burns and Holden (1995: 13) suggest that 'tourism is not so much about suntans as it is about being a major part of the globalisation of culture', they cannot be implying that all people are implicated in this cultural process, owing among other factors to the glaring inequalities among people both within the First World and between the First and Third Worlds. While the rapid increase in overseas travel is seen as symptomatic of globalisation's pleasure space, it is rarely contrasted with the massive increases in diasporic movements of the economic poor and politically refugeed from Third to First Worlds. At the same time that the frontiers of

Table 2.1 *International tourist arrivals and receipts from selected First World countries*

		1985	1990	1995	2000	2004
Australia	Arrivals (000s)	1,143	2,215	3,726	4,931	5,215
	Receipts ($m)	1,062	4,088	7,857	8,463	13,647
	000s Dollars received per arrival	0.9	1.8	2.1	1.7	2.6
Japan	Arrivals (000s)	2,327	3,236	3,345	4,757	6,138
	Receipts ($m)	1,137	3,578	3,226	3,373	11,265
	000s Dollars received per arrival	0.5	1.1	1.0	0.7	1.8
Spain	Arrivals (000s)	25,459	34,085	34,920	47,898	52,430
	Receipts ($m)	8,151	18,593	25,388	30,979	45,067
	000s Dollars received per arrival	0.3	0.5	0.7	0.6	0.9
UK	Arrivals (000s)	14,449	18,013	23,537	25,209	27,755
	Receipts ($m)	7,120	14,940	18,554	21,769	28,202
	000s Dollars received per arrival	0.5	0.8	0.8	0.9	1.0
USA	Arrivals (000s)	25,399	39,539	43,318	51,219	46,086
	Receipts ($m)	17,762	43,007	63,395	97,944	94,090
	000s Dollars received per arrival	0.7	1.1	1.5	1.9	2.0
World	Arrivals (millions)	326	456	567	687	764
	Receipts ($ billions)	116	261	407	484	633
	000s Dollars received per arrival	0.4	0.6	0.7	0.7	0.8

Source: World Tourism Organisation (various dates) *Compendium of Tourism Statistics*, Madrid: UNWTO

Table 2.2 *International tourist arrivals and receipts from selected Third World countries*

		1985	1990	1995	2000	2004
Ghana	Arrivals (000s)	–	–	286	399	584
	Receipts ($m)	–	–	233	335	466
	000s Dollars received per arrival	–	–	0.8	0.8	0.8
Grenada	Arrivals (000s)	52	76	108	129	134
	Receipts ($m)	26	38	54	93	83
	000s Dollars received per arrival	0.5	0.5	0.5	0.7	0.6
Guatemala	Arrivals (000s)	252	509	563	826	1,182
	Receipts ($m)	67	185	277	482	776
	000s Dollars received per arrival	0.3	0.4	0.5	0.6	0.7
Thailand	Arrivals (000s)	2,438	5,299	6,952	9,579	11,737
	Receipts ($m)	1,171	4,326	7,664	7,483	10,043
	000s Dollars received per arrival	0.5	0.8	1.1	0.8	0.9
Tunisia	Arrivals (000s)	2,003	3,204	4,120	5,058	5,998
	Receipts ($m)	551	953	1,393	1,682	1,970
	000s Dollars received per arrival	0.3	0.3	0.3	0.3	0.3
World	Arrivals (millions)	326	456	567	687	764
	Receipts ($ billions)	116	261	407	484	633
	000s Dollars received per arrival	0.4	0.6	0.7	0.7	0.8

Source: World Tourism Organisation (various dates) *Compendium of Tourism Statistics*, Madrid: UNWTO

tourism are dismantled through the liberalisation of the service sector, the *cordon sanitaire* against migration to the developed economic core is tightened, despite the fact that migrants' remittances are, in some cases, more significant than earnings from tourism (James 2001). We return to the issue of remittances in Chapter 11. This leads Bauman (1998) to conclude that globalisation is geared to 'tourists' dreams and desires', but that the unavoidable side-effect is the casting of many others as 'vagabonds, . . . travellers refused the right to turn into tourists . . . allowed neither to stay put . . . nor search for a better place to be' (Bauman 1998: 93).

In other words, globalisation is highly uneven too, with migration as a reminder of the global map of power, and it is a concept that has become oversimplified in its application. It is important to remember, therefore, that most accounts of globalisation are by westerners (as this is) and are essentially about western globalisation as a result of the expansion of western capitalism; clearly an exercise in power. As Robins reflects, for 'all it has projected itself as transhistorical and transnational, as the transcendant and universalising force of modernisation and modernity, global capitalism has in reality been about westernisation – the export of western commodities, values, priorities, ways of life' (Robins 1991: 25, quoted in S. Hall 1992a).

Table 2.3 *The world's top 25 tourism spenders (excluding transport), 2005*

Rank	Country	International expenditure (US$ billions)	Market share (%) 2004	
1	Germany	71.0	11.2	
2	United States	65.6	10.4	
3	United Kingdom	56.5	8.9	41.0
4	Japan	38.2	6.0	
5	France	28.6	4.5	
6	Italy	20.5	3.2	
7	China	19.1	3.0	
8	Netherlands	16.4	2.6	13.8
9	Canada	16.0	2.5	
10	Russian Federation	15.7	2.5	
11	Belgium	14.0	2.2	
12	Hong Kong (China)	13.3	2.1	
13	Spain	12.2	1.9	9.7
14	Austria	11.9	1.9	
15	Australia	10.3	1.6	
16	Sweden	10.2	1.6	
17	Korea, Republic of	9.9	1.6	
18	Singapore	9.6	1.5	7.4
19	Switzerland	8.8	1.4	
20	Norway	8.4	1.3	
21	Taiwan (PR of China)	8.2	1.3	
22	Denmark	7.3	1.2	
23	Mexico	7.0	1.1	5.2
24	Ireland	5.2	0.8	
25	India	5.1	0.8	
	World	**633**	**100**	

Source: World Tourism Organisation (2006b)

The notion that a 'new world order' has emerged, representing a triumph of western ideas also preoccupies the work of Francis Fukuyama (former Deputy Director, US State Department Policy Planning) who, following the end of the Cold War, proclaims we have reached the 'end of history' (Fukuyama 1989, 1992):

> The triumph of the West, of the Western *idea*, is evident first in the total exhaustion of viable systematic alternatives to Western liberalism . . . and can be seen also in the ineluctable spread of consumerist Western culture . . . What we may be witnessing is not just the end of the Cold War . . . but the end of history as such: that is, the endpoint of mankind's ideological evolution.
>
> (Fukuyama 1989: 3–4, quoted in MacCannell 1992: 62)

As we shall see shortly, the transposition of western concepts and ideas also saturates the history of 'development'.

In summary, globalisation may be an emotive and powerful idea, allowing each of us to invoke Big Macs, Coke and holidays in Thailand as our individual testimonies in support of that globalisation. But it can also be conceived as an attempt to place ethnocentric conformity over the way we perceive change and is a reflection of unequal development in its own right; an interesting story but a poor basis for analysis. Rather more ominously, it is a term that has allowed First World politicians, scholars, cultures, interest groups, business people and so on to impose an inevitability which is firmly premised upon the West – a global economy, global culture, global politics, a global environment. As such it obscures the varied processes of social and spatial changes to such an extent that globalisation defies both space and time; it is simultaneously aspatial and ahistorical (Daniels et al. 2001).

The reality of globalisation is considerably more complex and is characterised by uneven and unequal development. A careful examination of the economic, political and cultural processes is also central to a more in-depth analysis of the 'globalisation' of contemporary tourism. The next section advances and elaborates upon this discussion by considering the content and meaning of sustainability within the context of global change. Globalisation, it is suggested, is also useful in stretching our comprehension of sustainability.

Sustainability and global change

The second key word in our analysis of tourism is *sustainability*, a notion that at its most basic encapsulates the growing concern for the environment and natural resources, though sustainability has also had increasing resonance in social and economic issues. Just as processes of globalisation are implicated in the drawing of Third World destinations into the sphere of tourism, so too notions of sustainability are closely related and disproportionately reflected in the Third World as concern for the health of the planet has resulted in the emergence of a globalised environmental politics.

The end of the Cold War – the end of the East–West conflict between communism and capitalism – also meant that the so-called 'international community' (which is heavily imbued with the influence and power of the First World) has looked elsewhere for causes célèbres since the 1990s. In particular, the powerful nations of the First World have readjusted their attention on the Third World, and a concern for the global environment, 'development' and security now promises to become one of the principal focuses in political action and in the rhetoric and thinking of development studies (Adams 2001: 1). Just as globalisation is characterised by unequal relations, however, so too has the global environment debate already resulted in conflict between the First and Third Worlds: conflict that came to the fore at the Rio Summit in 1992 and the World Summit for Sustainable Development (WSSD) in 2002. The former was also known as the United Nations Conference on Environment and Development, UNCED, UNCED '92 and the Earth Summit and was held at Rio de Janeiro in June 1992. The latter is also often referred to as the Earth Summit, Rio+10 or the WSSD and was held at Johannesburg, South Africa, in 2002. Box 2.1 gives details of and opinions about both summits. As David Lascelles (1992) comments:

> The old East/West confrontation has shifted to one between North and South and environment and development have become fixtures on this new agenda. Some of the most acrimonious debates during the UNCED process focused on the North/South debate and the marked divisions of wealth and poverty between the two.
>
> (Lascelles 1992: 42)

Box 2.1 The Rio and Rio + 10 Summits

In 1989, the United Nations expressed deep concern at the 'serious degradation of the global life-support systems' (Resolution 44/228) and convened the United Nations Conference on Environment and Development in Rio de Janeiro in June 1992. It was attended by 178 governments including 120 heads of state.

The purpose and content of the conference were to 'elaborate strategies and measures to halt and reverse the effects of environmental degradation in the context of strengthened national and international efforts to promote sustainable and environmentally sound development in all countries'.

The immediate results – the Rio Declaration, non-binding treaties on climate change and biodiversity, forest principles, Agenda 21, and meagre financial commitments – fell far short of the envisaged aims of the conference. The declarations were vague enough to please everyone, and the commitment of resources was paltry ($2.5 billion compared with an estimated cost of programmes of $600 billion a year). Despite its size, the travel and tourism industry was not included as a separate item on the conference agenda.

The Business Council on Sustainable Development (BCSD) was formed to promote the international business community's standpoints on environmental issues. Carothers (1993) describes the BCSD as a coalition of some fifty multinationals, including some of the worst polluters on the planet, whose 'goals were predictable: "voluntary" rather than legislated reduction in toxic emissions, the right to corporate privacy and wholesale support for "free trade"' (Carothers 1993: 14–15). References to the over-consumption of the rich countries were removed from treaties, mention of corporate conduct was watered down, the poorest countries barely had a say, and, despite objections from all the environmental groups in attendance, the conference was used to endorse the General Agreement on Tariffs and Trade (GATT).

In September 2002 the UN World Summit on Sustainable Development (also known as Rio + 10) was held in Johannesburg, South Africa. Along with the International Chamber of Commerce, the BCSD formed an initiative called Business Action for Sustainable Development (BASD) whose aim was to 'ensure maximum participation from the world business community' (BASD 2001) in the preparation for and execution of the summit. Part of this preparation was the collection of narrowly chosen examples of good corporate citizenship, a tactic which Corporate Watch (2001: 1) refers to as 'greenwash'.

Before the conference, Friends of the Earth International (FOEI) declared that:

> Since Rio unsustainable development has continued and there has been a dramatic failure to implement the (already insufficient) commitments made. Governments have promised to address the reasons for this failure before Johannesburg. FOEI is appalled that action on the real causes, such as neoliberal economic global-isation and the excessive influence of corporations on policy have not been taken.
>
> (FOEI 2001)

FOEI called for the conference to commit itself to regulated action rather than voluntary action on corporate accountability, environmental rights, ecological debt, trade justice and environmental governance and to recognise the crucial role of the precautionary principle (FOEI 2001).

Of course, the polarisation is not so much geographical (a strict North/South polarisation) but a division between those with power and the relatively powerless (schisms that relate not just to the divisions between nation states but to divisions internal to nation states too).

It is necessary initially to trace how such debates over environment, development and sustainability are reflected in the issues of contemporary tourism. The first important point to establish about sustainability is that it is a word that is defined, interpreted and imagined differently between individuals, organisations and social groups. For the 'admen' of transnational corporations an interpretation of sustainability (how to entice customers to buy their product on the basis of their concern for the environment) is significantly different from that of those communities and activists claiming to resist the destruction of the countryside (how best to 'save' nature and Mother Earth). In this book, rather than taking sustainability as given or relatively easily defined (which is mostly done in terms of the highly ambiguous Brundtland definition),[2] sustainability is considered a contested concept, a concept (as Chapters 3 and 4 argue) that is 'socially and politically constructed' and reflects the interests and values of those involved.

This contest is well illustrated by the differing views of the proponents and critics of the Brundtland Report and definition. Protagonists of the report point out that it incorporates the essential principles of intra-generational and intergenerational equity and that it persuaded many governments to endorse the notion of sustainable development. Its critics would argue that it contains inbuilt assumptions about the need for continued expansion of the world economy and that it failed to stress the radical changes in lifestyles and society that would be required to overcome the problems inherent in the western model of development.

The second important point is that, in addition to acknowledging and assessing the different interpretations of sustainability, we are also interpreting the debate over sustainability in a broader context. This provides for a more free-ranging discussion, allowing us to consider the ways in which different ideas of sustainability are used, for example, to sustain profits in the tourism industry (Chapter 7), or are used by social classes to retain distinctive holidays (Chapter 5), or are used by 'host' communities to exclude outsiders (Chapter 8).

Sustainability, then, is a concept charged with power. We will be turning to the notions of ideology and hegemony a little later in order to help in the exploration of tourism and sustainability. The critical questions must remain: Who defines what sustainability is? How is it to be achieved? And who has ownership of its representation and meaning? It will be argued that, for the greater part, the answers to these questions are found in the First World: in businesses, governments, transnational institutions, scholars, environmentalists and new socio-environmental organisations.

Sustaining profits

Arguably, global economic restructuring and development are the most pertinent factors in the study of globalisation. Indeed the motor behind global economic change is the need for the growth of capitalism – new opportunities, new markets and, for tourism, new destinations – in other words the imperative for sustained growth and profitability. It is necessary to outline the most important features of global economic change before looking in more detail at why these changes have occurred and how they are reflected in contemporary tourism development.

The first feature has been a rapid growth in the world market, a process of inter-nationalisation that has resulted in the emergence of a global economic system (though it is necessary to contextualise this process as an uneven and unequal one, as we argued above). This is most clearly reflected in the expansion and reorganisation of the global financial market with the emergence, for example, of a global stock market. Economic globalisation is also reflected in 'footloose' capital and the growth of less nationally regulated industrial, banking and commercial sectors, a process that is also clearly represented in the global tourism industry's principal economic sectors with mergers and buy-outs between international airlines and hotels. It is a process that has been pursued fanatically through the General Agreement on Tariffs and Trade (GATT) and the General Agreement on Trade in Services (GATS) by the promotion of liberalised free trade by First World governments and that has been dominated by global agencies such as the World Trade Organisation (WTO). We shall return to this critical agenda in Chapter 10.

A second feature is the relatively rapid First World deindustrialisation, with an equally rapid growth in the service sector. Deindustrialisation has been necessitated by the long-term fall in manufacturing profit margins in the advanced capitalist economies (Daniels et al. 2001; Lash and Urry 1987), which have been forced to compete with the more cheaply manufactured goods in the so-called 'newly industrialising countries' such as the *little tigers* of Asia (Hong Kong, Singapore, South Korea and Taiwan). On the other hand, there has been a dramatic increase in service sector industries and varying degrees of reorientation in the First World to so-called service sector based economies. This is perhaps best understood as part of an ongoing process involving the international division of labour at a global scale. These shifts also provide important clues to the nature of attendant changes in the consumption of services (such as holidays), a grasp of which is essential for understanding the growth and development in the First World consumption of tourism.

Third, and very much interrelated with globalisation and the international division of labour, capitalism has increasingly penetrated the Third World and 'integrated' or drawn these countries into a global capitalist system. Some commentators argue that there are simply no viable alternatives to capitalism and it is inevitable that, given the right conditions, capitalism will finally triumph as a single global economic order (De Soto 2001; Easterly 2001; Fukuyama 1992). Consequently there has been a considerable increase in the number of countries which are implicated in capitalist production (Lash and Urry 1987: 6). We shall return to the consequences of economic interdependency at various points in this book, and in Chapter 3 will consider some of the parallels that have been drawn between cash crops (such as bananas, coffee or minerals) and tourism. It is important to note here that, while we are arguing that tourism must initially and principally be understood within the context of capitalist development and the dynamics of capital accumulation, it is nonetheless necessary to avoid deterministic reasoning and acknowledge that such development is locally conditioned and differentiated (Bianchi 2002a, 2002b; Massey 1995b). In other words, the form that tourist development takes and the respective roles of relevant agents (from 'local' people and entrepreneurs to external companies, central and local governments, and lending institutions) is not predetermined and will vary greatly from place to place.

Post-Fordism provides one way of capturing the processes of global restructuring and the qualitative changes in the organisation of both production and consumption (Allen 1992), changes which have been alluded to above and will be expanded upon below. The regime known as Fordism (taking its name from Henry Ford's assembly lines making mass-produced cars) characterised the major capitalist economies for the best part of the twentieth century. It expresses how economies of scale are ensured by goods which are mass-produced and mass-consumed. Under conditions of post-Fordism (or neo-Fordism), however, which many commentators suggest represent the current economic regime, there is a qualitative shift from mass production and consumption to more flexible systems of production (often referred to as economies of scope or batch production) and organisation (such as flexible work patterns). Post-Fordism also makes tentative links to changes in the way that goods and services are consumed, with rapidly changing consumer tastes and the emergence of niche and segmented markets.

The applicability of these ideas to the changes in tourism have now been acknowledged. It would appear that just as Butlins holiday camps or packaged holidays are indicative of services mass-produced and consumed under a regime of Fordism, the emergence of small group tours to Bolivia or truck journeys across sub-Saharan Africa is indicative of post-Fordism. Further examples of tourism types associated with post-Fordist consumption are given in Table 2.4. Lash and Urry (1994) in particular point to the demand for 'independent holidays' and the increasing environmental planning and control of tourism in places such as Belize or Bermuda as examples of new alternatives to mass tourism. They also recognise the importance of Third World countries in these changes, arguing that 'the development of "alternative tourism" in some developing countries' (Lash and Urry 1994: 274) is a clear example of post-Fordist tourism. We will return to this argument in Chapter 4 in discussing the shape of new tourisms that have emerged.

So far some of the most salient features in contemporary global change have been described. But what forces drive these global economic processes? The most penetrating analysis has been offered by the Marxist geographer David Harvey (1989b), who presents the concept of *time–space compression*, which is of considerable interest in unravelling the growth and development of Third World tourism. It is worth spending a little time considering his ideas, for they move an understanding of global change beyond the

Table 2.4 *Post-Fordism and tourism*

Post-Fordist consumption	*Tourist examples*
Consumers increasingly dominant and producers have to be much more consumer-oriented	Rejection of certain forms of mass tourism (holiday camp and cheaper packaged holidays) and increased diversity of preferences
Greater volatility of consumer preference	Fewer repeat visits and the proliferation of alternative sites and attractions
Increased market segmentation	Multiplication of types of holiday and visitor attractions based on lifestyle research – e.g. trekking, trucking, sport-based, bird-watching
Growth of a consumers' movement	Much more information provided about alternative holidays and attractions through the media, and increasingly the internet
Development of many new products, each of which has a shorter life	Rapid turnover of tourist sites and experiences because of fashion changes – countries join or are dropped off the list of 'places to go'
Increased preferences expressed for non-mass forms of production/consumption	Growth of 'green tourism' and of forms of accommodation which are individually tailored to the consumer (such as eco-lodges)
Consumption is less and less 'functional' and increasingly aestheticised	'De-differentiation' of tourism from leisure, culture, retailing, education, sport, hobbies

Source: Adapted from Lash and Urry (1994: 274)

descriptive analysis and prescription of *how* global capitalism works and seek instead to demonstrate *why* capitalism changes in the way it manifestly does. Furthermore, time–space compression provides an opportunity to overcome the problems of the geographical literature on tourism, criticised by Britton (1991: 456) as offering 'little more than a cursory and superficial analysis of how the tourism industry is structured and regulated by the classic imperatives and laws governing capitalist accumulation'.

Harvey presents the time–space compression thesis in his now widely cited *The Condition of Postmodernity* (1989b). The present phase of globalisation, Harvey argues, involves a marked increase in the pace of economic (and everyday) life and a phenomenal acceleration in the movement of capital and information. Time–space compression seeks to encapsulate this intensification as capitalists aim to overcome the barriers of distance and stretch their economic relationships to all parts of the globe. It is, in other words, part of an ongoing expansion of capitalist relations of production where the primary objective is to reduce the turnover time and to quicken the circulation time of capital, and to sustain profits. Both new markets and new products are sought in order to achieve this and the process is clearly reflected in the way in which an increasing number of holiday destinations are drawn into the global tourism industry.

Box 2.2 demonstrates the way in which information is moved more rapidly within global networks and the way that travel experiences are advanced to such a degree that it is no longer necessary to leave your home or office. Box 2.2 may appear an exaggerated example but it underlines the way in which places are increasingly drawn into (or excluded from) a global system. The telecommunications revolution allows teleworkers and computer users to take advantage of the global information 'super-highway', via email, bulletin

---

Box 2.2 Have mouse will travel

Dawn is breaking in the Costa Rica rainforest. Above me, a white hawk flits silently through the treetops. A pale-billed woodpecker alights softly on the side of a soaring trunk. The branches are swathed in mosses, ferns and orchids, and hung with a triangle of vines and lianas. The Braulio Carrillo National Park is less than an hour from the capital, San José, but it could be in another world.

Gazing up through the forest, I can make out in the distance the rows of small houses, running up and down the steeply sloping hills of Tegucigalpa.

Looking down, the wide blue waters of Montego Bay stretch out invitingly from the Jamaican shoreline, with the luxurious 'cottages' of the Sandals Hotel perched on the edge of their private beach. It is an inviting prospect, but one which, for the moment, I decide to pass on . . .

Yes, sad to say, it's not for real, it's the Internet – that sprawling worldwide computer network. Armchair travel has a whole new meaning when, with a laptop atop your lap and a modem by your side, you can skip across the globe on the World Wide Web from Abidjan to Zanzibar, alighting on everything from the Eurostar timetable to the latest malaria alerts for East Africa . . .

In some ways, it's the ultimate in eco-tourism, dropping in on distant lands without ripping holes in the ozone layer . . .

Source: M. Wright (1996)

boards, virtual conferences and the worldwide web (the web or www). Appendix 1 lists some of the travel-related bulletin boards and websites that can be accessed over the internet (p. 384). It is the stated objective of the telecommunications industry to provide access to the technology for all the population – a laudable aim. In practice, however, the promotion of its widespread use may serve to extend the uneven and unequal nature of access to information. In this respect, it may mirror the effects of access to travel. Furthermore, it is feasible that increasing regulation of the networks will serve to concentrate power in the hands of those who can afford it.

The need to accelerate the circulation of capital has necessitated an economic transition, or qualitative shift, and Harvey seeks to explain this in terms of the regime of capital accumulation, a central tenet of Marxist inquiry. The key, and perennial, question is how best to deal with the over-accumulation of capital and how it can be 'expressed, absorbed or managed' (Harvey 1989b: 131). Reflecting the post-Fordist debates discussed above, Harvey contends that in terms of production the problems encountered in achieving satisfactory productivity increase and the intense competition faced from, for example, 'newly industrialising countries' has forced the transition (in the First World) to a more flexible mode of accumulation in the post-1970 period, a 'new dynamic phase of capitalism' as he terms it (Harvey 1989b: vi). At the heart of this change lies flexibility. As Figure 2.2 demonstrates, a capitalist system based upon a Fordist mode of production has given way to more flexible modes of capital accumulation, a regime labelled 'flexible accumulation'. We will return to this diagram a little later. As Harvey argues, this comes into 'direct confrontation' with the rigidities characteristic of Fordism. Flexibility is introduced in a range of respects including labour markets and processes, products (new and different forms of tourism) and patterns of consumption (such as new tourism).

Harvey's discussion of capital circulation and the acceleration in the pace of our everyday lives has clear reflections in the rapid expansion of tourism in the Third World. It is not only capital that is circulated at an accelerated rate, but places too, as destinations come in and out of fashion and tourism moves on elsewhere. Indeed, the growth and development of Third World tourism may be another manifestation of time–space compression with the logic of capital accumulation driving the global spread and expansion of tourism. Equally, however, it is factors such as natural disasters, political instability, the environment and so on (and how these are represented and perceived in the First World)

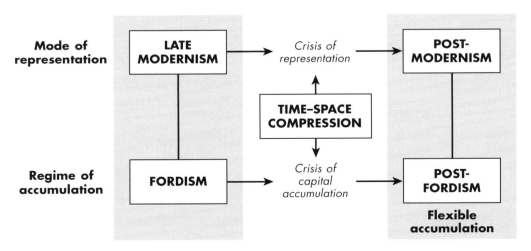

Figure 2.2 Transition in late twentieth century capitalism

Source: Adapted from Gregory (1994: 412)

which play a significant part in governing the ebb and flow of Third World destinations on and off the global tourism map. And it is to one of these other factors, sustaining culture and lifestyles, that we turn next.

Sustaining culture and lifestyles

The dramatic economic restructuring that has occurred under the latest and extremely intense phase of globalisation is, some argue, closely related to far-reaching cultural changes (Harvey 1989b; Jameson 1984). Figure 2.2 suggests that the fierce round of time–space compression reflected in an accumulation crisis and the emergence of 'flexible accumulation' are equally reflected in a crisis of cultural representation (our ability to make sense of a rapidly changing world). So, while Harvey's ideas are underpinned by economics, he also stresses the importance of cultural change within global restructuring:

> While simultaneity in the shifting dimensions of time and space is no proof of necessary or causal connection, strong *a priori* grounds can be adduced for the proposition that there is some kind of necessary relation between the rise of postmodernist cultural forms, the emergence of more flexible modes of capital accumulation, and a new round of 'time–space compression' in the organisation of capitalism.
>
> (Harvey 1989b: vii)

This argument provides some potential insights into how and why Third World tourism has become so popular. In other words, it is with changes in the First World, and with the factors that produce new forms of holidays and tourists, that analysis must focus first. Returning to Figure 2.2 for a moment, it will be noticed that the predicted outcome of the crisis of representation is *postmodernism*, and Harvey (1989b) refers to 'postmodernist cultural forms'. Very broadly, postmodernism refers to the emergence of new cultural styles (in art, architecture, music and the objects and experiences we buy and consume), and postmodernity to the idea that we now live in a new social epoch that has superseded modernity.

Postmodernism is a widely used and debated idea, though it is not our intention to provide a lengthy rendition on what postmodernism is and is not, or to provide an in-depth guide to protagonists' arguments. Rather, it is to accept that, like post-Fordism, this idea helps encapsulate the profound cultural changes that have been emerging, particularly since the 1970s (Featherstone 1991; Huyssen 1984). Practically, the changes we are experiencing are rooted in our everyday lives and 'can no longer be ignored' (Harvey 1989b). Moreover, postmodernism also helps to draw our attention to the important relationships between First World consumption habits and the capitalist imperative of increasing the turnover time of capital.

> Of the many developments in the arena of consumption, two stand out as being of particular importance. The mobilisation of fashion in mass markets provided a means to accelerate the pace of consumption not only in clothing, ornament, and decoration but also across a wide swathe of lifestyles and recreational activities (leisure and sporting habits, pop music styles, video and children's games . . .). A second trend was a shift away from the consumption of goods and into the consumption of services . . . The 'lifetime' of such services (a visit to a museum, going to a rock concert or movie, attending lectures or health clubs), though hard to estimate, is far shorter than that of an automobile or washing

machine. If there are limits to the accumulation and turnover of physical goods
... then it makes sense for capitalists to turn to the provision of very ephemeral
services in consumption.

(Harvey 1989b: 285)

It is clear that, for most people, contemporary lifestyles involve a dramatic increase in the
number of services consumed. (You may wish to consider how the shift to services is
related to the growth of tourism (a service) and how the circulation of fashion perhaps
relates to popularity of holiday destinations over time.) Box 2.3 summarises the shifts in
contemporary tourism predicted by this broader analysis.

But part of the postmodernism argument tends to lead to the conclusion that, just as
an increasingly globalised world has resulted in a global economy, so too there is an
inevitability that the same processes have resulted in the emergence of a global culture
characterised by Big Macs, Coke and the web. Such predictions (or forebodings) have led
to a succession of tourism commentators bemoaning the inability of Third World
communities to sustain their traditional lifestyles in the face of an imposition of western
values and beliefs, and the consequent erosion of cultural difference and authenticity.
Morrow (1995), for example, refers to tourism as a 'radioactive cloud of banalizing
sameness' which, he dramatically argues, 'threatens the earth; the sacred and beautiful
places, all the uniquenesses, have been invaded, desacralized, franchised for the masses,
dissolved into the United Colors of Benetton' (Morrow 1995).

While the ubiquity of baseball caps or holidays as evidence of global culture is
anecdotally appealing, it was briefly suggested earlier that it is also a highly simplistic
view of the world. It is a seriously deficient account in two important respects. First,
it discourages acknowledgement of the power and inequality reflected through cultural
globalisation (S. Hall 1992b). Not only is globalisation extremely unevenly developed,
affecting (or not as the case may be) different social groups in very different ways, but
also it represents an increasing interdependency between First and Third Worlds which is

---

Box 2.3 Shifts in contemporary tourism

OLD FORDIST	NEW POST-FORDIST
Mass	Individual
Packaged	Unpackaged/Flexible
Ss (Sun, sea, sand, sex)	Ts (Travelling, trekking, trucking)
Unreal	Real
Irresponsible (socially, culturally, environmentally)	Responsible
MODERN	**POSTMODERN**

highly unequal: 'the West and the Rest', as Stuart Hall (1992b) terms it. Crucially, in the context of tourism, the Third World consists ostensibly of tourist-receiving and not tourist-sending countries.

Second, postmodernism offers a position that tends to underplay the way in which commodities are used and understood by different social groups and communities. Much the same criticism could be levelled at the way tourists are received and perceived, and how their practices are adopted, rejected or adapted. In other words, because tourists are disliked (for a variety of reasons) in one locality, it does not automatically mean they are considered similarly elsewhere, and vice versa. Similarly, the charge that tourism bastardises cultures, ripping them from their traditional roots and subsuming them into a global culture, fails to offer a more sophisticated understanding of how this process of cultural change may be negotiated by communities in the Third World. We return to this process of 'transculturation' in Chapter 8.

This has resulted in a highly polarised and simplified debate in the First World concerning the most appropriate way of holidaying, which equates to 'tourists = mass tourism = bad' and 'travellers = appropriate travelling = good'. Again this simplistic view reflects the role of power with the prospects and problems of tourism calculated largely from a First World perspective. It is we in the First World who decide what is right (untouched primitive cultures) and what is wrong (Thai hilltribes listening to a radio or Iban longhouse dwellers with walkmans), what is appropriate and sustainable and what is not. Reflective of this, the American academic Tensie Whelan (1991: 3) proclaims of ecotourism, 'If we are to save any of our precious environment, we must provide people with alternatives to destruction'. In this sense, First World inhabitants seek to sustain their own ideas and work towards sustaining their own lifestyles, within which only certain types of 'holiday' are acceptable.

Finally, however, postmodernism does help to capture the high degree of difference and fragmentation that lies at the heart of contemporary cultural change and therefore, somewhat contradictorily, helps challenge the emergence of global homogeneity. It is a process of change that has had some important impacts. In terms of politics, for example, commentators have contrasted the disengagement of many from the formal political process and relative downfall of party politics to the emergence of new social movements focused upon issues such as women, minority rights, nuclear power or the environment and most notably in recent years the rise of anti-globalisation movements. The emergence of societies that are considerably more socially pluralistic in terms of class, culture, ethnicity, lifestyles and so on has also been noted. And it is worth considering this in a little more detail, for this too offers some important clues in the investigation of Third World tourism.

Postmodernism has been closely linked to the emergence and growth of the so-called new middle classes.[3] This has involved the relatively rapid expansion of social classes (and hence the plural, new middle *classes*), indicating that there is not one distinguishable middle class, especially in the First World.

An important feature of the new middle classes has been their identification as agents of the cultural change inherent in postmodernity, and they have been referred to as 'new cultural intermediaries' (Bourdieu 1984). Their ranks have swollen dramatically with the growth of the media (advertising, lifestyle research, and so on), caring personal services (from physiotherapy to reflexology) and the emergence of service sector-oriented economies more generally. They are, therefore, key social groups in initiating, transmitting and translating these new cultural processes and consumption patterns, of which holidaying is demonstrably a significant part. Later chapters of this book suggest that these social classes are not only important consumers of Third World holidays but also key groups in promoting and implementing notions of sustainability. This is not to claim that all Third

World tourism is the product of cultural consumption preferences and political leanings of members of the new middle classes; clearly this would be ridiculous. Rather, it is to suggest that they are major initiators and consumers of new tourism in Third World destinations. As Morrow (1995) observes, the 'great, global middle class is in motion'.

Crompton (1993) argues that much recent work on the growth and development of the middle classes is crucially linked to the growth of consumer capitalism and an emphasis on 'lifestyles'. Lifestyles range from the places in which people choose to live, to the things they eat, and, of course, to the types of holiday they take. With the rapidity of cultural and economic change, social groups are constantly attempting to identify and indicate to others their position in 'the new social and cultural order' (Shurmer-Smith and Hannam 1994); or in other words, attempting to sustain a distinctive lifestyle that has been matched by the emergence of new niche markets and commodities.

The new middle classes are not only significant as cultural intermediaries, but also heavily represented in the new political alignments (or new socio-environmental organisations) that have emerged focused on issues such as the environment, and that are concerned primarily with sustainability and sustainable lifestyles. Habermas (1981) interprets this as the emergence of a 'new politics' which includes and is centred on the anti-nuclear, peace, women's, environmental, minority liberation, religious and alternative lifestyle movements; to these we must add anti-globalisation. (Chapter 6 will develop this argument by assessing the relevance of environmental politics, and what might be termed the sustainable lifestyle movement, to the development of new forms of Third World tourism.) Drawing upon this recent social theory and acknowledging the significance of contemporary cultural processes will help to move the analysis beyond the somewhat restrictive surveys of tourist motivation and satisfaction and the rather static typologies of tourists that have characterised much work in the field of tourism in the Third World.

Sustainable politics

If globalisation is marked by profound economic and cultural changes, equally it also alerts us to important political processes and the emergence of what is commonly referred to as global politics. A more globalised political environment has become part of our everyday experience – the European Union (EU), the IMF, the WTO and international environmental organisations (such as Greenpeace) are all examples of organisations whose political relationships stretch beyond the boundaries of nation states. And the decisions of these institutions are felt thousands of miles from where they are taken, a fact that re-emphasises the power of unequal and uneven development. Broadly speaking, there are three groups of 'global' political bodies that interest us most here: supranational institutions, trans-national institutions and international non-governmental organisations.

Supranational institutions and agencies, which involve both a degree of political integration between states and a transcending of power of individual nation states, have grown in stature and influence in the post-1945 period. These institutions have had an important impact on global development, and their influence in terms of tourism is considered in later chapters.

The most widely cited supranational actors, and arguably the most significant in terms of their economic and social impacts, have been the IMF and World Bank. Together they have imposed so-called structural adjustment policies and programmes (more recently recast as Poverty Reduction Strategies) upon Third World states which have been forced to adjust their economies (mostly in terms of denationalisation, privatisation, a massive reduction in state services and a reduction or elimination of protective import tariffs) in order to secure further loans. Harvey (2005: 13) powerfully refers to this as a 'series of gyrations and chaotic experiments', through which these institutions stumbled towards neoliberalism. But these are by no means the only institutions involved. Equally important to the discussion are the policies of other supranational institutions such as regional development banks, the EU, the UNWTO or the United Nations which convened the Rio Summit in 1992 and the Rio + 10 conference in 2002.

Beyond these bodies there are other emergent intergovernmental structures that transcend the independence of the nation state in a variety of ways. Economic regionalism, for example, has resulted in a dramatic impact in this respect and it is necessary to consider the effects of trading blocs such as the North America Free Trade Agreement (NAFTA) and the Association of South East Asian Nations (ASEAN). Such arrangements can underscore and reinforce the inequalities between countries and regions and, in effect, foist certain choices upon weaker countries both within and beyond the agreement.

There are two important observations to be made here. First, not only have the political relationships and actions forged by these institutions increased in their global reach, but also they have been couched increasingly in a language of sustainability, of which the Rio and Rio + 10 summits are the most dramatic examples. Clearly, for whom these 'sustainable' policies, priorities and programmes are intended, what they entail and how they impact upon the Third World are important parts of our inquiry. Second, it appears that the product of globalised politics is the emergence of a euphemistically entitled international community, something of a globalised consciousness, a seemingly popularised and benevolent global collective or ombudsman that acts in the best global interests intervening to solve a myriad of problems from civil war, international crime cartels and terrorism to the killing of whales and the destruction of rainforests. We should add, however, that although there are key alliances between supranational agencies and Third World interests, these are not simple one-way relationships between First World and Third World. It is necessary to examine the composition of this international community, and enquire for whom it speaks, who it represents, who it is controlled by, who it seeks to influence and in what ways.

Transnational institutions have also emerged on the global stage and possess varying degrees of influence and power with variable impacts upon the Third World. Unlike supranational institutions they are composed of bodies other than governments (Allen and Massey 1995). Private global consortia, such as the World Travel and Tourism Council (WTTC) and the UNWTO, provide examples of such institutions and their relationship to tourism.

Established in 1990, the WTTC is a global coalition of 70 chief executives from all sectors of the travel and tourism industry. The WTTC's 'millennium vision' is to encourage governments, in cooperation with the private sector, to harness the industry's economic dynamism and increase overall growth and job creation. The WTTC aims to do this through the promotion of open markets, the elimination of barriers to growth, deregulation and liberalisation, while at the same time pursuing sustainable development through industry environmental initiatives such as the Green Globe scheme (see Chapter 7). Again, centred in the First World and with mostly First World members, such institutions help reflect global inequalities. Some of the WTTC's campaigns and work are examined further in Chapters 7 and 10.

There has also been a marked growth in non-governmental organisations (NGOs) both within nation states and transnationally. International non-governmental organisations (INGOs), including such household names as Amnesty International, Greenpeace and Friends of the Earth (FOE) (some of which have an indirect relationship with tourism) increased in number from 832 in 1951 to 4,649 in 1986 (McGrew 1995). Another notable fact is the way in which these organisations, especially those focusing on ecological and environmental problems, have invoked the global nature of the problems. Indeed, names such as FOE, the World Wide Fund for Nature (WWF) or Earth First explicitly emphasise the global nature of environmental concerns in the very titles of the organisation (Yearley 1995). In other words, since the 1980s we have witnessed the emergence of transnational environmental politics, a large part of which is devoted to saving Planet Earth.

It is necessary to consider how these INGOs have reacted to the growth of global tourism and equally how the activities of such INGOs have been received and are perceived by Third World countries and communities. Just as economic growth and change or cultural transformation are differentially experienced in different places, so too are global environmental politics. For example, conservation measures designed to maintain ecological biodiversity undertaken by organisations such as Conservation International have frequently been 'contradicted' by the priorities and aspirations of local communities attempting to secure their livelihoods. And yet it is the power of benevolence reflected by environmental INGOs that is the most striking feature of such cases. Through membership of such organisations or through a general empathy with their aims, the global concerns and consciousness of First World citizens are played out at a local scale; their 'will' is imposed upon communities thousands of miles away. Global environmental politics, therefore, also appear to confirm, at least in part, the contours of power that have already been mapped in the discussion of global economic and cultural processes.

In the preceding sections, it has been suggested that sustainability has a much broader currency than is acknowledged in contemporary academic literature. Sustainability is as much to do with ensuring continued profits through more flexible patterns of capital accumulation, or maintaining middle-class lifestyles in the First World and the ability of these social groups to experience (sustained) indigenous cultures while holidaying in the Third World, as it is to do with ecology and environment. Figure 2.3 provides a reminder of the focus of this book and the range of factors which must be brought to bear in a discussion of Third World tourism. It also re-emphasises the importance of the interrelationships of the themes of sustainability, globalisation and development. It is to the last of these key themes, development, that we now turn.

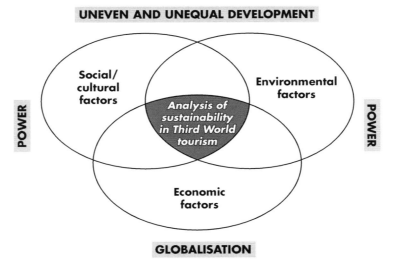

Figure 2.3 Tourism, sustainability and globalisation

Development and the Third World

Development is the last of our key themes and is of special significance, in that much tourism literature invokes the activity of tourism as a potential means of 'development'. We have already referred to the uneven and unequal nature of global capitalist development (Allen and Massey 1995; Daniels et al. 2001; N. Smith 1984) in the discussion above, and inferred that globalisation and sustainability are contemporary universalistic discourses. By this we mean they are global in their reach, and despite wide differences in interpretation, there is a broadly held powerful shared meaning and understanding of these processes and their implications for change. The same, we will argue, is true of development, and despite changes in the theory and approach to development, Third World change is contingent on, and reflective of, the First World: in the nature of economic change and expansion of capitalist relations of production and in the 'practice' and theorisation of development (Escobar 1995). In sum, we must consider the extent to which globalisation, sustainability and development are the standard bearers of western capitalist development and expansion, and gauge how far they are emerging as interchangeable ideas or projects. Our task is to try to understand the potential impact of such change globally, and the role and fate of tourism as part of this process. First, though, it is necessary to place development within a recent historical context. This is especially important as discussions of the relationship of development to tourism are relatively sparse (Reid 2003; Sharpley and Telfer 2002; Telfer 2002; Wall 1997).

It should be remembered that most theories of development emerge from Eurocentric thinking and analysis of western capitalist economic history. As Hettne (1995) contends, development is a product of the Enlightenment and is an unequivocal 'modern project':

> Once the first industrial nation had been born it provided the model to imitate . . . Not to imitate would mean permanent dependence . . . This basic dilemma was to be repeated more generally in the relation between the West and the decolonized world. In order to develop it was deemed necessary for the 'new nations' to imitate the Western model – it was a 'modernization imperative'.
>
> (Hettne 1995: 25)

In a similar vein Rist (1997) suggests that the hegemony of development involved a 'semantic conjuring' where the 'transformation of the term "underdeveloped countries" into "developing countries" simply reinforced the illusory promise of material prosperity for all' (Rist 1997: 238). As Escobar (1995: 53) argues in his deconstruction of development, the development discourse has achieved its success as a 'hegemonic form of representation' as it has unambiguously constructed the poor, poverty, the underdeveloped and the developing as a largely universal and homogenised take on 'reality'. (We will address the significance of concepts such as discourse and hegemony to the study of tourism and development in Chapter 3 in discussing the significance of relationships of power.)

Wolfgang Sachs (1999) suggests that US President Harry Truman's citation of large parts of the globe as 'underdeveloped' in his inaugural speech to Congress was the start of the race for the Third World to catch the First. On that day, 20 January 1949, two billion poor people were discovered and became underdeveloped (Escobar 1995; Esteva 1992). More significantly however, as Sachs (1999) continues, Truman redefined the world in economic terms – as a global economic arena – with Truman's vision conjoining the double imperative of global and 'economic development', and since that time, development, above all else, has signalled the need to escape the undignified confines of underdevelopment (Esteva 1992; W. Sachs 1999); a global strategy was mapped out in a few paragraphs (Rist

1997). Development may be best understood foremost as a hegemonic discourse that originates from, and is largely fashioned by, First World dominated global institutions, governments, agencies and academe. Above all else, there is a need to interrogate and scrutinise the 'aura of self-evidence surrounding a concept which is supposed to command universal acceptance but which – as many have doubtless forgotten – was constructed within a particular history and culture' (Rist 1997: 2).

The age of development

Figure 2.4 describes in broad terms the emergence of differing strands of development thinking from the 1950s onwards. This is not to suggest that development thinking falls into neat temporal periods or involves exclusive ideas within these periods. Nor is this an exclusive representation of the age of development; rather, it attempts to pick out the main components in the evolution of development. Indeed the history of development thinking and 'development studies' is marked by the infusion and overlap of ideas and of counter theories and approaches, all of which tend to work against a historical representation of their emergence. As Potter (2002: 63) comments, 'theories and strategies have tended to stack up, one upon another, co-existing, sometimes in a very convoluted and contradictory manner'. Nevertheless, Figure 2.4 attempts to describe the evolution of thinking and development discourses in a way that reflects on the economic and political context.

In the wake of the Second World War the first mainstream theories of development emanated from modernisation theory. Modernisation theory attempted to map the stages through which so-called 'traditional' or 'backward' societies would pass on a transitional path to development. The cornerstone of modernisation theory is born of a dualistic representation, as Larrain (1989: 87) argues, 'a dichotomy between two ideal types': the traditional society (otherwise referred to as underdeveloped, rural or backward) and the urbanised and industrialised modern society – a state of 'development'. While there is a range of academic approaches to modernisation including sociological and psychological explanations, it is the economic version that is most noteworthy for its practical formulation and impact on the juxtaposed 'undeveloped' world. The most widely cited approach is W. Rostow's (1960) *non-communist manifesto*, a five-stage model of economic growth that seeks to show how societies move from traditional to the preconditions for take-off, the road to maturity and the age of high mass consumption. This 'theory' must be understood within the political context of Rostow's capacity as chief adviser on Vietnam to President Johnson, and the need to secure US security interests through the defeat of communism by a process of support and protection to 'modernising' societies in Asia, Latin America, Africa and the Middle East (Larrain 1989). The politicisation of development should not be underestimated, from counteracting communism to the concerns about the links between poverty, terrorism and First World security. It is a one-directional, evolutionist representation of 'advancement' and 'maturity' that has biological overtones.

Modernisation theory was challenged from the late 1960s onwards as cryptoimperialist, by the emergence of neo-Marxist dependency theory which articulated the way in which development had been produced in Latin America (and by inference the Third World more generally) by the weak structural position of the Third World (Frank 1966, 1967). Notwithstanding the variance of approaches that emerged within the Dependency School, and critiques of such approaches (as Hettne (2002) argues, for example), in practice the *dependentistas* approach to development shared a focus on state-driven industrialisation with the modernisation school. Suffice to note that the Dependency School drew attention to the global *interdependencies* which aided an understanding of the *development of*

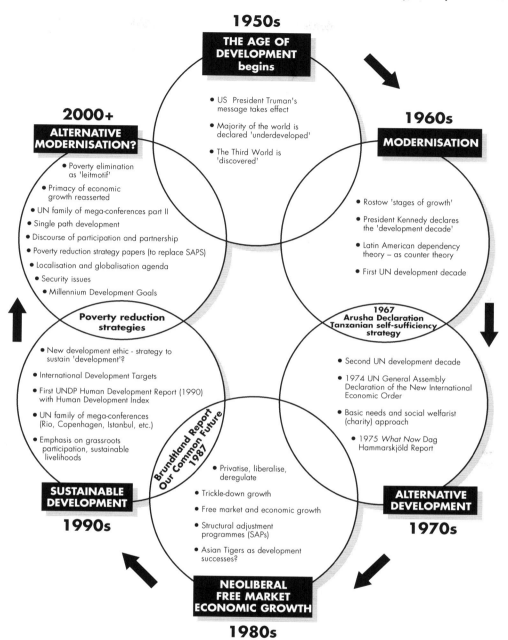

1950s

THE AGE OF DEVELOPMENT begins

- US President Truman's message takes effect
- Majority of the world is declared 'underdeveloped'
- The Third World is 'discovered'

2000+

ALTERNATIVE MODERNISATION?

- Poverty elimination as 'leitmotif'
- Primacy of economic growth reasserted
- UN family of mega-conferences part II
- Single path development
- Discourse of participation and partnership
- Poverty reduction strategy papers (to replace SAPS)
- Localisation and globalisation agenda
- Security issues
- Millennium Development Goals

Poverty reduction strategies

- New development ethic - strategy to sustain 'development'?
- International Development Targets
- First UNDP Human Development Report (1990) with Human Development Index
- UN family of mega-conferences (Rio, Copenhagen, Istanbul, etc.)
- Emphasis on grassroots participation, sustainable livelihoods

1960s

MODERNISATION

- Rostow 'stages of growth'
- President Kennedy declares the 'development decade'
- Latin American dependency theory – as counter theory
- First UN development decade

1967 Arusha Declaration Tanzanian self-sufficiency strategy

- Second UN development decade
- 1974 UN General Assembly Declaration of the New International Economic Order
- Basic needs and social welfarist (charity) approach
- 1975 *What Now* Dag Hammarskjöld Report

Brundtland Report Our Common Future 1987

- Privatise, liberalise, deregulate
- Trickle-down growth
- Free market and economic growth
- Structural adjustment programmes (SAPs)
- Asian Tigers as development successes?

SUSTAINABLE DEVELOPMENT

1990s

ALTERNATIVE DEVELOPMENT

1970s

NEOLIBERAL FREE MARKET ECONOMIC GROWTH

1980s

Figure 2.4 The age of development

underdevelopment, a global dialectic in which some parts of the globe are marginalised and dominated to fuel growth and development elsewhere. This juxtaposition of dominant cores and marginalised and dominated peripheries was taken up in political economies of mainstream mass tourism development in the Third World, which will be discussed in Chapter 3. In a similar fashion world systems theory (Wallerstein 1979, 2000) provides for an analysis and understanding of the role and function of the Third World as part of the increasing internationalisation of the world economic system, an approach that transcends the recent flurry of intellectual activity focused on the most recent and intense phase of globalisation (Klak 2002). Crudely then, both approaches share the fundamental need for a global analysis of development, an approach that has been reinforced in a range of subsequent critiques.

As Rist (1997) argues, despite the rhetoric, the UN General Assembly's declaration of a New International Economic Order in 1974 had all the hallmarks of business-as-usual in its emphasis on the fundamental need for continued economic growth, the expansion of world trade and increased aid expenditure by the industrialised countries. On the other hand, the 1970s also saw the emergence of the endogenous approach to development, also known as *another development*, which contrasted with mainstream approaches by an emphasis on self-reliance and, perhaps, served as the precursor to advocates of 'localisation', an approach that seeks to counteract (reverse) or counterbalance the trend towards globalisation (Hines 2000). This approach, which finds clearest expression in *What Now* (a report prepared by the Dag Hammarskjöld Foundation 1975), challenged the traditional approach to development. In particular it argues that development neither is simply an economic process nor follows the application of a universal formula, and that by contrast development must be geared to meeting the needs of the poor and that the situation of Third World countries needed to be understood within global structures of power and exploitation. As Escobar bluntly asserts, 'Europe was feeding off its colonies in the nineteenth century, the First World today feeds off the Third World' (Escobar 1995: 83). Development, therefore, was not something that happened solely to the Third World, but equally required adjustments in First World development, for example, through the reining back of economic growth or the reduction in levels of western consumption. At the very least it acknowledged the nascent power exercised through the hegemony of development. The implications and reflections for the growth and development of tourism are clear here, ultimately forgoing the individual choice to holiday in the Third World. Most interestingly perhaps, *What Now* had little official impact, in part, as Rist (1997) contends, because the considerable aggregated experience contributed to the report by (ex)-government ministers and international senior civil servants had momentarily thrown diplomatic caution to the wind with the report's contention of the 'gulf between economic theories and "development" practice' (Rist 1997: 157) and its 'bold, almost sacrilegious conclusion – that "the primacy of economics is over"' (1997: 156), an utterance that Rist (1997: 156) observes 'would never be repeated in any international declaration'. The primacy of economics to development and economic growth as the means of development remained.

The 1980s have been termed the 'lost decade for development' (Esteva 1992). The rise of neoliberalism symbolised by the Reagan–Thatcher axis, captured by Reagan's 'magic of the market' speech at the 1981 North–South Conference in Mexico, was characterised by the dominance and application of free market principles and trickle-down growth. Hettne (2002: 8) comments that this 'purified neo-classical discourse' produced a representation of development as an 'inherently universal and increasingly global economic process', so much so that globalisation (like sustainable development) has emerged as a – tautological – word for development. Whether real, or a manufactured and universalising discourse that legitimises neoliberal economic growth, globalisation has produced profound changes in the social sciences, including development studies (Schuurman 2002). As Wolfgang Sachs (1999: xii) concludes, since the 1980s, the age of development 'has given way to the age of globalization'. In this way we can begin to understand the emergence of debates about global tourism and its inferred development potential from the 1980s onwards.

Finally, the 1990s are seen by some as representing a convergence of thinking on development and on the significance of human welfare and human rights in the agenda

(Elliott 2002). The emergence of new systems of social and economic quantification – the Human Development Index (HDI) and the Human Poverty Index (HPI) compiled by the United Nations Development Programme (UNDP) – and the move towards people-focused and participatory approaches to development, could be considered as the return of *alternative development*. While the pro-poor development paradigm, however, is reflective of the Hammarskjöld report, the approach is considerably circumscribed in its premise of economic growth as the foundation of development. We provisionally refer to this era as 'modernisation' in an attempt to draw the parallels to previous approaches that have tended towards the production of universalistic discourses on development. One paradigm, that development is achieved through the linear progress to a modern economy, is displaced by another, that development is pro-poor, people-inclusive 'growth'. This is not an attempt to criticise such an approach to development but, rather, a critique of the way in which it is constructed (as ideology) and converted into practical strategies and approaches. For example, we consider the application of 'participatory development' in Chapter 8.

To bring us full circle, the 1990s was also a decade that witnessed the emergence of *sustainable development* following the publication of the Brundtland Report and the subsequent staging of the Rio Earth Summit. As we have argued, while the roots of sustainability are found in the western fad for ecology and environmental issues (Rist 1997), it has been interpreted in economic, cultural and social terms (in fact so much so that all aspects of contemporary life have been subjected to the gaze of development). Sustainable development ushered the ultimate oxymoron (the 'legitimizing camouflage' as Rist (1997: 174) suggests), a need to face up to the global ills of ecological meltdown and compounding poverty, but with a business-as-usual mentality to global economic growth; an approach, as we will argue, that is largely reflected in the growth and development of the tourism industry. The primacy of economic growth remains, albeit tempered by a social and environmental consciousness, and the logic of the Third World catching the First powered by international trade, is reaffirmed, with Rist (1997: 193) acerbically concluding that sustainable development is an 'enormous nothing'. Nevertheless, sustainability is an enduring concept and has found ample expression in a range of strategies (such as the ubiquitous poverty reduction strategies) and development concepts (such as sustainable livelihoods). Finally, and most significantly, in September 2000, the 'Millennium Declaration' produced by the UN General Assembly's Millennium Summit heralded the agreement of a set of time-bound and measurable targets known as the Millennium Development Goals (MDGs). The MDGs are an increasingly significant development framework primarily aimed at eradicating poverty, that is being pursued at varying speeds by Third World governments worldwide, supported by the UN system and multilateral and bilateral donors. Reflecting their significance (including their application to the tourism sector) the eight goals are summarised in Table 2.5. We will return to the MDGs in further detail in Chapter 11.

Table 2.5 *The UN's Millenium Development Goals*

GOAL 1	Eradicate extreme poverty and hunger
GOAL 2	Achieve universal primary education
GOAL 3	Promote gender equality and empower women
GOAL 4	Reduce child mortality
GOAL 5	Improve maternal health
GOAL 6	Combat HIV/AIDS, malaria and other diseases
GOAL 7	Ensure environmental sustainability
GOAL 8	Develop a global partnership for development

As Figure 2.4 suggests, there has been a tendency for the discourse on development to swing from one grand narrative to another (and reflecting the prevailing political economies that underpin the promotion of these ideas), so that, for example, major infrastructure is seen as an imperative for development within one era, but as anathema and anti-poor in another. Or as Parnwell (2002) argues, more mainstream approaches to development are ditched in favour of alternative models. But there are dangers to this ritual ditching and adoration of competing means of development, as Parnwell argues:

> Alternative models of development have tended to opt for the antithesis of the orthodox approach. Thus an urban and industrial bias is replaced with an emphasis on the rural and the agricultural; the top-down directionality and centralized character of development policy is challenged by decentralized, devolved and bottom-up initiatives; capitalism is superseded by socialist ideals; small-scale and particularistic development is seen as preferable to large-scale and universalistic approaches; and so on. In essence, we are presented with another binary, between the orthodox and the alternative. One danger of this is that the very real gains from orthodox development are seriously downplayed in favour of sometimes idealistic and neo-populist alternative visions of the future.
>
> (Parnwell 2002: 113)

This is of particular interest in the tourism field in the way that discussions of mass and alternative forms of development have been represented as binary opposites. But as we will see in Chapter 11, proponents of 'pro-poor tourism' (an alternative 'people-focused' approach to tourism development) have been quick to point to the benefits and necessity of mainstream tourism to coexist with 'alternatives'. Succinctly, we reject the notion that new – alternative – forms of tourism a priori constitute a mode of development, and that mainstream approaches to tourism do not. This discussion suggests that considerations of development are considerably more complex, not least because of the need to question at the outset the competing meanings and interpretations of development.

The myth of development?

To what extent then do sustainability, globalisation and development represent (in postmodern 'speak') a simulacra or three-way marriage and global hegemony on human advancement and order? Rist (1997) suggests that it is something of a 'postmodern illusion' or *trompe-l'oeil* within which the primacy of growth remains unchallenged and for which there is simply no alternative; and Escobar (1995: 4) refers to the history of development as a 'history of the loss of an illusion'.

Post-development approaches have surfaced as a radical reaction to, and dissatisfaction with, development, and more broadly with the intoxication of the 'West'. Bringing together a range of intellectual and political positions, they seek to criticise the project of 'development' rather than rejecting the possibilities of change (Rahnema and Bawtree 1997). As Nederveen Pieterse (2000) comments:

> post-development is a radical reaction to the dilemmas of development . . . with business-as-usual and standard development rhetoric and practice . . . Development is rejected because it is the new religion of the West . . . it is the imposition of power . . . it does not work . . . it means cultural Westernisation

and homogenisation . . . It is rejected not merely on account of its results but because of its intentions, its world-view and mindset.

(Nederveen Pieterse 2000: 175, quoted in J. Sidaway 2002: 16)

In the light of the experience of 'development' from the mid-twentieth century onwards, some commentators have turned to the demolition and revelation of the 'myth of development' (for example, see de Rivero 2001). As a powerful Platonic dualism in which development and underdevelopment give rise to a procession of easily worked, and often quoted binary opposites – civilised/uncivilised, free/enslaved, rich/poor, advanced/backward, modern/traditional – the complexity of both theorisation and practice has been bludgeoned into a series of unsustainable and inconceivable outcomes given the present global geopolitical and economic organisation: development is economically reductionist, has promoted rather than addressed inequality and is increasingly focused on issues of global security (W. Sachs 1999). Sachs concludes:

> So development has become a shapeless amoeba-like word. It cannot express anything because its outlines are blurred. But it remains ineradicable because it appears so benign. They who pronounce the word denote nothing but claim the best of intentions. Development thus has no content but it does possess a function: it allows any intervention to be sanctified in the name of a high evolutionary goal.
>
> (W. Sachs 1999: 7)

Former Peruvian diplomat Oswaldo de Rivero is equally critical. His critique is aimed directly at the globalisation gurus and the proposition that development will be achieved through 'worldwide competition with a totally unfettered global market' (de Rivero 2001: 5). This 'global financial casino', as de Rivero terms it, is created and maintained by the Washington Consensus, a set of economic principles that emerged from the US government, IMF and World Bank and an emergent international geopolitical regime reminiscent of Darwinism. The origins of the 'myth of development', argues de Rivero (2001: 110), 'lie in our Western civilization's ideology of progress', in the context of an 'ideology of happiness based on material progress' (2001: 111) and the brutal reality of capitalism's material accumulation. In a manner reflective of some protagonists in tourism studies who have contrasted the rapid global expansion of tourism with the unethical nature of much holidaying, de Rivero (2001: 141) argues the 'economy is being globalised, but ethics is not'.

Equally Rist's tentative conclusions are pessimistic, if somewhat depressing:

> 'development' appears as if suspended between the necessity and the impossibility . . . of its own realization. The talk about the 'world community' or 'global village' and the benefits of economic 'globalization' never ceases, but two-thirds of the planet is being increasingly separated off as the North patiently erects a wall to keep out the 'new barbarians'. Apartheid has been . . . reborn on a world scale. Does 'development' have a future in these conditions? Or should one hope, more simply, that there is a future after the end of the 'development era'?
>
> (Rist 1997: 210)

Ultimately, however, one universal discourse of development, as Esteva (1992: 9) suggests, continues to underscore and provide a 'global hegemony to a purely Western genealogy of history'; the Third World had and has no alternative than to follow a Eurocentric path to development: 'growth, evolution, maturation' (Esteva 1992: 10).

Development is real though and is unlikely to disappear as a system of thinking and instutionalisation of action. Development is an industry in its own right (with myriad global agencies, government departments, INGOs, consultants, experts, reporters, academics and so on), that despite good intentions nevertheless make livings from poverty; indeed, development has 'become a sector of economic activity in the same way as tourism . . . and too many interests are at stake for it to be simply wound up' (Rist 1997: 220).

Development as economic growth

As the discussion so far implies, at the heart of dissatisfaction, disquiet and disillusionment with development lie the intellectual and philosophical underpinnings that development is dependent on economic growth, *the* paradigm that has dominated and remained virtually unchallenged in the age of development. Escobar (1995: 94) forcefully argues that far from being an undisputed 'objective universal science' and neutral representation of the world, economics is a cultural discourse that socially constructs the world and which, by its very nature, is contestable. Invoking Edward Said's (1991) study of the power of cultural representation and domination from the West, to which we will return in Chapter 3, Escobar (1995: 62) argues: 'There is, then, an orientalism in economics that has to be unveiled – that is, a hegemonic effect achieved through representations that enshrine one view of the economy while suppressing others.' Similarly the Nobel Prize winning economist, Joseph Stiglitz, who was the World Bank Chief Economist between 1997 and 2000, and who provides an insider's perspective of the Bretton Woods institution is equally critical. He argues:

> discontent with globalization arises not just from economics seeming to be pushed over everything else, but because a particular view of economics – market fundamentalism – is pushed over all other views . . . it is not just opposition to the policies themselves, but to the notion that there is a single set of policies that is right. This notion flies in the face of both economics, which emphasizes the importance of trade-offs, and of ordinary common sense.
>
> (Stiglitz 2002: 220–1)

The significance and respect afforded economic growth and expansion will simply not go away. Two development commentaries are noteworthy and illustrative of the universalistic and growth-centred economic solutions and approaches to development. De Soto's (2001) work is particularly influential with key multilateral development agencies, including the World Bank, while Easterly's (2001) approach is interesting because it is, in part, an insider's critique of World Bank policy and programming. Moreover they are perhaps 'windows' into some of the most fundamental developmental debates (and could be read as manifestos for the continued growth and expansion of tourism as one of the key sectors of potential Third World development).

In accepting that capitalist economies provide the only demonstrable and serious option for development in the post-Cold War era, Peruvian economist Hernando de Soto (2001) argues in *The Mystery of Capital: Why Capitalism Triumphs in the West and Fails Everywhere Else* that the Third World is held back by an inability to produce capital. In a nutshell, de Soto argues that development is forestalled in the Third World because of the retardation of capitalism there, crucially 'not because international globalization is failing but because developing . . . nations have been unable to "globalize" capital within their own countries' (de Soto 2001: 219).

The centrepiece of de Soto's case rests on the production of capital and the fundamental significance of formal, legal property systems. In countries without formal property titling where the ownership of assets (the most significant of which is physical buildings) is difficult if not impossible to trace and validate, de Soto argues that most resources are financially invisible and cannot therefore be converted into capital. As de Soto (2001: 44) argues, it is 'formal property that provides the process, the forms and the rules that fix assets in a condition that allows us to realize them as active capital'. In effect, there are vast amounts of 'dead capital' which could otherwise be used to assist development. De Soto's research team estimate that the 'real estate held but not legally owned by the poor of the Third World and former communist nations is at least $9.3 trillion' (2001: 32), a figure that dramatically outstrips, for example, the total direct foreign investment in these countries between 1989 and 1999. By contrast, capitalism in the West has triumphed precisely because of the existence of 'one formal representational system' (de Soto 2001: 50) that records and regulates most assets.

Most stark is de Soto's unswerving belief that if capitalism has the chance to spread and develop in the Third World, it will deliver development. Although he acknowledges the inequitable nature of contemporary capitalism in developing countries, and indeed denies what might appear to be the credentials of a 'diehard capitalist', in his final analysis it is 'the only game in town' and the critical problems of economic and social apartheid are 'relatively easy to correct' (de Soto 2001: 241) if governments are willing to accept the 'property manifesto'. In an approach reminiscent of the pro-poor development paradigm, he confidently predicts from the 'perspective of the poor' that 'Everyone will benefit from globalizing capitalism within a country, but the most obvious and largest beneficiary will be the poor' (de Soto 2001: 201). This approach might be most readily translated to the field of tourism in the supposed processes of releasing capital (largely through formal property titling) to small producers and service providers linked to the mainstream tourism industry, or small-scale 'alternative' tourism providers.

William Easterly's (2001) *The Elusive Quest for Growth* is noteworthy as a critical inquiry of development economics and the application of theory and practice through World Bank programmes.[4] A former Senior Adviser in the World Bank's Development Research Group, Easterly's leitmotif (or the official motif of the book as he suggests) is that 'people respond to incentives' (Easterly 2001: 38) and that traditional economic models and practice, such as the pervasive financing-gap approach in the 1960s and 1970s (crudely where development aid plugs the temporary gap between national investment and savings in an attempt to support investment-led development) or the structural adjustment loaning of the 1980s, failed because they had violated this simple, universalistic rule. Needless to say, it is an approach wedded to economic growth and free market principles. Free trade, Easterly (2001: 230) argues, 'allows economies to specialize in what they are best at doing' with more open economies growing richer and faster, coupled with the underlying assumption that 'economic growth frees the poor from hunger and disease' (Easterly 2001: 8) and lifts them out of poverty. Statistical analysis, Easterly suggests, shows a one-for-one percentage increase between an increase in national average income and the income of the poorest 20 per cent of society.

Easterly's curious new right and economically deterministic account is vehemently opposed to debt relief as disincentivising and for throwing 'good money after bad' (Easterly 2001: 133), especially where government policy and behaviour is unchanged. His solution is, in part, to peg aid against a satisfactory policy environment and performance, in effect to 'tie aid to past country performance, not promises, giving the country's government an incentive to pursue growth-creating policies' (Easterly 2001: 118). One muted recommendation in order to enforce policy performance as a precondition for receiving aid is that 'countries should enter into "aid contests", whereby they would submit proposals for growth-promoting uses of the aid money', and as 'countries' incomes rise because of their favourable policies for economic growth, aid should increase in matching fashion' (Easterly 2001: 119).

The seductiveness of the approach taken by the likes of Easterly and de Soto lies in part in their conformity to economic tradition (that is, in their respect for economic growth and capitalist organisation) and in part in their simplicity, universality and supposed promotion of the economic 'missing link'; they are to development economics what self-help books are to western consumers and for the most part are decontextualised and dehumanised (with poverty and 'underdevelopment' largely homogeneous and indistinguishable geographically, socially or culturally). Both share a belief that incentives are capable of overcoming the barriers to development. For de Soto (2001: 228) this is that formal property offers the 'means to motivate people to create real additional usable value', though even the right leaning *The Economist* is pessimistic about the prospect of the 'silver bullet' (Economist 2006d: 66) of land titling: 'although land titling is certainly a good thing', it argues, 'it does not provide a single theory of development, as many *Sotistas* have in the past implied. Poverty, alas, is itself a barrier to risk-taking and enterprise' (Economist 2006c: 12). For Easterly (2001: 38), the crucial point is that 'people respond to incentives'. It is important that concepts of economic development in tourism avoid similar, supposedly universally applicable models and approaches.

Both analyses also present relatively flat, globally homogeneous, economic maps. Neither account traces the contours of the inherent uneven and unequal manifestation of the capitalist development process. For de Soto, the implication is that capitalism has produced equitable and healthy societies in the West. And neither demonstrates a particular interest in globalisation nor geopolitical questions. The historical context of colonialism and imperialism, the hegemony of global capitalism and the seemingly inherent inequalities in the development and application of global trading relationships are relative silences in these works. Despite Easterly's own critical observation of the apolitical nature of much development economics, both works appear to suffer from a lack of political analysis at the intra-nation state level; politics, like the economic analysis, appears to be the confine of the nation state. Such accounts fundamentally require a counterbalancing geopolitical critique (ÓTuathail 1994).

Easterly's functional and static analysis suggests pro-growth policies are most likely to work when 'class conflict and ethnic tensions are absent' (Easterly 2001: 279) and that by inference such countries are those with the biggest consensual middle classes. He contrasts the likes of Japan to Guatemala. And so the general pattern deduced is that 'countries with a high middle class share and low ethnic heterogeneity . . . are rich; those with a low middle class share and high ethnic heterogeneity are poor' (Easterly 2001: 279). This is both dangerous and crude. It is dangerous because it hints at a revisionist geopolitical history. Guatemala, to take one example, is a 'development failure' because of a 'fatal mixture of ethnic and class hatreds' (Easterly 2001: 276). The story of Guatemala's place in the history of imperial expansion and a civil war prompted and funded by successive US governments that has left 200,000 dead (Chomsky 1985; Painter 1989; J. Pearce 1982; Weinberg 1991) is not told (Blum 2002; Jonas 1991; J. Pearce 1982). It is crude because it offers an ahistorical and largely undifferentiated account of nation state formation and development, and tends to view the relationships between nation states as somewhat benign. De Rivero's critique of 'quasi nation-states' is noteworthy here for, as he argues, the majority of Third World countries experienced nation state formation in reverse where 'political authority . . . the state, emerged before the nation, before the national cultural identity and before the development of a true middle class and unifying national market' (de Rivero 2001: 21). De Rivero (2001: 23) concludes that 'quasi nation-states have been stabilised in underdevelopment for many years . . . they are not viable with their own resources' and that unlike the 'shocks' that are applied to Third World economies in an attempt to kick-start economic growth, First World capitalist economies achieved growth through protection.

It is also curious that while the contemporary analysis and critique of development increasingly situates itself within the context of globalisation, and more particularly economic forms of globalisation, the economic accounting entity, so to speak, remains the nation state. There is little analysis above or within the rigidity of the state. Quantitative data and achievement of targets at the state level has become an end in itself, and the fixation of the territorial boundedness of sovereign nation states has, more dangerously, signalled an unchallenged belief that all nation states have, in theory, viable economies capable of sustained economic growth given the right economic and political environment. But, as Easterly (2001) asks, is this reality of neoliberal free market economic specialisation an economic prospect for all nation states? Or does it retire some states to the dustbin in the quest for growth? Furthermore, few economically deterministic accounts of development appear to have questioned the dominance of realist interpretations of the primacy of the nation state. In contrast, de Rivero suggests 'developing countries' are 'dysfunctional, ungovernable chaotic entities', that are 'slipping towards the status of non-viable national economies' (de Rivero 2001: 143, 9). Sachs too has misgivings about the economist's 'blind eye' that fails to account for social relations in any other than economic terms (a shortcoming with a strong resonance in tourism studies), and the manner by which 'whenever development strategists set their sights on a country, they see not a society that has an economy, but a society that is an economy' (W. Sachs 1999: 17).

As suggested above, the primacy of economic growth has found fertile ground in multilateral and bilateral development policy. As an example, it is interesting to briefly consider the UK government's approach to overseas development. The government's official policy on development has been set out in three White Papers by the Department for International Development (DFID 1997, 2000a, 2006). The first, published in 1997 and entitled *Eliminating World Poverty: A Challenge for the 21st Century* (DFID 1997) addresses how the international development targets can be achieved with an emphasis on building partnerships and consensus, the need for consistent policies (on trade, for example) and the building of support for and understanding of the government's stance. Overall, the government appears anxious to recast overseas development as an approach to securing 'sustainable international development . . . [and] a new global society' (DFID 1997: 80), rather than a process of 'aid' giving. Such has been the pace of change in the global economic system that the second White Paper, *Eliminating World Poverty: Making Globalisation Work for the Poor* (DFID 2000a), specifically addresses the issues of development and globalisation. The Secretary of State for International Development contends that by analysing the nature of globalisation, the government 'sets out an agenda for managing the process in a way that could ensure that the new wealth, technology and knowledge being generated brings sustainable benefits to the one in five of humanity who live in extreme poverty' (DFID 2000a: 7).

The degree to which the faith in economic globalisation is well placed or the ability to successfully redistribute or channel its benefits to the poorest (the so-called pro-poor growth) is realisable, will only become clearer with time. What cannot be underestimated is the scale of the international development challenge nor the seemingly unconditional faith in the opportunities that globalisation is said to provide. For example, Tony Blair, the then UK prime minister, asserts that: 'Globalisation creates unprecedented new opportunities . . . If the poorest countries can be drawn into the global economy . . . it could lead to a rapid reduction in global poverty – as well as bringing new trade and investment opportunities for all' (DFID 2000a: 6).

There is a neoliberal consensus that poverty reduction induced through globalisation is contingent on economic growth and wealth creation, 'an indispensable requirement for poverty reduction' (DFID 2000a: 18). The starting point, therefore, is the almost universal acceptance that 'efficient markets are indispensable for effective development' (DFID 2000a: 23) and that growth is dependent on a 'continuation of market-based policies which promote investment in the context of low inflation and effective macro-economic management' (DFID 2000b: 9). The approach is pinned against continuing market liberalisation and opening of economies, led by the private sector. But so-called pro-poor growth also requires equity, and the White Paper argues, poverty reduction is fastest where inequality is also declining. Of course, while the subject of analysis is global, the strategies and outcomes are confined to nation states. This development platform accepts that the current levels of growth and consumption in the First World are not only acceptable but necessary for 'global international development', and that all Third World nation states have viable national economies. The benefits of globalisation can be attained by the right policy choices of governments, international institutions and business. (Chapter 11 will discuss how this policy approach is being translated into the promotion of pro-poor tourism initiatives.)

The good intentions of wishing to slash the numbers of those in absolute poverty aside, within the context of an international division of labour and increasing global inequality and unevenness, some critics will find it difficult not to conclude that the necessary precondition for this approach to development will result in *making the poor work for globalisation*. Even if there is an appreciable reduction in extreme poverty, and notwithstanding the real and important difference that this makes to individual lives, there is little or no evidence to suggest that there will be a deceleration of increasing levels of global inequality, arguably the barometer of a 'sustainable global society'. In the final analysis, and very much part of Eurocentric discourse, development is primarily, and is preconditioned by, an economic process. Economists are driving the contemporary development agenda, and the power, prestige and wealth characterised by First World economies is non-negotiable.

One is left to ponder how far economic analysis is short on meaningful social and geopolitical assessment. The degree to which such analysis is oversimplified and decontextualised, and whether it is too focused on deducing causal relationships from the experience of the First World and 'richer' Third World states where statistics are practically easier to collect,[5] may suggest it has serious limitations in theoretical construction and policy formulation. As David Harvey (2006) concludes:

> I cannot convince anyone by philosophical argument that the neo-liberal regime is unjust. But the objection to this regime of rights is quite simple: to accept it is to accept that we have no alternative except to live under a regime of endless capital accumulation and economic growth no matter what the social, ecological or political consequences.
>
> (Harvey 2006: 56)

New directions in development?

Even Nobel Prize winning economists sound a note of caution when it comes to the power of economics in providing *the* key to development. Joseph Stiglitz, former Chief Economist at the World Bank between 1997 and 2000, launched a stinging rebuke of the Washington Consensus and the special role of the IMF, describing the policies pursued by the IMF, World Bank and WTO as founded on a 'simplistic model of the market economy' and the 'outworn assumption that markets, by themselves, lead to efficient outcomes' (Stiglitz 2002: 74, xii). Trickle-down economics, Stiglitz (2002: 78) argues, 'was never much more than just a belief, an article of faith'. Of most concern to outsiders is the unsophisticated, single-track and unquestioning approach of blue-chip international agencies that wield such enormous influence. Stiglitz's frustration is clear:

> There was a single prescription. Alternative opinions were not sought. Open, frank discussion was discouraged – there was no room for it. Ideology guided prescription and countries were expected to follow the IMF guidelines without debate.
>
> (Stiglitz 2002: xiv)

Similar caution is sounded by economist Jeffrey Sachs, Special Advisor to the former United Nations Secretary General Kofi Annan. Sachs advocates both the will and means to address absolute poverty globally through assisted economic development (including global measures to cancel debt and institute global trade and aid policies) largely at the national and international level. He too recognises the excesses of economic primacy, a context within which 'free market ideologues took the argument to extremes that are utterly unsupportable by evidence or good economic reasoning' (J. Sachs 2005: 328), the results of which the political commentator Mike Davis (2006: 174) evocatively refers to as the 'brutal tectonics of neoliberal globalization'. Sachs heads for the centre ground, dismissing the extremism of 'free-market ideologues' and convinced that capitalism with a 'human face' (J. Sachs 2005: 318, 357) – or 'enlightened globalisation' – is capable of delivering mass salvation to the global poor. Sach's most important message (and one offered in response to de Soto's focus on housing, land and property) is a seemingly obvious one that development is a complex and messy business requiring careful local analysis (a spurious analogy to modern medicine he coins 'clinical economics'):

> all single-factor explanations fail the scientific test of accounting for the observed diversity of development experience. Dozens of recent statistical studies have shown that difference in economic growth rates across countries depends on a multiplicity of factors . . . The real challenge is to understanding which of these many variables is posing particular obstacles in specific circumstances – what I mean precisely by 'differential diagnosis'.
>
> (J. Sachs 2005: 322)

Nevertheless, some critics argue that far from challenging the prevailing development orthodoxy, Sachs' (2005) approach is a back-to-basics development 'top-down, expert-designed and run approach that has dominated development thinking and practice for over half a century' (Breslin 2007: 63) – an approach reminiscent of Rostow's stages of growth theory; evidence perhaps of the growing popularity of alternative modernisation (see Figure 2.4). Moreover, the increasingly public duelling between Sachs and Easterly (Easterly 2006) is perhaps a barometer of the lamentable state that the world's self-publicising development gurus, and indeed development as an 'industry' itself, are actually in. Sachs (2005) is focused very much on the basics of human life (in fact a matter of life and death involved in food and disease, for example, and largely African-centred), while Easterly (2006) is engaged with longer term, locally rooted, trajectories of change and growth. The former advocates the need for more centralised (and global) systems of provision; the latter backs home-grown solutions. Different sides of the same development coin, some might argue.

The approach taken by Amartya Sen (a fellow economics Nobel Prize winner) in *Development as Freedom* simultaneously provides a counterweight to the tendency of economics to reduce human agency to statistics and individuals to units of the labour force. Sen (1999) dismisses the arguments of post-development commentators as conceptually bankrupt. His approach is of particular interest in that it focuses on the causality of underdevelopment and looks at the role of individual agency rather than institutional and state structures (though he recognises both are an important element in ensuring individual freedoms). We will return to a discussion of approaches to development based on human needs, rights and securities in Chapters 10 and 11. Sen's approach is founded on 'freedom', and removal of unfreedoms he argues is both the means and the ends of development and the role of the individual is central to this; development *is* freedom. While the range of freedoms is diverse and varied in the local context, Sen focuses on five so-called 'instrumental' freedoms: political freedoms, economic facilities, social opportunities, transparency guarantees and protective security. For the analysis of tourism, Sen's approach is of particular interest in that it brings a critical consideration of those involved in tourism ('host' communities, tourists and the industry) and the degree to which tourism is able to support development as freedom. Thus, it implies an analysis of tourism not just as a purely economic effect (as is often promoted by advocates of tourism). It also provides a potentially powerful framework for considering the ethical justification for new tourism in countries with poor human rights records such as Burma; we return to this example in Chapter 10.

Sen's arguments are centred on the notion that while low incomes, poor health, inadequate education, violence and insecurity and so on may be the *symptoms* of under-development, it is the lack of freedom that is the *cause*. A focus on freedom is important for two reasons, one centred on 'processes', the other on 'opportunities'. The first emphasises the distinctiveness of the '*processes of decision making*' (Sen 1999: 291) underlining that development cannot focus solely on economic performance, and must see processes such as political participation as 'constitutive parts of the *ends* of development in themselves', not just a means to an end. The second advocates that development as freedom must gauge the degree to which people have the opportunities to achieve desirable outcomes. While rising real income levels may be important in giving people the opportunity to buy goods and services that improve their lot, 'income levels may often be inadequate guides to such important matters as the freedom to live long . . . or the opportunity to have worthwhile employment, or to live in peaceful and crime-free communities'. As Sen concludes, these 'non-income variables point to opportunities that a person has excellent reasons to value and that are not strictly linked with economic prosperity', and 'require us to go well beyond the traditional view of development' (Sen 1999: 291). For Sen, therefore, poverty is a measure of deprivation in processes and opportunities, not a result of low levels of income. Alternative ways of understanding and promoting development should be found therefore, as freedom 'cannot yield a view of development that translated readily into a simple "formula" of accumulation of capital, or opening up markets, or having efficient economic planning (though each of these particular features fits into the broader picture)' (Sen 1999: 297).

What then can we conclude from the decades of theorising and intellectualising about development, and of the development policies of governments and supranational institutions over fifty years of formulation and practice? The only point of agreement between the representative divergent views ranging from Easterly (2001) to de Rivero (2001) is that there is no agreement on how and if 'development' can be achieved. For Easterly (2001: 289), somewhat contradictorily given his account, there are 'no magical elixirs' in the quest for growth and development; for de Rivero (2001: 115), the reality is 'nobody knows how to reach El Dorado'. And on a practical level, as Rist (1997: 239) argues, far from 'bridging the ritually deplored gap' between rich and poor nations, development 'continues to widen it'. One thing is clear: there remains a widely held belief that economic growth is fundamental to the well-being of the poor and the progress of development, and there is little questioning of the need to slow down and redirect growth, with an attendant scaling down of our comparatively luxurious, free and consumptionist western lifestyles.

Conclusion

This chapter has introduced a number of ideas and concepts which will be elaborated and expanded in the rest of the book. It is best to think of these concepts as a bag of tools that are useful for building up an understanding of the complexity of tourism in the Third World and most especially for dealing with the new forms of tourism that find particular expression there.

There are three notions that are fundamental to the analysis of new tourism: globalisation, sustainability and development. In a number of important respects we have argued that these are not only the leitmotifs for this book, but also key to commencing an analysis of contemporary global change and capitalist expansion. Moreover, these three discourses (or ideas) are rapidly becoming one – interchangeable and a powerful structuring of First World geographical imaginations. The starting point is the accelerated pace of global change. We examined three aspects of globalisation – economic, cultural and political – in order to explore the key changes in late twentieth and early twenty-first century capitalism. In terms of economic globalisation, it was argued that an intense phase of time–space compression has necessitated a move to post-Fordist modes of production and a regime of flexible accumulation. The emergence of a global economy as a result of time–space compression was linked with the expansion of capitalist relations of production, especially to the Third World, where these changes are seemingly reflected in the rapid increase in tourism and the rise of different and more flexible forms of new tourism.

Time–space compression was also invoked in the discussion of cultural globalisation. Here the focus was on the emergence of so-called postmodernist cultural forms and of new patterns of consumption and the relationship of these forms to the growth of the new middle classes. It was suggested that these classes are major consumers of Third World tourism and that political globalisation serves to highlight the emergence of a range of global institutions and organisations, especially concerned with issues of sustainability, that are likely to have a major impact upon the development of Third World tourism.

Finally, despite the implicit suggestion within the idea of globalisation that everywhere is becoming the same, the discussion of globalisation alerted us to the possibility of increasing differences and to the uneven and unequal nature of development.

An attempt was also made to fix the notion of sustainability in a broader analytical framework. Again, this was traced through economic, cultural and political processes, ranging from the need to sustain profitability and economic growth through 'sustainable development', to the sustaining of cultural lifestyles and the politics and programmes of global institutions and organisations. Overall it has been argued that while a discussion of environmental or ecological notions of sustainability is central to the arguments about Third World tourism, on its own it provides too narrow a focus, and the importance of social, cultural and economic applications of sustainability must be acknowledged.

Finally, we emphasised the uneven and unequal nature of global capitalist development and the way in which this is inherent in the development of Third World tourism and examined the ways in which globalisation and sustainability are 'invested' in western concepts of development. Tracing the 'maturity' of development since the 1950s, it was argued that despite changes in approaches to development theory and practice, it remains indelibly marked by universalistic western discourses.

3 Power and tourism

Throughout Chapter 2 the uneven and unequal nature of development was emphasised and it was argued that an effective analysis of tourism must acknowledge the importance of relationships of power. In this chapter we begin to consider the way in which power is reflected through tourism in more detail. We start with a consideration of concepts of power and how these can assist a critical comprehension of tourism development; ideology, discourse and hegemony will be discussed in turn. The chapter then reviews the most systematic attempt to explain the unequal nature of tourism development – the political economy of Third World tourism that seeks to emphasise the dominance and control of tourism from the First World. The discussion moves on to trace other ways in which power has been implicated in the analysis of Third World tourism, particularly through the use of imperialism and colonialism. It is argued that these relationships of dominance have also emerged in new forms of tourism with the citation of 'neocolonialism' and 'eco-colonialism'. This chapter also provides a review of the importance of 'authenticity' to the study of tourism. It is argued that a consideration of authenticity is a further way in which relationships of power can be traced.

While the political economy of Third World tourism is of considerable interest and applicability (and indeed remains an important framework for understanding unequal development, especially of mass tourism) it is argued that it does not provide such a penetrating critique of new forms of tourism in the Third World. Indeed, political economy approaches suggest that the dominance of the First World over the Third World can be overcome, in part, by the creation of new 'alternative' forms of tourism. We challenge this suggestion.

The final section of the chapter suggests an alternative critique through four key characteristics of much new tourism. The first emphasises that all forms of tourism are tied into the growth and expansion of capitalist relations of production. We call this characteristic 'intervention and commodification'. It builds upon the economic aspects of globalisation from Chapter 2 and stresses the way in which holiday destinations are either drawn into a global system of interdependency, or are bypassed by it or in some cases are expelled from it. Given the context of global inequality and unevenness of development, the second characteristic stresses the 'subservience' that critics have argued characterises much tourism in the Third World, regardless of the form it takes. The final two characteristics seek to provide a more nuanced critique of new forms of tourism, referred to as 'fetishism' and 'aestheticisation', which seek to demonstrate the way in which the reality of the Third World is either hidden or is used to create a special aura of travelling in Third World regions.

Power play

Although we invariably associate tourism with pleasure and a certain playfulness, Indian academic, Nina Rao, reminds us that 'Tourism takes place in the context of great inequality of wealth and power' (quoted in Gonsalves 1993: 8), and power relations are central to our discussions in this book: we have already indicated in Chapter 2 that power is crucial to a critical understanding of development (see also Bianchi 2002a, 2002b). In pursuing this argument, we are seeking to address an identifiable weakness in much work on tourism. On the one hand, much tourism analysis has played down relationships of power, which remain either implicit or are absent. Such studies have largely consisted of identifying structural and deterministic models of tourism. These are examined more appropriately in Chapter 4. On the other hand, where power is invoked in a discussion of tourism it has tended to be in passing; references to ideology, discourse, colonialism, imperialism and so on appear in a rather unstructured, even anecdotal, fashion. Although such analysis is commendable in signalling the importance of power in the study of tourism, the treatment of power needs to be approached more thoughtfully. As Crick (1989) concludes from a wide-ranging review of social science literature, there is an inadequate representation of the complexities of tourism. Although some of the gaps have been filled since Crick's observation, the issue of power is still skirted around in many tourism analyses rather than placed at the heart of the analysis.

Initially, it is necessary to consider concepts of power that may assist a critical understanding of contemporary tourism and the themes we introduced in Chapter 2 (globalisation, sustainability and development). The discussion below presents three useful concepts that will be employed at various points in the book. These are ideology, discourse and hegemony, and factors of relevance to each are summarised in Figure 3.1. In addition, the discussion suggests how relationships of power are embodied in the 'project' of sustainability. In short, we are arguing that we require what Massey (1995b) refers to as a 'geography of power' to make sense of Third World tourism development.

HEGEMONY
- New tourism and new middle-class values
- WTTC and UNWTO promotion of tourism
- IMF and World Bank promotion of tourism
- Sustainability as a contested concept
- Globalisation

IDEOLOGY
- Environmentalism
- Travel advisories as reflectors of politial ideologies
- Sustainability as an ideology in its own right
- Eco-colonialism

DISCOURSE
- Traveller versus tourist
- Tour operators and tour brochures
- The tourist industry
- Environmental issues
- Assuming objectivity
- Intellectualisation of travel

Figure 3.1 The power jigsaw

Ideology

While ideology is a complex term, one profound trait stands out: namely, its concern with the 'bases and validity of our most fundamental ideas' (McCellan 1986: 1, quoted in Dobson 1995). In using the term ideology, we will be referring not only to the sustaining of relationships of domination in the interest of a dominant political power (the USA as the only global superpower, for example) or social thought (the supposed significance of religion to 'civilisation', for example), but also to interests that are opposed to dominant power (the anti-nuclear movement, environmentalists, feminists and so on) that are themselves capable of forming ideologies in the pursuit of power. Although, as Eagleton (1991) notes, this may signal a degree of contradiction in the meaning of ideology, it is nevertheless fundamental to the notion that ideology is about the way relationships of power are inexorably interwoven in the production and representation of meaning which serves the interests of a particular social group. Referring back to Chapter 2, for example, Stiglitz (2002) casts the IMF's operations as ideological:

> a set of beliefs that are held so firmly that one hardly needs empirical con-
> firmation. Evidence that contradicts those beliefs is summarily dismissed. For
> the believers in free and unfettered markets, capital market liberalization was
> obviously desirable: one didn't need evidence that it promoted growth.
>
> (Stiglitz 2002: 222)

As Dobson (1995: 7) concludes, ideologies 'map the world in different ways', and it is the intention of this book to map the way in which different interests are implicated in the uneven and unequal development of tourism.

Sustainability is ideological in the sense that it is largely from the First World that the consciousness and mobilisation around global environmental issues have been generated and in the sense that sustainability serves the interests of the First World. Adams (2001), for instance, refers to the ideology of sustainable development, and in the context of tourism in southern Mexico, Daltabuit and Pi-Sunyer (1990) refer to the 'ideology of environmentalism'. The power implicated through First World environmentalism has led increasingly to the 'charge' of eco-imperialism and eco-colonialism.

Implicit in these criticisms is the idea that sustainability is ostensibly ethnocentric. Reconsider for a moment the quote from Robins (1991, see p. 16) where he talks about the export of western values and priorities. Such observations can also be applied to the current debate on sustainability. For the most part, it is a discussion framed in the West and imposed on the 'Rest', and hence the acrimonious debates between First and Third World countries at the Rio Summit, Seattle trade talks, the G8 Summit in Genoa and the collapse of the Doha Development Round of trade talks, to name just a few.

Discourse

The second key concept, discourse, is closely related to ideology. Ideology is perhaps best thought of as a discriminator between power struggles which are central to a 'whole form of social life' (socialism, feminism, ecologism perhaps) and those which are, for whatever reason, relatively less holistic. Prioritising the most important forms of struggle may be an exercise of power itself, but it is important to signal which struggles are ideological and which are not.

Discourse can be considered as complementary to ideology. Indeed ideology is a matter of 'discourse', a 'question of who is saying what to whom for what purposes' (Eagleton

1991: 9). But discourse can also be non-ideological; in other words, it is not reducible to ideology.

The French philosopher Michel Foucault (1980) suggests that discourse expresses how 'facts' can be conveyed in different ways and how the language used to convey these facts can interfere with our ability to decide what is true and what is false. Discourse, Foucault argues, is so much more than 'mere' words; words are not 'wind, an external whisper, a beating of wings that one has difficulty in hearing in the serious matter of history' (Foucault 1972: 209, quoted in Escobar 1995); words as a discourse provide the conditions, practice, rules and regulations on thought. As such, 'development' and 'sustainability' are powerful discourses as our earlier discussion suggested. For example, as Chapter 4 considers, the term 'carrying capacity' (an important tool in the study of sustainability) can be subdivided into different types: ecological, social, economic, physical, real, effective, aesthetic; all of these can be interpreted and measured in different ways by different people at different times and in different circumstances. But carrying capacity is often treated as if it were a 'neutral' ecological term. Zaba and Scoones (1994) challenge this neutrality:

> most of us have no problems with the notion of the carrying capacity of Botswana (pop. 1.3 m, area 567,000 sq km), but would be incredulous at the idea of calculating the carrying capacity of Birmingham (pop. 1.1 m, area 300 sq km).
>
> (Zaba amd Scoones 1994: 197)

Not only does the notion of ecological sustainability bring some kind of scientific validity with it, but also it suggests that some places (in this case Third World environments) are more suited to its application than others. In this way, carrying capacity as discourse transmits and translates power.

There is no agreement over the exact nature, content and meaning of sustainability. It is a contested concept in all senses of the word. Different interests – supranational and transnational organisations, INGOs, socio-environmental organisations, social classes and so on – have adopted and defend their own language (discourse) of sustainability. The new socio-environmental organisations mobilised around issues of environment, for example, are not in power, and yet their ability to influence the meaning of sustainability for our everyday lives has been marked.

Similarly, consider the power to interpret and represent the Third World through travel books and brochures. On the one hand, we have the highbrow, intellectual accounts of best-selling travel writers such as Paul Theroux and Eric Newby, authors noted for the 'authoritativeness of their vision' (Pratt 1992: 217; see also Wilson and Richards 2004), and the serious travel pages of broadsheet newspapers. On the other hand, we have glossy high street tourist brochures selling destinations from the Caribbean to Thailand, and which are the subject of much highbrow, intellectual criticism. These are simply different ways of outsiders representing and interpreting the Third World to their audience, each claiming authenticity and truth, albeit in very different ways. Equally, there is the perpetual discursive battle among tourists themselves (especially 'anti-tourists') in effectively providing social space between mass consumption and more individualised forms of new tourism (Jaworski and Pritchard 2005; Richards and Wilson 2004).

Foucault's ideas may lead to the conclusion that knowledge in tourism is produced by competing discourses. Discourse, therefore, is a useful concept in emphasising how a certain subject or topic is talked and thought about and how it is represented to others. Most importantly, discourses are 'part of the way power circulates and is contested' (S. Hall 1992b: 295).

Hegemony

Discourse is also an essential property of hegemony, our last concept, in the power jigsaw. Hegemony was a concept developed by the Italian Marxist, Antonio Gramsci, to emphasise the ability of dominant classes to convince the majority of subordinate classes to adopt certain political, cultural or moral values; a more efficient strategy than coercing subordinate social groups into conformity (Jackson 1992). Hegemony, therefore, is essentially about the power of persuasion and is immediately differentiated from ideology, which by contrast may be imposed forcibly (Eagleton 1991), as in the former apartheid system in South Africa or through the imposition of IMF structural adjustment policies. The best way to conceive of hegemony is as a 'broader category than ideology' which '*includes* ideology, but is not reducible to it' (Eagleton 1991: 112, emphasis in original).

The real innovativeness of Gramsci's thinking is the conclusion that hegemony is never fully realised in capitalist societies – that it is continually contested (Jackson 1992). As Williams concludes, hegemony must be 'renewed, recreated, defended, and modified' (quoted in Eagleton 1991: 115) and is 'inseparable from overtones of struggle' (Eagleton 1991: 115); a relationship that does not necessarily hold true for ideology.

The concepts of the Third World, development, sustainability and tourism are examples of hegemony in practice. Tourism, as we shall see in later chapters, is replete with examples of hegemonic strategies ranging from tourism codes of conduct to the advocacy of more responsible, appropriate or sustainable forms of tourism. It is also evident in the way in which tourism is contested between different social groups (traveller versus tourist, for example) and between different places (Thailand versus Chile, for example). Hegemony is especially useful for its dynamism and practical usage encompassing and focusing attention on a wide range of practical strategies that are adopted by a variety of interests. Such characteristics place notions of struggle and contest at the centre of the enquiry. A useful example is provided by Hutnyk's critique of travellers-cum-volunteers in Kolkata (India). Considering questions of cultural hegemony, Hutnyk (1996) argues that travellers' ability to engage in and promote the complexities of Kolkata are compromised by: '(a) the insularity of traveller culture and traveller style; (b) the cultural and class background of western travellers; (c) the hegemony of western versions of Calcutta in "traveller lore"; and (d) the hegemonic effects of the traveller "gaze"' (Hutnyk 1996: 44).

The advocacy by environmentalists of the need to act globally, for example, is an interesting aspect of the persuasiveness of sustainability and how it ties in both the global and local dimensions and stresses the interdependency of places. Residents of distant places are asked to 'consider' other places; in the dictum of Friends of the Earth, 'think globally, act locally'. Conservation measures in southern Africa and rainforest preservation in Central America can be lobbied for and financed from the First World; and a degree of control and influence over Third World affairs is exercised through First World conscience-prodding. Sachs (1993) refers to this as the 'hegemony of globalism':

> Until the 1980s, environmentalists were usually concerned with the local or national space . . . But in subsequent years, they began to look at things from a much more elevated vantage point: they adopted the astronaut's view, taking in the entire globe at one glance. Today's ecology is in the business of saving nothing less than the planet.
>
> (W. Sachs 1993: 17)

Testimony to the hegemonic properties of sustainability, perhaps, is the rapidity with which the word has entered public usage on a seemingly global level since its use by Brundtland in 1987 (World Commission on Environment and Development 1987), along with the large number of texts that are devoted to dissecting, interpreting, defending or reclaiming the idea of sustainability. For some it is a means of sustaining much more than just environment. It is about 'sustainable development' and incorporates indicators such as income, employment, health, housing, human welfare indicators that are concerned with a 'more rounded policy goal than "economic growth"' (Jacobs and Stott 1992: 262). For others sustainability is to be reclaimed within a far more radical agenda of political ecology where ecological issues and questions of social justice are paramount (for example, see Hayward 1994; Lipietz 1995; Shiva 1988). Characteristic of hegemonic positions, sustainability is contested within a continuum of viewpoints (Adams 2001) ranging from 'reformism' (often referred to as light green, conservationist or environmentalist) to 'radicalism' (referred to variously as dark green, deep ecology or, in Dobson's (1995) phraseology, ecologism).

Similarly, sustainability and its application to tourism should not be considered a once-and-for-all position – a neutral, scientific term to which techniques can be applied and upon which policies and programmes can be implemented and evaluated and blueprints, ideal types and models catalogued and advocated. Rather, it constantly changes as the broader influences and interests change, reflecting a dynamic situation and concept.

In the next section we turn to the most concentrated analysis of power in tourism: that offered by the approach known as political economy which is derivative of the neo-Marxist dependency theory discussed in the previous chapter.

The political economy of Third World tourism

By the early to mid-1970s it was already acknowledged that tourism did not necessarily offer a panacea to Third World countries struggling for economic growth (Turner 1976; Turner and Ash 1975). A number of highly critical studies focusing, in particular, on the fate of the small island economies in the Caribbean (Bryden 1973; Hills and Lundgren 1977; Perez 1974, 1975) began to highlight the unequal economic and social impacts associated with tourism. Of special importance was the observation that Third World economies drawn to tourism as a way of earning foreign exchange witnessed the leaking of much of the money made, straight back out of their national economies. This leakage, as it is now commonly known, was seen to arise primarily as a result of the First World ownership and control of the tourism industry in the Third World: from hotels to tour operators and airlines – see the section in Chapter 7 on size and structure of the tourism industry.

These early studies also began to hint at the relationship between tourism and 'underdevelopment'. It was not until Stephen Britton's analysis of Fiji, however, that a more thorough attempt was made in applying dependency theory to the study of tourism (Britton 1981a, 1981b, 1981c, 1982). The importance of Britton's analysis is that he stresses the need 'to place tourism firmly within the dialogue on development' (Britton 1982: 332) and investigate why tourism so often perpetuates uneven and unequal relationships between the First and Third Worlds.

As discussed in Chapter 2, the theory of dependency is best understood as a riposte to the laissez-faire (free market economics) approach to economic development and international trade. The global expansion of capitalism has drawn the Third World into increasingly tight economic relationships with the First World, and tourism, now the largest global industry, has been a significant component in this process. Dependency theory has sought to demonstrate how and why these tightening relationships are highly unequal.

Dependency theory argues that western capitalist countries have grown as a result of the expropriation of surpluses from the Third World, especially because of the reliance of Third World countries on export-oriented industries (coffee, bananas, bauxite and so on) which are notoriously precarious in terms of world market prices. The theory uses the notion of centre–periphery (or core–periphery) relationships to highlight this unequal relationship, where the core is the locus of economic power within a global economy.

The most widely cited of the dependency theorists, André Gunder Frank (1966, 1969, 1970), takes matters one step further in his notion of the 'development of under-development' which stresses that it is the underdevelopment of the structures in Third World countries created by First World capitalist development that creates dependency (Kay 1989; Rodney 1988). Above all else, theories of dependency are in general agreement that the interdependence resulting from global economic expansion and the suppression of autonomous growth results in unequal and uneven development.

Britton applies this body of theory to tourism. Centrally, he argues, dependency involves the 'subordination of national economic autonomy' (Britton 1982: 334) as a direct result of the unequal relationships inherent in the world economy and that within the present structure of international tourism, Third World countries can assume only a passive role (Britton 1981a). Britton summarises his approach as follows:

> Underdeveloped countries promote tourism as a means of generating foreign exchange, increasing employment opportunities, attracting development capital, and enhancing economic independence. The structural characteristics of Third World economies, however, can detract from achieving several of these goals. But equally problematic is the organisation of the international tourist industry itself.
>
> (Britton 1982: 336)

But Britton's research and narrative are very much part of the analysis of the mainstream – mass – tourism industry (see Bianchi (2002b) for a broader discussion of new political economies of global tourism). As such, he argues that tourism in Third World economies is best conceptualised as an enclave industry (referred to by Turner (1974, 1976) as the 'golden ghettos' and by Krippendorf (1987) as 'holidays in the ghetto') where tourists only occasionally venture beyond the bounds of their hotel compounds (referred to as an 'environmental bubble'). While Britton's critique is widely cited and provides valuable insights into the unequal structure of Third World tourism, we must ask how useful his analysis is for a critical understanding of new forms of Third World tourism that seek to escape the 'ghettos'. We return to this consideration a little later.

Tourism as domination

Given the arguments advanced by an increasing number of tourism commentators that the Third World is structurally dependent on the First World, there is little surprise in finding a wide range of references to the principal forms of global domination: colonialism and imperialism. While these terms are often used loosely and interchangeably, colonialism is best conceived as a special form of imperialism (that is, the imposition of power by one state over another) involving the occupation of territories. The following sections begin to build up a picture of how these relationships of power are reflected in the analysis of tourism.

The significance of colonialism and imperialism to theories of underdevelopment and dependency has a special appeal to writers on tourism. Both the characteristic First World ownership of much Third World tourism infrastructure and the origin of tourists from the First World have for many become an irresistible analogy of colonial and imperial domination. Indeed, the distinction drawn in dependency theory between a First World core and Third World periphery is part of a more general theory of imperialism. Nash (1989) argues that tourism exists only in so much as the metropolitan core generates the demand for tourism and the tourists themselves. He concludes: 'it is this power over touristic and related developments abroad that makes a metropolitan center imperialistic and tourism a form of imperialism' (Nash 1989: 35). Similarly, van den Abbeele (1980) laments tourism as doubly imperialistic both in turning Third World cultures into a commodity and providing hedonistic practices for wealthy First World tourists. Clearly, this is more than just an academic concern or critique. Take, for example, Box 3.1, which provides the background to Survival International's campaign on tourism and tribal peoples.

Box 3.1 The 'new imperialism'

The majority of the world's tourists are from the industrialised countries: 57 per cent from Europe, 16 per cent from North America; 80 per cent of all international travellers are nationals of just 20 countries. Thus it is largely the tourist industry in the affluent tourist-generating countries that determines the nature and scale of tourism. These tour operators are interested primarily in short-term benefits and realising a return on capital and investments . . .

Much of the money generated by tourism is sent abroad; 60 per cent of Thailand's $4 billion a year tourism revenues leave the country. Some critics of the tourist industry have called it the 'new imperialism'.

While tourism usually promises to provide employment to the local community, the jobs are most often unskilled, menial and poorly paid . . . More often than not, the needs and rights of indigenous peoples are ignored. For example, in west Nepal, the Chhetri people were moved from their lands to make way for Lake Rara National Park . . . In Kenya's Shaba reserve, scarce water is diverted from the spring once used by local Samburu herdsmen to water their cattle, in order to fill the swimming pool of the Sarova Shaba Hotel . . .

Survival does not claim to be able to resolve the debate surrounding ecotourism. However, when tribal communities are the destination for tourists, it is right and appropriate that the wishes of these communities are respected. The key word is control. Not only do tribal peoples have a right to their lands, they also have the right to decide what happens on their lands, to determine their future and way of life . . .

Ecotour operators are busy selling 'rainforest tourism' to the environmentally interested traveller by promoting the image of tribal people as 'noble savages'. Rather than patronising tribal peoples . . . we need to see them on their own terms as dynamic and complex societies . . .

The need to bring in foreign currency is used to justify this abuse of tribal peoples' rights and denial of their dignity. Clearly, tour operators and governments are often willing collaborators and perpetrators of this form of exploitation. This can be stopped if tribal peoples are given control over the . . . development of tourism in their communities.

Source: Extracts from Survival International (1995) 'Tourism and tribal peoples', information sheet.

Note: Survival is a worldwide organisation supporting tribal peoples, and more examples of its research are given in Chapters 6 and 8.

For Third World critics in particular, as Gonsalves (1993: 11) observes, it is the very presence of tourists that leads to the 'view that modern tourism is an extension of colonialism (with all the attributes of a master–servant relationship)'. It is an increasingly widely shared opinion within the Third World. Chung Hyung Kyung's (1994) observations are illustrative of the passion and conviction with which these are expressed: 'Colonialism has many faces. Third World tourism, an advanced form of "post-colonialism", is a disease which destroys people's bodies and souls . . . Third World tourism carries a major symptom of colonialism: "Domination and Subjugation"' (Chung 1994: 21).

It is this notion that tourism is implicated in the maintenance of neocolonial states that is so important here. Perez (1974: 473), for example, argues that 'Travel from metropolitan centres to the West Indies has served historically to underwrite colonialism in the Caribbean'. Bruner (1989: 439) insists that, however much we attempt to deny or evade the relationship, 'colonialism . . . and tourism . . . were born together and are relatives'. They are, Bruner contends, driven by the same social processes involving the occupying of space (by tourist infrastructure and ultimately by tourists) opened through the expansion of power.

It is not just academics that are drawing parallels between tourism and colonialism. Srisang (1992), a former Executive Director of the Ecumenical Coalition on Third World Tourism (ECTWT), the world's largest NGO on tourism, suggests that:

> tourism, especially Third World tourism, as it is practised today, does not benefit the majority of people. Instead it exploits them, pollutes the environment, destroys the ecosystem, bastardises the culture, robs people of their traditional values and ways of life and subjugates women and children in the abject slavery of prostitution. In other words, tourism epitomises the present unjust world economic order where the few who control wealth and power dictate the terms. As such, tourism is little different from colonialism.
>
> (Srisang 1992: 3)

The ECTWT itself is equally outspoken, referring to the majority of Third World tourism as 'an expression of neo-colonialism contributing to racism, erosion of moral values, economic impoverishment and cultural degradation' (ECTWT leaflet, undated).

For other writers, however, the relationship between colonialism and tourism to which they allude amounts to little more than a casual or anecdotal observation, often on the tourists themselves. Hence, the analogies between the affluent middle classes and 'scavengers' (MacCannell 1976), the 'easy-going tourist' and the 'conqueror and colonialist' (E. Cohen 1972), the suggestion that 'for many tourists, aggressive – almost colonialist – behaviour becomes a norm while on holiday' (G. Shaw and Williams 1994: 80), and the charge that tourists are the 'terrorists of cultural expansion' (Iyer 1989, quoted in Hutnyk 1996: 13). As Krippendorf (1987: 56) concludes, in the absence of changes, tourism will remain for the host 'a special form of subservience'.

While such observations are understandable, even justified in the way in which tourism seems to reawaken memories of a colonial past (Crick 1989), they represent a reaction to tourism based more upon an emotional response. In this vein, as Allen and Hamnett (1995: 252) conclude, it 'can be argued . . . just as some "Third World" countries have thrown off the yoke of colonialism, they have taken up the yoke of tourism'.

Two observations arise from this review. First, is the rather ambiguous fashion in which the charge of imperialism and colonialism is often made. Because tourism is a conduit for relationships of power, it has been easy for authors to use terms for these forms of domination to describe a vast array of relationships involved, from multinational hotel chains to a waiter–diner exchange. It has, therefore, become an attractive comparison to make, with the words imperialism and colonialism immediately invoking certain images and responses in our minds. Second, as we observed with Britton, a good deal of the critique arises from observations of the mainstream mass tourism industry. It is somewhat blunt or crude in dealing with new forms of tourism whose claim is to escape these very relationships of domination. In the section 'Alternative critiques for alternative tourism?' (pp. 60–81), therefore, we attempt to analyse (or disaggregate) these forms of power. This provides a clearer picture of how relationships of domination are manifest and suggests how these observations might be applied to new forms of tourism.

One way in which the discussion can be reframed is through the reference to 'neo-colonialism'. As Nicholas Thomas (1994: 1) argues, although colonialism as a pervasive moment in history has all but gone, 'the persistence of neocolonial domination in international and inter-ethnic relations is undeniable'. In the context of tourism, the charge of neocolonialism has already emerged as a principal way of describing the retention of former colonies in a state of perpetual subordination to the First World, in spite of formal political independence, a view reflected in de Rivero's (2001) analysis of non-viable national economies. Hence, Britton (1981c) refers to Fiji as a neocolonial economy and seeks to demonstrate why tourism reinforces the pattern of spatial organisation which evolved during colonialism. The tourist industry he argues is a 'neocolonial extension of economic forms present in pre-independent Fiji' (Britton 1981c: 149). Similarly, Shivji (1973) argues of Tanzania, 'Since the success of tourism depends primarily on our being accepted in the metropolitan countries, it is one of those appendage industries which give rise to a neo-colonialist relationship and cause underdevelopment'.

While such analysis is clearly significant in constructing a broader critique of Third World tourism, the discussion has too often been restricted to a consideration of economic impacts. So, for example, the complexities of class and race have been largely neglected and the spectre of neocolonialism engendering a subtle, but pervasive racism has remained largely unexplored. Again, we will return to such considerations below. Most notably then, a discussion of neocolonialism allows us to think in terms of the existence of discourses of 'colonialisms' (N. Thomas 1994), and explore the many different ways in which power is spatially and socially expressed in a so-called postcolonial world.

One way in which we can immediately see the relevance of thinking in terms of different forms of neocolonialism is its application to sustainability and environmentalism, and ultimately to development. On the one hand, there are emerging critiques of environmental organisations themselves, not far removed from the critical attacks launched on other supranational agencies such as the World Bank and IMF. John Phillipson, an internal auditor of the WWF, for example, accuses them of egocentricity and neocolonialism (as cited in Fernandes 1994). Central to these criticisms is the way in which organisations seek to impose policies and programmes on Third World countries. On the other hand, a more thorough critique of environmentalism and ecologism as a movement has begun to emerge. Such criticisms have tended to focus on the morality vested in the 'environment' (an entity that must be protected and saved) and the crusade-like fashion with which environmental issues are pursued. To advance the analogy, there is a sense in which an army of eco-missionaries, or as some would argue, ecofundamentalists, have fanned out across the Third World to green the Earth's poor.

As noted earlier, the moral basis for environmentalists' claims has emanated from the symbol of interdependence and 'oneness' of the Earth which is founded upon the notion of a global ecosystem. Wolfgang Sachs (1992a: 22) neatly draws out the relationships of power and domination from this 'systems language' that is committed to 'regulation and control', arguing the 'terms "ecosystem" or "global system" cannot shake off the legacy of engineering'; Sachs (1992b: 31) also argues that the 'concept ecosystem that gave to the ecology movement a quasi-spiritual dimension and scientific credibility at the same time'. Sachs (1992b: 31) concludes that for many environmentalists 'ecology seems to reveal the moral order of being . . . it suggests not only the truth, but also a moral imperative and . . . aesthetic perfection'. Central to Sachs' concern is the way in which the univer-salistic discourses on development (as discussed in Chapter 2), or the 'hegemony of globalism' as he refers to it, have imposed a system of global resource management that undermines nature and undercuts local autonomy, difference and diversity. Sachs concludes:

> In the face of the overriding imperative to 'secure the survival of the planet', autonomy easily becomes an anti-social value, and diversity turns into an obstacle to collective action. Can one imagine a more powerful motive for forcing the world into line than that of saving the planet? Eco-colonialism constitutes a new danger for the tapestry of cultures on the globe.
>
> (W. Sachs 1992b: 108)

From Sachs' (and others') writings, we are quickly led to question the intention and outcome of much environmentalism and the way it is advanced through notions of sustainability. As the environmental critic, Vandana Shiva, asks rhetorically: 'Global environment or green imperialism?' (Shiva 1993: 151). The importance, once again, is the way in which such discussion reflects back on the global changes considered in Chapter 2. Globalisation drives us towards the logical conclusion that there is only one world: a global economy, a global culture, a global environment. It is the violent imposition of this idea from the First World, an imposition clearly reflected in the Rio Summit, which creates the 'moral base for green imperialism' (Shiva 1993: 152).

These critiques of environment and ecology have clear and wide-ranging ramifications for the study of tourism too. It is the power invested in the concepts such as sustainable tourism and environmental tourism that has been central to critical responses. Reflecting the discussion of discourse, Herman (1992) argues that 'sustainable tourism is rooted in much "double-speak"'. This double-speak needs to be recognised as 'The misuse of words by implicit re-definition, selective application of . . . words, and other forms of verbal manipulation' (Fernandes 1994: 28, quoting Herman 1992). Similarly, writing of tourism development in Quintana Roo, southern Mexico, Daltabuit and Pi-Sunyer (1990: 10) refer to environmentalism as a 'powerful rhetoric'.

Beyond dependency?

The discussion so far argues that relationships of power have been at the heart of critiques of Third World tourism. Consequently, traditional responses and suggested solutions to these problems (exemplified in Britton's approach) have been couched in the need to move beyond dependency and to break off the subordination and domination that is said to characterise mass tourism. As we have suggested, much political economy is situated firmly within a typical mass packaged tourism environment. Predictably, prescriptions for a reform of such tourism rest on the need to move beyond mass tourism – or 'conventional' tourism – and develop alternative forms (Britton and Clarke 1987). Among other things, this shift responds to the problems Britton identifies, promotes the local ownership of tourism resources, creates local employment and helps stem the haemorrhaging of foreign exchange from the economy.

A second, and very different tack, is offered by Auliana Poon (a UNWTO and UNDP consultant attached to the Caribbean Tourism Organisation; we return to the policies of these organisations in Chapter 10). Like political economists, Poon (1989a) sees dependency as the key question. But for Poon 'dependency' is defined in a very different context and she challenges the general agreement that small island states are compelled to participate in a world economy beyond their control. Of the Caribbean, Poon (1989a: 74) argues, 'The innovativeness of many indigenous tourism enterprises . . . coupled with new developments in the world tourist industry, warrants a re-conceptualisation of the world question of dependence'. The conclusion drawn by Poon is that it is not the metropolitan core singled out by political economists (and others) nor the lambasted activities of multinational and transnational companies upon which Third World countries are dependent. In a linguistic (and conceptual) twist reminiscent of free market discourses, Poon argues that tourism in the Third World is dependent on 'innovation', the 'fostering of indigenous skills, creativity and innovativeness, rather than perpetual reliance on MNCs [multinational corporations], which hold the key to the future survival' (Poon 1989a: 74). It is a conclusion that appears to lay both onus and blame squarely on the shoulders of tourism providers in the Third World and exonerates the mass tourism operators from culpability in the highly uneven and unequal development of global tourism. Moreover, it is an interpretation of dependency that anticipates the policies and programmes of donor bodies. These agencies impose their own vision of Third World tourism on the countries and projects to which they lend.

Others have followed suit in advocating a range of tourisms prefixed with descriptions that indicate their qualitative difference from current forms of mass tourism and which in some cases seek to express the goal of tourism as an activity: community-based tourism, ethical tourism, pro-poor tourism, responsible tourism and so on. We review these approaches later.

Similarly some authors have chosen to advocate the role of independent travellers in promoting socially just forms of tourism. Kutay (1989: 35) rather eulogises travellers, referring to them as 'Peace corps-type travellers looking for a meaningful vacation', and Jeremy Seabrook (1995: 22) sings the praises of those dissatisfied with the packaged form of travel: 'the success of the Lonely Planet Guides testifies to the hunger of young people for a deep exploration of the countries they visit and that includes a growing curiosity about social and living conditions'. Seabrook moves towards what could be argued is at best wishful thinking, at worst an apology for the acknowledged problems of all forms of tourism. His approach is problematical not only in assuming to speak for Third World communities, but also to speak on behalf of tourists themselves. His new-found activity is social tourism:

> Social tourism can never be a mass movement of people . . . It is a response to a growing number of people who wish to deepen their understanding of north-south relations . . . What they wish for is contact with rootedness, true diversity, and an extension of understanding, rather than more escapism, with which their own culture amply supplies them.
>
> (Seabrook 1995: 23)

While, therefore, mass tourism has attracted trenchant criticism as a shallow and degrading experience for Third World 'host' nations and peoples, new tourism practices have been viewed benevolently and few critiques have emerged. But do these alternatives offer viable responses and solutions to the existing problems as so many authors imply? Or are these new tourism practices further evidence of the way in which the ebb and flow of tourism is conditioned and controlled from the First World?

Returning to Poon's analysis, we find that a transformation and metamorphosis is predicted from an old tourism which was 'not only mass, but . . . standardised and rigidly-packaged', to new tourism based 'upon a new "commonsense" or "best practice" of

Flexibility, Segmentation and Diagonal Integration' (Poon 1989a: 74, 75); a point that will be expanded upon in Chapter 4. Box 3.2 summarises Poon's position.

In an approach that lacks both explanatory power in assuming an oversimplistic binary division and a critique of the impacts (both positive and negative) of tourism, Poon (1993) suggests that mass tourism will fade into relative unimportance though not disappear altogether. Drawing upon an analogy of typewriters and computers she concludes:

> While there will continue to be a market for typewriters (mass tourism), the growth of new computers (new tourism) will be far greater. Having used typewriters and then been exposed to the power of computers, users will be unwilling to go back to the old way. This exact logic holds for old and new tourism.

> (Poon 1993: 23)

As discussed in Chapter 2, what Poon (1993) describes as new tourism is referred to as 'post-Fordist consumption' by Lash and Urry (1994: 274). Although Poon's analysis is less conceptually elaborate, both approaches strike the same chord. Lash and Urry (1987, 1994) build their ideas upon a transition from one stage of tourism to another, although their ideas are set within a deeper theoretical seam that runs through their two principal works, which set out a political economy of late capitalism. For Lash and Urry the development of capitalism takes place in three main phases: liberal capitalism, organised capitalism and disorganised capitalism. In terms of this last stage, they do not contemplate a shift to some form of 'random disorder' but rather a 'fairly systematic process of disaggregation and restructuration' (Lash and Urry 1987: 8) of capitalism and capitalist countries.

So how do these stages relate to the emergence and development of tourism? Liberal capitalism, they argue, is characterised by individual travel by the wealthy, usually

---

Box 3.2 Tourism in metamorphosis?

Old tourists	*New tourists*
Search for the sun	Experience something new
Follow the masses	Want to be in charge
Here today, gone tomorrow	See and enjoy but not destroy
Show that you have been	Just for the fun of it
Having	Being
Superiority	Understanding
Like attractions	Like sport and nature
Reactions	Adventurous
Eat in hotel dining room	Try out local fare
Homogeneous	Hybrid

Source: Poon (1993: 10)

stereotyped by the Grand Tour of the eighteenth and nineteenth centuries. The second phase is best understood as one of organised mass tourism that stretches from the development of Blackpool to the emergence of the costas (in Spain) and package destinations elsewhere. Their last phase of disorganisation is a little more difficult to justify in claiming the 'end of tourism'. Sociologically, they argue that tourism loses its specificity as an activity and that people become 'tourists most of the time', an argument that stems from the 'analysis of the social relations actually involved in tourism' (Lash and Urry 1994: 259, 270). While this is a rather convoluted approach to the debate over contemporary tourism, their observations are useful in setting the emergence of alternative (and by inference sustainable) tourism in a broader conceptual context. Box 3.3 summarises the changing features of tourism with the development of capitalism. The key point here is that these transitional frameworks appear to predict a positive qualitative shift in the nature of tourism and tourists, and that by inference new types of consumption and consumers may help to create alternative and beneficial forms of Third World tourism that help to break dependent relationships.

Alternative critiques for alternative tourism?

The preceding discussion has highlighted what we consider to be the main constraints in the analysis of Third World tourism. It is evident that much criticism has centred upon the traditional mass forms of tourism. While a critical analysis of the activities of supranational

---

Box 3.3 Tourism and the development of capitalism

Mass consumption	*Post-Fordist consumption*
Purchase of commodities produced under mass production	Consumption rather than production dominant
High or growing rate of expenditure on consumer products	New forms of credit and indebtedness
Individual producers dominate particular industrial markets	Almost all aspects of social life become commodified
Producer dominant	Consumer dominant
Little differentiation between commodities	Greater differentiation of purchasing patterns
Relatively little choice and producers' interests reflected	Consumer movements and politicisation of consumption
	Consumers react against the 'mass' and producers more consumer-driven
	Many more products and shorter lives
	New kinds of specialised commodity emerge

Source: Urry (1990a: 14)

institutions, such as the IMF and World Bank, and multinational companies is evidently necessary, we have suggested this fails to tell the whole story. It is not just the actions of the 'big players' that we must place under investigation; they are in a sense the obvious suspects. It is equally necessary to provide a critique of the actions of environmental organisations or the armies of backpackers whose actions are largely seen as benign or benevolent. As Hutnyk (1996: 13) argues, it is necessary to 'listen attentively to the echoes of power'. His critique of alternative tourism seeks to trace the relationships between the micro-centred hegemony of 'travellers' to the 'whole deal' – the 'global world disorder' (Hutnyk 1996: 23). Thomas touches on this need for a more sophisticated discussion because it is

> easy to denounce government policies and bodies such as the IMF, but perhaps more difficult to explore constructions of the exotic and the primitive that are superficially sympathetic or progressive but in many ways resonant of traditional evocations of others.
>
> (N. Thomas 1994: 170)

In particular, it is essential that we challenge the tacit assumption that the emergence of new forms of tourism are both designed for, and will result in, surmounting the problems that have been identified. Returning to our starting point, exactly what is sustainable tourism seeking to sustain and for whom? We must then attempt a response to Paul Gonsalves' call: 'Wanted: A Third World Political Economy of Tourism' (Gonsalves 1993: 11).

The following sections provide one way of beginning to assess and criticise the other – alternative – parts of tourism, to which reference will be made later in the book. They set out what we consider to be the principal processes through which power is conveyed and reflected in tourism. This framework assists in a more critical understanding of new forms of tourism, although it is not meant as a definitive, critical account of Third World tourism. The framework is split into four elements (intervention and commodification; subservience; fetishism; and aestheticisation) with each section anticipating the main arguments that will be developed in later chapters. Box 3.4 provides a short definition of each of these elements and it may be useful to refer back to this box as you read through the sections. Each element is not independent of the others. On the contrary, they are overlapping and interrelated.

Intervention and commodification: controlling the goods

In Chapter 2 we suggested that the spread of capitalist relations of production throughout the Third World is one of the most notable global economic processes in the period from the 1960s. It was suggested that time–space compression provides a conceptual understanding of why this expansion has taken place and why services, such as tourism, have become increasingly attractive as capitalism attempts to speed up the turnover time of capital. This is of considerable significance, to repeat Britton's observations, because 'geography texts on tourism offer little more than a cursory and superficial analysis of how the tourism industry is structured and regulated by the classic imperatives of and laws governing capitalist accumulation' (Britton 1991: 456). And some critical observers of this transition of capitalism and neoliberalism, such as David Harvey (2006: 44), contend that the 'commodification (through tourism) of cultural forms, histories and intellectual creativity entails wholesale dispossessions'.

Intervention, however, has not only involved the activities of capitalists. Equally significant, as Truong (1991: 99) notes in her study of South-East Asia, 'there has been a

Box 3.4 Elements of a new tourism critique

Intervention and commodification

These ideas attempt to capture the rapid expansion of capitalist relations of production in the Third World (a critical factor of economic globalisation as discussed in Chapter 2) and the way in which the spread of tourism has led to destinations, local cultures and environments (such as national parks, wildlife, flora and fauna, and so on) being transformed into commodities to be consumed by tourists. Examples of commodification are the way in which an Amboseli lion is calculated to be worth $27,000 a year in tourism revenue, or the way in which cultural traditions and ceremonies are packaged and sold to tourists, and the timing of rituals is altered to fit tourist schedules (see, for example, MacCannell 1992).

Subservience (domination and control)

As First World tourism expands and commodifies Third World destinations there is a tendency for Third World communities and individuals to assume unequal or subordinate relationships to both First World tourism interests and the interests of 'local elites' (which is discussed further in Chapter 8). It is a reflection of unequal and uneven relationships of power and development.

Fetishism

The fetishism of commodities (or commodity fetishism and the associated concept of reification) is a concept that embodies the way in which commodities hide the social relations of those that have contributed to the production of that commodity (be it a good or a bad experience) from the consumer (such as the tourist). In a nutshell, tourists are generally unaware of the conditions of life experienced by the waiters, cooks, tour guides and so on, the people who service their holidays and the other people who form part of their tourist gaze. There are limitations to arguments that promote new tourism as a means of getting closer to the reality of life in tourism destinations.

Aestheticisation

This represents the process whereby objects, feelings and experiences are transformed into aesthetic objects and experiences (of beauty and desire). Aestheticisation is a notable characteristic of the way in which the new middle classes construct their lifestyles and is well represented in the ascendancy of new forms of tourism as important cultural goods (discussed further in Chapter 5). But aestheticisation must be interpreted broadly. Not only is there a desire to experience primitive cultures, environments and wildlife, but also there is a desire to experience 'real' poverty and really dicey situations that new tourism sometimes presents.

high degree of external influence on the formation of tourism and leisure policies', a pressure exerted by foreign governments, global financial institutions such as the World Bank, and by 'development' agencies.

These processes have been most obvious in the development of international mass tourism. With the spread and intervention of capitalism into Third World societies, tourism has also had the effect of turning Third World places, landscapes and people into commodities. In other words, we consume these elements of a holiday in the same way as we consume other objects or commodities. As Drew Foster, chair of one of the leading UK tour operators to the Caribbean (Caribbean Connection) puts it: 'The Caribbean is a great product' (author's transcripts 1995).

This process of intervention and commodification continues as mass tourism ventures explore new Third World opportunities, but is now joined by a considerably more subtle and superficially benevolent form of external influence. As we have suggested, support for mass tourism developments has tended to give way to new, 'softer' forms of tourism that appeal to notions of sustainability. The smaller and more diversified tour operators, together with new social movements mobilised around issues of environmental and cultural difference, that are now commonplace in the First World are seeking out rather different experiences from their excursions into the Third World. In part they are seeking to ensure the maintenance of these experiences and attractions (or commodities), by translating their desire for environmental conservation and cultural preservation to Third World communities. Tourism is being upheld as both the cause and effect of environmental, and to a lesser extent, cultural preservation. Although the emergence of new, alternative forms of tourism are less discernible as the hard edge of economic globalisation, they are nonetheless of increasing significance and 'tinker at the edges of capitalist expansion into new market niches' and represent the 'soft edge of an otherwise brutal system of exploitation' (Hutnyk 1996: x, ix).

Through a range of initiatives, from the 'debt-for-nature' swaps pioneered by the US government (Enterprise for the Americas Initiative Act, Dept of Treasury press release, 27 June 1990) to the phenomenal growth in 'green tourism' operators and to lending agencies placing environmental conditions and caveats on their loans and grants, a *greening* of social relations is being promoted. It is a kind of *eco*-structural adjustment where Third World places and peoples must fall into line with First World thinking. For example, the Domestic Technology Institute (a US-based non-profit organisation), which planned a 'multinational organisation to develop and co-ordinate low impact tourism in Third World countries', had warned, 'These [Third World] countries can be forced into establishing natural and cultural resource policy before they can get World Bank loans' (Pleumarom 1990: 14).

As we have already suggested and will see in later chapters, there is a range of ecological concepts that are capable of spearheading and justifying interventionist policies. *Biodiversity*, for example, a concept adopted by organisations such as the WWF, the International Union for the Conservation of Nature and Natural Resources (IUCN) and Conservation International (see Chapter 6), seeks to impose conservationist regimes on Third World countries. Some would argue that their efforts to link up Third World communities with US corporate buyers accelerate the commodification process and dependencies of capitalist relations of production and, arguably, undermine the ability of Third World producers to trade fairly – a point to which we return in Chapter 7.

At their most powerful, donor agencies and environmental organisations join forces to pursue conservationist policies in which tourism is a major economic factor. As Gordon (1990: 7) argues of Namibia, 'Adjusting the sails of tourism Nature Conservation began to explore the viability of elite expeditions to "natural and wild" parts', a context Gordon refers to as little more than 'welfare colonialism'.

Other interventionist methods have also emerged among environmental organisations and are being successfully employed, such as appeals to adopt animals and parts of nature (see, for example, Conservation International and the World Land Trust). In what would appear to be the antithesis of local control, the multinational World Land Trust (WLT) explains the origins of its private reserves in Belize which it acquired by selling certified acres of rainforest:

> The World Land Trust (WLT) was originally set up to raise funds for the Programme for Belize [1989], and subsequently raised over $2 million for land purchase and research into sustainable use of the forest. By 1996 all the loans had been paid off and the WLT has established Friends of Belize to encourage supporters to make a further commitment to helping the long-term conservation of Belize's wildlife and natural resources.
>
> (WLT website, accessed 16 November 2007)

Independent tour operators have also sought to support or become actively involved in conservationist measures. (In the following pages the tour company brochures quoted are from the years 1992 to 2007.) Many operators cite their membership of environmental organisations (such as the WWF) and contribute financially to their work. Other operators have become more proactive in intervening in conservationist measures; Worldwide Journeys and Expeditions in 1992, for example, claimed management involvement in a 'privately' managed national park in Africa, at Kasanka (Zambia). In any case the ultimate objective is the same: the need to protect their principal product: 'pristine environments' and 'wildlife'.

In Chapter 5 it is argued that individual representatives of the new middle classes (travellers, backpackers and so on) are no less interventionist. The scale and scope of their activities may be qualitatively different from the power wielded by tour operators or large environmental organisations, but their role in the process of commodification is as noteworthy. For such tourists it is the supposed desire to experience 'indigenous cultures' – the Third World otherness – that is a major driving force of their travels and results in the search for 'off-the-beaten-track' or 'lesser visited areas'.

Speaking of trekking in northern Thailand, Trailfinder's travel brochure warns that as 'the popularity of this type of holiday experience increases it is more difficult to find . . . "untouched" or "traditional villages"'. Of course, solutions to such problems are always on hand and Trailfinders promises treks to villages 'which are only visited a couple of times a month. These groups are welcomed as a refreshing diversion to normal village life'. Such activities underscore a nostalgic desire for an imagined, 'real' and 'authentic' primitiveness and support the drive for cultural preservation.

The whole language of new forms of tourism is also premised upon intervention and commodification. 'Welcome to South America', the opening paragraph from tour operator, *Passage to South America*, expresses a neocolonial process of discovery, penetration and expropriation – of last and ultimate frontiers. The early colonists are celebrated by many tour operators and Figure 3.2 illustrates some of the ways in which this colonial past is reproduced offering holidays to the adventurous and relatively affluent new middle classes. For tours to Africa, Dr Livingstone is frequently invoked to assist in these hedonistic discoveries. For example, of Malawi, J. & C. Voyageurs promised: 'Following in the footsteps of Dr Livingstone, gliding quietly up the Shire River, unchanged since the first appearance of the white man' (1992). Or Africa Exclusive in expressing their thoughts on Zimbabwe: 'when we first discovered this beautiful country – like the great missionary and explorer Dr David Livingstone 150 years earlier – we were very excited' (1992). Jacobs, writing of Mali, testifies: 'Mungo Park, the late 18th century explorer, would have recognised the view', a land he tells us is 'bypassed by history'.

Figure 3.2 Neocolonial discovery?

Even against the backdrop of the five hundredth year of exploitation since the arrival of Christopher Columbus in South and Central America, those presenting a supposedly more enlightened form of travel still find room wryly to celebrate his conquest and assure us that the age of discovery has not passed:

> Given the right advice and [a] . . . Travel Round the World ticket, Christopher Columbus might have saved himself a lot of trouble. He might, for example, have tried both eastern and western routes around the world. But the age of exploration is not dead. There's the whole world waiting to be discovered afresh.
>
> (STA Travel 1995)

It is not just the language that is used, however, but the manner in which some new forms of tourism have attempted to capture the aura and mystique of colonial forms of travel and holidaying. There has, for example, been a resurgence in the popularity of colonial rail travel, train cruises as Eames (1994: 88) describes them, 'elegantly packaged tours with a taste of tradition', and luxury safaris: 'Classic Kenya' – 'an escorted private safari in the old style tradition' in 'large, walk-in tents with en suite private bathrooms' (Worldwide Journeys and Expeditions 2003). Again, Figure 3.2 illustrates some of the images found in tour brochures and promotional material.

It is a romanticism for travel modes of the colonial periods which, unwittingly perhaps, re-creates the subordination of Third World peoples in an invidious aura. And it has invoked a nostalgic longing for untouched, *primitive* and native peoples who are there to meet the demands of tourists, both in terms of service and as an object to be enjoyed and photographed. It is not the issue of product development that is of relevance here, but the question of who has the control and power to create these images.

Subservience: enslaving the goods

In the 1970s Turner wrote that a 'significant part of tourism involves shipping rich white pleasure seekers into some of the world's poorest black societies' (Turner 1976). Of course, nothing much has really changed in this, as Tables 2.1–2.3 suggest. But as has been argued, there has been a widely held belief that changes in the forms of tourism will produce qualitatively different experiences. But different experiences for whom, we must ask?

As already suggested, conventional analysis of the political economy of Third World tourism has tended to focus on the three principal components of the industry: international airlines, tour operators and hotel chains. Critical commentary has confined itself largely to observing that all three components are owned and controlled from the First World. As the political economy of tourism demonstrates, Third World countries are subordinated to the flows of tourists, tourism capital and resources from and to the advanced capitalist economies. Logically, such analysis has reached the conclusion that 'alternatives' that avoid such ownership characteristics represent either a considerable step towards the creation of a much fairer basis for international tourism or, for some, the answer to unequal and uneven tourism development. In part this may be true (and we consider alternative tourism's pro-poor credentials in Chapter 11). However, it also represents an oversimplistic conclusion in two important respects.

First, it is a conclusion that is based upon too narrow an analysis of tourism within a global context. In this book we argue that, despite changes to the ownership of tourism resources, tourism in the Third World will remain a special form of domination and control, although this is not to argue that there will be no individual success stories (of places and even countries). Rather, it is the global inequality between First World and Third World in terms of power that is of such significance as to warrant an analysis of much tourism as 'a special form of subservience' (Krippendorf 1987: 56; see also Fennell 2006); or that the widely cited Ss of tourism – sun, sea, sand and sex – are matched by the Ss of the content and outcome of tourism – subjugation, servility and subservience. Critically we argue that tourism is one among a number of 'symptoms' (not a *cause* in itself) of unequal and uneven development – as much as the restriction of the reverse flow (immigration from Third to First World countries) is a symptom. It is important that we walk a tightrope here, mindful that too global an analysis ignores local lessons and too local an analysis ignores global questions.

Second, existing studies have tended to blind commentators to the coalface of tourism: the relationship between tourists and those they are visiting. It is this basic encounter that remains the most significant activity in tourism. Rao (1991) captures this and reflects on the way in which many First World commentators refer to this as a relationship between host and guest:

> Can one really describe the encounter between the tourist . . . and the Other in the so-called voluntary relation of guest and host? Such a relation is again dictated by the tourism discourse which seeks to sweep away the basic commercial nature of the encounter.
>
> (Rao 1991, quoted in Gonsalves 1993)

Thinking critically about this 'relationship' opens new doors in our analysis. For example, it questions the holiday consumption practices and preferences of the new middle classes and should help us understand processes external to many Third World countries where tourism has produced a helpful, smiling and servile tourism class, serving the interests and economic preferences of business and local elites.

Even after training courses and 'tourism weeks', Jean Holder (the Secretary General of the Caribbean Tourism Organisation) complains of Caribbean tourism, 'aggressive attitudes . . . often emerge' which 'have their basis, to some extent, in the fact that a significant number of employees are not proud of what they do, and harbour resentments rooted in the inability to distinguish between service and servitude' (Holder 1990: 76). Arguably it is easy for those in positions of influence and power and those professional local elites assisting in the running of the industry to be able to make this distinction, but perhaps far harder for the army of labour that is required to service the industry and whose structural disadvantage (in terms of class, gender, race, income and so on) is compounded.

Whereas under industrial capitalism the workforce confronted its subordination in the industrial structure, within tourism workers in many cases must daily confront the symbolic representation of their servility – the tourist. Holder (1990: 76) continues: 'there appears to be a deep-seated resentment of the industry at every level of society – a resentment which probably stems from the historic socio-cultural associations of race, colonialism and slavery'. This is an important point, for it strikes at the heart of the perception of tourism and tourists that reflects back on deep-seated historical inequalities, and a number of Caribbean writers have drawn the parallels to the dawn of a 'new slavery' (see Fanon 1967; Kincaid 1988; Naipaul 1962; for an up-to-date discussion of the Caribbean, see Pattullo 2005).

For new tourism to claim that these forms of servility and subordination are overcome may prove disingenuous. The number of tourists involved and whether they are conventional mass packaged tourists or an off-the-beaten-track type of traveller is secondary with subservient relationships tending to persist, albeit manifest in different ways – witness the scepticism exhibited in the headlines shown in Figure 3.3, referring to both mass tourism and new forms of tourism. We have already noted the neocolonial aura that is re-created through eco-safaris, for example. Box 3.5 shows a single black-and-white

Figure 3.3 To be or not to be a tourist?

Box 3.5 Global porters

You have to live off the land when on safari in the Selous Game Reserve. This is Africa as it used to be . . . The walking safari normally lasted for 6 days, and a group of 6 guests is supported by 35 porters . . . The crew undertake the domestic chores, such as cooking and laundry, etc., and pitch and break camp each day . . . A typical day would be: up at six in the clear cool dawn, walk until the midday heat builds up, then lunch and a siesta in the shade; finally, after a cup of tea, an evening walk, with the countryside serene and the animals out in force.

Source: J. & C. Voyageurs brochure

photograph offered in a J. & C. Voyageurs brochure of a trail of porters (thirty-five we are told) shown tramping through an 'eco-colonial landscape', carrying the supplies for the group of six: 'most of the comforts of home – iced drinks, spacious sleeping tents, loos and showers'. It is an image and aura that is re-created by many so-called new small independent tour operators, whether it be luxury safaris or treks and expeditions for the young and adventurous; an army of global porters trot behind or ahead (via different routes) to ensure that these new, ethical tourists are regularly refreshed.

Kutay (1989), former director of the International Ecotourism Society, poses the critical question: 'Those who view tourism as the final, most humiliating stage of human domination condemn it . . . But is modern tourism any more dominating or humiliating than religious imperialism or European mercantilism?' Indeed, it is not, and this suggests we must consider whether tourism in its new forms is still a highly interventionist and subordinating activity.[1]

Fetishism: hiding poverty

The concept of fetishism is a useful way of getting us to think about the relationships that lie behind the things we buy and consume. A central precept of Karl Marx's thinking ('historical materialism') expresses the way in which commodities hide or veil the social relations embodied in their production. Think of the last piece of fruit you ate that may have come from the Third World. There will have been a multitude of people and relationships responsible for getting this fruit to your grocer or supermarket shelves. And you are able to consume your fruit without considering, for example, the conditions faced by the workers who picked it and what aspirations and dreams these hidden workers have (Cook 1993). It is necessary to get behind these surface appearances and unveil the fetishism of commodities.

When we talk of the fetishism embodied in the production of commodities, we are also implying that the same processes are identifiable in the production of services, such as tourism – a task of applying Marx's ideas first undertaken in the context of leisure by Veblen (1925). There are two points at which notions of fetishism are particularly interesting in the study of tourism.

The first is the way in which tourism is an issue so hotly contested between different social classes (a point discussed at greater length in Chapter 5). Travel acts like any other commodity in expressing who we are, what we consider our status to be, what we believe in, and so on. It is a commodity through which we are able to express ourselves culturally. Take this example from Borzello (1991), who attempts to tell us what the difference is between tourists, travellers and those who choose to take a 'truck' expedition (passengers Truck Africa prefers to call 'would-be backpacker[s]'): 'Tourists just graze through a country. Travellers live and integrate into it. Truck travellers are not travellers but a very peculiar sort of tourist.' The implication is that 'travellers' seek to roll back the fetishism that Borzello (1991) accuses 'tourists' and 'truck travellers' of experiencing (see Box 3.6). And yet at the same time the search (or competition) for the most virtuous way to take a holiday becomes a fetishistic cultural game in its own right; it is culture as the 'supreme fetish' (Bourdieu 1984).

Box 3.6 Focus on truck travels

Truckers

A number of companies now offer expeditions to most regions of the Third World travelling in 'trucks'. They are operationalised versions of overland trips that have become increasingly popular for 'a generation well used to looking for services on the internet', as Truck Africa's (2007) website puts it, and are aimed at the 18–45 age bracket. Exodus stipulates: 'This style of travelling is more suited to younger people . . . We therefore apply an age limit of 17–45 on all trips of 4 weeks' duration and over. We will occasionally waive this for people who are older, and have demonstrated to us that they have the right attitude and have completed similar journeys before.'

Risk and adventure

An essential ingredient of such trips which are 'more likely to appeal to the adventurous in spirit' (Truck Africa 2007). As Dragoman contends, this is for 'those who want the thrill of "real travel" . . . We will be crossing areas of the world that do not adhere to western safety standards and may have inherent political and economic instability.'

But civil wars, revolutionary insurrections and suppression become part and parcel of the travel experience; obstacles to be got round and overcome. Exodus simply states: 'we have seen and coped with numerous coups, wars and revolutions in all three continents, and our leaders have proved themselves to be pretty good at finding their way round the problems that politics can place in their way.' A serious interrogation of the meaning of the civil struggles experienced in many Third

World countries is replaced by macabre celebrations that destinations are back in business: 'For many people Peru is the most interesting of all Andean countries. With the leaders of the Shining Path guerilla movement now firmly behind bars, Peru is back to business as usual, and once again on our list of top destinations for the discerning traveller' (Exodus newsletter, April 1994).

Exploration

Invoking the aura of exploration (see Figure 3.2), it is the ease of penetration and a truck's panoptic qualities that become pre-eminent selling points. Exodus refers to its trucks as 'particularly suitable . . . where contact with the outside world is of paramount importance'. It must provide for optimum surveillance so the prey of this postcolonial gaze is captured and greeted. Encounter Overland ensures of its vehicles, 'With the sides up, the all-round views are superb and there is a great sense of being in close with the world around.' Exodus claims: 'The open sides . . . give the best views and the best contact with the people and countries that we visit . . . They allow all-round vision, ease of entry and exit, good off-road capability.'

Paradoxically, others point out, it is the ridiculed, enclave, mass tourist infra-structure and resorts, from which truckers attempt to distance themselves, that is mimicked by the truck. The charges of environmental bubbles from which tourists need not leave are replicated by the truck. Simultaneously it allows both a 'close-ness with nature' but guarantees 'everyone can remain well within the framework of the vehicle body' (Encounter Overland).

Critics will suggest that travel brochures present the neocolonial consumption of black landscapes by white truckers. Where natives are able to assimilate whiteness, it is in ridiculing, fleeting and patronising representations; natives momentarily peering through a camera, or situated directly within their colonial legacy pulling an ailing truck from muddied waters. Natives drawn into the 'white man's' world. 'Lunch with the Maasai', for example, captions a photograph illustrating the ease with which this other world can be consumed. Lunch is prepared beside the truck and the Maasai have their trinkets examined by truckers in the foreground.

Second, travel in many cases brings the tourist in direct contact with labour (hotel staff, tour guides and so on). Superficially then, the fetishistic 'mist' that enshrouds, masks and objectifies labour dramatically clears within much tourism, as the 'guest' is faced with the 'servant'. Indeed, as suggested above, it is the claimed ability to transcend the fetishism characteristic of mass tourism and to meet real people in real places producing real things, that lies at the heart of new forms of tourism. It would not be unreasonable to expect new practices to achieve this and dismantle such fetishism. However, we have argued that travel is a commodity, and one that is consumed principally as an image. This has led to new, more ingenious and intellectualised ways of creating an aura of travel (a point Figure 3.3 seeks to convey) and a reordering of the fetishism involved.

So it is that a range of less savoury realities of some parts of the Third World today – inequality, poverty and political instability – are also there to be enjoyed as part of the tourism experience (as tours of urban slums perhaps testify, an issue that we discuss further in Chapter 9). They are called upon to both titillate and legitimise travel, to help distinguish these experiences from mere mass tourism and packaged tourists. It is perhaps an extreme method of seeking authenticity through travel (a point we consider below) with many backpackers, for example, simultaneously bemoaning and celebrating the existence of such characteristics.

Exodus exemplifies the glib juxtapositions. Of Latin America, for example, it states: 'some of the most sophisticated and wealthy cities in the world . . . but along with the wealth there is appalling urban poverty.' Predictably, no more is to be said on urban poverty and instead Buenos Aires is sold as a 'huge, sophisticated capital' with 'plenty to see' and where 'carnivores can wash down one of the giant steaks with excellent local wine'. Similarly, *Backpacker's Africa* (Bradt 1994) sweeps aside a politically repressive environment; Kenya, it is argued, is 'blessed with outstanding geographical and cultural variety and her relatively stable government and capitalist economy is an added attraction for visitors' (quoted in *In Focus* 2 1991). Or as Tucan, a company specialising in adventure tours to South America 'to meet the needs of younger travellers', suggests (in 1992) of the 'real' Latin America experience: 'Enormously wealthy people in their impressive mansions and estates live virtually cheek by jowl with the very poor in their squalid adobe huts and yet the populace as a whole show a special friendliness coupled with a real sense of hospitality.' Widescale repression of human rights, deeply rooted racism and intense class political struggle are null and void in the brave new world of adventure travel, and provide for a rather different itinerary of attractions in regions such as Central America – see Figure 3.4.

Arguably, then, new forms of tourism seek to penetrate the less visited parts of the Third World and commodify what is there. It is a form of commodity racism (McClintock 1994), which has its roots in the diaries (or travelogues) of nineteenth-century Victorian 'explorers'. The desire to consume these strange, other worlds that they had 'discovered' became a fetishistic ritual which tourism has maintained. And it is this discovery of these other people and places that is such a striking feature, especially, of new forms of travel today.

It is with the stream of touristic images, the trophies of these discoveries, that fetishism is most visibly maintained. *Out*, or at least marginalised, are the images of long, sandy, palm-fringed beaches. *In* are the images of unlimited wildlife, adventuristic landscapes and painted indigenous cultures. In fact the treatment of animals and peoples has become dramatically inverted with Cara Spencer Safaris proclaiming: 'A bonus for us was the warmth, concern and courtesy of the people . . . And their approach to wildlife conservational land management is highly progressive.' At best they are, it would seem, of equal value and are there to be photographed and embraced. The search and expropriation of authenticity, it appears, has become a mission (Wheat 1994) of neocolonial proportions:

> Both explorers and dragomen journeyed through far-flung lands in search of long-forgotten civilisations and empires. They explored hostile deserts to find nomadic tribes and ancient cities. They went in search of legendary mountains deep in the heart of jungles and brought back stories of fabulous wildlife.
>
> (Dragoman brochure 1995: 1)

To return to the observation that Said (1991) so forcefully expresses in his study of *Orientalism*, exploration, travel writing and tourism are means of representation. But they are ways of representing the world that also amount to, and maintain, a 'formidable structure of cultural domination': a system of truths, as Said's contested argument runs, that has 'rarely offered the individual anything but imperialism, racism and ethnocentrism for dealing with "other" cultures' (Said 1991: 18).

Figure 3.4
An alternative
tourist's guide to
an exciting region

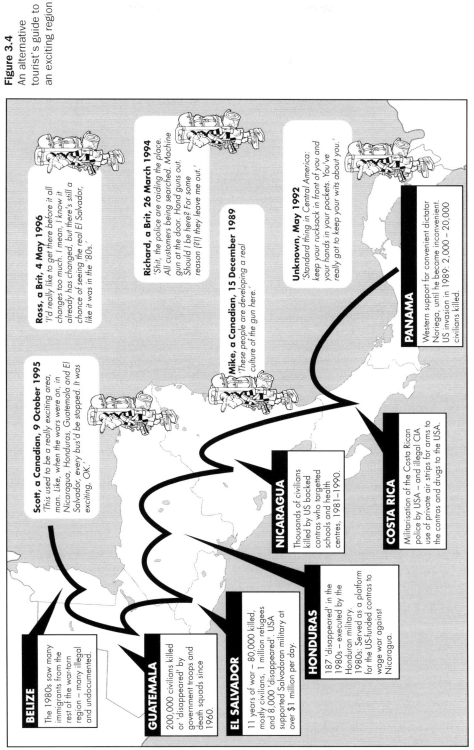

Aestheticisation: enjoying poverty

Not only is the consumption of tourism fetishistic, it is also intensely *aestheticised*. By this is meant the way in which travel and tourism is used to express 'good taste', a cultural accoutrement indicating to others what it is we find of beauty or note, what we are seeking to get from a holiday and, significantly, an indication of our social position and 'belief system'.

We need to start here with a consideration of 'authenticity', a critical concept that has a rich heritage in tourism studies, and one of the most enduring debates in the sociology of tourism (for example, see E. Cohen 1989; MacCannell 1973, 1976, 1992; P. Pearce and Moscardo 1986; Turner and Manning 1988); Box 3.7 provides an introduction to the concept's evolution. It is now commonplace to hear references to authenticity in everyday usage, people who refer to their holiday experiences as real or unreal, authentic or fake, for example, or the range of programmes and publications – the *Real Holiday Show*, the *Independent's Guide to Real Holidays*, *Rough Guides* and so on – that are motivated by the desire for real, authentic experiences. As Poon contends, tourists are moving away from 'tinsel and junk' in the search for 'more real, natural and authentic experiences' (Poon 1989a: 75). Authenticity is central to our discussion, therefore, because it is fundamental to much of the debate about the content (real, ethnic, off-the-beaten-track and so on) and appropriateness (eco-, alternative, sustainable) of new forms of tourism in the Third World. As Daltabuit and Pi-Sunyer (1990: 11) argue, for example, '"The environment", as commodity or experience, is no less a fantasy than any other image elaborated by the leisure industry. That it has so broad an attraction probably owes much more to the postmodern quest for authenticity.' As we have suggested, authenticity also has strong resonance in the debates over sustainability – environmental and cultural.

Among the new middle classes, the 'real' appears to indicate a desire for authenticity, for 'honesty', however circumscribed that may be in reality. Kestours, a private UK travel agent formed in 1956, specialising in Caribbean holidays, exemplifies this and is indicative of the transformation in contemporary travel. The company's two-brochure package symbolises the transition from a derided mass tourism to a new, possibly postmodern, and more truthful tourism that characterises certain segments of the industry. The first brochure, a traditional glossy with large-format coloured photographs of sun-drenched beaches, shady palms, luxury hotels and loving couples, contains minimal text. These images say it all, this is paradise. The second brochure, printed on manila, seemingly recycled, paper, contains only line drawings (and two passport-size black-and-white photographs on the back cover). This is reflective of new tourism and it is here that the Caribbean paradise is introduced:

> The Caribbean is a tropical destination . . . Ants, mosquitoes and cockroaches thrive in hot climates and while usually harmless are sometimes a nuisance. In the long hours of sunshine, lack of rain can mean erratic water and electricity supplies – really hot water is rare. The Caribbean is a long way away. Your flight takes at least 8 hours and even in modern wide-bodied jets, it can be tedious as can protracted entry formalities when you arrive . . . The Caribbean is very different to Britain. To enjoy your holiday to the full, you must accept its shortcomings as a challenge and as an enriching experience rather than as a reason for complaint.
>
> (Kestours, *The Caribbean Holiday Guide* 1991/2: 1)

As we indicated in Chapter 2, however, discussions have tended not to explore authenticity within the context of wider cultural consumption practices. This is especially

Box 3.7 Travels in authenticity: the evolution of the concept

Pseudo-event

Daniel Boorstin (1961) attempts to chart the shift from individual traveller to the emergence of mass tourism and tourists. In *The Image* he argues that contemporary Americans are unable to experience 'reality' directly and instead thrive on what he terms 'pseudo-events'; *inauthentic* contrived attractions. After a while this pleasure becomes a self-perpetuating system of illusions – *an environmental bubble*. This concept of tourists being cocooned from reality in a bubble is a widely held idea today. In their famous *The Golden Hordes*, Turner and Ash (1975) follow Boorstin's lead and place tourists at the centre of a strictly circumscribed world, with travel agents, carriers, hotel managers and so on taking the role of surrogate parents. Analyses of this kind are underlined by issues of class (see Chapter 5).

Staged authenticity

MacCannell detects the class bias in Boorstin's analysis. For MacCannell the essence of all tourism is the quest for authenticity that he considers absent from everyday modern life. In this quest for authenticity local communities – *hosts* – become cheesed off with being gawped at, but also see the chance to make a quick buck! Applying the work of the Canadian sociologist, Erving Goffman, MacCannell argues that front and back regions (lavish dining room and grimy kitchen, for example) can be identified in tourism spaces. This results in contrived tourist spaces where front regions are made up to look like the 'real' thing, and hence *staged authenticity*.

Communicative staging

Erik Cohen is in general agreement with MacCannell but disagrees that all tourists seek authenticity: a highly idealised global view of tourists. He identifies four types of 'touristic situations': authentic, staged authenticity, denial of authenticity, contrived. Cohen suggests we need to consider the role of *communicative staging* where tourist sites have not been transformed but the sites are presented and interpreted as authentic to tourists by their guides.

The gaze

John Urry concedes that authenticity may be an important component of tourism but only because there is some sense of a contrast with everyday experiences. He agrees with Cohen that some tourists delight in inauthenticity (what he terms post-tourists). The gaze, Urry argues, 'is as socially organised and systematised' and in any historical period is constructed in relationship to its opposite (that is, non-tourist forms of social experience and consciousness). A particular type of tourist gaze will therefore depend on what it is contrasted to. Tourist sites can be classified in terms of three dichotomies: historical/modern; authentic/inauthentic; romantic/collective.

Useful sources: Boorstin (1961); E. Cohen (1974, 1989); MacCannell (1973, 1976); New Internationalist (1984: 24–5); Urry (2001)

so among the new middle classes who are major consumers of new forms of Third World holidays. (We attempt this exploration in Chapter 5.) On the one hand, there are rhetorical exchanges over the invidious tourist–traveller continuum, coupled with the more rigorous attempts to typologise different types of 'tourists' and tourist experience. On the other hand, there are those accounts that rely ostensibly on motivation, attitude or satisfaction surveys (see, for example, Dearden and Harron 1991, 1993). Consequently, broader issues such as how social class and tourism embody class differentiation tend to be overlooked.

Arguably the most significant point to note about authenticity in tourism is the manner in which it reflects the wider global processes already discussed in Chapter 2. The laments of the fake, for example, can be seen to reflect the suggestion that we are moving towards an undifferentiated global culture and the accelerating desire to experience real cultures through travel. MacCannell (1992) captures the importance of authenticity within this broader framework:

> By now, almost every individual on Earth has been informed that he or she is related to every other individual in 'The New World Order'. The message is usually accompanied by a sinking feeling that all our actual relationships, even with former intimates, are falling apart. A secondary effect of the alleged globalisation of relations is the production of an enormous desire for, and corresponding commodification of, *authenticity*.
>
> (MacCannell 1992: 169–70)

But while this debate is of interest, it has tended to mask other, perhaps more invidious, processes that are at work. Authenticity, it will be argued, must be understood within a broad frame – it is not just about tourism providing access to 'real' tribes in Thailand, Kenya or Bolivia; it is about the ability to witness and consume 'real' lives too, and this includes poverty, civil struggle, and so on.

A second critical concept in understanding the manner in which tourism aestheticises is the process of othering. Through a brilliant analysis of western literary representations of the Middle East, *Orientalism*, Edward Said (1991) has demonstrated how western intellectuals lovingly created the countries to which they were devoted in an insidiously romanticised fashion and that those scholars were 'inseparable from the political structures within which and for which they wrote' (Hutnyk 1996: 7). Through their writings, they portrayed how these 'other' cultures were exotic, sensuous, erotic (to which we could add today: simple and sustainable), at one with nature. This process of creating, aestheticising and interpreting other peoples and places is referred to sociologically as a process of othering and it is especially significant in an analysis of tourism in that, for many, it is seen as the 'orientalism today' (Hutnyk 1996: 11).

There has been a marked growth of interest in mass and minority (non-western) cultures, religious traditions, ethnicity, and environment and ecology: aspects of otherness that find special representation in the Third World and are reflected in the First World in alternative markets, charity shops, and so on. Short (1991) refers to some homes in the First World resembling 'ethnographic museums' and Third World literature, food, cinema and music are increasingly demanded, especially by the new middle classes. And this has clear reflections in the idea of the creation and representation of authenticity. As the satirical quote with which Said opens *Orientalism* encapsulates: 'They cannot represent themselves; they must be represented.'

In addition, it is a process, as Stuart Hall (1992b) argues, that involves representing an 'absolutely essentially different, *other*; the Other'. In Hall's analysis of the West (or First World) and the Rest (essentially the Third World), othering produces a series of sharp binary opposites: for example, the West as democratic, free, developed, peaceful and the

Rest as despotic, undeveloped, violent, barbaric, fundamentalist, and so on. Rather than aestheticising, such an approach demonises the Third World, although it will be argued that the representation of regions as risky also plays a part in the aestheticisation and attractiveness of 'new tourism destinations'.

Othering, therefore, clearly involves a process of reflection. Other cultures and environments are everything that *our* cultures and environments are not. Thus western lifestyles can be denigrated as empty, culturally unfulfilling, materialistic, meaningless, while, on the contrary, Third World cultures are bestowed with meaning, richness, simplicity and, of course, authenticity. Indeed, the search for authenticity implied here has been seen very much as a response to the dissatisfaction with 'modern' living – a postmodern reaction to the decline of difference (different cultures, places and so on) and diversity. Both MacCannell (1976) and Cohen (1979a, 1985) refer to 'postmodern travellers', or what Erik Cohen calls *experimental* and *experiential* tourists.

It is not difficult to see how this vein of analysis can be applied to contemporary Third World travel. In tourism othering is a key process of socially constructing and representing other places and peoples. As Explore, a small company formed in 1982 offering 'exploratory holidays' to those 'keen on discovering real qualities and real places', promises, the intention is to know the 'other' side so rarely seen. One specialist tour operator entices its readers to South America as follows:

> The adventurous traveller longs to visit faraway places and discover the unknown. South America is, for many, the last frontier – a wild, exotic corner of the world mostly unexplored and largely unspoilt. From the Caribbean coast in the north, through the rich rain forests of Brazil, the high plains of Bolivia and the boundless prairies of Argentina, to desolate Tierra del Fuego, South America stretches nearly five thousand miles with the bristling Andean Mountain range straddling the entire continent from north to south. Sheer canyons, crystal clear lakes, majestic volcanoes, ice-blue glaciers and cascading waterfalls combine to make it a land of unparalleled beauty and splendour.
>
> (Passage to South America brochure)

Shurmer-Smith and Hannam (1994: 19) argue that such an approach 'informs much of the genre of travel writing and people's choice of holidays'. Equally, it also demonstrates the power to bestow and expropriate meaning. Veteran war correspondent and travel writer, Martha Gellhorn (1990), describes the ethnically diverse Belizeans as lovely, 'untainted by tourism', the Mayan 'Spanish-speaking bird-like women, all married at 14, all giggly and happy with hordes of children', and the Garífuna as 'matt black, with sharp strong Indian features, and a reserved ungiggly manner'. Belize is, Gellhorn (1990) concludes, an achievement of colonialism: 'self-confident people . . . loyal subjects of the Queen . . . They don't know how lucky they are past and present.'

Writing in the UK *Guardian* newspaper, Ros Coward (1996) alludes to the relationship between the middle classes, authenticity and otherness:

> the middle classes smugly believe that . . . problems are not created by their sort of holidays. They travel independently, [and] visit ever more remote places. The moral superiority of this tourism comes from the idea that it provides an experience of the authentic culture of the host country rather than its destruction. The . . . problems are blamed on the kind of holiday taken by less affluent members of the affluent west . . . who do not understand that the true purpose of travel is to experience otherness . . . As a result, the discerning have to travel further afield.
>
> (Coward 1996: 11)

Otherness and authenticity are united in a desire to ensure that culture and ethnicity are preserved and aestheticised. It is the promotion of primitiveness within which authenticity becomes the principal commodity (E. Cohen 1979a, 1979b, 1989; Errington and Gewertz 1989). This is perhaps best conceived, albeit in a rather different context, as part of wider postmodern nostalgic yearnings (Jameson 1984, 1991). In the context of Third World tourism, this is translated through a range of new tourisms, from colonial tourism in the Caribbean to the excitement of *discovering* Zimbabwe, 'like the great missionary and explorer Dr David Livingston [*sic*] 150 years earlier' (Africa Exclusive). Not only is there nostalgia for ancient traditions and environments, most visibly evoked in nature documentaries, but also a nostalgia for the travel styles of yesteryear. But cultural preservation and sustainability can also be understood as a reaction to indigenous or real Third World cultures under siege from global capitalism and western values.

In Chapter 2 it was argued that sustainability can again be regarded as a transmitter of power relationships concerned, in part, with the sustaining of Third World environments and cultures: the preservation of the other. Leading alternative tour companies are indicative of this strand of sustainability. Dragoman, for example, extols Survival International as the leading authority on tribal peoples and donates £2.00 for every direct booking. Encounter Overland actively supports WWF, Save the Gorillas, Project Tiger and rainforest preservation movements in South America. Guerba Expeditions cites over £12,000 to be contributed to WWF as well as its exploration with WWF to lessen the impacts of tourism. High Places highlights its support of the Rio Mazán Project, and Himalayan Kingdoms' pitch is:

> Being 'green' is a popular bandwagon, but we do whatever we can to minimise the impact of tourism. We have been instrumental in working out a code of conduct with Tourism Concern, and are currently spearheading a new concept of 'mountain friendly' climbing expeditions.

To dismiss these expressed environmental concerns as merely marketing ploys is to understate not only a valuable research resource but also the importance of these concerns expressed both within certain social groups and by the companies themselves; many small companies have an undeniably genuine commitment to environmental issues. But for some commentators the intimate relationship between tourism and environment has amounted to hegemony, or what Daltabuit and Pi-Sunyer (1990) refer to as an ideology of environmentalism. At the height of ecological reductionism, tourism itself is considered an 'ecological phenomenon' (Attenborough 1986).

In addition, as Chapter 5 will suggest, the 'environmental other' both reflects the need of the new middle class to sustain travel experiences that are capable of maintaining cultural capital (and the conferring of social status) and represents the manner in which social groups (as well as political institutions and businesses) establish boundaries of power. The travel company Explore exemplifies this in its 2007/8 brochure:

> Explore was founded on the principle that travel should take you beyond the magnets of mass tourism and enable you to experience the real wonders and cultures of our world, but without compromising the integrity of their culture and environment.

Aestheticisation also demands the transmission of what, and how, experiences have been consumed, how they can be verified; and the power relationships involved in these forms of representation and construction are noteworthy. The use of photography and its relationship to tourism has been the subject of much critical commentary (Albers and

James 1988; Barthes 1981; Sontag 1979; Urry 2001); it is as if things have come full circle, for as Sontag (1979: 57) argues, 'From the beginning, professional photography typically meant the broader kind of class tourism'. While holiday 'snaps' are supposedly symptomatic of shallow tourist experiences, photography (especially monochrome) supposedly captures both the historical ambience and the closeness of tourists to the other.

It is this need to accumulate authentic images through the stream of portraiture of 'natives' or the embracing of the tribal child that has become an enduring image. Figure 3.2 portrays the innocence, authenticity, naturalness and 'nativeness' of the Third World at the same time as aestheticising the inequality of development. The momentary encounters captured in some of these photographs symbolise the inherent power that travel embraces. The passivity with which 'natives' and wildlife are portrayed in photographs has become indistinguishable. Passive, they are to be discovered, sighted, viewed and, ultimately, 'shot' (Sontag 1979).

But it is not just places and peoples that are the subject of aesthetic representation. Our discussion has pointed towards a process where it is suffering and poverty that become aestheticised, so much so that Hutnyk(1996: 11) claims: 'Third World tourism participates in a voyeuristic consumption of poverty.' Consider the 'alternative' tourism guide to Central America in Figure 3.4. Travel to 'dangerous' Third World countries, to regions suffering civil war and insurrection has become attractive to the bearers of new tourism who have become increasingly preoccupied with the need to distance themselves from tourists.

> Michael, a personable 28 year old on a three month leave from a New York finance business is out of sorts over breakfast. Things are not going according to plan. He has come to Calcutta to find poverty – and, like so many travellers to India, to find himself. Both are proving elusive. Sure, there are beggars, he says. Sure, there are people sleeping rough. But, quite frankly, it's not real poverty, is it? It's not – he toys thoughtfully with his poached egg – it's not swollen-bellied poverty . . . In the dining room, Michael from New York is a changed man. After breakfast, he told a taxi-driver to find him real poverty. After half-an-hour they found it. Real squalor. Real swollen bellies. Michael beams over the mulligatawny. He is well on the way to finding himself.
>
> (McClarence 1995: 55)

Story-telling or myth-making, a process of rumour-mongering (Hutnyk 1996), plays an instrumental role in much new tourism. Miles Warde (1992), writing in the UK *Guardian* newspaper, captures the way in which romanticism has seeped into contemporary Third World tourism. A romantic – aesthetic – aura feeds into the retelling of these travel accomplishments.

> Someone produced a bottle of whisky, and the stories began . . . It is a custom among travellers in South America to put at least two hours a day for telling stories. They are rarely pleasant. Some are cliched . . . A Canadian begins. I heard this in Belem, about an English guy, first day in Brazil, clean off the plane from London, and he was on this bus looking for somewhere to stay when four guys jumped him by the turnstile. Pinned him down on the ground – and get this – they rammed a fork in his arse and cut his pack off his back and threw it out the window to another guy. Jeez, can you believe it? A fork – they robbed him with a fork. More whistles of admiration. Even I liked that story, and was very impressed with the way it had remained so accurate as it travelled 6,000 kilometres up the coast from Rio . . . You see, it was my story, and I still have

the fork marks to prove it . . . Finally, a note of encouragement. Just before I flew home from Colombia I was in a bus queue. The bus arrived and the queue surged forward. I put both hands in my pockets – it becomes second nature – and waited for someone to try something. Then I noticed a loss of weight from the shoulder around which my camera had been hanging, hiding on the inside of my jumper. Wheeling round I caught the first throat which came to hand. It belonged to a young lad; he was holding my camera in one hand and a pair of wire snips in the other. I pulled my knee up in his crotch and he gasped, dropping the camera into my outstretched hand. The whole queue broke into applause. It was a sweet moment.

(Warde 1992)

This piece also suggests that the term aestheticisation incorporates the brutal reality, rather than stepping aside from it; that reality is then glorified and becomes intrinsic to the uniqueness and quality of adventure travel, a bitter twist for those young adventure travellers visiting unstable regions such as Central America, who can now include army checkpoints, state repression and civil strife on their list of things to tick off.[2] As Morrow (1995) gestures:

As Vietnam undergoes a new invasion from the West, you can be sure that returning travellers will have to endure the incessant cries of 'You should have been there a few years ago', from the pioneers, each claiming to have been there before it was spoilt by tourism. 'Vietnam? Oh, it's not the same without the smell of napalm.'

(Morrow 1995)

In arguably the most brazen of aestheticisations, the extracts from the travel article below, entitled 'Weekending in El Salvador', appeared in the UK *Independent on Sunday* newspaper accompanied by a Foreign Office travel advisory, 'Areas to avoid in Central America', of which it is advised that El Salvador is 'one of the most dangerous countries to visit in the region'; advice, as Simon Calder argues, some 'backpackers interpret . . . as a challenge rather than a warning' (Calder 1994a). Reading like the script to Oliver Stone's *Salvador*, it begins to express the aestheticisation that is supported by new forms of tourism.

The tension-filled journey to El Salvador was not one that we wished to prolong . . . Apprehensiveness hung in the air like smoke . . . Five hours later we arrived at the Salvadoran border . . . We had been careful not to bring in anything remotely *subversivo* – Graham Greene, Joan Didion and Patrick Marnham left reluctantly behind . . . Finally satisfied that we weren't revolutionaries or agents of the left Farabundo Martí Front for National Liberation, they let us through . . . we started to count all the men in round helmets and pale green uniforms . . . Everywhere we looked there they were: American M16 machine guns slung casually round shoulders . . . A station-wagon pulled out of a side road, seven plywood coffins strapped on top. At a cafe in the bus terminal we sat next to a young man with two shining claws instead of hands. He was the first of many amputees we were to see over the next couple of days . . . We took a yellow taxi . . . 'How is San Salvador these days?' we asked the driver innocently, aware that most taxi drivers in El Salvador moonlight as paid informers. 'Muy pacífico,' came the confident reply. But every time he stopped at a red light, his head would turn slowly to the left and then the right, and his steady gaze was invariably returned . . . Salvadoran television news showed reports of 'guerilla activity' in Chalatenango, an area I had hoped to see. Later I learnt that the Salvadoran army . . . had fired mortars into the nearby village . . . and machine-gunned three women and two small babies. Unable to get to Chalatenango we went north to Tonacatepeque . . . Emerging into the bright sunshine, we found a young man face down in the gutter. We thought he was dead, in El Salvador a fair assumption. But he was merely dead drunk. We then drove southwards, to the beach at La Libertad, to see El Salvador at play . . . Just half a mile away was the high black lava field known as El Playón, favoured for years by the death squads as a convenient dumping ground for mutilated bodies . . . We went back to San Salvador . . . Top on our list of essential visits was the tomb of Archbishop Oscar Romero, gunned down by the army while saying mass in March 1980.

(Wolff 1991: 59, 61)

How then can we make sense of this ultimate aestheticisation of reality, from which racism and class struggle actually seem to be enjoyed? Or as Kincaid (1988: 15) has commented, people 'visiting heaps of death and ruin and feeling alive and inspired at the sight of it'. It is not that people do not, or cannot, 'see through the clichés', Krippendorf (1987) contends, but their complicity in being 'seduced by them, again and again'. Perhaps the answer should be sought in the nature of tourism itself as a commodity, especially in its new and thin disguise as a more ethical and moral activity.

Conclusion

This chapter has sought to place power at the heart of tourism analysis. Initially, it reviewed the ways in which concepts of power – ideology, discourse and hegemony – could be deployed in an understanding of contemporary Third World tourism and elaborate the relationship between tourism and uneven and unequal development. It examined the various ways in which critics of Third World tourism have placed relationships of power at the heart of their analysis. It argued that the political economy approach still provides the most systematic critique of Third World tourism. An increasing number of subsequent commentaries have drawn analogies between tourism and colonialist and imperialist forms of domination as a way of expressing the unequal power relationships between the First World and Third World.

But it has also enquired how effective such critiques are in handling the new forms of tourism that have begun to emerge in the Third World. Forms of tourisms with prefixes such as 'alternative', 'appropriate', 'responsible', 'low impact' and, of course, 'sustainable', attempt to challenge the notion that all forms of tourism necessarily draw the Third World into a highly unequal relationship with the First World. Indeed, it was argued that the political economy of Third World tourism implies that such forms of tourism (smaller scale as opposed to mass) can overcome this unequal relationship. It is our contention that such a notion is too simplistic.

The final section of the chapter sought to demonstrate the ways in which new forms of tourism can actually maintain unequal power relationships, albeit in less obvious ways. In doing so, we suggested an alternative critique of the new forms of tourism, a critique which seeks to take the analysis beyond the characterisation of tourism as a form of colonialism. It is not suggested that such a characterisation of the relationships involved in tourism is wrong or inappropriate or irrelevant; rather, that these relationships cannot be fully explained simply by the notions of dependency and domination.

The approach presented suggests we can usefully build upon the former notions of power and capitalist relations in asking how far new tourism spreads capitalist relations of production in the Third World and what impacts this may have. Equally there is a need to reassess how far new tourism can alter the distribution and control of ownership and power. Critically, we add a final set of factors, fetishism and aestheticisation, which are necessary to explain many of the recent features, interpretations and characteristics of the new forms of tourism.

The next chapter examines the relationships we have set out by focusing more specifically on the association between tourism and sustainability.

4 Tourism and sustainability

Chapters 2 and 3 have argued that the term 'sustainability' is now an essential item in the vocabulary of modern political discourse. It is an ideological term. It can be used, and is used, by many if not most businesses, all the mainstream political parties, and those outside the mainstream, to illustrate and describe policies, the implications of which can be shown to be anything but sustainable in a number of ways. And it is widely used in a meaningless and anodyne way.

Sustainability is perceived and described as an essential part of the ideology of the New World Order and all the trends and tendencies that are associated with it. These tendencies, almost movements, include a 'new' consumerism, whose semantic ally is sustainability. The two notions have developed hand in hand to give mass consumption a more acceptable justification to the new middle classes who can afford to consider sustainability. But as Selwyn (1994) remarks:

> supported by a chorus singing of the joys of economic plenty in a world scarred by scarcity, the tourist often appears as a shining hero of the most pervasive myth of our time: that which tells of the omnipotence and untrammelled sovereignty of the individual consumer.
>
> (Selwyn 1994: 5)

In the field of tourism, the term 'sustainability' can be and has been hijacked by many to give moral rectitude and 'green' credentials to tourist activities. And it is by no means just the tour operator and other profit-making companies standing to gain from the activity who have used the term for their own ends. Conservationists, government officials, politicians, local community organisations and tourists themselves all manipulate the term according to their own definition. Illustrations of such instances are given in Chapters 5 to 11.

But whereas the paradigms of political discourse have changed little as a result of the new word 'sustainability', the study of tourism has had to adapt itself to the creation of a whole new branch of the discipline, 'sustainable tourism'; this term is used in conjunction with many other descriptive labels, which are discussed later in this chapter. The practice of tourism is also modifying itself to take account of the new forms of tourism; and opportunities to indulge in 'sustainable tourism' have sprung up all over the world. These too are considered in Chapters 5 to 11.

This chapter examines the notion of sustainability as it is applied to tourism. It begins with a consideration of the development of the mass consumption of tourism and its lead into a new form of consumerism in the industry. A brief survey of the terminology of the new forms of tourism is followed by an analysis of a range of definitions of these new forms. This leads on to examinations of, first, a number of principles often applied to sustainability in tourism and, second, the tools and techniques commonly used to measure and describe sustainability. Finally, we speculate on the intertwined futures of sustainability and tourism.

The growth in mass tourism

The rise and rise of the importance to our lives of holidays taken collectively at a distance from our home is well documented in many texts – see, for example, Krippendorf (1987), Lavery (1971) and Murphy (1985). The early association of this feature of our lives with industrial capitalism is also well noted.

The stimulus given to the holiday industry by technological developments in the field of transport is clear from the histories of the railways (Great Britain in the nineteenth century), the motor car (widespread ownership in the First World after the 1950s) and the wide-bodied jet (post-1960s).

A potted and humorous history of some of the major historical influences on the tourism industry is given in Figure 4.1. We do not wish to make light of these developments and influences, but, rather than chronologically documenting the history of the growth of tourism, Chapters 2 to 4 are used to offer a critique of the current political and developmental contexts of the tourism industry.

The models

Such a critique cannot be made without at least a brief examination of the explanatory models of tourism development. We believe that these are generally deficient as they fail to account for the distribution of power, and the need to set an understanding of tourism in the context of its structures of power has already been presented in Chapters 2 and 3.

Very broadly, explanatory models can be grouped into those which explain the tourist's motivation, those which explain the role of the tourist industry, and those which explain the development of the destination community. Some models such as Butler's Product Life Cycle Model (Butler 1980) attempt to explain the behaviour of both the industry and of the destination community. But such categorisations are simplistic. Moreover, these categories and the models which follow them fail to explain the relationship between the different elements of the industry (tourist, service provider, and local populace at its simplest) and the wider context of development processes. Chapter 2 has already examined these processes in some depth.

Probably one of the simplest models, which serves also as a definition of tourism, is the equation provided by Valene Smith (1989):

$$T = L + I + M$$

where:

T   =   *tourism*

L   =   *leisure time*, which has increased for the majority of workers in the industrialised, technocratic, western, capitalist nations since the Second World War

I   =   *discretionary income* (often referred to as surplus income), which is now commonly used up in the pursuit of instant happiness rather than in the savings for future security associated with the work ethic which was prevalent in industrialised nations in the first half of the twentieth century

Figure 4.1 History of the tourism industry

Source: Biff Products, *Guardian*, 25 July 1994

> M   =   *positive local sanctions* (or *motivation*), which are those factors that prompt the tourist to tour – these are many and varied, but often spring from an escapist motivation. (Thus the local sanctions or conditions may be positive in the sense of prompting travel, but negative in the sense that they reflect unchosen or unhappy aspects of the life of the traveller.)

The model should more correctly be expressed as a functional relationship which varies with a range of factors rather than as a mathematical equation, but it is obviously not intended as a precise representation. Despite its intention, however, its use, even as an aid to understanding, is restricted by its vagueness, and it offers no explicit recognition of the importance of the structure of power over the activity.

Jost Krippendorf (1987: xiii) also referred to the surge in the importance of leisure time in modern western life: 'Most people in the industrialised countries have been seized by a feverish desire to move. Every opportunity is used to get away from the workday routine as often as possible.' His simplistic Model of Life in industrial society (Work–Home–Free Time–Travel) reflects the historical change in the balance of work and leisure in the lives of industrialised and urbanised populations. Again, a historical inevitability underlies the model rather than a political perspective. As has already been argued, the latter is crucial to an understanding of the nature of tourism. And it is no less crucial to an understanding of the new forms of tourism.

Murphy (1985) cites the three crucial growth factors of motivation, ability and mobility as explanations of the evolution of tourism. His characterisations of each of these factors in each of four chronological eras of development are given in Table 4.1. While these factors are clearly germane, neither the historical developments themselves nor tourism can be understood without an analysis of their relationship with the prevailing power structures. And all of these models fail to offer such analysis.

Albeit a little flippant, one model offered recently that does explicitly base itself on the prevailing structure of power and that also relates specifically to Third World countries is that offered by Chang Noi (a pseudonym) in the Thai newspaper *The Nation*. The anger with which it was devised shines through, and in this manner it clearly reflects the local power structure on which the development is based. Chang Noi (2001) suggests that three stages of tourism development can be viewed throughout Thailand:

> Stage 1: Start with a place of outstanding beauty . . . Impose absolutely no controls. Allow get-rich-quick entrepreneurs to encroach on the beach, blow up the rocks, scatter garbage and pour concrete everywhere.
>
> Stage 2: The resort is now popular but rapidly losing its natural charm. Add large quantities of sex and comfort. Build large, luxurious hotels. Import lots of girls.
>
> Stage 3: By now the natural beauty is totally obliterated. The seafront is an essay in bad architecture. The hinterland is a shanty town of beer bars. Develop the remains as a male fantasy theme park. Add anything with testosterone appeal – big motorbikes, shooting ranges, boxing rings, archery. Bring in more and more girls (and boys).
>
> (Noi 2001: 6)

This model is clearly based more on personal observation and political interpretation than on academic research, but for all that it may be as applicable in some Third World countries as many other models.

An attempt to summarise the major features of studies in tourism is offered in Box 4.1. This simple summary emphasises two crucial points. First, there are many textbooks that

Table 4.1 *Murphy's growth factors in the evolution of tourism*

Era	Motivation	Ability	Mobility
Pre-industrial	Exploration and business Pilgrimage/religion Education Health	Few travellers; those involved were wealthy, influential or received permission	Slow and treacherous
Industrial	Positive impact of education, print and radio Escape from city Colonial empires	Higher incomes More leisure time Organised tours	Lower transport costs Reliable public transport
Consumer society	Positive impact of visual communication Consumer society Escape from work routine	Shorter work week More discretionary income Mass marketing Package tours	Growth of personal transportation Faster and more efficient transport
Future	Vacations a right and necessity Combined with business and learning	Self-catering Smaller families Two wage earners per household Demographic trends favour travel groups	Alternative fuels More efficient transport Greater use of public transport and package deals

Source: Murphy (1985: 22)

describe in detail the existing approaches to tourism analysis. Second, it is striking how few concepts there actually are in the tourism debate and the hold that these ideas have retained in directing subsequent research. Many of them are endlessly repeated or contested in case study material.

The 'ethics'

Models can explain tourist behaviour patterns through the contexts of ideology and/or political developments. One example of this is the work ethic. The notion of the work ethic has been a contentious point in political discourse for many decades (see, for instance, Harvey 1973; Marx 1965), but as an explanation of the work patterns and practices of industrialising and industrialised societies it has been and is commonly used and referred to, even if not accepted by all. In essence, it relates the pursuits of moral rectitude and economic survival as the principal motives for people's actions. In this sense the work ethic is used both as a support for and justification of capitalist economic development and ideology. As Fennell warns, however, 'ethics too can be wrong in its support of ideologies and utopias that have more to do with the agendas of a few at the expense of the many' (Fennell 2006: 13).

The waning of the strength of the work ethic as a variable explaining the development of western capitalist trends in behaviour has been shadowed by the appearance on the scene of the notion of the leisure ethic. The two ethics are naturally closely related, for an important justification for the leisure ethic in many minds is that it is not work. The leisure

Box 4.1 Studies in tourism

Structure of the tourism industry

Attempts to identify the main actors and structures in the tourism industry usually take the form of a flow diagram. The most widely cited is Mathieson and Wall's (1982) conceptual framework of tourism. Other texts worth consulting include Burns and Holden (1995) and Shaw and Williams (1994).

Impacts of tourism development

Seemingly a favourite among academics, who list as many impacts as they can under three headings: environment, economic, social-cultural. Mathieson and Wall (1982) started the trend which many others have followed (see, for example, Burns and Holden 1995; D. G. Pearce 1989, 1995; Shaw and Williams 1994). The best and most accessible work of this nature dealing with the Third World is John Lea's (1988) *Tourism and Development in the Third World*.

Models of tourism development

Another 'method' with a vice-like grip on academics is to look at the way in which holiday destinations move from boom to bust. The most famous examples are Doxey's (1976) *index of irritation* (with destination communities moving from 'euphoria' to 'antagonism') and Butler's (1980) *resort life cycle model* (with destinations moving from initial 'exploration' to 'stagnation'). An army of others have followed in testing these models out.

There is also an increasing number of texts devoted to providing blueprints – or models – for appropriate tourism development. These offer both methods and case studies. For example, see Gunn (1994), D. Hawkins et al. (1995) and Whelan (1991).

Tourist typologies and motivational characteristics

Interesting attempts, usually from anthropologists and sociologists, to place tourists into boxes. The leaders have been Erik Cohen (1972, 1979a), whose tourists range from 'recreational' to 'existential', and Valene Smith (1989), whose tourists range from 'charter' to 'explorer'. Many others have set off 'tourist spotting' too.

ethic grows in importance as the work ethic wanes when the pursuit of hedonism replaces the pursuits of moral rectitude and economic survival associated with the work ethic. But it is important to note that the two ethics are not mutually exclusive.

It is also important, however, to point out that the leisure ethic appears to have two distinctly different faces at present – the face of the urban salaried westerner and the face of the local people who have to cater for the leisure ethic of the former. In 1991 Butler wrote of the 'leisure ethic which is at least parallel to, and in some cases more powerful than, the work ethic' (Butler 1991: 201). The work ethic still holds strong in many societies. The leisure ethic has not yet overtaken it, save in a few wealthy societies. But it is gaining ground, not so much as a result of the spread of wealth as something personally experienced by populations around the world, but rather more as a result of the spread of service to the wealthy around the world.

Since the early 1980s, the leisure ethic as an explanation of First World tourist behaviour has been joined by the notion of a conservation ethic which has begun to have a bearing on patterns of travel and tourism, especially in the Third World. The two recent ethics, leisure and conservation, are closely associated. The former reflects the economic power of the individual tourist, and the latter reflects their ability or desire to impose that power on the areas, communities and populations that they wish to visit.

In a Third World which is constantly faced with the extraction and degradation of its natural assets and resources, it is imperative for the new middle classes who travel that those assets and resources which they wish to travel to are preserved for that purpose. Hence, the conservation ethic: introduced by environmental and conservation organisations (such as the WWF, IUCN, Friends of the Earth International (FOEI), the International Ecotourism Society, the Audubon Society and private organisations such as the Caribbean Conservation Corporation (CCC) and the National Association for the Conservation of Nature (ANCON, in Panama)), which suggest debt-for-nature swaps, and entice northern populations to donate towards or to buy an acre of tropical rainforest in the belief that they will be contributing towards the conservation of the planet's biosphere. According to Ryel and Grasse (1991), the conservation ethic

> provides the framework within which all marketing and travelling should take place and includes several basic components: increasing public awareness of the environment, maximising economic benefits for local communities, fostering cultural sensitivity, and minimising the negative impacts of travel on the environment.
>
> (Ryel and Grasse 1991: 164)

In such sentiments Jim Butcher (2003) would detect a whiff of what he refers to as New Moral Tourism. Referring directly to Poon's work (1989a, 1989b – see our Chapter 3) rather than that of Ryel and Grasse (1991), he suggests that

> These people perhaps constitute a new school of 'ethical' tourism – the *New Moral Tourism*. The key features of their moralized conception of leisure travel are a search for enlightenment in other places, and a desire to preserve these places in the name of cultural diversity and environmental conservation.
>
> (Butcher 2003: 8)

David Fennell points out, however, that in tourism studies, 'there are virtually no underlying ethical principles – in a theoretical context – that might act in guiding a comprehensive vision for the importance of human values in tourism decision-making' (Fennell 2006: 191).

Figure 4.2 represents the three ethics and associates them with types of tourism and prevailing economic and cultural power relationships. The models outlined earlier say little of the relationships of power between different elements of society with respect to the tourism industry, while it is these relationships of power which underlie the different 'ethics'. The structures of power in the industry are a crucial explanatory variable of the growth, development, patterns and types of tourism practised, and these are alluded to in an understanding of the ethics.

Cultural trend	Economic trend		Tourism	Power
Modernist	Fordist	**THE WORK ETHIC**	Mass and package	Merchants and new service providers
		THE LEISURE ETHIC	Package, exploration adventure	Transnational corporations (TNCs) + lending organisations
Postmodernist	Post-Fordist			
		THE CONSERVATION ETHIC	Nature and sustainable	Socio-environmental organisations + TNCs + lending organisations

Figure 4.2 Ethics and the industry

The growth

There is of course no doubt that the facility to take a holiday (that is, leisure) has spread, especially since the 1960s. The linking of the package holiday with increasing opportunities for large numbers of people to travel overseas has had profound effects on many areas of the world which now serve as receivers of tourists. The growth and nature of the package holiday business are conveniently and humorously summarised in the UK *Guardian*'s Pass Notes in Box 4.2.

Box 4.2 Package holidays

Pass Notes

No 223: Package holidays

Born: In modern form, sometime in the 1950s, although the original Thomas Cook hatched a rudimentary version in the mid-19th Century.
Father: Fascist dictator.
What? General Franco hit on tourism as a short-cut to economic prosperity; fishing villages filled with concrete hotels and BEA supplied the Brits.
Distinctive features of a package holiday? Watney's, chips with everything, pot-bellied men with tattoos . . .
You know what I mean: All right, the distinctive features of a package holiday are, first, that flight, hotel and everything else are bought in one deal off the shelf and . . .

What about the reps? . . . second, that a holiday company representative is on hand to provide assistance at the resort itself.

Don't these reps drag people out of bed and make them go cycling in the Atlas mountains and take part in hokey-cokey competitions? The more pro-active rep is, indeed, a legendary figure.

So the package holiday boomed? Yes. By the seventies it was an integral part of the British way of life, inspiring its own pop songs, like . . .

No, please! . . . Soleil Soleil, Y Viva Espana, The Birdie Song . . .

Enough! Why did the Brits go such a bundle on Spain? Same reason posher Brits went a bundle on France and Italy.

Which is? The uniquely civilised European way of life.

Pardon? Sex and low-duty liquor.

But now package holidays are finished? Far from it. True, Thomson, Thomas Cook and others are discounting 1994 holidays, but that's recession. This year saw 12.5 million packages sold, matching the 1988–90 peak years.

Where are they going? The old favourites – Spain, Greece, France.

What about Thailand, Egypt, South America? They're not package holiday destinations. They're experiences.

Sorry? A middle-class package holiday, as in "The Nile Experience".

What's the difference? None, except (a) the price, (b) the respectability and (c) you come back on Orient Express.

How is the package holiday business run? Like the rest of British industry. Manic expansion one minute, psychotic cutbacks the next.

That's why so many have gone bust? Intasun et al.? Yes, 'fraid so.

Their epitaph?: Y Vi-va! Esp . . .

Stop! What's a surviving package-holiday tycoon most likely to say? It's a great place for the kids, and a great place for you.

Where is? Anywhere

Least likely to say? You'd have a better time walking in Yorkshire.

Source: Guardian (October 1993: 3)

This global spread of tourism is explained by Prosser as the 'tidal wave of the pleasure periphery' (Prosser 1994: 24–25). Assuming an origin of most tourists in western Europe and eastern USA, Prosser identifies five peripheral regions of the world which have been successively commodified for the tourist industry over the past hundred years or so. Beyond the origin region, which represents the first periphery, these are successively: the western Mediterranean and Florida; the eastern Mediterranean, North Africa, California and the Caribbean; Africa, Asia, Latin America, the Pacific Basin and Australasia; and finally Antarctica and remote areas of all other continents and oceans.

This is purely a descriptive model of the spread of tourism. But Prosser (1994) offers an analytical model to explain the changing consumption patterns and trends through the concept of successive class interventions:

over time, a particular mode of consumption, fashion or lifestyle will spread downwards through the socio-economic class structure of a society. An admired elite inspires or propagates a fashion which is then aspired to by progressively broader sections of society, who as they become able, attempt to emulate the behaviour and style of the perceived elites . . . As this process continues, the discoverer and elite groups, driven by the desire for novelty, uniqueness and exclusivity of experience, seek out fresh destinations and move on.

(Prosser 1994: 24)

UNWTO projections for the year 2020 predict a continuation of this spread and growth for the industry into the foreseeable future. Figure 4.3 shows a 26-fold increase in global tourist arrivals from 1950 to 2000, and they are further projected to increase by more than double from 2000 to 2020. It should be stressed that these figures include all international arrivals using tourist visas. This includes a proportion, which probably differs with space and time, of people who travel for reasons other than tourism. Furthermore, we would draw your attention once again to the reservations concerning the UNWTO statistics expressed in note 1 of Chapter 2. Also shown in Figure 4.3 are the actual global tourist arrivals from 2000 to 2004, and from these it can be seen that the UNWTO's projections (as given in their 'Tourism: 20:20 Vision' report) may be rather optimistic (in the UNWTO's terms). Nevertheless, the figures are clearly illustrative of the trend.

The WTTC claims that travel and tourism 'is one of the world's largest industries, employing approximately 230 million people and generating over 10 per cent of world GDP' (WTTC website 2007). The WTTC includes more than just the tourism industry in its figures, and it is in effect a lobby group for the industry; but even though their claims may be exaggerated, they are nevertheless in agreement with all other sources that tourism is a large-scale and fast-growing industry.

This growth in the importance of tourism has not passed by the Third World. Box 4.3 illustrates this using the examples of Tanzania and the region of Eastern Africa. This point

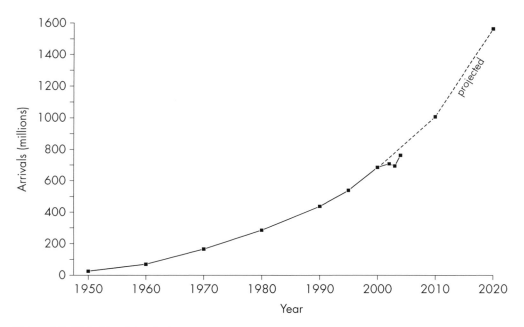

Figure 4.3 Global tourist arrivals

Sources: World Tourism Organisation (UNWTO)
– *Compendium of Statistics 1987– 1991*
– *Tourism: 2020 Vision: Executive Summary*
– *Tourism Market Trends 2005: World Overview and Tourism Topics*. Madrid: UNWTO

Box 4.3 The growth of tourism in Tanzania and Eastern Africa

Figure A shows a gradual growth in tourist entries into Tanzania from the early 1990s to 1997. In just two years after 1997, the number of tourist arrivals into the country almost doubled. Receipts showed the same sharp increase after 1997, again almost doubling in just two years. The effects of the global slump as a result of the terrorist attacks in New York are shown in 2001 and the following two years.

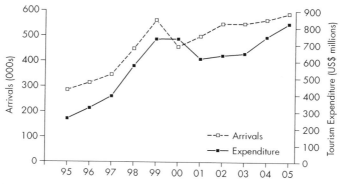

Figure A International tourist arrivals and expenditure in Tanzania, 1995–2005

Although not as spectacular as Tanzania's growth, Figure B shows a steady growth in both international tourist arrivals and monetary receipts from tourism for the region of Eastern Africa as a whole. This occurred despite the problems of politics and violence in Burundi and Rwanda and their overspill into neighbouring countries.

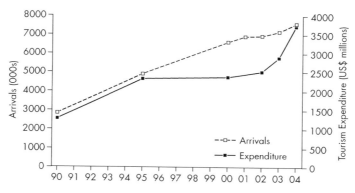

Figure B International tourist arrivals and expenditure in East Africa, 1990–2004

Eastern Africa comprises:
Burundi, Comoros, Djibouti, Ethiopia, Kenya, Madagascar, Malawi, Mauritius, Réunion, Rwanda, Seychelles, Tanzania, Uganda, Zambia, Zimbabwe.

Sources: Tanzania – UNWTO Compendium of Tourism Statistics
E. Africa – UNWTO Tourism Market Trends. 2005 edn, Africa

is further emphasised in Table 2.2, which gives data on the growth of tourism from and to a number of Third World countries. Regionally, from 2000 to 2004, the average annual rate of increase in tourist arrivals to Africa was 4.4 per cent, to Asia and the Pacific 6.9 per cent, and to the Middle East 9.5 per cent. The world average annual rate of increase over the same period was 2.7 per cent (all figures from World Tourism Organisation 2006b). Europe (1.8 per cent) and the Americas (–0.5 per cent) were regions which fell below the world average annual rate of increase (2000–2004).

Mass tourism, then, has increased remarkably in recent years and, despite economic recessions and terrorist attacks, most projections show a continuation of this trend. (The effects on global tourism of the 11 September 2001 attacks have proven to be relatively short-lived and are discussed in greater detail in Chapter 11.) The ability to holiday anywhere in the world has become an essential part of modern professional life in the wealthy world.

Not surprisingly, the growing middle class in the middle-income economies of the world are also increasingly keen to participate in this pursuit of hedonism, and domestic tourism within these countries is becoming or has already become a sizeable and significant sector of the tourism market. In particular, the recent growth in outbound tourist numbers from India and China (see Table 2.3) is remarkable. The UNWTO reported that in 2005 more than 31 million Chinese travelled outside mainland China and forecast that this number would rise to 50 million by 2010 and 100 million by 2020. As *The Economist* magazine put it, 'across the world, hotels, shops, restaurants and travel agents are salivating at the prospect' (Economist 2006b). The prospect prompted Wolfgang Georg Arlt (2006) to ask in his keynote speech to the Beijing International Travel and Tourism Market seminar 'Will China change the face of global tourism?' It should be noted that 'around 90 per cent of [these] visits are still to other Asian destinations, including Hong Kong . . . But tourism agreements reached with EU countries in the past couple of years have boosted travel to Europe' (Economist 2006a). Regarding India, the Pacific Asia Travel Association (2007) reports that 'International arrivals, outbound and domestic travel continue to grow dramatically'. The potential for more growth is, therefore, great.

This recent growth in domestic tourism in Third World countries does not alter the theses pursued in this book. With some exceptions, much of this growth can be characterised as affecting the mass or traditional sectors of tourism rather than the 'new' forms of tourism. Additionally, as we have already noted, there are pockets of the First World in the Third World (as there are pockets of the Third World in the First), and we believe that in general the relationships of power and wealth that we have described (especially in Chapter 3) still hold true.

But can the planet sustain this growth? Is the current practice of tourism suitable for us to pass down to future generations as a model of economic development which will guarantee them a source of income without the destruction of the environment from which they make it? The next section outlines just a few of the growing litany of social, cultural, economic and environmental problems created by the industry and its practices and conduct.

Resulting problems and the rise of new forms of tourism

The phenomenon and growth of mass tourism has led to a range of problems, which have become increasingly evident and well publicised over recent years. In fact, impact analyses have been one of the major outputs of tourism research. The problems include environmental, social and cultural degradation, unequal distribution of financial benefits, the promotion of paternalistic attitudes, and even the spread of disease. These have been described in many books (Bugnicourt 1977; P. Harrison 1979; Hong 1985; Krippendorf 1987; Lea 1988; Pattullo 2005) and other publications (*Cultural Survival Quarterly* 1982, 1990a, 1990b; *Equations* (various); *In Focus* (various); *New Internationalist* (various); the Tourism Investigation and Monitoring Team of Thailand 2000, 2001, 2002), and by a variety of other organisations. The problems have also resulted in a range of campaigns run in recent years by NGOs: Tearfund, Voluntary Service Overseas (VSO), Action for Southern Africa, the WWF and others have all run tourism-related campaigns as well as those NGOs such as Tourism Concern, whose principal motive is action on tourism.

Some of these problems have become matters of global concern, as in the cases of, for example, the state of the Mediterranean Sea, deforestation and consequent soil erosion in various regions of the Himalayas, litter along Nepalese mountain tracks, the disturbance of wildlife by Kenyan safari tours and most recently of course the contribution of air travel to climate change. It may be most illustrative to outline the general problems associated with mass international tourism through the words of the Reverend Kaleo Patterson, a pastor on the island of Kauia, Hawaii:

> I have counselled the prostitute, the desk clerk, the maid and the bartender. I have had to counsel and pray with the whole housekeeping section of a major resort development consisting of over a hundred Filipinos and Hawaiians. I have been involved in hundreds of re-burials of ancient Hawaiian grave sites because of a new resort development or existing resort renovations. I have witnessed the desecration of our sacred places and cried over the senseless pollution of our reefs and rivers. I have held picket signs in protest and given testimony at public hearings. I have organised workshops and forums on tourism. I have even chased an obstinate tourist into the sanctuary of a local pizza restaurant in an attempt to vent my anger in confrontation. I have seen the oppression and the exploitation of an 'out-of-control' global industry that has no understanding of limits or responsibility or concern for the host people of a land . . . All is not well in paradise.
>
> (Patterson 1992: 4)

The problems illustrated by Patterson are tangible and identifiable and can be solved in practical ways. They are often and commonly associated with mass tourism, although there is mounting evidence from impact studies to suggest that new forms of tourism also create similar problems.

It has often been claimed that, in part, the development of alternative forms of tourism has resulted from the need to address these problems. Other factors have also been cited as responsible. For instance, the rise of a population of tourists who are becoming increasingly sophisticated and aware in their leisure pursuits; socio-economic trends in northern countries (Krippendorf 1987) and increasingly in Asian countries (Ghimire 2001); the replacement of the work ethic with a leisure ethic (Butler 1991); and the post-Fordist production trends and postmodern cultural trends which were discussed in Chapter 2. To a greater or lesser degree all these factors explain the rise of new forms of tourism.

While we do not deny the existence of these factors, we believe that the association of the growth of new forms of tourism with the problems arising from conventional mass tourism is misplaced. They may indeed have been used at times as an excuse for this growth in new tourism. But we believe that this growth has come about more as the 'natural' continuation of the historical inequalities between First and Third World countries (Fernandes 1994; Munt 1995). As Fernandes (1994: 4) argues, much of what are now seen as new forms of tourism have arisen because 'the mainstream tourism industry has in fact merely tried to invent a new legitimation for itself – the "sustainable" and "rational" use of the environment, including the preservation of nature as an amenity for the already advantaged'.

Whatever the reasons for their growth, there now abound forms of tourism apparently attempting to grapple with the negative impacts of mass tourism and claiming to be alternative, different or sustainable. As Frank Barrett reports in the UK *Independent* newspaper:

> When the Independent was launched in 1986, there was some debate in the Weekend department of the paper as to whether we should call the travel section 'Independent Traveller'. Hard to believe now, but as recently as the mid-Eighties, independent travellers were still considered in some quarters to be socks-and-sandals wearing, knapsack-toting, five star eccentrics. At that time most people taking a foreign holiday bought some sort of inclusive package from a tour operator . . . As the Eighties continued, people no longer thought about package holidays being fun or good value; instead they became associated with lager louts and unseemly behaviour. Resorts like Torremolinos, Benidorm and Palma Nova emerged as the modern equivalent of a music hall joke . . . By the start of the Nineties, independent travellers were no longer considered oddballs . . . there is a demand from sophisticated travellers for information about sophisticated travel . . . about slow boats to China, express trains to Ulan Bator, coaches across America and rambles through the Himalayas.
>
> (Barrett 1994)

Within the industry, tourism to protected areas and pristine wilderness is one of the most rapidly growing sectors. This is what Boo (1990: 2) describes as the ecotourism sector which 'has rapidly evolved from a pastime of a select few, to a range of activities that encompasses many people pursuing a wide variety of interests in nature'. It also goes under several other names or descriptors which will be examined in the following section.

Published data on the increase in the importance of new forms of tourism are difficult to come by. Where they exist, they do so for specific sites, parks, or tours, and their overall significance in the tourism industry is still difficult to measure. For Costa Rica, a country renowned for its national parks and its promotion of ecotourism, the Costa Rican Institute of Tourism's annual surveys of visitors have regularly shown that around 70 per cent of its tourists (both national and international) visit its protected areas. Costa Rica may not be representative of Third World countries in this sense, although it is often held out as a model of tourism development for others to follow – but see the case study of the country given in Chapter 10. Evidence seems to suggest an increasing share of the tourism market for types of tourism which may (or, as we shall see, may not) be referred to as 'responsible', 'sustainable', 'alternative' or 'environmentally friendly', and an increase in holiday journeys to Third World countries. In the UK in 2001, Tearfund (2002) reports that one in ten holidays taken by British people were to Third World countries (that is, 4.3 million holidays), and points out that many of the favourite new tourist destinations are among the poorest countries in the world. It is noteworthy that many estimates are generally not dissimilar to the estimated 10–12 per cent of the First World population interested in the issues which concern the socio-environmental movement (World Tourism Organisation 1995), and this link is explored again in Chapter 6.

It is clear that one of the difficulties in measuring this growth is the uncertainty of what is being measured. The terminology associated with the types of tourism and the different definitions of these types varies, as does the debate about their degree of sustainability. The terminology and definitions of the new forms of tourism are discussed in the next two sections.

Terminology

Box 4.4 presents a range of terms associated with the new forms of tourism. These have been culled from the vocabulary of relevant academic papers, journals, advertisements and tour operators' brochures.

It would be tempting to dismiss the terminology as insignificant and of little consequence to the notion of sustainability except inasmuch as it provides us with descriptive labels. But the use of these terms represents an attempt to distance the activities associated with the new forms of tourism from what are presumed to be the unsustainable activities pursued by the mass. Frank Barrett of the UK *Independent* newspaper calls this 'a reaction to the naffness of package holidays' (Barrett 1989). But, as seen in later chapters (especially Chapter 7), sustainability is a goal and/or claim of various sectors of the mass tourism industry as well as the sector of the industry which can be described as 'new'.

With this burgeoning list of new terms has emerged a new range of travel agents and tour operators which offer their clients individually centred, flexible, personalised holidays. Phrases employed to appeal to the tourist's desire for something different and exclusive include 'designer' tourism from Cara Spencer Safaris and 'bespoke' itineraries from Journey Latin America. The markets associated with this are referred to as 'individuated' or 'specialised', as distinct from 'mass'.

It is also necessary for operators to differentiate themselves from 'conventional' travel operators. Magic of the Orient, therefore, claim that their holiday brochure is 'like no other', for it is a 'collection of ideas', and Roama Travel (specialising in treks and climbs in India and Nepal) establish that they are not 'a travel agent but a specialist tour operator'. Travel consultancies, a middle-class transformation of the travel agent, have also appeared. Marco Polo Travel Advisory Service, for example, offers a consultation service to the 'imaginative and independent traveller looking for an extra dimension' to their holiday. The small specialist operators catering for the new middle classes who form an increasingly significant market segment can translate their desire to be a twenty-first-century adventurer, explorer or 'traveller'. Urry (1990) argues that this represents a consumer reaction against being part of a mass; and, as we have discussed, the emergence of these specialised markets is a feature of a post-Fordist mode of consumption. In the same way

---

Box 4.4 An A–Z of new tourism terminology

The following list is not exclusive but is indicative of the types of terms used as descriptors of the new forms of tourism.

Academic	Contact	Ethnic	Soft
Adventure	Cottage	Green	Sustainable
Agro-	Culture	Nature	Trekking
Alternative	Eco-	Risk	Truck
Anthro-	Ecological	Safari	Wilderness
Appropriate	Environmentally friendly	Scientific	Wildlife
Archaeo-			

In addition, terms used to describe markets include:

Bespoke	Designer	Individuated	Personalised
Customised	Flexible	Niche	Specialised

that the new middle classes assume control of the 'new' activities through their exclusiveness, so the operators assume exclusivity, and therefore status, for themselves on the grounds of their specialised, individualised offerings.

The messy word de-differentiation (a key feature of postmodernism) is used to convey a straightforward idea. It involves the way new tourism practices may no longer be about tourism per se, but embody other activities. On the one hand this means combining a variety of 'activities' such as adventure, trekking, climbing, sketching and mountain biking. More significantly, on the other hand, it means the marriage of different, often intellectual, spheres of activity with tourism (that is, academic, anthropological, archaeological, ecological and scientific tourisms – see Chapter 7).

These latter forms of tourism are becoming increasingly important to the small-scale tour operator and travel agent. School, college or group tours combining elements of fieldwork and vacation offer a way of catching larger numbers of clients on a package which has to be especially designed, but for the group rather than an individual. Some examples of this type of development are given in Chapter 7.

In this book we use the phrase 'new forms of tourism' generically to cover the range of terms given in Box 4.4. Other terms, such as 'sustainable', 'alternative', 'ecotourism' and the like are used either because they appear in quotations or because they refer to a specific type of tourism that is appropriately described only by that word.

Defining the 'new' tourism

Because the study of the forms of new tourism is still young, there is no clear agreement on their definitions and conceptual and practical boundaries. This lack of consensus is at its most conspicuous between those who study new forms of tourism and those who operate tours. But disagreements are also evident among others in the field, the conservationists, government officials and service providers. The new tourisms are truly contested ideas and the tourism literature is peppered with claim and counterclaim, with mainly academics and interest groups advocating and defending particular terms and definitions. A little like tourism destinations themselves, the terminology of new tourism experiences a relatively rapid circulation as terms come in and fall out of fashion (although there is often little to distinguish one term from another).

As with any activity which involves many groups, the terms mean different things to different people, according to the role they have within the activity. Protagonists perceive and portray the activity they are involved in as 'sustainable', 'no-impact', 'responsible', 'low-impact', 'green', 'environmentally friendly' or use some other suitable term to convey the message. We do not intend to get sidetracked into a lengthy discussion of the many and varied definitions of new tourism types. Suffice it to say they share, in varying degrees, a concern for 'development' and take account of the environmental, economic and socio-cultural impacts of tourism. They also share an expressed concern, again with varying levels of commitment, for participation and control to be assumed by 'local people' and the degree to which they engage and benefit the poor. Box 4.5 summarises some of the major features of the front-runners in the struggle for supremacy of terms. Most of these terms and their meanings underlie or echo the notion of sustainability in its varying guises and again suggest an important link between sustainability, development and new tourism.

The extent to which these terms can be used interchangeably is itself a debatable point. But more significant is the extent to which the claim of sustainability can be justified, a point examined later in this chapter under the heading 'The principles of sustainability in tourism' (p. 100) and in later chapters.

Box 4.5 Tourism and development: defining new tourisms

The seemingly endless list of new tourism terms apart, perhaps the most noteworthy activity has been the definitional battle between different forms of tourism, tourism advocates and protagonists, in defining the most ethical way to take a holiday. Here we briefly review the front-runners that seek to define themselves in relationship to development and sustainability.

Ecotourism

Fared badly during the UN's International Year of Ecotourism. Considered by Third World protagonists as an elite form of western-defined pleasure and by First World proponents as a means of protecting ecologically valuable Third World destination habitats. Unashamedly focused on the environment, with largely incidental benefits for local host communities.

Sustainable tourism

Sustainable tourism focuses on environmental issues – relabelled from ecotourism. Although pro-poor tourism advocates would agree with much that sustainable tourism stands for, the overall objective of sustainable tourism is not to reduce poverty, though this may happen as a result of sustainable tourism development.

Community-based tourism

Seeks to increase people's involvement and ownership of tourism at the destination end. Should initiate from and control stay with the local community, but sometimes arising from operator initiative as, for example, in South Africa (see Simpson 2007). Has some resonance in the other types of tourism reviewed here.

Fair trade and ethical tourism

Fair trade tourism policies seek to create social, cultural and economic benefits for local people at the destination end and minimise leakages. Such policies adhere to national laws; establish strong First World and Third World consultation structures; are transparent; involve open trading operations (such as social accounting); are ecologically sustainable; and respect human rights. The key focus is on changing consumption patterns in the First World. Limited take-up, but growing.

Pro-poor tourism

Out to capture the emerging development consensus on poverty reduction by generating net benefits for the poor. Set to become the developmentalists' favourite;

packed with the most up-to-date technical development-speak. Proponents argue that pro-poor tourism puts 'the poor at the centre of analysis', 'focuses on tourism destinations in the South' and is 'particularly relevant to conditions of poverty' (Ashley et al. 2001: 2–3).

Sources: Ashley et al. (2001); Cleverdon and Kalisch (2000); Mann (2000); Simpson (2007); Tourism Investigation and Monitoring Team, *Clearinghouse for Reviewing Ecotourism* (various)

One further point of relevance to the sustainability of tourism and to definitions of new forms of tourism is the acknowledgement in some quarters that mass packaged tours may be just as sustainable as some of the new forms of tourism (see, for instance, Table 7.1), which makes a qualitative comparison of the sustainability of a package sun-sea-and-sand holiday and a trekking tour. Acknowledgement is made by organisations such as Tourism Concern, Green Globe and by some in the industry that sustainability is not the exclusive concern of new forms of tourism. But the attempt to subsume sustainability is reflected in the language and terminology of the new forms of tourism.

There appears in fact to be a recent marriage of the mass and the new forms of tourism with the emergence of the notion of mass ecotourism. As early as the mid-1990s Gerardo Budowski (1995) discussed the idea of 'a mass ecotourism towards natural areas'; certainly the first few years of the new millennium have witnessed a massive growth in demand for visits to areas such as the Galapagos Islands, Costa Rica's national parks, whale watching, African safari tours and the Antarctic. The new demand comes not only from the rich, industrialised nations, but also from the now significant (although still minority) middle-class sectors in the emerging economies of India, China, and elsewhere in South East Asia and Latin America. Martin Jacques (2005) believes that 'the world is being – and will be – reshaped by the rise of China and India', and perhaps this is – and will be – true for the world of tourism too. As reported earlier in this chapter, the forecast of potential growth in this sector is large. Domestic tourism (that may be categorised as among the new forms of tourism) is also now significant in more than a few Third World countries (Ghimire 2001), and perhaps we and others have erred in stereotyping new forms of tourism as solely the preserve of First World middle-class sectors.

Although the scale and growth rate of new forms of tourism may currently be an open question, the issues surrounding their sustainability remain similar, as do the definitions. These definitions, however, differ according to the different individuals and groups involved, the position or role of those involved who make the definition and according to the context in which the definition is made. With such a diverse range of standpoints, it is not surprising that there is little agreement on the definition of the term. In these circumstances, it cannot be surprising that there is such a plethora of terms, as Box 4.4 illustrates.

The principles of sustainability in tourism

Given that there is no unarguable, comprehensive and all-encompassing definition that is accepted by all, an approach more appropriate than devising definitions is to examine and assess tourist activities according to whether they satisfy a number of criteria of sustainability. Whether, for instance, a specific tour, lodge or wildlife reserve is operating sustainably might be assessed by reference to the criteria listed in Figure 4.4. The list in Figure 4.4 is not presented here in a prescriptive manner; rather, it has been culled from observed practice, especially the practice of those organisations which attempt to publicise

Figure 4.4 Criteria often used for sustainability in tourism

lists of environmentally and ethically sound companies. As such, it stands as descriptive of these practices. It is not our view that these principles represent a 'correct' or absolute version of the meaning of sustainability. Indeed, we believe there to be no absolutely true nature of sustainability and, as the last section has illustrated, it is not definable except in terms of the context, control and position of those who are defining it. As Figure 4.4 points out, the notion of sustainability has many ramifications. These are briefly examined in the following subsections and discussed further at various points in later chapters.

Ecological sustainability

The condition of ecological sustainability need hardly be stated as it is often the only way in which sustainability is publicly perceived. The need to avoid or minimise the environmental impact of tourist activities is clear. Maldonado et al. (1992) suggest that the calculation of carrying capacities is an important method of assessing environmental impact and sustainability. In an important work on carrying capacity, they calculate the capacities for seven different tourist foci in Costa Rica, and in so doing take the notion of sustainability beyond a rather fatuous interpretation by so many users of the term. Box 4.6

Box 4.6 Carrying capacity calculations for the Guayabo National Monument, Costa Rica

The following calculations and conclusions were made by Calderón and Madriz Díaz of the Geography Department of the University of Costa Rica, and were reported in *Análisis de Capacidad de Carga Para Vistación en las Áreas Silvestres de Costa Rica* by Maldonado et al. (1992).

The Guayabo National Monument covers 217 hectares and is the most significant area of archeological interest in Costa Rica. Its structures date back to 500 AD.

The map shows its location in the centre of the country. Within its area there are three trails along which visitors are chanelled. In 1990 the monument received 12,356 visitors, 92 per cent of whom were Costa Rican nationals, and 80 per cent of whom arrived in their own vehicles.

Among the problems of the area, identified by the team of geographers who conducted the study on which the carrying capacity calculations were based, were inadequate government funding for protection of the area, increasing visitor numbers, tree felling in a buffer zone around the monument, indiscriminate hunting of birds, deterioration of the archaeological remains due to visitors, and the lack of a management plan.

Carrying capacity calculations

Three types of carrying capacity were calculated for each of the three trails separately, as it was not possible to apply the three different concepts to the area as a whole. The three different types are:

- physical carrying capacity
- real carrying capacity
- effective or permissible carrying capacity.

They are defined by the variable used in their measurement.

Physical carrying capacity (PCC) is calculated according to the space necessary for one person to move freely in a specified time and assumed to be 1 square metre per person. The average width of the trails is 1 metre, so each visitor uses 1 linear metre of the trail at any given moment. For one of the trails, Sendero Los Cantarillos, other relevant assumptions made are:

- visitors follow the trails in groups of no more than 25 (each group with a guide)
- a distance of at least 100 metres is maintained between groups
- the trail has a length of 1,100 metres
- an average time of 1 hour is required for a visitor to complete this trail
- the monument and trail are open to the public for seven hours per day and 360 days per year.

PCC   =   length x visitors/metre x daily duration (hrs/day)
          = $1,100 \times 1 \times 7$
          =   7,700 visits per day
          = $7,700 \times 360$
          =   2,772,000 visits per year

Real carrying capacity (RCC) is the physical carrying capacity 'corrected' to allow for the following factors: precipitation (FP = 1.39%), vulnerability to erosion (FE = 38.28%), degree of slope (FS = 38.28%). Correction factors were calculated for each of these and expressed as percentages. The calculations are based on survey data, for details of which the reader should consult the original work. The real carrying capacity calculations are:

RCC   = $PCC \times (100 - FP)/100 \times (100 - FE)/100 \times (100 - FS)/100$
          = $7,770 \times 0.9861 \times 0.6172 \times 0.6172$
          =   2,892 visits per day
          = $2,892 \times 360$
          =   1,041,276 visits per year

Effective carrying capacity (ECC) is the real carrying capacity 'corrected' to allow for the difference between the actual management capacity and the ideal management capacity, and is represented as FM. The actual management capacity of the monument is given by the number of personnel (administrative staff, park guards, and guides) employed (in this case 10). The ideal management capacity is given by the number that would be required to fulfil all functions allocated to the staff of the monument (39).

FM   = $(39 - 10)/39 \times 100 = 74.36\%$
ECC   = $RCC \times FM$
          = $2,892 \times (100 - 74.36)/100$
          = $2,892 \times 0.2564$
          =   741.5 visits per day
          = 741.5×360
          =   266,943 visits per year

For work on carrying capacities, see also Box 8.5.

gives an outline of their calculations of the carrying capacity for the Guayabo National Monument, a small area of archaeological significance in Costa Rica.

While the work of Maldonado et al. (1992) is undoubtedly a valuable contribution to the measurement of carrying capacity, it is important to understand that the notion of carrying capacity may be used to wrap a social or economic constraint in a cloak of scientific jargon. Where exclusivity is promoted by the operators, a low carrying capacity is likely to be publicised. Conservation organisations involved in the promotion of new forms of tourism are more likely than most to foster imaginary maximum capacities in pursuit of conservation and economic gain. Several examples of the variability of carrying capacity calculations are given in Chapter 8.

Moreover, despite the progressive nature and importance of this work, Box 4.6 illustrates that the calculations are dependent on assumptions which are in some cases arbitrarily chosen (such as the maximum number in a group and the ideal management capacity) and in others widely variable (such as the degree of slope). Other assumptions and conditions affecting the physical and management capacity of the area (such as the availability of guides, maps, rest spots and the incidence of low cloud cover) might be thought of as relevant, but are not included. Furthermore, a change in one value allocated to one assumption or input could have a substantial effect on the final carrying capacity calculated.

Social sustainability

Social sustainability refers to the ability of a community, whether local or national, to absorb inputs, such as extra people, for short or long periods of time, and to continue functioning either without the creation of social disharmony as a result of these inputs or by adapting its functions and relationships so that the disharmony created can be alleviated or mitigated.

Some of the negative effects of tourism in the past have included the opening of previously non-existent social divisions or the exacerbation of already-existing divisions. These can appear in the form of increasing differences between the beneficiaries of tourism and those who are marginalised by it, or of the creation of spatial ghettos, either of the tourists themselves or of those excluded from tourism. Stonich et al. (1995) provide a clear example of these social divisions on the Bay Islands of Honduras (see Box 4.7).

If we accept the premise that tourism sets up an intrinsically false and fabricated social division between the server and the served in the first place, it is of course inevitable that tourist developments (resorts, enclaves, condominia) will create such divisions. It is one of the purposes of the tools of sustainability, such as carrying capacity calculations, environmental impact assessments, and sustainability indicators, to minimise the effects of these divisions to a point at which they can be excused. To this end, John Clark (1990) has suggested the possibility of calculating social carrying capacity.

Cultural sustainability

Societies may be able to continue functioning in social harmony despite the effects of changes brought about by a new input such as tourists. But the relationships within that society, the mores of interaction, the styles of life, the customs and traditions are all subject to change through the introduction of visitors with different habits, styles, customs and means of exchange. Even if the society survives, its culture may be irreversibly altered. Culture of course is as dynamic a feature of human life as society or economy; so the processes of cultural adaptation and change are not assumed by all in all cases to be a negative effect. But cultural sustainability refers to the ability of people to retain or adapt elements of their culture which distinguish them from other people.

Cultural influences from even a small influx of tourists are inevitable and may be insidious; but the control of the most harmful effects, emphasis on the responsible behaviour of the visitor, and the prevention of distortion of local culture might be assumed to be essential elements of sustainable tourism. Pratt's (1992) notion of transculturation encapsulates the way in which marginalised or subordinated groups select and invent from materials transmitted to them by dominant 'metropolitan' cultures – see Chapter 8 for further discussion of the notion of transculturation. Cultural adaptation occurs in this way, and may result in change towards the wishes of the dominant culture.

Cultural impacts are more easily seen over the long term and are therefore more difficult to measure, although the cultural subversion of many local communities has been well documented, especially but not exclusively by anthropologists – the work of de Kadt

Box 4.7 Social divisions in the Bay Islands of Honduras

An examination of the costs and benefits of tourism developments on the island of Roatán, off the Caribbean coast of Honduras, was made by Stonich, Sorenson and Hundt and reported in the *Journal of Sustainable Tourism* (Stonich et al. 1995).

Inhabitants on the island were categorised by:

Community	*Ethnicity*	*Gender*
West End	islanders	Male
Sandy Bay	ladinos	Female
Flower's Bay		

The degree of tourism development differs considerably in each of the three settlements. Variables describing income, household economic strategies and demography were analysed for each of the above categories.
Among other findings and conclusions were:

- increased social differentiation as a result of tourism developments
- the assignment of the majority of ladinos and islanders to low-status, low-paid, temporary jobs
- reduced access for local people to the natural resources on which they depend for their livelihoods
- escalating prices
- land speculation
- increased outside ownership of local resources
- deterioration of the biophysical environment.

(1979), Plog (1972), Valene Smith (1989) and Smith and Brent (2001), for instance, illustrates the cultural ill-effects of tourism. Organisations such as Survival International and Tourism Concern have also documented the cultural subversion of indigenous groups.

Economic sustainability

The condition of economic sustainability is no less important than all others in any tourist development. Sustainability in these terms refers to a level of economic gain from the activity sufficient either to cover the cost of any special measures taken to cater for the tourist and to mitigate the effects of the tourist's presence or to offer an income appropriate to the inconvenience caused to the local community visited – without violating any of the other conditions – or both. As expressed thus, it may appear as if the other aspects or conditions of sustainability are being 'bought off'. In other words, regardless of how much damage may be done culturally, socially and environmentally, it is perfectly acceptable if the economic profitability of the scheme is great enough to cover over the damage, ease the discontent or suppress the protest.

Economic sustainability, we would argue, is not a condition which competes with other aspects of sustainability. Rather, it can be seen as equally important a condition in its own right. On the other hand, it is not the only condition of sustainability, as might appear to be the case from the thoughts of numerous active agents of the industry. The condition of this as an element of sustainability in no way reduces the significance or level of acceptance or tolerance of the other conditions. Nor does it cloud the importance of the contextual issue of power over tourist activities. With this in mind, the question of who gains financially and who loses financially often sets the power and control issue in sharper and more immediate focus than all other facets of sustainability.

The educational element

It is often stated that an important difference between the new forms of tourism and conventional tourism is found in an element of educational input into the activity. This does not mean that it is necessary to reach high academic levels in order to be a sustainable tourist; but a greater understanding of how our natural and human environment works is often a goal, if not always stated, of the activity. At times, however, it is stated as a goal without being practised. Pressure of business may render this so, but cynicism may also explain it – the flimsiest pamphlet of information for the tourist can be used as evidence of an educational input, and therefore of the 'genuine' motives of the operators and the real desire to aim for sustainability.

Again, it is important to refer this principle back to its context of power and development. Who is the beneficiary of the educational element? Does this enhance their degree of control over the activity and its distribution of benefits? This element, we would argue, has the potential to further widen the inequalities of tourism developments.

At conferences on the subject (First World Congress on Tourism and the Environment, Belize 1992; Ecotourism – A Sustainable Option, Royal Geographical Society 1992; Managing Tourism, Commonwealth Institute 1995; Sustainable Tourism, San José, Costa Rica 1995, to name just a few) education in this respect is generally taken to mean one of two things: first, the enlightenment of the new tourist in the cultural ways and norms of those they are visiting – an education for its own sake; and second, the training of the 'hosts' so that they are better able to cater for the wishes of the new middle classes who visit them.

There are very few acknowledgements of the need to educate the local populace of the destination communities about the tourists. One notable exception to this is Krippendorf (1987), who encourages the dissemination of information about the tourists to those they are visiting:

By supplying the host population with comprehensive information about tourists and tourism, many misunderstandings could be eliminated, feelings of aggression prevented, more sympathetic attitudes developed and a better basis for hospitality and contact with tourists created . . . Such information should aim at introducing the host population . . . to the tourists' background: their country, their daily life (working and housing conditions, etc.), their reasons for travelling and their behaviour patterns.

(Krippendorf 1987: 143)

Another form of educational input into sustainable tourism is the provision of 'technical information on how to do ecotourism *right*' (Whelan 1991: 4, our emphasis). Arrogance like that betrayed in this paternalistic attitude expresses the idea that 'we' know how to do it and 'the rest' just need to be educated in our ways.

Local participation

The importance attached by many parties to the inclusion of the local populations is considerable. Indeed, there is more debate about the degree of inclusion or control to be exercised by destination communities than about the need for their involvement at all. Six different types of participation are identified by Pretty and Hine (1999: 6), ranging from 'passive participation' ('people participate by being told what has been decided or has already happened') to 'self-mobilization' ('people . . . taking initiatives independently . . . [and] retain control over how resources are used') – see Table 8.1. Chapter 8 discusses the issue of participation further.

This debate is thrown into sharp contrast by the two standpoints of 'host' communities as objects of tourism or as controllers of tourism. Again, this matter is often considered to be at the heart of the difference between conventional mass tourism and supposedly sustainable new forms of tourism. But it is argued here that the issue of control is the same whether it refers to mass tourism or any of the new forms of tourism. Indeed, there may be something in the idea that the local authorities and local service providers of a mass tourism clientele have a greater degree of control and power over their activities than do those of the new forms of tourism.

The conservation element

It is often argued that new forms of tourism assist or should assist in the conservation of specific aspects of the biodiversity or culture of a given area, and hence that an essential element of new forms of tourism is or should be such conservation. This criterion has the tendency to divide the conservationists into two distinct camps. On the one hand, we have the proponents of the benefits of specific new forms of tourism who cite examples such as the Annapurna Conservation Area in Nepal, private nature reserves in Costa Rica and selected rainforest areas in Brazil in order to illustrate the relationship between tourist money and the conservation of natural or cultural phenomena by placing a value on their retention rather than their extraction. Gerardo Budowski, for example (former Director General of the IUCN, former Head of Ecology and Conservation at the United Nations Educational, Scientific and Cultural Organisation (UNESCO) and former President of the International Ecotourism Society), believes that 'ecotourism cannot survive without conservation and a symbiotic relation must therefore be established' (Budowski 1996).

On the other hand, we have those who believe strongly that the disbenefits of tourism outweigh the benefits, who see the only valid form of conservation as that which excludes the malign influence of human visitors, and who claim that the former group focus exclusively on species preservation at the expense of local people. This view sees ecotourism as a new form of ecological imperialism in which western cultural values override local cultural values and thereby oppose the principles of sustainability which ecotourism claims to support. These ideas are discussed further in Chapter 6.

The aspects of sustainability discussed above are not presented here as prescriptive. We are not suggesting that a given lodge, tour or reserve (or indeed a regional or national tourism strategy) can be assessed through these criteria for sustainability. In fact, it should be clear that no establishment would be able to meet all these criteria. If they were universally used for making judgements about whether a given practice was sustainable and if all criteria had to be satisfied, then clearly nothing would be judged as sustainable. But this raises the point that sustainability should perhaps be seen as a continuum, and should be assessed on a scale similar to that of probability, offering differing degrees of sustainability. Such an idea opens the concept up to distortion and misuse, but as we have seen and as we shall see, it is indeed misused already.

In any case, it is our contention that sustainability is contested and is not reducible to a series of absolute principles. If principles can be applied to the notion, then it can only be in a relative way, relative to each other without contradiction, relative to the varying perceptions of those who use them, and relative to the values, ideological and moral, of those who apply and interpret them. 'Good' and 'bad' are relative terms, as is sustainability. With this in mind, it is worth considering the priorities for sustainable development set out by Agenda 21.

Agenda 21 and sustainable development in tourism

Agenda 21 is a global action plan endorsed by the 1992 Rio Summit in Brazil (see Box 2.1) and 'reaffirmed at the World Summit on Sustainable Development (WSSD) held in Johannesburg, South Africa' (United Nations Division for Sustainable Development website 2007) in 2002. It sets out the priorities for sustainable development into the twenty-first century. Stancliffe (1995) provides the following summary of the points of relevance in Agenda 21 for the tourism industry:

> Agenda 21 impinges on tourism in two ways. First, tourism is specifically mentioned as offering sustainable development potential to certain communities, particularly in fragile environments. Second, tourism will be affected by Agenda 21's programme of action because its many impacts may be altered by the legal framework, policies and management practices under which it operates. Among other priorities given in Agenda 21, governments are urged to:
>
> - improve and reorientate pricing and subsidy policies in issues related to tourism;
> - diversify mountain economies by creating and strengthening tourism;
> - provide mechanisms to preserve threatened areas that could protect wildlife, conserve biological diversity or serve as national parks;
> - promote environmentally sound leisure and tourism activities, building on . . . the current programme of the World Tourism Organisation.
>
> Business and industry, including transnational corporations, are urged to:
>
> - adopt . . . codes of conduct promoting best environmental practice;
> - ensure responsible and ethical management of products and processes;
> - increase self-regulation.
>
> (Stancliffe 1995)

In its widest sense, tourism is a form of trade, not of goods perhaps, although the commodification of tourist destinations and talk of the 'tourist product' is now firmly established and accepted. Shortly after the Rio Summit, Arden-Clarke (1992: 13) argued that 'the whole of the Agenda 21 section dealing with trade amounted to an evasion of key trade and environment issues, rather than a basis for their solution'.

Arden-Clarke's (1992) arguments about Agenda 21's treatment of the general area of trade are applicable to the field of tourism. Essentially, his criticism is based on two particular features of the Agenda: first, it endorses the GATT rules which encourage the externalisation of environmental and social costs; and second, it endorses the idea that only trade liberalisation will bring about sustainable development.

The first of these endorsements stems from GATT's agreement that the degree to which a country internalises its costs is left to choice. This effectively fixes the externalisation of environmental costs as the norm and makes clear that those countries which deviate from this will lose short-term competitiveness. The second endorsement on trade liberalisation implicitly depends on:

> the 'trickle down' mechanism to solve environmental problems – free trade leads to increases in per capita income through the economic growth it engenders, which in turn creates wealth, some part of which can then be invested in environmental protection . . . The argument essentially says that you must first dirty your own backyard to generate the wealth to clean it up . . . [This] ignores the facts that:
>
> (a) there is no automatic mechanism which guarantees that 'trickle-down' wealth is invested in the environment;
> (b) environmental damage is cheaper to prevent than cure, and in many cases is irreversible.
>
> The flaws in this argument are being learned painfully around the world, but most notably in developing countries.

(Arden-Clarke 1992: 14)

Arden-Clarke's (1992) critique highlights the ideological values which underpin the priorities of Agenda 21 and reinforces the arguments about the importance of relationships of power. The principles of sustainability are not absolute and immutable. In any tourism analysis there is a need to examine the questions of who is stating the principles, priorities and policies, who will benefit from related action and who will lose, and the arguments around Agenda 21 illustrate this point. The ability of the tourism industry to take advantage of the trickle-down effect is investigated and further questioned by Mowforth et al. (2008). The significance of Agenda 21 for the tourism industry is again analysed and discussed in Chapter 10.

Sustainability, or certain elements of it, may be measurable; it may be judged according to given yardsticks. But in the same way that the principles of sustainability may be contested, so too may be its measurement. As with the principles of sustainability, it is necessary to examine the exercise of power in its measurement. Those who employ the tools used to measure sustainability (covered in the following section) may also exercise power over its definition.

The tools of sustainability in tourism

Box 4.8 lists the techniques available for use in assessing or measuring various aspects of sustainability under ten major groupings.

Area protection

As applied to the field of tourism and for the purposes of this chapter, we use the term 'tools or techniques of sustainability' in a general sense. Even the designation of an area of land as a national park or as some other category of protected area can be seen as a tool of sustainable tourism. Those countries with high proportions of their land area under some

Box 4.8 The tools of sustainability

The following listing of techniques is not exhaustive. Each group of techniques is briefly discussed in the main text.

1 Area protection

Varying categories of protected area status:

- national parks
- wildlife refuges and reserves
- biosphere reserves
- country parks
- biological reserves
- areas of outstanding natural beauty (AONBs)
- sites of special scientific interest (SSSIs)

2 Industry regulation

- government legislation
- professional association regulations
- international regulation and control
- voluntary self-regulation
- corporate social responsibility

3 Visitor management techniques

- zoning
- honeypots
- visitor dispersion
- channelled visitor flows
- restricted entry
- vehicle restriction
- differential pricing structures

4 Environmental impact assessment (EIA)

- overlays
- matrices
- mathematical models
- cost–benefit analysis (COBA)
- the materials balance model
- the planning balance sheet

- rapid rural appraisal
- geographic information system (GIS)
- environmental auditing
- ecolabelling and certification

5 Carrying capacity calculations

- physical carrying capacity
- ecological carrying capacity
- social carrying capacity
- environmental carrying capacity
- real carrying capacity
- effective or permissible carrying capacity
- limits of acceptable change (LACs)

6 Consultation and participation techniques

- meetings!
- public attitude surveys
- stated preference surveys
- contingent valuation method
- the Delphi technique

7 Codes of conduct

- for the tourist
- for the industry
- for the hosts
  - host governments
  - host communities
- best practice examples

8 Sustainability indicators

- resource use
- waste

- pollution
- local production
- access to basic human needs
- access to facilities
- freedom from violence and oppression
- access to the decision-making process
- diversity of natural and cultural life

9 Footprinting and carbon budget analysis

- holiday footprinting
- carbon emissions trading
- personal carbon budgets
- carbon offsetting

10 Fair trade in tourism

form of legislated protection might be considered as practising more sustainable tourism than those with low proportions of their land protected. This assumption can of course be questioned. Some governments, for instance, have designated large areas of land as national parks or wildlife reserves but have failed to provide the resources required to afford an appropriate level of protection on the ground. Guatemala and Brazil may be cited as examples here, but they are not alone. It is difficult to blame such governments – they simply do not have the capital resources to pay for land protection, which after all has become a fashionable policy to pursue only in recent years.

Indeed, the very idea of protected areas begs the questions of who is protecting the area for whom and from whom. Many such areas have been so designated as a result of the tide of environmental consciousness that has been promoted, especially by environmentalists and conservationists since the 1970s (see Figure 6.3). In 1994, for instance, WWF International began a fund-raising and recruitment campaign with the patronising slogan: 'He's destroying his own rainforest. To stop him, do you send in the army or an anthropologist?' The advertisement that followed was, as Survival International (1996: 1) observed, 'glibly "pro-nature" and implicitly anti-people'. This example is detailed further in Box 6.1.

This consciousness portrays areas of natural beauty as wilderness areas, unspoiled by contact with humans, and reserved for visits by the 'discerning' and appreciative urban dweller in need of rest and recuperation. This view conveniently ignores both the indigenous inhabitants of such areas and the proportion of the national population in search of and in need of land for survival.

Lorenzo Cardenal, a Nicaraguan environmentalist, has characterised this approach as 'parquismo' (Cardenal 1991). He suggests that a progressive, integrated approach should replace it, referring to the integration of humanity and nature rather than their separation or compartmentalisation as typified by parquismo.

Industry regulation

Regulation of the tourism industry can come from local governments in the form of planning restrictions, national governments in the form of laws relating to business practice, professional associations in the form of articles of affiliation, and international bodies in the form of international agreements and guidelines to governments.

It is axiomatic that government legislation is intrinsically political in multi-party democracies. International agreements may also be explicitly or implicitly political, especially when they stem from a body such as the World Tourism Organisation

(UNWTO) whose 'mission is to promote and develop tourism as a significant means of fostering international peace and understanding, economic development and international trade' (UNWTO 2007). Other international agreements and guidelines, especially those stemming from the work of the scientific community, such as agreements to reduce carbon dioxide emissions, may suffer from a lack of commitment without statutory legislation on the part of national governments and a difficulty in enforcement.

Regulation imposed on the industry by industry associations is normally promoted (at least by industry personnel) as a more effective way of preventing unethical or illegal activity than is government legislation. In 1986 the American Society of Travel Agents (ASTA) produced a set of Principles of Professional Conduct and Ethics. They added weight to these with the threat of disciplinary action against those failing to live up to the responsibilities embodied in the set of principles.

With this type of discipline, the industry tries to promote voluntary self-regulation and to fend off what it sees as restrictive government legislation. On the one hand, it seems to be an intrinsic part of the doctrine of monetarism that any form of regulation should be voluntary and conducted by the industry itself. This accompanies other planks of the doctrine, such as the right to corporate privacy, reduction in public expenditure, transfer of national assets from public to private hands, deregulation of industry, and wholesale support for the notion of free trade 'as the key to planetary prosperity and environmental protection' (Carothers 1993: 15). On the other hand:

> It has to be appreciated that tourism is an industry and, as such, is much like any other industry . . . There is no more reason to expect tourism, on its own accord, to be 'responsible', than there is to expect the beer industry to discourage drinking or the tobacco industry to discourage smoking – even though many agree that such steps would be socially desirable.
>
> (Butler 1991: 208)

The tool of regulation is clearly one which allows specific groups to take control of the industry. The argument around regulation represents a power struggle between different interest groups. So should the industry be regulated, presumably by a branch of government? Or should it be left to regulate itself voluntarily? This issue is addressed again in more detail in Chapter 7.

Visitor management techniques

A range of visitor management techniques exist for use by those who cater for and control the movements of tourists. Some of these are listed in Box 4.8. There are several texts which outline these in depth (Ceballos-Lascurain 2001; Elkington and Hailes 1992; Lavery 1971; Lindberg and Hawkins 1993; Witt and Moutinho 1994).

Worthy of particular note is the current trend towards the restriction of motorised vehicles in areas normally attractive to lovers of nature. On the premise that the motor car as currently run is inherently unsustainable, this trend would seem like a move which the scientific community, the hosts, and the planners could all agree works towards the goal of sustainability.[1] This particular issue is currently of topical concern in countries like the UK and USA where levels of car ownership and use are high. It is also a topical issue in many cities in Third World countries, although in national parks and protected areas in the Third World the problems have generally not yet prompted the same level of concern as those of the national parks in the developed countries. There are exceptions to this, as, for instance, in the case of the highway through the Metropolitan National Park in Panama

City, the largest area of tropical rainforest within the boundaries of a city. Wildlife safari vehicles in East Africa have also created problems sufficient to be widely noticed and publicised in recent years.

Another interesting visitor management technique is that of differential charging for foreign and national visitors. Such a policy is not always understood by the visiting tourists from the north (see Chapter 5), but it promotes the condition of local participation as an intrinsic aspect of sustainability.

Environmental impact assessment (EIA)

A technique which has attained fashionability and respect relatively recently is that of environmental impact assessment (EIA).[2] It has been described as 'among the foremost tools available to national decision makers in their efforts to prevent further environmental deterioration' (Sniffen 1995). But it can be used at more than just the local scale: 'the EIA process is seen as a means not only of identifying potential impacts, but also of enabling the integration of the environment and development' (H. Green and Hunter 1992: 36).

According to Tourism Concern (1992: 34–35), 'For the tourism industry to develop and survive in a sustainable and responsible manner, an anticipatory approach is essential'. This is eminently sensible, as is Goodall's (1992: 62) pronouncement that 'Only where the result of the EIA clearly demonstrates that the development will be environmentally responsible and sustain the destination's primary tourism resources should planning permission to proceed normally be granted'.

But EIAs are not an exact science and can be manipulated like most other techniques. Indeed, it is now standard corporate practice to present an EIA that will allow the building of a development under consideration rather than impede it. Results of EIAs are responsive to those factors used as inputs and assessment can be qualitative and quantitative, and hence subject to degrees of subjectivity. The choice of input, then, is crucial, and it is vital that we recognise that 'If we are to account for environment . . . then the idea of a politically neutral social science has to be dropped' (Mulberg 1993: 110). Mulberg's statement is a reference to the externalisation of unquantifiable factors by the practitioners (accountants) of capitalist economics. For 'externalisation' we could read 'ignoring', a practice which has indirectly led to many of the world's worst environmental catastrophes.

Environmental auditing of specific activities or processes has arisen out of the practice of EIAs, and in turn environmental auditing has more recently spawned the practice of eco-labelling and the certification of products and activities. The ways in which this practice affects tourism are examined further in Chapter 7.

Carrying capacity calculations

Carrying capacity calculations have already been briefly discussed in this chapter (pp. 101–104), and an example given in Box 4.6. It is worth adding, however, that Mulberg's (1993) point about political and social neutrality, which can be extended to include commercial neutrality, is as applicable to carrying capacity calculations as it is to EIAs. Calculations can be manipulated by, for instance, tour operators, protected area officials, officers of conservation organisations, or government officers, to promote either a destination's exclusivity (a low carrying capacity) or its ability and potential to absorb more visitors (a high carrying capacity). Both these strategies might be seen as in the interests of different parties under different circumstances. Circumstances may also change and lead to an increase or decrease in the carrying capacity of a particular place – as Butcher

(2003: 52) points out, 'development itself, including tourism development, can *transform* the carrying capacity. Better transport, drainage, facilities . . . would enable greater numbers of people to visit if this was deemed appropriate.' And it is interesting to note how upper thresholds of visitors, arrived at as a result of carrying capacity calculations, have increased as time progresses. Calculations for the carrying capacity of the Galapagos Islands provide an example of such changes over time – see Mowforth et al. (2008) for further details of this example. Other examples are discussed later, along with the technique of limits of acceptable change which is becoming more fashionable than carrying capacity.

It should also be remarked that the notion of carrying capacity reflects the prevailing relationships of dominance between First World and Third World. Management of the carrying capacity of a particular national park or other protected area gives considerable power to those who have that control. And control of the technique itself offers academics their own degree of power in the debate.

Consultation and participation techniques

Stewart and Hams (1991) argue that 'Sustainable development must be built by, through and with the commitment of local communities. The requirements of sustainable development cannot merely be imposed; active participation by local communities is needed.' As we have already discussed, participation is regarded as central to people-focused and pro-poor approaches to 'development'. In the field of tourism, those who speak of sustainable development almost always include participation of the destination communities as one essential element or principle of that sustainability. For this reason, techniques for promoting public participation and involvement in development projects are included in Box 4.8 as one type of tool available for the measurement of sustainability. Furthermore, techniques for measuring public perceptions, attitudes and values (such as contingent valuation techniques and stated preference surveys) are seen here as a necessary stage in the measurement of sustainability. They too are included here, although for details of the application of these techniques, see David Pearce (1993) and Pearce and Moran (1994).

Techniques which allow for consultation and participation (of those people affected) are still young in their development and subject to problems of definition and interpretation. They are vulnerable to the type of distortion and bias which is introduced in the selection of inputs. They can also be hijacked to give an appearance of consultation with local people while in reality there is only consultation with so-called 'experts'. (Such difficulties are illustrated with the use of the Delphi technique in Chapter 8.) Additionally, consultation with local people has become fashionable, especially among conservationists, but as Survival International (1996: 2) point out in relation to the role of indigenous people in managing protected areas, 'This looks good on paper, but they are hardly an adequate substitute for land ownership rights and self determination'.

Attempts to value social costs and benefits include surveys of public perceptions of, expectations of, and attitudes towards a range of social problems and manifestations such as shopping opportunities, access to recreational facilities, noise levels, litter, standard of living, and vandalism. Measurements of perceptions are placed on a scale, generally extending from highly positive (strongly in favour of) to deeply negative (strongly opposed to). Perceptions, attitudes, expectations and values, however, vary from person to person and from group to group. Difficulties therefore arise with the interpretation of results, which often appear weak and ambiguous, but which are nevertheless used and excused as participation and consultation.

Codes of conduct

The 1990s saw a rising tide of codes of conduct for use in the tourism industry. Their design, promotion, contents, relevance, uptake, effectiveness and monitoring have become important features of the industry and are all worthy of attention.

There are two general points that can be made about almost all codes. First, they attempt to influence attitudes and modify behaviour. Second, almost all codes are voluntary; statutory codes, backed by law, are very rare (Mason and Mowforth 1995).

Many codes of conduct are very impressive in their range of issues and in their depth of discussion and information. But they can be abused by the industry as marketing ploys or as veils extending over many of its impacts. There exist a number of problems associated with the use of codes of conduct which can be summarised under the following descriptions: the monitoring and evaluation of codes of conduct; the conflict between codes as a form of marketing and codes as genuine attempts to improve the practice of tourism; the debate between regulation or voluntary self-regulation of the industry; and the variability between codes and the need for coordination. These are discussed in greater depth in Chapter 7.

Sustainability indicators

The development of sustainability indicators arose from the Rio Summit of 1992 (see Box 2.1 and p. 18). It is now commonly accepted that conventional indicators of 'well-being', such as gross national product (GNP), give a restricted, partial and one-sided view of development. It is the search for indicators which show the linkages between economic, social and environmental issues and the power relationships behind them which has given rise to the development of so-called 'sustainability indicators'. Such indicators have been developed mainly as trials and are currently applied largely at local authority level, although exceptions such as the indicators developed as part of the Association of Caribbean States' Zone of Sustainable Tourism for the Caribbean can also be found (Quintero and Vega 2006 – see also the website of the Association of Caribbean States, given in Appendix 1).

One important aspect that has been built into these indicators from their inception has been the participation of local community members in their formulation. There is no doubting here the genuine and diligent attempt to promote such participation as part of the development of sustainability indicators. There is also no doubting that it is precisely this participation which has led to the use of indicators which are much less remote and much more comprehensible to people than are nationally and internationally derived measures such as GNP, gross domestic investment and the like.

But their acceptance and utility faces an uphill struggle. The measures most frequently used at the level of the national economy relate precisely to that: the economy. Other relevant factors are ignored (or, as economists call it, 'externalised'). Moreover, their use is well entrenched and perpetuated by conservative media which accept new ideas with great reluctance unless they are forced to do so by a public that has already moved ahead. The need to include the social, cultural, environmental and aesthetic factors which our commercial world and controllers normally externalise has not led to a quick redress for such factors, despite public debate of the issue.

Furthermore, it has yet to be proved that these more locally accountable, more relevant and less remote indicators are less likely to be subject to bias and manipulation to suit the ends of those who use them.

Footprinting and carbon budget analysis

The 'ecological footprint' provides a means of quantifying environmental impacts in a single easily understandable indicator. It also provides a means of identifying opportunities for cost savings.

It is calculated on the assumption that the earth is a reserve of natural capital, each year producing interest in the form of renewable natural resources such as fish, soil, fresh water and many more. Ecological sustainablity requires that we live off this interest rather than eat into the underlying 'capital'. The interest is quantified in units of area. At present, there are about two units of area available per person on the planet per year. The WWF–UK has developed the tool of ecological footprinting and estimates that on a global scale humanity is currently eating into the earth's underlying capital by annually consuming around a third more resource than the earth produces, which, if accurate, is clearly unsustainable.

Colin Hunter and Jon Shaw have applied the technique of ecological footprinting to journeys from five of the G8 countries to a range of Third World countries (Hunter and Shaw 2006).[3] Despite seeking conservative estimates of relevance to the ecotourism sector, 'our results suggest that ecotourism holidays involving air travel are likely to produce an absolute demand on global natural renewable resources. The magnitude of this demand may be very substantial' (Hunter and Shaw 2006: 7). And later in their work, 'a higher EF [ecological footprint] at the destination area merely reinforces our central conclusion that ecotourism experiences involving international air travel will normally exert an absolute (and substantial) net demand on global natural resources' (Hunter and Shaw 2006: 8). Regarding the uncertainty surrounding the oft-debated comparison between the effects of ecotourism and those of mass tourism, Hunter and Shaw (2006: 9) comment that 'it would only be logical to infer ecotourism as having a greater impact than mass tourism if, at a global scale, ecotourism products generate more air passenger km. than mass tourism products'.

Hunter and Shaw's (2006) work is of considerable significance for the development of ecological footprinting and holiday footprinting, tools of sustainability measurement that may increase in importance in the future. But as they acknowledge, the technique needs to be made more robust and at the same time more refined before results from its application can be treated with confidence. For such a methodological tool to enter into common practice, it will have to become acceptable to the travelling public, for few tour operators and travel agents will begin to describe their holidays in terms of their holiday footprints unless the idea is already widely understood and accepted by the public. Even if such a situation is reached in the future, it would also be necessary to overcome the industry's resistance to change and new ideas by persuading them that the technique can genuinely work in their financial interests.

A sustainability technique which is already employed by some travel companies despite its newness involves the trading of carbon emissions. Under the European Union's Emissions Trading Scheme, participating companies are allocated allowances (in tones of carbon dioxide equivalent) of the relevant emission. A company can emit in excess of its allocation by purchasing allowances from the market, and companies that emit less than their allocation can sell their surplus on the market. 'In contrast to regulation which imposes emission limit values on particular facilities, emissions trading gives companies the flexibility to meet emission reduction targets according to their own strategy' (Department for Environment, Food and Rural Affairs (DEFRA) 2005). An example of one company which has joined the scheme is British Airways, details of whose involvement are given in Chapter 7, where the company's environmental policies are highlighted.

Effectively, emissions trading schemes give polluters the opportunity to avoid taking action to reduce their own emissions by 'purchasing' the clean record of others as a counterbalance to their own dirty record. To offset their own high emissions in this way does not imply that they are reducing them, which is widely perceived as the only solution to global warming. Whether this argument against emissions trading schemes is justifiable is open to debate; certainly proponents of trading would argue strongly that this is a major new technique which allows the world to begin to combat global warming. These arguments against emissions trading are not raised here in order to cast doubt on the motives and meaning of the companies concerned; rather they are presented in order to illustrate different facets of the debate.

A sustainability technique of the future – if current research bears fruit – may be the development of a personal carbon budget. A few years ago this may have sounded a little like science fiction, but in July 2006 the UK government Environment Minister (David Miliband) announced his ministry's intentions to examine schemes for implementing and managing tradable personal carbon allowances (DEFRA 2006).

Such ideas bring the responsibility for sustainable human development on the planet down to the level of the individual and ultimately ask the individual to restrict their behaviour patterns (including holiday travel) in order that global sustainability may be achieved.

To leave the combat against global warming up to the individual is wantonly irresponsible. At the corporate level, however, there are a number of companies (such as Climate Care and the CarbonNeutral Company) now established to sell carbon offsets. They provide funding for projects that reduce carbon dioxide (CO_2) in the atmosphere, for example, through forest restoration and energy efficiency schemes, in order to offset flight emissions. Companies offering this service receive funding from a range of transport and tour companies which pay a nominal amount for each client who uses their services (such as flights) which add greenhouse gases to the atmosphere. There is no compulsion on companies to support such schemes and so it can be assumed that those that do are genuine in their motive to reduce their contribution to global warming. There is an important question, however, not over the motives of the companies involved, but over the effectiveness of the measures taken to offset emissions. Tree-planting schemes in particular are notoriously unreliable and subject to fire, land takeover and change of use. The *New Internationalist* magazine of July 2006 exposed several carbon offset refor-estation schemes which lasted for only short periods of time and managed to offset only a small fraction of the CO_2 emissions caused by the journeys of those who contributed towards the schemes (New Internationalist 2006).

Criticism of such schemes now appears to be mounting, although travel pages of the mainstream media still regularly offer advice to travelers to offset their flights in articles on 'How to be a responsible tourist' (Guardian 2006) and similar. We need to ask whether the savings of energy made by these schemes really offset the emissions produced by the clients of tour companies and whether an absolute reduction in emissions could be achieved only by reducing the number of flights sold. Ultimately we should ask if it is possible to reconcile a tourism industry which encourages people to fly as frequently and as far as possible with attempts to reduce carbon emissions and global warming.

Fair trade in tourism

Fair trade is a challenge to traditional economic theory and practice in that it seeks to set a price for a product based on principles other than those of pure profit maximisation and on practices other than seeking the lowest cost of production in so-called efficient markets.

It can be seen as a technique of sustainability in that it seeks to redistribute the benefits of an activity or production and thereby to eliminate any resulting disadvantages accruing to a given sector of the population concerned. In theory it should reduce the uneven and unequal development which we have noted and discussed.

The neoliberal economic model, however, is not designed to operate on principles which work against the principle of profit maximisation, and one of the institutions most strongly embodying this principle, the World Trade Organisation, is well known for 'its incapacity to be inventive when it comes to trying to marry sustainable production, fair trade and ethical consumption with international trade policy' (Banana Link 2004).

One of the few attempts to provide a consolidated understanding of fair trade in tourism is given by Cleverdon and Kalisch (2000), who observe that 'Little or nothing is known about fair trade in services, let alone fair trade in the hospitality sector'. Cleverdon and Kalisch's work and the notion of fair trade in tourism are discussed further in Chapter 7.

Whither sustainability in tourism?

It is worth restating the point that the notion of sustainability has to be taken beyond its current bland usage and interpretation, as best illustrated by politicians and daily media pundits. If it remains a 'buzzword' which can be so widely interpreted that people of very different outlooks on a given issue can all use it to support their cause, then it will suffer the same distortions to which discourses such as 'freedom', 'democracy' and 'development' are commonly subjected. All of these terms, and others, are frequently and regularly distorted by most of our politicians (for analyses of such distortions, see Beder 1997; Chomsky 1989, 2000; Curran et al. 1986; Postman 1985).

We move on, then, to the question of how sustainability should be taken beyond its current usage and how it should be given a substantial, tangible and unequivocal meaning. This further leads us on to the question of whether we should be promoting the principles and tools of sustainability, as outlined above, towards this end.

It has already been pointed out that the principles of sustainable tourism are open to manipulation in the service of operators and others in the industry. This is not to say that the principles are not worthy of attention by all those involved in the industry, but it does suggest that the motives of those who apply them should also be scrutinised.

On the assumption that the use of the techniques of measurement and description will help a move towards a clearer, workable and meaningful analysis of sustainability, awareness of the limitations and immaturity of the techniques is also necessary. This means that they are susceptible to manipulation for partisan purposes. In turn, this raises the need to politicise the tourism industry in order to promote its movement towards sustainability and away from its tendency to dominate, corrupt and transform nature, culture and society. The politicisation of the tourism industry would require a clarification and emphasis of the associations between the prevailing power structures and the control of tourism developments, and a clear linking of the goal of reducing uneven and unequal development with the policies pursued by the tourism industry and the governments and international institutions which promote it.

Without this politicisation, sustainability will continue to be hijacked by the prevailing model of development, capitalism, and will increasingly fall into the service of the controllers of capital, the boards of directors of major transnational companies and other organisations which manage the industry. This tendency has already become apparent, and as the new forms of tourism gather ground and increase their share of the tourism market, as seems likely, the current power structure and the processes by which power is held and retained will attempt to subsume them, as has already been shown.

Concurrent with this trend in many areas of the Third World, however, is a grassroots groundswell to take control of, and exploit, tourist opportunities at the community level. Currently this tendency seems to assume automatically that 'sustainability' is their prerogative, and use of the term is as loose as it is in other tendencies. Automatic assumptions are often used to cover over awkward questions. The existence of these different tendencies highlights the debate between a tourism that is industry controlled and one that is community controlled.

The obstacles to change towards sustainability briefly mentioned above and the struggle for power over the definition of the concept are themes which run throughout the remainder of the book. The discussions on global changes, power relationships, dependency theory and sustainability in the first part of the book inform and fuel the analyses of the roles of different sectors of tourism that now follow. Specifically, the key themes (uneven and unequal development, globalisation, relationships of power) and key words (new tourism, sustainability and the Third World) of the book will reappear as the bases of discussions in each of the following chapters.

In Chapter 5 the role of the new tourist class of relatively wealthy westerners is considered, and we ask how and why new tourism has a distinct class dimension. In Chapter 6 the high moral ground of the socio-environmentalist organisations which attempt to influence the full range of human development, not simply that of the tourism industry, is examined. Chapter 7 looks at the often criticised operations of the agents of the tourism industry, the tour operators, travel agents and airlines, and asks how they are adapting themselves to the new forms of tourism and the principles which these embrace. Chapter 8 views the concept of sustainability from the point of view of the populations at the receiving end of tourism, especially at the community level, and makes use of a range of different types of participation. Chapter 9 investigates new forms of tourism in urban areas of the Third World, a largely neglected field of study. Chapter 10 seeks to explain the response of governments to the myriad factors affecting the development of the tourism industry, its relationship with sustainability, and the power and influence of supranational institutions such as the World Bank and IMF. Chapter 11 examines one of the most recent developments in new forms of tourism, namely pro-poor tourism. Finally, the issues and debates briefly alluded to in this last subsection are discussed again in Chapter 12, in the book's conclusions.

5 A new class of tourist: trendies on the trail

As Crick (1989) argues, 'human beings', the tourists themselves, are only infrequently the object of consideration in much tourism research, and yet it is upon tourists that the industry is founded. In part this has resulted from academic prejudices of analysing production rather than consumption (Lash and Urry 1994) and the tendency to focus upon the impacts (economic, environmental, cultural) of tourism where tourists are at best a homogeneous and undifferentiated group and at worst are deemed immaterial.

It has been suggested at various points in earlier chapters that tourists need to be taken more seriously, and campaigns in recent years on the ethical forms of travel and tourism from development agencies (such as VSO and Tearfund) seem to suggest that they are. To this point in the book, a number of key factors have been stressed. First, Chapter 2 sought to demonstrate the significant economic changes that have resulted in post-Fordist modes of both producing and consuming goods and services. In part, these changes can be traced through to the emergence of new and varied forms of tourism in the Third World. It was also argued that these changes have tended to invest more power in us, the consumers, or the tourists: we have more choice. Second, it has been suggested that the influence of the new middle classes can be linked to the emergence of what may be referred to as postmodern cultural forms. The importance of these social groups in both producing new forms of tourism in the Third World and forming a significant role in taking such holidays was highlighted. It may be worth taking another look at Figure 2.2, Table 2.4 and Box 3.3 to recall this framework.

Both points stress the increasing significance of tourists, with the second factor highlighting the importance of social class. Of course this is not to say that class is the only factor in studying tourism; far from it. But it is a significant factor and is especially important in the analysis of new forms of tourism, in that the 'world of tourism is rife with the class distinction in our everyday world' (Crick 1989: 334). Yet an analysis of the significance of class and tourism is only weakly developed. It was argued in Chapter 2 that analysis of tourists has centred around either classifying tourists or carrying out motivation and attitude surveys. Although such approaches are of interest, they have tended to limit the scope of tourism analysis.

This chapter explores what use an analysis of class offers in developing a critique of Third World tourism. As elsewhere in the book, this is not a definitive statement of how tourists should be analysed and understood, but an attempt to broaden our thinking and approach to the field of tourism. Initially, the chapter discusses the importance of social class as a concept and how it is reflected in travel and tourism. In particular, the relationships of contemporary global change to the way classes are formed will be considered. The second half of the chapter identifies a number of crucial factors evident in the formation of new middle-class tourism, and it is argued that travel and tourism have an increasingly important symbolic role to play as social classes seek to define and distinguish themselves from other social classes. The final sections of the chapter indicate why social class analysis is of importance in considering the impacts of Third World tourism.[1]

Class, capital and travel

Of all social science concepts, arguably, it is class that has been subjected to the most thoroughgoing marginalisation since the 1980s. Politicians began to talk in terms of classless societies, academics became preoccupied with the fragmentation of traditional class lines (working class, middle class and upper class) and more and more people have begun to think in terms of their individual, as opposed to class, status. This is a lure of individualism that led Crompton (1993: 8) to conclude that the 'retreat from class . . . is becoming the sociological equivalent of the new individualism'.

However, a recognition of the role of social class in contemporary capitalism has begun to re-emerge. A number of authors have begun to illustrate the relationships between cultural changes and the development of new middle classes (Betz 1992; Featherstone 1991). In common, these authors agree that these significant and numerically expanding class fractions have an instrumental part to play in new cultural forms. Indeed, Lash and Urry (1987, 1994) refer to postmodernism as a 'hegemonising mission' for the new middle classes, and elsewhere they are considered as both producers and consumers of post-modernism – par excellence (Featherstone 1991). In Box 2.3 we identified a number of key characteristics of postmodernism and highlighted their relevance to the analysis of Third World tourism. Briefly these involved the emergence of specialist agents and tour operators (and its adjunct, more individually centred and flexible holidays), the de-differentiation of tourism as it becomes associated with other activities, and the growth of interest in Other cultures, environments and their association with the emergence of new social movements.

Nevertheless, the different ways in which different social classes consume tourism is vastly under-researched. This is significant in that we have argued above that much new tourism is based on the notion of individual travel. As Figure 5.1 conveys, a culture of travel (as opposed to tourism) that is shamelessly hedonistic has emerged: chill out, the world's a breeze, so experience it. But most of the existing commentary amounts to little more than the rhetorical and polemical ranting of middle-class writers bemoaning the scale, effects and popularity of mass tourism. The analogy of tourism to pollution is commonplace (Brooks 1990; Guerrin 1991; Turner and Ash 1975) and is often expressed in socio-cultural terms where it is the swarms or tidal waves of tourists that constitute a problem that must be contained. It is a view that appears to be prevalent across the arts, academia, the media and the industry. The renowned French photographer, Henri Cartier Bresson, contends: 'I'm an adventurer . . . The Africa I experienced was literally that of the "Voyage au Bout de la Nuit". It was a time when people were travellers, not tourists . . . I travelled on a shoestring. Regimented tourism is a contagious form of pollution' (quoted in Guerrin 1991). Rosenthal (1991: 2) admits that perhaps it is timely to 'accept that some areas will remain "sacrificial resorts" which continue to attract tourists who do not seek an experience of "deeper" quality'. As an operator contends: 'With the days of the package holiday . . . drawing to a close, we are about to discover very special holidays for very special people: like you' (Sherpa Expeditions 1992 brochure).

Equally limiting is the fact that the role of geography, the spatial dimension, is all but ignored, despite its fundamental role both to understanding tourism as a process and how social classes are constituted as part of that process. As Thurot and Thurot (1983: 182) note, the ability of the middle classes to maintain their 'social distance' from others has become more difficult because 'great distance no longer offers substitutable destinations'.

Figure 5.1 The culture of travel

As Box 4.1 suggested, the circulation of tourist areas has tended to focus either on the deterministic cyclical nature of tourist development or the cyclical character of tourists themselves. In the former approach, tourist areas are modelled as moving from discovery to overcrowding (Butler 1980; Doxey 1976). In the latter approach, a taxonomy of visitor characteristics are noted (E. Cohen 1974, 1979a; V. Smith 1989). Although this approach offers greater scope for thinking about the relationships between social class and tourism, the links remain weakly developed.

Bourdieu and 'habitus'

By acknowledging the importance of cultural consumption in the study of social classes, the French sociologist, Pierre Bourdieu (1984), has provided a productive way for thinking about the relationships of class and tourism. A number of authors have noted the applicability of his influential work to tourism (Bruner 1989; Errington and Gewertz 1989; Lash and Urry 1994; Urry 1990).

Briefly, Bourdieu (1984) argues that social classes wage 'classificatory struggles' seeking to distinguish themselves from each other by education, occupation, residence and so on, and, of course, through commodities, which is taken to include both objects (cars, furniture and so on) and experiences, such as holidays. They achieve this, Bourdieu (1984) suggests, by constructing 'lifestyles'. We have already alluded to this notion of lifestyle ('Sustaining culture and lifestyles', Chapter 2) as a useful way of considering the individuals' uses of a range of objects, experiences, hobbies and beliefs to, sociologically speaking, 'mark their territories'.[2] These lifestyles, Bourdieu concludes, are the products of what he terms habitus, and it is worth spending a few moments considering the significance of this concept. Habitus represents the ability and disposition of individuals and social classes to appropriate objects and practices, to act in certain ways (J. Thompson 1991), that differentiate them from others. A knowledge of 'foreign' food, good wine, classic literature or Latin American film, for example, may all assist in differentiating from others without such knowledge or appreciation. Habitus is, therefore, a 'cognitive structure' (Jackson 1992) which 'gives people a sense of their place in the world' (King 1995: 28). Or as Painter (2000) suggests,

> Habitus gives individuals a sense of how to act in specific situations, without continually having to make fully conscious decisions. It is this 'practical sense', often described as a 'feel for the game', that Bourdieu's theory of practice seeks to understand.
>
> (Painter 2000: 243)

For Bourdieu it is the necessary starting point for a theory of strategies that 'aims to account for the logic according to which groups, or classes, form and break up' (Bourdieux 1986: 119). Box 5.1 provides an application of the concept.

Two important points emerge from Bourdieu's work. First, social classes are in constant struggle to ensure that differentiation from class fractions above and below is maintained. Travel has always had an important role to play in this process of differentiation, and as tourism has become more widespread the struggle has intensified. It is difficult to deny that Marbella, Kos, Phuket and 18–30 provide a very different set of meanings and representations from Tikal, Chiang Mai, Kilimanjaro and Explore. So different places, the different experiences to be had, and even different operators, add up to what we could describe as a 'symbolic system'; it is the way in which we represent objects and experiences and then communicate this to others. We can also argue that these different components

Box 5.1 Ecotourists: a personal profile

Age 58, retired World Bank agricultural economist

Meet a member of the new 'whoopie' club – wealthy, healthy, older people. Like many ecotourists, they are from the rich industrialised countries, aged between 44 and 64 years of age, and this growing elitist club has 'been there and done that'.

Not for them the packed beaches of the mass tourist resort. According to a survey carried out by the US-based financial company American Express, these travellers have exhausted traditional destinations and are now in search of original, preferably pristine destinations. As the gap widens between retirement and death, and an ageing population grows, these rich people are destined to make up the core of future eco-travellers.

Age 33, teacher, partner also in work, vegetarian and eco-friendly, donates to Third World charities

Meet the north's sensitive soul. A member of the liberal middle classes who have grown more environmentally aware over the eighties. For these people various eco-travel operators seem to offer the exotic destination, but in a way that can help the local economy and not be too disruptive to the environment. And in a busy, do-gooding life, a laid-on tour is convenient.

Age 20, student

Meet what some observers call the 'curriculum vitae builders' or 'ego-tourists'. Young northerners who may be more mindful of environmental issues and who take off for longer holidays in the Third World, searching for a style of travel which reflects an 'alternative' lifestyle. Many will backpack, others will join tours that explore the last great wildernesses. What they have in common with all other ecotourists is the necessary capital to embark on the adventure.

Age 36, director of southern-based NGO

Meet the new affluent southerner. Works for an organisation which receives most of its funds from individual and corporate donations and grants from the north. Intensely concerned with his work, which allows him to travel to other areas in the Third World. Is keen to network with northern environmentalists. Spends holidays with family visiting national parks in own country and neighbouring countries.

Source: Panos Institute (1995)

'each embody particular class "tastes"' (Allen and Hamnett 1995: 240). Habitus, therefore, represents a certain 'class culture and milieu' (Zukin 1987: 131) and provides the basis for class reproduction and differentiation. The traveller/tourist distinction, although highly stereotyped, is reflective of a wider debate over social class differentiation.

These battles are an example of hegemony in practice (as discussed in Chapter 2). Of course, this also suggests that social classes are themselves constantly formed and hence the term class fractions to distinguish from the 'once-and-for-all' idea of the working class, or the middle class. Social class formation is, therefore, a dynamic process, 'a flame whose edges are in constant movement, oscillating around a line or surface' (Bourdieu 1987: 13), and of conceptual importance in helping to understand the dynamic nature of tourist consumption: today's cool spot is tomorrow's classificatory dinosaur.

Second, a discussion of class also suggests the importance that exploitation (of one class by another) should play in the analysis; in 'order for a social collectivity to be regarded as a social class it has to have its roots in a process of exploitation' (Savage et al. 1992: 5). Moreover, an exploitation-centred analysis of class provides for a more coherent strategy for considering the characteristics of the middle classes in contemporary capitalism (E. Wright 1985: 131); hence, the straightforward observation 'one person's welfare is obtained at the expense of another' (E. Wright 1985: 65). Traditionally, class analysis has been undertaken with a production-side bias with the focus on the exploitation of the workforce by capitalists (E. Wright 1985), and a principal focus on economic measures (occupation, income, and so on) as indicators of class position. The critical part of Bourdieu's work is that 'domination remains, but it must be reconceptualised in a world of consumption. Domination is now mediated by taste' (A. W. Frank 1991: 66). Reflecting this power, Eagleton (1991) refers to habitus as a 'practical ideology', or what we might call an 'everyday hegemony', whereby the need to dominate is translated into routine social behaviour. The importance for our discussion of tourism is that the struggles inherent in social classes have both a social element (ecotourism is better than package tourism) and spatial element (the Brazilian rainforest has more kudos than a Gambian beach). In other words, tourism can be used to differentiate and exploit both socially and spatially.

So exactly how are these struggles waged and differentiation achieved? An important part of the answer lies in the accumulation of capital. But in this case it is not the finance capital necessary to sustain profits or individual financial assets, but cultural and symbolic capital linked to certain kinds of tourism.

Cultural and symbolic capital

As a number of authors have contended, consumption has become more skilful in recent years (Lash and Urry 1994: 260). Not only do consumers ask more questions of the commodities they buy (the least polluting car to buy or holiday to take?), but also they are more sensitive to the symbols they are consuming. The consumption of holidays has assumed an increasingly significant role in the cultural differentiation characteristic of other forms of consumption and the Third World has an important role within this and represents a sizeable cultural asset. It is a way of investing in and accumulating what Bourdieu (1984) refers to as cultural capital.

Unlike economic capital, cultural capital is not something that can be strictly bought, but instead relies on the ability of individuals to join the game of being able to 'know' and 'appreciate' what to eat, drink, wear, watch and what types of holiday to take. In other words, it requires the skill of reading the cultural significance of certain types of consumption; for example, the significance of 'original features' in the process of gentrification (Jager 1986) or the cultural significance of certain types of holidays, whether ecotourism in Kenya or backpacking in South East Asia. In the former case, 'it is not simply that a desire to see ecological attractions in far-off destinations conveys a sense of superior status; rather it is the actual ability to appreciate ecology which indicates cultural competence' (Allen and Hamnett 1995: 240). In the latter case, it will be argued that cultural capital is also accumulated through journeys to remote and hazardous regions (recalling the discussion of fetishism and aestheticisation in Chapter 3). It is these young travellers who often emerge as 'figures of admiration' (Jardine 1994). As Bob Samuel, information manager at Trailfinders, observes: 'There is a lot of kudos in coming back and telling your friends you have travelled overland in Cambodia. It gives you a certain cachet' (quoted in Jardine 1994).

If tourism has increasingly taken on a symbolic meaning (in part due to the dramatic increase in tourism), it has also meant that tourism as a commodity embodies 'sign-value', an important means of stocking up on cultural capital. It is symbolic in the way in which travel and tourism embody certain attributes: personal qualities in the individual, such as strength of character, adaptability, resourcefulness, sensitivity or even 'worldliness'. In a somewhat traditional sense, travel is widely regarded as 'character building'.

In this way tourism not only is something to be enjoyed but also represents a strategy for building a reputation that can be converted into economic capital. In other words, it transforms travel into a commodity with exchange-value and can be traded in, so to speak, for material benefits; and hence the phrase 'curriculum vitae builders' in Box 5.1. Trailfinders' 'long haul travel consultants' must be aged 22–30 and 'have travelled extensively'. But of course it is not only the travel industry that recognises the 'benefits' of travelling and there is a number of occupations where overseas travel is an increasingly significant component or, indeed, prerequisite. Curriculum vitae building starts young, as Sansom (2001) contends, with universities and employers agreeing 'that a constructive year out embellishes your CV'. Myriad agencies have sprung up to offer opportunities for a 'year out' – in the UK, this includes Teaching & Projects Abroad, Group Activity Projects, Frontier, The Year Out Group, Operation Wallacea, Greenforce, VSO's Youth Volunteering and many others. In fact, an industry has grown up around this segment of the travelling market, and many companies now exist principally to make money out of catering for the gappers and voluntourists – see Box 5.2. In a 'Review of Gap Year Provision' made for the UK Department for Education and Skills, Andrew Jones (2004) identified '85 specialist organisations whose primary "market" represented gap year participants' and suggested 'they provide at least 50,000 placements in the UK to gap year participants annually' (Jones 2004: 14). As Tourism Concern (2007: 6) describe them, there is now 'a rapidly growing industry champing at the bit to create placements to fulfil this new market'.

How are the sign and exchange values of these travel commodities maintained or enhanced? Before we try to answer this question, consideration will be given to how class fractions may be conceived in relation to tourism. This is not to suggest that class fractions are fixed and immutable as described below; rather, it suggests a heuristic approach – a style of analysis – that may help our understanding of tourism and class.

Ecotourists

The first principal grouping Bourdieu terms the new bourgeoisie. They are located firmly within the service sector with jobs involving finance, marketing and purchasing, for example. They are also class fractions that are well paid and are correspondingly high in terms of economic capital (finance). They can afford expensive holidays that are exclusive

Box 5.2 Voluntourism and gappers

Two new (rather awkward) words have entered the common English language since the turn of the century: voluntourism and gappers. They are not unrelated. **Gappers** are generally young people of university age who through their own efforts or with the financial backing of their parents and family choose to take a year out of their continuing studies in order to gain experience of the world. Increasingly this young segment of the population is being joined by members of the working population who want to take a break from their chosen career and by retired people who feel that they still have energy enough left to 'do good work' and travel a bit too. Some of them simply travel as backpackers, but many of them volunteer to work on projects abroad, projects which are associated with some form of development – building or painting a school, teaching English, beach clean-ups, wildlife censuses, national park path maintenance, and so on. For their efforts and the experience they gain, they have to pay; so they are volunteers and tourists at the same time – hence, **voluntourists**. Jones (2004: 8) defines a gap year as 'any period of time between 3 and 24 months which an individual takes "out" of formal education, training or the workplace, and where the time out sits in the context of a longer career trajectory'.

Potentially, they are current and future ego-tourists and some may eventually join the ranks of the ecotourists, or even the travel professionals who guide the ecotourists. But for the moment they have to find the money to pay for their good deeds in Third World countries and for the experiences they gain, and in many cases the price is not cheap. Indeed, in an *Education Guardian* article, Jessica Shepherd (2007: 1) suggests that with respect to gap year activities there is 'a widening gulf between middle-class students and their less well-off peers'. There may be some argument about the degree of association between voluntourism and specific class fractions, but that it is identified with a broad middle-class movement and with the motive of 'doing good works' is hardly contestable. By implication, then, in its relationships of power between those who do the good and those who receive it, or hopefully benefit from it, voluntourism is political. The fact that this is not perceived by many leads Hutnyk to despair:

> It is not enough just to raise questions about the moral propriety of First World youth taking holidays among the people of the Third World; it is not enough to encourage discussion of such contradictions in cafes along the banana-pancake trail. Nor is it sufficient to reflect critically upon the politics of charity, while working – because something must be done – at a 'sound' street clinic.
>
> (Hutnyk 1996)

Tourism Concern (2007) points out that there is no control and no regulation over the activities of the companies which cater for the gappers and voluntourists, and anecdotal evidence suggests that some (but an uncertain proportion) of the experiences offered have not resulted in the benefits expected for the voluntourists concerned. More worryingly, an increasing number of reports tell of work projects that construct unwanted buildings, take jobs which would otherwise be taken by locals, promote projects which are opposed by some segments of the local populations, plant saplings which will not be tended, and leave 'white elephants' which cannot be sustained or maintained by the local communities involved.

There are many potentially positive aspects of such work opportunities, and not all the reports tell of projects gone wrong. But as a result of its research, Tourism Concern has felt the need to produce a code of practice for volunteer-sending companies to ensure that the industry provides a beneficial experience for the volunteer and host community alike (Power 2007; Tourism Concern 2007).

in terms of price affordability and the numbers of tourists permitted: private game reserves with luxury accommodation and limited capacity, for example. They are class fractions, therefore, which can take holidays in environments which, by virtue of price, are exclusive. For shorthand, we shall refer to this group as ecotourists, older and professionally successful tourists (Errington and Gewertz 1989) or as Wood (1991) refers to them, the 'cream of Manhattan and the City'. 'Ecotourist' has a double meaning, however, for not only does it signal an interest and focus of this type of tourist on the 'environment' (ecology), but it also indicates the ability to pay the high prices that such holidays command (economic capital).

It is not a coincidence that the 'taste' of which Bourdieu speaks is actually translated into culinary metaphors for the holidays aimed at this social class, for luxurious eating is another of this fraction's cultural characteristics. Thomson Holidays' 'Travel à la Carte' urges you to 'choose from the menu or dream up a dish of your own' and Barrett (1989) refers to 'delicatessen' travel agents. It embraces the delicate, discriminating and generally luxurious holidays to an off-the-beaten-track, out-of-the-way place. And there are apparently no expenses spared on the level of service and culinary provision. As Africa Exclusive, a tour company specialising in East African safaris, put it, 'gastronomy is taken as seriously as zoology'. It carries a level of cultural capital as well. It is a way of buying into a shared meaning that the environment has an acknowledged value: 'ECO-TOURISM . . . It's not in any dictionary yet, but if you want to impress at your next dinner party, it's a dead cert' (Wood 1991). So, not only are the new bourgeoisie high on economic capital, but also they are high in terms of their ability to accumulate cultural capital.

Ego-tourists

The second important class fraction that Bourdieu identifies, and one that is potentially more important, is the 'new petit bourgeoisie'. This social class is characterised by the increasing numbers working in employment involving 'presentation and representation' (Bourdieu 1984) such as media and advertising, but it also encompasses a wide range of service jobs (such as the guides working for small tour operators) and the growth in the number of people working for charities, for example.

You may recall that some commentators have emphasised the significance of this social class in interpretation and representation. This not only includes opinion-forming in terms of the latest film to see, clothes to wear or designer alcohol to drink, but also how we should perceive issues such as homelessness or international development: in other words, a guide to the things we should be 'into' (consider the profiling of tour leaders in Box 5.3 on p. 133).[3] Chapter 2 made reference to this group variously as taste-makers, trend-setters, new cultural intermediaries or cultural brokers, labels which seek to capture the structural significance of such groups in postmodern culture.

Unlike the new bourgeoisie who are high on both economic and cultural capital, however, the new petit bourgeoisie are not so economically well endowed and must seek out cultural capital in order to establish their social class identity. In this way the burden of class differentiation weighs most heavily on this social class who must differentiate themselves from the working classes below (mass packaged tourists) and the high spending bourgeois middle classes above (ecotourists). Part of the reaction has been for the new petit bourgeoisie to deem themselves unclassifiable, 'excluded', 'dropped out' or perhaps, in popular tourism discourse, 'alternative'; anything other than categorisation and assignment to a class (Bourdieu 1984). Yet, as Bourdieu concludes, the A to Z of their practices, from 'aikido' to 'Zen' (or from alternative tourism to wildlife tourism) exudes classification, being nothing other than 'An inventory of thinly disguised expressions of a sort of dream

Figure 5.2 Self-discovery and the social field

of social flying, a desperate effort to defy the gravity of the social field' (Bourdieu 1984: 370) – see Figure 5.2.

This slide into individualism is an expression of what Neil Smith (1987) refers to as a new ideology founded upon the 'pursuit of difference, diversity and distinction'. Indeed, the pursuit of difference is of critical significance in understanding the expansion of tourism in the Third World. Contradictorily, while the assertion of individualism becomes a frame of action of the new middle classes, it also reaffirms their class status and position. As already argued, the struggle and, by inference, exploitation, are 'quasi-inevitable' (Bourdieu and Eagleton 1994).

The idea of 'classlessness' or 'individualism' is an indelible mark among these class fractions, resulting, some argue, in the arrival of ego-tourists (Munt 1994a; Wheeller 1992, 1993a, 1993b) or as Box 5.1 suggests, 'curriculum vitae builders'. Ego-tourism is more characteristic of less formalised forms of travel, such as backpacking (see, for example, O'Reilly 2005; Richards and Wilson 2004), overland trucking (Truck Africa refer to their customers as 'would be backpackers') or small group travel, which often involve longer holidays overseas, especially in Third World regions. 'Going-round-the-world' or 'taking a year off to travel' has become a latter-day equivalent of the Grand Tour and an important component of the culture of travel. Ego-tourists must search for a style of travel which is both reflective of an 'alternative' lifestyle and which is capable of maintaining and enhancing their cultural capital. Furthermore, it is a class fraction that attempts to compensate for insufficient economic capital, with an obsessional quest for the authentication of experience (E. Cohen 1979b, 1979c; Errington and Gewertz 1989). MacCannell (1976) portrayed the middle classes as systematically scavenging the Earth in search of new experiences. Errington and Gewertz (1989: 42) note, 'For travellers, the encounter with what was seen as "primitive" – the exotic, the whole, the fundamentally human – contributed to their own individuality, integration and authenticity'. Stressing the link of ego-tourists to environmental issues, Gordon (1990) associates tourists he observed in Namibia to 'an emergent, urban-based, alienated petit bourgeoisie'.

Ultimately it is a competition for uniqueness (E. Cohen 1979b, 1989) with which ego-tourists engage. This is bound with the knowledge, difference and distinction that certain forms of tourism imply – a struggle, so to speak, to stamp the hallmark of individualism in the traveller's passport. While individualism has always been a key characteristic of bourgeois culture, which regards travel as an 'activity for the stimulation and development of character' (Rojek 1993: 114), the rate at which individualism is sought now has significant consequences and impacts for places, especially in the Third World.

Individualism also underscores why ego-tourists feel they are exempt from the criticism levelled at much Third World tourism. First, as Sherpa Expeditions contends of its customers, 'They are involved and interested in the world around them, and in their turn are interesting people'. Second, individualism provides the foundation for claiming the moral high ground in the battle over 'what kind of tourism?' (see Figure 3.3). These are people travelling on a shoestring, and as individuals their practices are not harmful to environment or society. Ego-tourists conveniently disaggregate their actions – they are individuals and their actions therefore have no significant impact. In other words, they refuse to acknowledge their part in a larger entity or mass. By contrast, some argue their travels are actually beneficial (for example, Kutay 1989; Seabrook 1995), supporting local, indigenous businesses, or forging new cultural understandings with Third World communities. Hutnyk's (1996) study of travellers working in Kolkata refers to the 'hegemonic understandings of how those mostly middle-class children of privilege who are able to travel – even, or especially, as "alternative travellers" giving "aid" as people who "care" – see their relation to the . . . "Third World"' (Hutnyk 1996: ix). As we shall see below, however, moral claims are by no means sufficient to secure victory in this battle. The new middle classes have adopted strategies to remedy this by, for example, creating jobs and professionalising not just occupations, but consumption practices as well.

In effect, the boundaries between these new middle classes – between ecotourists and ego-tourists – are blurred: they share their ability to make relatively expensive journeys overseas and their willingness to wage hegemonic struggles to define and differentiate their forms of tourism. In this way, tourism has become part of that critical assemblage of goods, practices and experiences that are taken up as social 'bridges and doors', so important in bonding some and excluding others (Featherstone 1991: 111).

A new class of tourist?

Tourism cannot be understood as just a means of having some enjoyment and a break from the routine of everyday, an entirely innocent affair where there are some unfortunate, incidental impacts or some fortunate incremental benefits to 'host' communities. To recap, so far we have argued that we need a deeper understanding of tourism to appreciate fully its content and expression and, of course, its potential impact. There are two principal components to this argument. First, tourism must be understood as a commodity with both a symbolic or sign value and an exchange value. Second, much so-called new tourism is an expression of the new middle classes' hegemonic struggle for cultural and class superiority (Stauth and Turner 1988).

Travel, of course, has always been an expression of taste and a way of establishing class status (Adler 1989). But, with the rapid growth in the numbers of people taking holidays, it has never been so widely used as at present. Put simply, the democratisation of tourism has created a social headache when it comes to classes attempting to differentiate themselves from one another.

Two fundamental questions arise from the discussion. How are the sign and exchange values of these travel commodities maintained and enhanced? And how do social classes pursue differentiation and to what effect? In answering the first question, we shall concentrate on three key methods that are used to define and legitimise contemporary tourism: the appeal to intellectualism and professionalism, together with the discourse that has arisen to describe travel today. In addressing the second question we will identify what we have labelled 'hegemonic spatial struggles' pursued by the new middle classes. Underlying this discussion is the acknowledgement that much new tourism focuses on the Third World.

The new intellectuals

It has been argued that the new petit bourgeoisie have sought to intellectualise new areas of activity and expertise (Bourdieu 1984; Featherstone 1987). These 'new intellectuals' attempt to ape and popularise intellectual lifestyles and transmit intellectuals' ideas to a wider audience – arguably the ultimate in a cultivated, even scholastic and 'romantic flight from the social world' (Bourdieu 1984: 371). Tourism is a particularly important component in this respect.

In Box 4.4 the 'de-differentiated' characteristic of much contemporary Third World tourism was noted, in which academic prefixes are increasingly apparent. Holidays have moved beyond sheer relaxation towards the opportunity to study and learn, to experience the world through a pseudo-intellectual frame. With the acknowledgement of tourism as an environmentally, socially and culturally problematic activity among certain fractions of the new middle classes, tourism and tour companies catering for the intellectual demands of these class fractions are of increasing importance in the legitimation of travel. Thus we find companies such as Academic Travel (part of STA, an ostensibly student-oriented travel agency), Adult Education Study tours (established in 1982), who offer 'special interest travel for thinking people', and Temple World, offering 'exclusive journeys into wildlife, archaeology, natural history and ecology', whose guest leaders are academics such as the geographer, Andrew Goudie.

Moreover, the intellectualisation of travel has been enhanced by interlinkage between academia and tour companies. It is convenient for companies such as Himalayan Kingdoms (1992–93) to prioritise places on its Bhutan visit (a country that has imposed severe restrictions on the number of tourists admitted each year) to Fellows of the Royal Geographical Society, and for Temple World to run a series of 'Royal Geographical Society African Tours'. At the height of what we might term 'intellectual mimicry', the travel operator Voyages Jules Verne offers tour leaflets (or information sheets) that are indistinguishable from weekend broadsheet newspaper supplements and are appropriately entitled Travel Reviews. Intellectual legitimation is sought in this instance by interspersing descriptions of holidays and itineraries with other considered features: 'Islam', 'The Turkish Inner Man' and the 'Shepherds of Shanghai', for example.

Among the younger new petit bourgeois tourists, with less economic capital, intellectualisation is sought in other ways. Most frequently this involves the claim of cultural superiority, of true and real contact with indigenous peoples, pursued through organised tours, such as 'overlanding', or 'individual' travel informed by travel guides such as the Rough Guide or Lonely Planet Guide series. As Guerba (world travel company) put it, this means more than 'scratch[ing] the surface', it requires getting participants 'personally involved in the country, its people and its wildlife'. In addition, as noted earlier, longer periods of travel overseas have in themselves evoked the spirit of intellectualism and are increasingly referred to as sabbaticals among the new middle classes. But intellectualising is insufficient in itself, and has necessitated the use of other forms of classification and legitimation.

Travel professionals

In the context of socio-economic and cultural change discussed earlier (Chapter 2) coupled with the associated growth of the new middle classes, there have been two notable processes at work. First, professions have become relatively more open following the educational 'revolution' of the 1960s, and the aura (medicine, law, the City and so on) with which they had been traditionally conceived has, at least in part, been dismantled. Second, however, with still relatively limited access to the established professions, the new middle classes have been busy legitimising new licences and certificates (Featherstone 1991; Lash and Urry 1987): a classic process of establishing new professions (with professional education and qualification periods, codes of ethics and accrediting bodies or institutes), or professionalisation as Bourdieu (1984) notes. This dual process may indicate that the infallibility of professional distinction is beginning to falter and surprisingly little analytical consideration appears to have been focused on a logical extension (or migration) of the struggle for distinction through professionalisation, into consumption practices.

Professionalism is a key characteristic assigned to company tour leaders, who in many cases are introduced personally: 'dedicated professionalism' as one brochure says (Abercrombie & Kent) indicating that a tour manager not only has a 'deep love of and insight into the areas visited', but also has 'academic credentials and expertise in a specialised field'. This citation of 'experience' and academic qualifications is commonplace. The following pages cite and quote many tour company brochures dating from 1992 to 2007. In many cases the quoted passages are repeated from year to year.

In tour companies specialising in overland trucking and travelling for ego-tourists, this also means that tour leaders are 'experienced international travellers in their own right' (Hann Overland). At Journey Latin America (Small Group Escorted Trips) even the reservation staff are 'South American experts' and have 'travelled extensively in Latin America'; the tour leaders are all graduates (some in Latin American Studies), a portfolio similar to Trailfinders. Tour leaders in these companies also embody, and are ambassadors of, new middle-class lifestyles, as the tour leader profiles in Box 5.3 testify.

In addition, the spirit of professional dedication is also widely cited, where work in this part of the tourism industry is more akin to a vocation. Encounter Overland refer to their trips as 'projects' and talk of their leaders in glowing terms: 'ordinary men and women – often previous trip members – who have elected to put promising careers on hold and devote half a dozen years or more to what they like doing best'. Of course, with greater flexibility in the service sector a higher proportion of new middle-class employees are able to take longer periods off between jobs or contracts, or are able to negotiate relatively long periods of absence. All in all a more dedicated, professional, and avaricious tourist class has emerged.

The point we anticipated earlier is that 'travelling' has emerged as an important informal qualification, with the number and range of stamps in a passport acting, so to speak, as a professional certificate; a record of achievement and experience. Not only is travel a professional prerequisite for employment in parts of the tourism industry, but it is also an important attribute in many new professions, such as international development work; it has become a rite of passage into certain occupations.

Box 5.3 The Himalayan Kingdoms Team: tour leader profiles

Cathy Woodhead

Cathy has trekked in Nepal, Tibet, Turkey, Peru, Ecuador and elsewhere. She has been a rock climber most of her life and now combines climbing and trekking trips with her freelance job as a consultant in the oil industry. She is a qualified mountain leader and first-aider.

Mike Rutland

Mike, who has known Bhutan since the early Seventies when he was Tutor to the present King, has an encyclopaedic knowledge about its fascinating history and culture. He founded the UK-based Bhutan Society, which is dedicated to increasing knowledge about the kingdom and raising funds for worthwhile projects, including schools and village schemes.

Beetle Seymour-Williams

Beetle has lived and worked around the world, including a spell running tiger safaris in India's Kanha National Park. She has travelled extensively, including eight months in China and Tibet, and undertaken several treks and climbs in Nepal. She has also led three treks for us in Bhutan and two in Sikkim.

Catherine Darjaa

Catherine is the operations manager for the local Mongolian agents we use for this trek and has lived and worked around the world. She worked for the UN in Bhutan for three years and Mongolia for two years, during which time she travelled extensively around Asia. She speaks Mongolian.

Mike Ford

Mike has spent much of the past fifteen years travelling across the Himalaya, either independently or researching for his work as a Rough Guides writer. Mike has used his experiences in Nepal, Tibet, Thailand and India not only in writing various guide books, but also in giving slide presentations and teaches English as a foreign language at Bristol University.

Source: Himalayan Kingdoms website www.himalayankingdoms.com (accessed November 2007)

There are also indications that professionals working in other disciplines have begun to diversify into travel. Many of the occasional tour leaders, for instance, are professionals in other fields. For example, the two directors of Papyrus Tours, a company established in 1984 with an aim to provide tours to East Africa, who were supportive of conservation efforts are also a senior officer in a large local authority and a self-employed professional landscape architect, respectively. Of course, we should also note that Third World tourism has created a huge range of job opportunities for the new middle classes in the First World. It is not just operators in the industry but consultants, journalists, tourism commentators, academics and charities focusing on tourism issues; take The International Ecotourism Society (TIES), for example, whose founder and president, Megan Epler Wood, explains:

> It was decided that The Ecotourism Society should be an organisation for professionals. It was clear in 1991 that a broad array of professionals from a variety of disciplines was needed to make ecotourism a genuine tool for conservation and sustainable development.
>
> (Epler Wood 1994)

More recently, TIES's website describes the organisation as 'Representing a worldwide network of 1,700 members from 55 professions and 65 countries . . . We'd like to invite you to the growing ranks of ecotourism professionals who are working with TIES' (TIES website 2002).

It is these professionals who are the opinion formers, the teachers, the advisers, even the ones who take decisions, and a number of publications have sought to prescribe 'technical information on how to do ecotourism *right*' (Whelan 1991: 4, our emphasis). But the degree to which these 'benevolent' organisations are somehow value free is questionable. It is clear who these professionals are: in TIES's case, middle-class Americans, many of whom represent other powerful NGO interests (WWF, Conservation International, IUCN and so on – see listing of board of directors and advisers). We must ask what vision of the world they are pursuing and the degree to which such visions are imposed from the First World on the Third World, a question considered in the context of the activities of these organisations in Chapter 6.

The commencement of professionalisation processes in consumption are also detectable. As noted earlier, consumption has become more skilled. First, there are clear signals that the distinctions between occupational professionalism, and consumption and leisure, are beginning to blur. Illustrative of this is the growth of outdoor management and 'team' building exercises, especially popular among the new middle classes. High Places, for example, which offer mountain holidays in a number of Third World countries, also offer 'another type of HOLIDAY! . . . training programmes for people at work in industry, commerce and the public sector'. This is an operation appropriately called 'HIGH PROFILE', an experience for people like 'YOU, the sort of people who come on our holidays', described elsewhere as 'intelligent and discerning people who wish to retain a taste of independent travel'.

Second, and more significant, tourists themselves are attempting to professionalise travel, a process encouraged, it would appear, by tour operators and environmental organisations. While the emergence of a formal travel qualification, such as the need to support an application for a Himalayan Kingdom expedition 'with your climbing CV', remains the exception rather than the rule for the time being, the number of tourist codes addressing the ethics and conduct of travel have exploded (see Chapter 7). A tourist 'Code of Conduct' established by the Ecumenical Coalition on Third World Tourism has been reproduced in many places, especially by the network of organisations concerned with the effects of tourism, but also, for example, among travel agency associations and tour operators, who have formulated their own versions. Area-specific codes with emphasis on ecological and cultural issues such as the 'Himalayan Tourist Code' formulated by Tourism Concern (in the UK) have also appeared. Organisations such as these have attracted increasing support from members of the new middle classes and from new independent tour operators like High Places, which reproduces the Himalayan Code in full and which claims to strive to 'adhere to the ethics' of Tourism Concern. It is the emergence of an ethical, if not professional, approach to tourism reminiscent of Kutay's observed 'Peace Corps-type travellers looking for a more meaningful vacation' (Kutay 1989: 35).

These codes of ethics form the backbone of a hegemony of travel (or ecotravel as it is now known in North America), which is advanced by the new middle classes, the small independent tour operators and the vanguard of this hegemony, tourism organisations. Together they begin to represent a new tourism social movement with organisations such as Tourism Concern, the Campaign for Environmentally Responsible Tourism (CERT) and the US-based International Ecotourism Society, emerging as symbolic 'institutes of travel'. These institutes now provide ethical yardsticks against which the activities of operators and tourists can be measured and classified. With an overwhelming concern for environmental ethics (such as the US National Audubon Society's 'Travel Ethic for Environmentally Responsible Travel'), it is in the ecological and indigenous heart-lands of the Third World that these hegemonic struggles are most readily detected and played out.

The discourse of 'new wave tourism'

'Out goes the "'ere we go, 'ere we go" Spanish Costa-style, in comes a more thoughtful middle class approach' (Barrett 1990: 2). This was how Frank Barrett of the *Independent* newspaper, the UK's champion of new middle-class values, describes the shift to 'new wave tourism' (Wheeller 1992).[4] There can be little disputing that the attributes and qualities of the supposedly more alternative, individualised and sensitive forms of travel are generally unspoken but appear to be deep within the new middle-class psyche. As The Travel Alternative, a specialist company offering craft and textile holidays, comments: 'On a recent tour a passenger said "You can either be a tourist or a traveller." We would like to think that our tours allow everyone the opportunity to be travellers' (Newsletter 7 1994).[5]

Of course, it is not only through the classificatory distinction between tourist and traveller that differentiation is pursued via discourse, but it is illustrative of the battle being waged to put into words exactly what kind of holidaymaker we are and what we stand for, and, conversely, state exactly what we are not for; as Barrett argues, 'tourist' is the 'worst kind of insult' (Barrett 1990: 3).

Some observers have momentarily reflected on the highbrow nature of traveller and the intellectual snobbishness within which the whole debate is framed, and have attempted to rescue 'tourist' from middle-class derision. Tourist has become prefixed by benevolent adjectives such as 'The Good Tourist' (Wood and House 1991) or 'The Reluctant Tourist', title to a series run by the UK broadsheet *Independent* which culminated in *The Reluctant Tourist's Handbook*. There are also references to ethical tourists, alternative tourists and so on. Arguably though, this is further confirmation that contemporary tourism is charged with the classificatory struggles in and between class fractions (Jaworski and Pritchard 2005; Walter 1982; Welk 2004). As Culler (1988: 158) observes: 'Once one recognises that wanting to be less touristy than other tourists is part of being a tourist, one can recognise the superficiality of most discussions of tourism, especially those that stress the superficiality of tourists.'

By contrast, the term 'traveller' assumes that it is no longer a process of tourism with which the individual is engaged, but a considerably more de-differentiated, esoteric and individualised form of activity; travel is to tourism, as individual is to class. It is a strategy for seeking differentiation or a 'paper-chase aimed at ensuring constant distinctive gaps' (Bourdieu 1984: 481). Most importantly, the discourse adopted by these class fractions is a further reminder that consideration of the way in which power is expressed and imposed must lie at the heart of the analysis of tourism (Munt 1994a; Urry 2001; Wheeler 1993a).

Many small tour companies and individual travellers have attempted to maximise the distinction. As already argued, the practices of travellers are perhaps best conceived as part of the 'cult of individualism' (Pels and Crebas 1991), though it is deeply ironic that they are largely indistinguishable from each other by virtue of their discourse, dress codes and the informal 'packages' they follow through travel guides. Whole regions have become travel circuits (in popular travel discourse 'doing' South East Asia, Central America and so on). Errington and Gewertz (1989: 40) reproduce excerpts from a Papua New Guinea guest house visitor book to illustrate the fundamental traveller distinction, with two US travellers advising: 'Explain difference between tourist and traveller . . . Be a traveller, not a tourist. It makes a big difference.' Travellers, therefore, attempt to create their own aura, while attempting to prevent the encroachment of 'tourists' in their quest for authentication:

> Momentarily alone in one of the wildest places on earth, you feel exhilarated, exhausted and scared. Unfortunately the feeling is unlikely to last more than a few moments. If you are lucky, all that will happen is another traveller will appear on the ridge to exchange pleasantries and wonderment. If you are unlucky . . . your reverie will be interrupted by half a dozen Londoners swearing and shouting.
>
> (Edwards 1992: 19)

As Box 5.4 suggests, mass tourism and tourists have also become the target of independent tour operators. Intellectualism oozes through, with package tourists 'invading'

Box 5.4 On tourism and travelling

Tourism	Travelling
Invasion, rape, poisoning, tidal wave, pollution, swarms, juggernaut	Discovery, exploration, understanding, peaceful contact
Benidorm, Torremolinos, Kos	Tikal, Phnom Penh, Zanzibar
Sun, Sand, Sea and Sex	Sensible, Sensitive, Sophisticated, Sustainable and Superficial (Wheeler 1993a: 122), or Intelligent, Inquisitive, Independent and Idealistic
Unadventurous, narrow-minded, undiscerning, unenergetic, inexperienced, unimaginative, unintelligent, boring, unreal, false	Adventurous, broad-minded, intelligent, discerning, energetic, experienced, keen, imaginative, independent, intrepid, real, true

backpackers' discoveries and the celebration of areas where the invasion has been forestalled: at the height of aestheticisation, Cambodia is lucky for its 'fascinating culture bruised by war' and even luckier with a culture 'unscathed by tourism' (Calder 1994b). Companies refer to the 'anonymity and inflexibility' (J. & C. Voyageurs), of 'run of the mill' (Abercrombie & Kent), 'conventional' (Explore) and 'impersonalised' (Africa Exclusive) mass packaged tours, and celebrate their demise (Magic of the Orient). Ultimately, it is an appeal to the idea that travel is 'individual enough to be a sustainable alternative to the juggernaut of mass tourism' (Explore).

In its place, and by contrast, we have the emergence of African destinations – where 'nothing is packaged' (Africa Exclusive), and the 'Unpackaged Caribbean' (Kestours). For luxury tour operators and their new bourgeois clientele (ecotourists) exclusiveness is sufficient enough a distinction, with Cox and Kings taking 'a select few to Africa', and operators speaking of 'a small and select clientele' (Abercrombie & Kent) and 'limited edition holidays' (EcoSafari).

For tour operators catering more for the young and adventurous ego-tourists where economic capital is clearly insufficient to confer taste, it is the invidious distinction of participant travellers from tourists that has become critical. It advances the distinctions beyond a mere reference to (or 'charge' of) packaging and instead focuses on the qualities and practices of travel and contrasts these to tourism. This is particularly notable of overlanding operators such as Dragoman specialising in trucking for young travellers, a world 'shunned by the masses who prefer the resorts and beaches', or, as Explore put it, travel for 'people who want more out of their holiday than buckets of cheap wine and a suntan', a stereotype also used by Exodus Adventure. In similar vein Journey Latin America contrast their ('Small Group Escorted Tours') participants to those who prefer 'two weeks in Torremolinos' and where preference is for a 'cold beer after a three hour walk in the jungle, not nightly booze-ups'. As Tours to Remember warn, if you are looking for 'two weeks in the eastern equivalent of Benidorm', look elsewhere.

Logically, tour companies also promise 'tourist-free' zones (Naturetrek), 'few lager and litter louts' (Cara Spencer) and avoidance of 'tourist haunts' (High Places). In a more positive sense, Encounter Overland celebrates the traveller-cum-wayfarer: 'today's custodian of the ancient relationship between traveller and native which throughout the world has been the historic basis of peaceful contact'.

So what of the qualities and practices that are claimed to constitute the distinction and embody the traveller? The simple listing of adjectives applied to travellers in travel brochures (Box 5.3) begins to map the key coordinates. By implication, if not explicitly as noted above, the counter-distinctions constitute tourists. If, for any reason, the citation of these qualities proves insufficient, however, there are other prerequisites which help define travellers. High Places tours of India, for example, demand 'patience, stamina, humour and adaptability', and Journey Latin America cite 'essential qualifications for all trips' as 'emotional balance, maturity, a spirit of adventure, and a desire for good companionship'.

There are two other notable features of travelling and travellers. First, with an overwhelming emphasis on a mixture of often physical activities, especially among ego-tourists, bodily fitness is important. High Places offers the symbolically new middle-class holiday: mountain biking. Second, and most important, there is need for the fellowship of other like-minded travellers, whether it be small intimate group tours (with generally between six and twenty participants) or individual backpackers meeting at predestined travel guide recommended hotels, restaurants or sites. In both cases, albeit in rather different ways, the emphasis is on participation, on 'action and involvement' (High Places) and, ultimately, on accomplishment (a recurrent theme in tour brochures). It is not that travellers are unreflective; just the opposite, as Hutnyk (1996) observes of backpackers in

Kolkata (India) where a 'popular alternative critique of travel' emerges from their reflections, and manifests itself in:

> a) the search for 'authentic' experiences; b) dismay at the effect of tourism; and c) condemnation of other tourists and sometimes themselves. The correlates of these three moments are a) claims to the 'once-in-a-lifetime' experience; b) nostalgia for the days when such and such-a-place was not so well known; and c) 'of course I'm doing it differently' stories.
>
> (Hutnyk 1996: 9)

As Hutnyk (1996) concludes, not only is there an intense struggle in classifying and legitimising this notional traveller–tourist distinction, but also ego-tourists vie among themselves over what actually constitutes a traveller in the first place. As was seen in Chapter 3, Borzello (1991) considers that 'Truck travellers are not travellers but a very peculiar sort of tourist'. And it is an intense classificatory struggle that has pronounced spatial reflections, a point to which we turn below.

Tales of the unexpected

> Both explorers and dragomen journeyed through far-flung lands in search of long-forgotten civilisations and empires. They explored hostile deserts to find nomadic tribes and ancient cities. They went in search of legendary mountains deep in the heart of jungles and brought back stories of fabulous wildlife.
>
> (Dragoman brochure 1995: 1)

These fabulous stories or 'travellers' tales' (S. Hall 1992b; Massey 1995b) have a deep history. There are noted discussions and a resurgence of interest in ethnographic travelogues and travel writings (see, for example, Blunt 1994; Fussell 1980; Pratt 1992; Wilson and Richards 2004). But debates around the contemporary nature of tales, myths (Borzello 1994; Calder 1994b; Selwyn 1994, 1995), or rumour-mongering and 'traveller lore', have remained sketchy. This is of particular significance for, as Massey (1995b) argues, tales are a culmination of the ongoing process of travel which fill out our geographical imaginations, and story-telling is the archetypal 'traveller's medium' (Hutnyk 1996: 64); 'rumour-mongering is the architect. Rumour is the stuff of the social' (Hutnyk 1996: 29). Travellers' tales are a modern (or perhaps postmodern) continuation of the stories of colonial encounters. They are a useful way of demonstrating how discourse is used to represent or recreate a reality and impose meaning, in much the same way as the fantastic tales retold by colonialists in the nineteenth century. As Hutnyk (1996: 42) shows, 'word of mouth, the say-so of a friend of a friend, and rumour and gossip, operate to orientate and produce experience for travellers. This discourse must be read as a text.' In this way discourse can be conceived as part of that everyday hegemony.

Latter-day travellers' tales are a powerful way of sustaining the aura and mystique of the Third World (reminiscent of Said's Orientalism) and in turn sustaining the value of travel as a commodity. Their construction is often complex and draws upon a range of images and representations, including film (as suggested in 'Weekending in El Salvador' (Wolff 1991): see pp. 80–1), travel writing and colonial explorations (K. Adams 1984, 1991).

Tales also have an important part to play in the reproduction of social class. It has already been suggested that tales have an important symbolic currency as a form of cultural capital; the example used suggested that travelling on a 'shoestring' simultaneously denies the existence and need for substantial lumps of economic capital and confers status, uniqueness, worldliness and resourcefulness. As Simon Calder records of travelling in rebel-held Nicaraguan territory, by the time you reached the 'safety' of neighbouring Costa Rica 'all that backpacking bravado returns and you have a couple of cracking stories with which to trump fellow travellers' tales' (Calder 1994a: 35). Travel and the construction of tales, therefore, represent an informal credentialism.

The construction of tales is also intimately bound with the quest for authenticity (Box 3.7). It is a way of creating an aura in an effort to remove this experience from the tourist's world. These are feelings that cannot be 'snapped'. Take this example from High Places:

> The trip begins with a short drive from Nairobi down into the Great Rift Valley, to our camp set among the trees, overlooking grassy plains where zebra and giraffe graze. The only sound is the breeze and a kettle boiling on the stove. Time to relax as the trip unfolds . . . Walking through the Loita Hills, in Maasai territory, following little-used trails and using local warriors as guides . . . watching a majestic herd of elephants cross the Mara plains, or a cheetah stalk its prey; sitting round the fire, swapping stories with Maasai warriors; relaxing, as the sunset turns the empty plains red.
>
> (High Places brochure)

Of all modes of travel, it is trekking in areas of solitude that most gives rise to a romantic gaze and the construction of tales almost spiritual in their content. The phenomenal growth of trekking (see Chapter 8) in South East Asia, Latin America and Africa (Brockelman and Dearden 1990) underlines its importance in new middle-class travel. It is the authenticity of this form of alternative tourism that lies at the heart of trekking (E. Cohen 1989), and the accumulation of tales relies upon authenticity to verify its uniqueness and currency: as High Places promote, 'Our trips are unpretentious, authentic and adventurous', and Trailfinders contrast 'staged tourist "shows"' with the experience of 'untouched' and 'traditional' villages in the highlands of Thailand.

In this case authenticity is expressed in a somewhat different manner, conveyed through an emotional response that is either unique or intensely difficult to achieve. It is an exclusionary experience, that even if the struggle to dominate and control a spatially discrete area is lost, tourists are still unable to share the aura. Frank McCready (Sherpa Expeditions Managing Director) comments on the realisation of dreams: 'When the ordinary person reaches a place like Annapurna Base Camp it changes their whole view of life . . . Some of the places we take people are so fantastic I've seen grown men entirely overcome with emotion.' It is this experience of exhilaration and exhaustion (Edwards 1992) that culminates in the eulogising of trekking. Himalayan Kingdoms comments: 'The blend of grandeur and timelessness defeats all superlatives and such a place should be visited at least once every lifetime.'

To return briefly to the theme of aestheticisation: in order to legitimise and authenticate the travel experience, tales must be aestheticised in two important respects. In the first place, there is the dual requirement to make trips both purposeful and distinguishable from those of the average tourist or, indeed, traveller. Travel is purposeful, tourism is not. For example, one travel writer notes of a visit to Nepal, 'We had tea and increasingly battered chocolate to deliver to a friend of a friend working in a health clinic there' (Norris 1994). Sabbaticals, 'research' work, the presence of friends, colleagues and relatives, the emergence of work brigades and the visits promoted by international development charities – a range of work – and activity-centred holidays, are all used by the new middle classes to signal that this is more than just a holiday. It must be sufficient in distancing itself from supposedly inactive or inert forms of tourism. Exodus ('Overland Newsletter', summer

1994, 'What the customers say' section) notes, 'I think that was the best trip I've ever done, partly because of, rather than despite the difficulties and challenges we faced.'

Second, as suggested in Chapter 3, there has always been a nagging inadequacy around the assertions that 'one cannot sell poverty, but one can sell paradise' (Crick 1989; Rojek 1993). The quest for authenticity, like travel itself, has had to seek out new experiences. No longer can it rest upon the cultural preservation and discovery of highland tribes and deep forest natives. It has migrated to the quest for danger and risk and the emergence of contemporary travelogues and diaries: '"I'm keeping a journal; maybe I can make it into a book," says one traveller to me as we sit in a bar in a small village' (Gordon-Walker 1993: 19). Tales legitimise travel by sensationalising it, making it appear daring and risky, while at the same time peripheralising danger (Jardine 1994). Of Vang Tau (Vietnam) the STA Travel Guide notes: 'a two hour boat ride from Ho Chi Minh City and a million miles away from conflict'. Similarly, Encounter Overland's philosophy remains: 'With true adventure there is always an element of chance and of risk. This fact is not regrettable. It is often the fact upon which the best travel experiences are based.' Suffering can often enrich the travel experience, as extracts of Morrison's (1996) 'Tales of the unexpected' suggest:

> A glass of fiery rice liquor was thrust into my hands, and I sat down with them and joined in their infectious laughter, without a single word of common language between us . . . Seven years on I'm none the wiser about the purpose or origins of that strange harvest festival in northern Thailand . . . As it was I could find no mention of the festival or town in anything I read, which probably accounted for the fact that I was the only westerner on the scene. But it wasn't this smug one-upmanship that made it so special, but rather the very act of discovery. For one impetuous moment I dared to stray from the charted territory . . . and it was bliss . . . To me, surprise is the primary reason to travel. All those other time-worn clichés about broadening the mind and stimulating an appreciation of the world are merely justifications, not motivations . . . Think about those special moments from your last few trips, the ones that you bore your friends rigid with at parties. Are they accurate reports of the splendour of the temples or the excellent cuisine? I hope not. No, it's those unexpected events when your plans are interrupted by that out-of-the-blue experience . . . To my mind, if there's one thing worse than a holiday that fails to meet your expectations it's one that does . . . And let's be brutally honest here, if in the second week of your big adventure you haven't had at least one near-death experience you're going to want your money back. Right? . . . do we really want a boil-in-the-bag cultural experience? No we bloody well don't – we want an unpolished, unbleached wholegrain adventure with bits of stone and fingernail in the bottom!!!
>
> (Morrison 1996)

Most invidious, however, is to utilise tales to aestheticise risk and boost the cultural capital accumulated in travel. As suggested in Chapter 3, while the hazards of travelling in particular regions may act as a warning, it simultaneously signals the degree of legitimacy or coolness to be attributed to particular destinations. As Gordon-Walker (1993: 19) notes of Peru, 'travellers who are disappointed not to have captured a Sendero Luminoso flag raise their spirits . . . and swap stories about friends of friends who have been robbed'. Similar experiences were recounted by Warde (1992) and Wolff (1991) in Chapter 3. Travelling in potentially dangerous regions, being hoisted from a bus and frisked at midnight or braving certain urban areas are experiences to be enjoyed and admired by other travellers. Risk titillates, even eroticises, adventures in the Third World.[6]

And it is this genre of tale-telling that has come to the fore in discussion among fellow ego-tourists, in travel guides and reviews and within television coverage. In short, it is now widely adopted discourse focused towards social differentiation and the insurance, or hope, that certain places will stay unexplored by other groups of tourists.

The scramble for Third World destinations

We have argued that a pseudo-ethical and moral infrastructure underpins the growth of Third World tourism (Lea 1993), with new middle-class tourists (or travellers) contrasting their morally justifiable means of travel with the morally reprehensible practices of tourists. There is a temptation to conclude from this, as Gordon-Walker (1993: 19) does, that you 'cannot help suspecting that the campaign for "sustainable tourism" is little more than a rationalised desire to keep the Third World a cheap place to visit'. But such a conclusion would result in underselling the importance of how First World tourists impose their desires on Third World destinations. If the new middle classes are forced to wage this hegemonic classificatory struggle founded on taste, then it might be reasonable to expect that such struggles will also be reflected spatially. So at the very heart of this campaign lies, not just the preservation of a nice cheap place to visit, but the playing out of the battle of social class differentiation.

The professionalisation and intellectualisation of travel, together with its associated discourse, have been insufficient in themselves to ensure social differentiation and, more importantly, the creation of physical distance between mass tourists and travellers. The new middle classes must adopt strategies of exclusion, to seek and protect the new travel commodities. Extracts from veteran correspondent Martha Gellhorn's (1990) piece, 'Too good for tourists', illustrate the process by which this is achieved. It is an attempt to persuade and to impose the idea that this First World view is correct:

> The airport terminal at Belize City is very encouraging. It is an overgrown shack . . . After each passport has been examined as if passports were new-fangled inventions, you enter a square, wooden room where the baggage has been dumped in the middle of the floor. A free-for-all follows . . . My heart rose like a lark. This is how it used to be before the Caribbean was homogenised by tourism. Tourism, even . . . minor, modest tourism, corrupts. I cannot exaggerate the pure physical pleasure of underpopulation and empty space . . . Half a century ago I thought this ghastliness of packed humanity was the peculiar doom of China: now much of the world I knew and loved is ruined because there are too many of us and we move everywhere. For my taste, Belize is ideal.
> A young American in the shortest possible frayed shorts was questioning the American Rum Point owners about the effects of tourism on the population and the environment, his thesis for the University of California at Berkeley. We agreed that tourism is a destroyer and Belize is far too good for it. None of the inn owners I had met, English, French, American, Belizean, Irish, want to expand, though, between them, they could take care of less than a hundred perceptive guests. They love the country as it is. How much better if oil is found in viable quantities in the north, and tourists go somewhere else. It is astonishing at my age to stumble on a new country. I feel astonished.
>
> (Gellhorn 1990)

As some authors contend, the near-absence of other tourists to which Gellhorn alludes is most efficiently achieved by the consumption of 'positional' goods (Hirsch 1976), with dominant class fractions attempting to impose scarcity (Featherstone 1987; Leiss 1983). At the simplest level this involves claiming that certain areas 'receive very few tourists' (Detours Travel 1992), are 'tourist free' (such as national parks visited by Naturetrek), or that even whole countries are for the 'traveller rather than the tourist', as Meon described Ecuador in 1992.

But this amounts to little more than marketing, and alternative strategies and claims must be pursued, such as active searches for 'off-the-beaten-track' (or, as Asian Affairs claims, 'even further off-the-beaten-track') and 'lesser visited areas', areas 'rarely visited' (Himalayan Kingdom) or just 'secrets' (Okavango Tours and Safaris). Exclusiveness, uniqueness, romanticism and relative solitude are central to the philosophy of these tour operators.

This philosophy both embodies the contradictions in contemporary travel to the Third World and reveals the protracted and increasing difficulties which companies and travellers have in spatially defining separated practices from other like-minded travellers. As tourism spreads to more and more destinations in the Third World, the distinction of a mass packaged tourist from the individual character of travel is spurious. Ego-tourists crowd into cheap guesthouses, basic bus stations and out-of-the-way beach resorts, while ecotourists crowd into game reserves and national parks. So how far is travel little more than a figment of wishful middle-class thinking? It is worth considering the views of Budi (aged 28), an Indonesian guide from Bali, interviewed by Tourism Concern's Sue Wheat:

Q. Do you think there is a difference between a 'tourist' and a 'traveller?'
A. The traveller thinks they know everything about the local people and the country. But it's usually because some other traveller told them before. But they do whatever they like – some travellers are good, but 90 per cent are not, they can be very impolite. With the tourist, everything is organised, so they don't destroy as much. The traveller wants to see something new, and wants it to be cheap and then tells others about it. I prefer tourists . . . they go to specific places, it is more professional. But the traveller is uncontrolled – they won't go to the places already prepared for them; they want to go to other places and then they spoil it – and don't spend any money! Travellers always talk the same and say: 'Don't go to Kuta because it's spoiled.' Then they go to a new place.

(Wheat 1994: 9)

The example of the island of Bali is taken further in Box 5.5, which alludes to this blurring of tourist distinctions with an examination of the effects of tourism policy on the island.

It is a context, therefore, within which new tourism must not only do battle with mass tourism and tourists, but contend within itself both spatially and qualitatively for the most virtuous practice. This is particularly notable in safaris. For example, Papyrus Tours ensure that they spare you 'any involvement in safari bus "rat runs"', referred to by Africa Exclusive as the 'herds of landrovers that . . . plague some other parts of Africa' (but which do not, of course, afflict this tour operator's safaris in Zimbabwe). Even vehicles themselves are a point of differentiation among safari operators and overland trucking companies (as shown in Box 3.6), with Into Africa claiming that 'landrovers do less damage' than larger trucks and are more sensitive to both wildlife and local people.

Similarly, trekkers face increasing difficulties in spatially legitimising their experiences, as Edwards' (1992) lamentations have already indicated (E. Cohen 1989). With the sharp increase of trekking in northern Thailand, for example, it is more difficult to find a trek which visits 'untouched' or 'traditional' villages, we are warned by Trailfinders, a company with which it is possible to visit 'more remote villages less exposed to western

Box 5.5 Paradise lost: Bali and the new tourist

Tourism is no recent phenomenon on the Indonesian island of Bali. Colonised by the Dutch, the Balinese held on to a unique blend of Hindu, Buddhist and animist religions that imbued their cultural and social life. Tales of this exotic society living in tune with art and nature attracted foreigners as long as a century ago, and since independence, Bali has been the jewel in the crown of government tourism policy, hosting ever larger numbers of visitors (see graph).

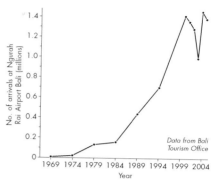

Contemporary tourists to Bali display a wide mixture of motivations. The island accommodates seriously wealthy hedonists and conventional package tourists (many on long-haul stopovers or multi-centre holidays), alongside large numbers of young Australians who arrive on charter flights and regard the island as a cheap and cheerful holiday spot and a surfing mecca. There is a huge pool of homestays and guesthouses, known locally as *losmens*, catering for this budget market.

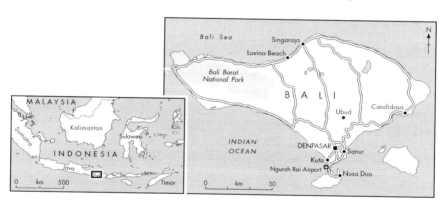

Bali's enduring image as a cultural paradise still draws independent travellers who also patronise the *losmens*, and it features frequently in the brochures of companies appealing to the new tourist as a land where it is possible to get off the beaten track while never straying far from creature comforts: 'the industry has been controlled so that hotels are primarily located in the southern peninsula, leaving the rest of the island essentially unchanged and uniquely Balinese' (Mortlock, 1988).

The village of Kuta, close to Bali's international airport, has been a favourite haunt of budget tourists since the 1970s, when its residents woke up to the lucrative possibilities of turning their homes into *losmens* and offering other services from bike hire and restaurants to massages on the huge crescent-shaped beach, which soon became a haven for novice surfers. *Losmens* began appearing in nearby Sanur, and on the north and west coasts too – for example at Lovina Beach patronised by the backpacking fraternity as a restful retreat from an overdose of culture or socialising in southern Bali: 'It's a good place to meet other travellers, and there's quite an active social scene' (Wheeler and Lyon, 1994).

The inland village of Ubud, renowned for its artists and lush beauty, has also grown rapidly. From a scattering of *losmens*, restaurants and artists' studios attracting backpackers and longstay visitors, Ubud has mushroomed into the alternative inland base for anyone not locked into a standard beach hotel package deal. Nowadays, new tourists wishing to escape the mass market experience still favour Ubud, but may well avoid the commercialised southern coastal resorts, opting instead for accommodation in less developed spots.

However, if the entrepreneurs of outposts such as Candidasa and Lovina Beach are counting on increased custom from this source, they may be disappointed. Like the first visitors to the island, new tourists are specifically interested in Bali's unique harmony of culture and landscape and they still tend to visit the prime sites on every tourist's itinerary. As many commentators (see, for example, Noronha 1979; Picard 1991) have remarked, Bali's cultural life has so far been remarkably resilient. But as tourist sites become increasingly crowded and as its distinctive art forms become increasingly commercialised, new tourists may well decide not to bother with Bali at all, taking up instead the Indonesian Tourist Board's invitation to explore the rest of the republic's giant archipelago and leaving this particular 'paradise' to the surfers and conventional package tourists.

Regardless of the tourism type, however, in October 2002 and October 2005 the major tourist town of Kuta suffered terrorist attacks that had dramatic impacts on the tourism industry as well as inflicting much loss of life, injury and damage. In the two weeks after the 2002 attack, visitor arrivals dropped by 80 per cent, hotel occupancy plunged from 73 per cent to 14 per cent and more than half of the workforce in the industry were threatened with or experienced unemployment (Bali Tourism Authority 2004). From a tour operator perspective, the attack immediately removed Bali and other Indonesian resorts from the marketplace. Other Asian Pacific resorts such as Fiji, however, gained. In particular, Great Keppel Island noted a large increase in students (mostly Australian) seeking the type of uninhibited partying for which the tourism of Kuta had become famous. This immediate effect on Bali is clear from the graph of annual tourist arrivals included here, although the remarkable bounce-back in 2004 is just as clear. The terrorist attacks in Kuta three years later also had an immediate effect on tourism to the island, but we believe that, once again, the bombings will be felt largely in the short term and that the long-term analysis of the island's tourism industry given in this box will remain relevant. (We revisit the effects of terrorism on the tourism industry in Chapter 11.)

(This analysis of Bali's tourism industry was prepared especially for the first edition of this book by Alison Stancliffe and amended by Alison for the second edition. Extra notes and data were added for this third edition by the two authors on account of the bombings in Bali in 2002 and 2005.)

influences', where 'Groups are welcomed as a refreshing diversion to normal village life'. At possibly its most advanced, spatial legitimation involves promises of carefully researched itineraries which avoid the 'overcrowded trek routes, often exploring untravelled or Restricted Areas, or visiting cultural festivals unknown to most westerners' (Himalayan Travel 1992). Ultimately, legitimation for this company is sought through the pages of the *Geographical Magazine*, with the co-director writing of their trek to Mustang:

An understanding and appreciation of both cultural values and ecological balance is essential when visiting such remote and unspoilt regions. This expedition was the first in a series which it is hoped will help turn the tide of 'tourist pollution' to remote areas of the world.

(Brooks 1990)

This form of tourism therefore means the need to ensure the absence of other tourists. Some areas are considered especially attractive by virtue of the exclusionary nature of state policies which limit the number of tourists, and the political scientist, Linda Richter, identifies a key issue in tourism planning as class versus mass tourism (Richter and Richter 1985).

The small, land-locked, Himalayan mountain kingdom of Bhutan, to the north of India, is undoubtedly the most unequivocal example of this phenomenon. Bhutan has chosen to concentrate its tourism on high-spending, 'classy' tourists. The country allowed entry to its first tourists in 1974, but only at the rate of 1,000 per year. Since then the rate has increased and in 2006, 18,000 tourists were allowed entry into the kingdom. But as the website of the kingdom states, 'Visitors to Bhutan must either be guests of the government or tourists. All tourists (group or individual) must travel on a pre-planned, prepaid, guided, package tour, or custom designed travel program. INDEPENDENT TRAVEL IS NOT PERMITTED IN BHUTAN' (Bhutan Tourism Corporation Limited 2007, emphasis in original). In 1988 the government closed some monasteries to foreigners because of 'growing materialism' among monks who accepted trinkets such as sweets and pencils together with money from visitors (V. Smith 1989: 14–15). A glance through several of the trekking brochures shows that Bhutan is an expensive destination as a result of the weekly charge levied on all foreign tourists. This restrictive entry has been utilised by some 'alternative' travel companies to demonstrate the exclusivity of their holiday destinations (Naturetrek; Himalayan Kingdoms; Coromandel). In 1986, a study made by the UNWTO praised the Bhutanese system. UNWTO planner, Edward Inskeep, agreed that 'selection' can be achieved: 'the limited tourism approach can be applied, through selective marketing techniques, to attract tourists from any socio-economic groups who respect and do not abuse the local environment and culture' (Inskeep 1987).

Trekking companies thus enter an intense competition for the most authentic locations, and that means destinations with the least tourists and, by implication, the most difficult to enter. Take, for example, the sample of statements from the UK-based Himalayan Kingdoms' 2008 brochure given in Box 5.6.

But it is with environmental conservation that tour companies and tourists have discovered perhaps the most effective method of exclusion, or 'inclusiveness'. Many tour companies are now supportive of conservation measures and cite membership of both global (WWF, CARE, FOEI and so on) and the more localised environmental organisations (such as Elefriends and Mountain Gorilla Project). Such projects are essential to both ecotourists and ego-tourists in restricting the number of visitors to such areas and in so doing securing some form of exclusiveness, however circumscribed in practice. Beyond this, some companies have actively pursued conservation policy. Examples are Papyrus Tours, where the director is a founding trustee of the Kenya Wildlife Trust (a UK-based trust seeking to support conservation initiatives in Kenya), or the Ultimate Travel Company which is 'wholeheartedly committed to the continued financial support of the invaluable work of the Galapagos Conservation Trust, Lewa Wildlife Conservancy in Kenya, and Tusk Trust' (2008 brochure: 11). Similarly, the certificated purchase of an acre of rainforest has also been encouraged by multinational environmental organisations such as Programme for Belize (now run as part of the World Land Trust) and the Nature

---

Box 5.6 Himalayan Kingdoms Treks and Tours 2008 brochure

'This is a collection of "classic" and "off-the-beaten-track" walking and adventure holidays.' (Inside front cover)

'How many people, for instance, have even heard of *The Plain of Jars* in Laos, let alone visited it?' (p. 2)

'Vast views of the hitherto largely unexplored mountains of the eastern Himalaya.' (p. 12)

'. . . extremely few Europeans have visited it at all.' (p. 13 – The 'Lost Pass' Trek, Arunachal Pradesh)

'. . . in 2001 we were the first British company to take a party to visit this unique culture.' (p. 13 – Nagaland)

'The five-day trek to reach the least-visited of all the Himalayan capitals, Thimphu.' (p. 16)

'Only a small handful of westerners have ever managed the whole route.' (p. 18)

'Even today there are still parts of Bhutan that remain closed, and this was the case for the Ha Valley until 2001. Himalayan Kingdoms was proud to take the first British group that year to visit the main township and fortress.' (p. 20)

'. . . this newly opened trek in the little visited central area of Bhutan.' (p. 21)

'. . . the number of groups allowed is strictly controlled.' (p. 21)

'After its annexation by India in 1975, Sikkim became a Restricted Area, requiring special entry permits . . .' (p. 22)

'[Tibet] has always been a difficult country to enter.' (p. 36)

Conservancy; these areas are often reserved for the consumption of discerning First World ecotourists.

Sustainability has proved the perfect ally of the new middle classes, with the social construction and application of ecological concepts such as carrying capacity proving the ultimate justification for natural exclusiveness. As already argued, notions of sustainability are hegemonic in their own right and are readily translated spatially. The inflated ranks of academia have been quick to seize upon this, using 'research holidays' to impose their own ethnocentric analysis and recommendations. Of Thailand, Brockelman and Dearden (1990: 146) (Dearden is a former president of TIES) argue that the number of trekkers be kept relatively low, concluding that 'nature trekking should . . . be promoted for special tours and not among general tourists'. The social exclusiveness of their proposals, the class-laden nature of this debate, and the cultural capital that is to be accumulated and invested, are plain to see:

> It should be designed for serious naturalists or Nature enthusiasts and others wishing to experience genuine local cultures. Most such persons are reasonably affluent and educated and read a variety of Nature, conservation, and botanical

or zoological society publications . . . Such clientele is likely to get the most out of the experience, and probably also leave behind the best impressions.

(Brockelman and Dearden 1990: 146)

Conclusion

In this chapter we have attempted to show how the contemporary socio-cultural changes outlined in Chapter 2 are reflected through the growth and popularity of travel in many parts of the Third World. While the discussion does not imply a determinism that social class lies at the heart of all tourism, it has sought to show that it is an important and frequently overlooked factor in attempts to explain why tourism spreads and seeks out new frontiers. In this respect, it is a further demonstration of why the boom-to-bust cycles of tourism, which are most frequently used to demonstrate the circulation of destinations, amount to descriptions rather than explanations.

The chapter has argued that a sizeable proportion of new forms of tourism in the Third World can be related directly to the burgeoning new middle classes of the First World. This group is being joined increasingly by new middle-class tourists from the Third World, most particularly from South East Asia. This growth was also related to the increased interest in otherness, with a particular concern for ecology and ethnicity, both of which are found in plentiful quantities in the Third World. In addition, it was argued that central to the swelling ranks of the new middle classes is the necessity to differentiate socially from other social classes. This is most readily achieved through the construction of lifestyles, of which holidays are undeniably an important part.

The chapter explored how both ecotourists and ego-tourists are able to accumulate what has been identified as cultural capital and the strategies – professional, intellectual and discursive – that are employed to ensure that the accumulation of such capital is sustained and enhanced. In particular, we have argued that this provides a rather different understanding of the current debate over sustainability. This debate can be recast, in part, as the drive towards sustaining the opportunity and ability to consume authenticity and exciting experiences, which simultaneously necessitates the exclusion of other types of tourists. Ultimately, this represents a cultural and social reaction of the new middle classes to the crassness with which they perceive tourism and a craving for social and geographical distance from tourists.

6 Socio-environmental organisations: where shall we save next?

Much of the discussion so far has been underlined by the uneven and unequal nature of tourism development. This is most vividly expressed at a macro-scale as the inequality of First World and Third World tourism. In particular it has been suggested that questions of power which underlie the way in which tourism is owned and controlled largely from the First World, for example, should not stop with the analysis of the tourism industry itself. This is especially true in an analysis of new tourism.

In this chapter we consider the involvement and activities of the socio-environmental movement and the role of socio-environmental organisations; although, as we examine below, these are very broad terms. Chapters 1 to 4 emphasised that a good deal of the debate concerning the emergence of new forms of tourism has centred upon questions of environmental harm attributed to traditional forms of tourism and has sought ways in which to prevent or mitigate these negative effects. Equally, however, it has been argued that such activity can also be traced to the desire to preserve environments – the areas of so-called wilderness and virgin territories where nature can be experienced by 'discerning' new tourists. This also reflects two facets of sustainability: on the one hand, an environmental or ecologically centred meaning of sustainability as protecting and enhancing resources and biodiversity, and, on the other hand, an attempt to sustain cultural products for the benefit of (predominantly) First World new middle-class tourists, or, in other words, retaining places that are free from mass tourism and tourists.

As might be expected, the socio-environmental movement has spearheaded the advocacy and implementation of programmes and policies centred upon sustainability. This movement is, therefore, a key interest in the analysis of new tourism, and the Third World has become a major focus for environmental concerns partly as a result of its spectacular environments. There are two introductory observations that can be made, each of which embodies a paradox. First, in general, environmental issues and environment-centred organisations have been treated as benevolent causes and have attracted widespread interest and support, even in the light of quite interventionist policies. While this support reached its height during the 1980s and has subsequently waned somewhat, it nevertheless remains a significant pull for the new middle classes. Second, tourism very much represents a double-edged sword for the socio-environmental movement in that it is an activity which is both reviled and revered. It has become a focus for both criticism, as a result of its impacts, and promotion, as one means of achieving sustainable development. With the concerns expressed primarily in terms of the environmental and cultural impacts of tourism, the new socio-environmental movement has rounded upon some forms of tourism (that is, mass tourism) and promoted others (alternative, appropriate, sustainable and so on). Concerns have centred on the need to protect endangered habitats, maintain biodiversity and promote minority rights. An example of the latter is Survival, an INGO supporting and campaigning on behalf of tribal peoples. It presents us with an interesting contradiction in that its advocacy has tended to result in exclusionary policies that subsequently lead to a reduction in visitor numbers and the inevitable outcome of elitism in tourism. (Survival and its tourism campaign, 'Danger Tourists', are covered in some depth in Chapter 8.) Where new tourism projects exist they are often lauded for their vision and sensitivity, while mass tourism stands accused of crassness and environmental genocide.

This chapter attempts to introduce these debates and to reflect the complexity of the positions taken, the range of environmental concerns and how these relate to tourism. Initially it reflects back on the previous discussion to suggest the links between new tourism, the new middle classes and so-called new socio-environmental movements (of which the environmental movement is a key part). It then traces the different streams of the environment and tourism debate through a continuum of environmentalism. In doing so we are seeking to avoid a simplistic reading of environmentalism as an undifferentiated whole. Following the central thrust of our argument, we also seek to demonstrate how and why power is a fundamental component of tourism analysis.

New socio-environmental movements

Chapter 2 sketched out the most noted social, economic and political changes in the late twentieth century and suggested ways in which the growth of new forms of tourism were related to these factors. In particular it was argued that there are apparent relationships between the growth of the new middle classes and a concern with 'other-ness' which includes an interest in minority cultures, religion, ethnicity and, arguably most significant, a concern with environment and ecology. Notably, the 10–12 per cent of all tourism that is attributed to new tourism activities is similar in proportion to that of the First World populations interested and concerned with these other issues (UNWTO 1995: 28).

The growth of interest in sustainable lifestyles was singled out as the most momentous of these movements and an important component of a so-called 'new politics' which stand in contradistinction to 'old' style party politics. The emergence and growth of a 'new social movement' (or what we refer to here as socio-environmental organisations) that lies at the heart of the new politics and which has been responsible for campaigning and politicking on single issues (anti-nuclear and world peace, anti-globalisation, anti-racism, environment and so on) has in some ways 'transformed the political scene' (Crompton 1993: 16). The socio-environmental movement in its many guises has become one of the most enduring images since the 1980s and has captured the public imagination in a way that has far surpassed other movements. As Eckersley (1992) comments in her study of environmental political thought:

> The environmental crisis and popular environmental concern have prompted a considerable transformation in Western politics over the last three decades . . . Whatever the outcome of this realignment in Western politics, the intractable nature of environmental problems will ensure that environmental politics (or what I shall refer to as 'ecopolitics') is here to stay.
>
> (Eckersley 1992: 7)

One of the most prominent organisations, both globally and within, for example, the UK, is Greenpeace which reflects these significant changes. Figure 6.1 charts the growth in Greenpeace UK's membership numbers from its early days. The most dramatic rise in its popularity came in the second half of the 1980s, reaching a peak in the early 1990s and falling off somewhat to the mid-1990s, since when it has been relatively stable.

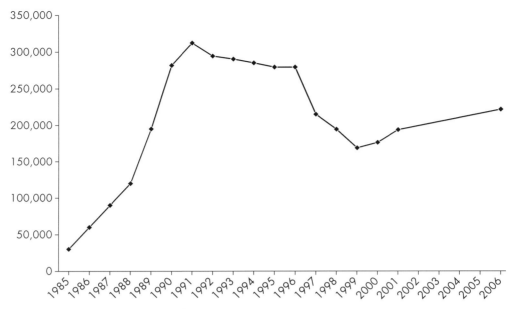

Figure 6.1 Annual membership of Greenpeace UK

The growth accompanied a variety of other factors, including an increasing visibility through the media of environmental disasters and problems, and a concomitant rise in the public perception of these problems. Growth can also be attributed to concern about the increasing importance of the role of supranational organisations such as the World Bank, and a rising intrusion of transnational companies into most ways of life, and possibly a growing disenchantment with the potential for bringing about change through increasingly distant, but powerless, democratically elected representatives. The debacle surrounding the Kyoto agreement on the reduction of emissions of greenhouse gases, together with the differing perspectives on the Rio + 10 Conference in Johannesburg in 2002 (the World Summit on Sustainable Development), have underlined the significance of a globalised environmental politics. More recently, the uncertain and contested outcome of the climate talks at Bali in December 2007 have emphasised to many in the broad environmental movement the disappointment that they feel with the globalised power of transnational corporations and international financial institutions and their chances of bringing about progressive change in these organisations and institutions.

But many of the major socio-environmental organisations have themselves become transnational and now campaign on issues of global concern. These are the BINGOs (Big International NGOs) such as Greenpeace, World Wildlife Fund, Friends of the Earth and Conservation International. Box 6.1 gives an idea of the growth of BINGOs and the relative importance of their distribution among the continents. In becoming global in their reach themselves, some of them have taken on characteristics associated more with economies of scale than with local concerns and have accepted compromises in their moral stances and policy promotions as a result of their search for the funding necessary to maintain their operations at such a scale.

The recent falling-off of membership numbers is a little less easy to link with other factors (perhaps because of the lesser benefit of hindsight), but may well be traced back to the demise of the Soviet Union and its geopolitical counterweight to the First World. This has given rise to both complacent attitudes of 'victory' and equally despair and apathy

Box 6.1 Big International Non-Governmental Organisations (BINGOs)

Some international NGOs have 'consultative' status with the UN Economic and Social Council (ECOSOC). They are divided between large 'general' ones whose work covers most of the ECOSOC agenda, and special ones that do not – though some of these are also quite big. The number of both grew slowly until the 1990s, then very much faster. This was partly due to new, and often more open policies at the UN, but also a growing number of BINGOs want this status.

Number of NGOs with consultative status at the UN

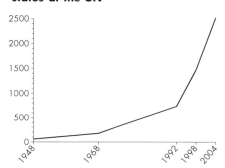

Representation of NGOs at the UN, 2004

Between them, Europe and North America account for more than two-thirds of the NGOs with consultative status at the UN – and for a much greater proportion of the largest, richest and most influential of these NGOs.

Sources: www.un.org; New Internationalist (2005)

in opposition, both of which may well have led to a disillusionment with the purpose behind involvement and activism.

A second factor of relevance here is the fact that the environmental lobby, initially perceived as a single issue movement in itself, has managed to raise its public profile high enough to promote a range of single issues under the general umbrella of the environment. Environmental issues are still widely equated with ecology rather than society or culture, although this broadening of issues under the general term of the environment should be acknowledged. This division of issues has also been associated with a rise in the number of organisations now dedicated to more justifiably described single issues, especially within the field of tourism. Tourism Concern, for instance, was first established in the UK in 1989 with 200 members. By 2002 it had over 1,500 members and had gained a respectable reputation among the relatively highbrow sector of the country's media. It had also become an organisation whose blessing was sought by some of the new and specialised tour operators (which are described in Chapters 3, 5 and 7). Similarly, by 2002 the US-based International Ecotourism Society (TIES) could boast 1,700 members in 65 countries and had changed its name from the less worldly 'The Ecotourism Society'. TIES describes itself as an organisation for professionals from a variety of disciplines who are 'needed to make ecotourism a genuine tool for conservation and sustainable development'

(TIES website 2002). TIES was heavily involved with the World Tourism Organisation and the United Nations Environment Programme in the organisation of events to mark the United Nations' controversial International Year of Ecotourism in 2002 (see Chapter 10).

Appendix 1 lists the websites of many NGOs concerned with the impact of tourism. The list is far from comprehensive, but serves to highlight the recent burgeoning of organisations focused on tourism. Few of the organisations listed have existed for longer than twenty years or so. Many have their headquarters in First World countries and see the issues of global tourism or tourism to Third World countries as their major focus.

Another organisation whose remit can be considered to be a single issue is that of Survival. It is of particular relevance to the tourism industry partly because it has conducted research and produced reports on the effects of tourism on tribal peoples and partly because of the nature of these effects – an issue taken up again later in this chapter and in Chapter 8. Figure 6.2 indicates not only the growth of Survival International since 1987 but also its predominantly European membership. It has its base and was founded in the UK, but since 1990 has opened satellite offices in France, Spain and Italy.

This growth in membership of and support for socio-environmental organisations with their bases in First World countries is likely to result in the wider dissemination of their particular interpretation of the role of tourism in the political, social, economic and cultural systems which prevail in Third World countries. As support for such organisations grows, so will their ability to influence the relevant and topical debates. Whether this influence will act as a counterweight to the geopolitical forces currently wielded by transnational companies, supranational institutions and First World government interests or, on the contrary, as an additional weight to these forces, is still a matter for debate. There can be little doubt that many organisations representing the socio-environmental movement rail strongly against the activities of supranational institutions and their effects. But in different

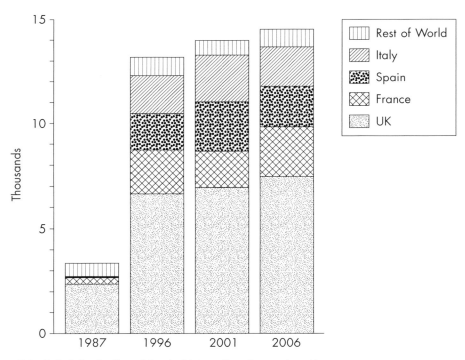

Figure 6.2 Global distribution of Survival International's membership

ways and in differing degrees they also represent a significant hegemonic interest in their own right, in convincing their constituents of the positions they adopt on tourism. The debates also reflect the contested nature of sustainability as competing interests struggle to legitimise their own definitions of the concept. How this debate will evolve is a matter for speculation, but it is already clear that all the interests involved in this struggle represent different facets of First World power.

While new socio-environmental organisations are of special interest in the way in which they defy, as it were, the traditional left/right political ideologies, they also tend to be closely related to class in that their memberships are drawn largely from the middle classes.[1] As Dobson (1995: 154) asserts, 'there is plenty of sociological evidence to show that the environmental movement is predominantly a middle class affair' with its ranks drawn from the educated, 'intellectual', and 'socially aware' elements. As Jonathon Porritt (1984) argues:

> One must of course acknowledge that the post-industrial revolution is likely to be pioneered by middle-class people. The reasons are simple: such people not only have more chance of working out where their own genuine self-interest lies, but they also have the flexibility and security to act upon such insights.
>
> (Porritt 1984: 116, quoted in Dobson 1995)

The reason for the predominance of new middle-class involvement in the new socio-environmental movement (as both supporters, members and employees) and their participation in 'green politics' is a matter of considerable debate and disagreement and one that is reflected in the degree of involvement of the new middle classes in new tourism. As Eckersley (1989: 210) explains, on the one hand there are the selfishness and self-interest arguments that posit that 'it is mainly the new class that is involved . . . because it is a means by which it is able to further its own class interest'. On the other hand, there are those who argue that the new middle classes are the vanguard of postmaterial values, the harbingers of sustainable lifestyles. Eckersley (1989: 222) argues in favour of the importance of education and the new middle class awareness of the 'scale and depth of the social and ecological problems'. This is what Inglehart (1977, 1981) refers to as the empathy and sympathy that the new middle classes have with such issues and the new socio-environmental organisations that mobilise around them and which produce, he argues, an 'ideology of the new middle classes'.

Although Eckersley strenuously defends the socio-environmental movement against charges of elitism, it is difficult to deny the piety and sacrificial overtones which are apparent in much promotion of new middle-class ideals. They are reminiscent of the 'Peace-corps type tourists' (Kutay 1989) and the Explore leaders who put 'promising careers on hold' (see Chapter 5). As Eckersley (1986) suggests:

> the assumption that environmentalists will necessarily defend their own level of affluence simply overlooks the fact that many committed environmentalists have deliberately forsaken the material lifestyles and career opportunities of their fellow class members (which they argue are wasteful and devoid of purpose) in favour of a more frugal and socially responsible lifestyle which is encapsulated in the slogan: 'To live more simply that all others may simply live' (Porritt 1984: 205).
>
> (Eckersley 1986: 28)

While seeking to avoid a crude determinism and correlation between these various factors of social, economic and political change, we can nonetheless detect some important apparent relationships. As already argued, many authors have traced the connections between the expansion of the new middle classes and the emergence and growth of new socio-environmental organisations (Gouldner 1979; Offe 1985). In turn it has been suggested that there is a significant, although clearly not exclusive, relationship between these phenomena and the growth of new forms of tourism. We are not able to present in full and unravel this complex debate. But it is interesting to begin to think about and question the dynamics of new tourism, new middle-class involvement and the campaigns and foci of new socio-environmental organisations.

Environmentalism and new tourism

Environmentalism, or ecopolitics as Eckersley (1992) refers to it, is a useful way of exploring the political globalisation and global political reordering that Chapter 2 touched upon. Yearley (1995) identifies several reasons for assessing environmentalism as a vehicle for examining 'global political re-ordering'. First, the socio-environmental movement has repeatedly stressed the global nature, or 'worldwide-ness', of environmental problems and concerns embodied in the much-used, catch-all phrase 'think globally, act locally'. Second, environmental issues more than any others have given rise to international summits (such as the Rio and Rio + 10 summits) and agreements and the emergence of international and supranational bodies focused upon the environment.

Tourism adds another facet to this global environmental politics and has become a focus for debates over the environmental impacts of tourism; a focus that has intensified as tourism has grown and spread. In addition, of course, much new tourism has become intimately associated with the environment, environmentalism and debates over sustainability on how to achieve less environmentally harmful forms of activity.

But it is far too crude to imply that environmentalism is one undifferentiated issue and movement. Far from it. Environmentalism embraces a broad range of views, policies and actions, and this is clearly reflected through tourism. The continuum or spectrum of environmentalism set out by Eckersley (1992) provides one way of exploring this relationship with tourism more systematically. The remainder of this chapter will use this framework to explore a range of positions.

Eckersley (1992) identifies five major strands of environmentalism: resource conservation, human welfare ecology, preservationism, animal liberation and ecocentrism. This spectrum is summarised in chart form in Table 6.1. For the purposes of applying this typology to the debate over new forms of tourism we have omitted the 'animal liberation' category, which is of less significance to our discussion (though clearly there are questions about the promotion of voyeurism of wildlife with the increasing number of game parks and use of animals as an added attraction in some forms of new tourism). Eckersley (1992) suggests that most environmental causes and movements fall within the two poles of resource conservation and ecocentrism, and that, in general terms, as you move from left to right in the table you are moving from anthropocentric views of environmentalism to ecocentric ones. Anthropocentrism means that these strands of environmental political theory are human-centred and seek to offer 'new opportunities for human emancipation and fulfilment in an ecologically sustainable society' (Eckersley 1992: 26). By contrast, ecocentric environmental politics takes the emancipatory goals one stage further by acknowledging the rights of non-human life forms; in other words, saving nature for nature's sake and not for human exploitation. Eckersley concludes:

> This spectrum represents a general movement from an economistic and instrumental environmental ethic toward a comprehensive and holistic environmental

ethic that is able to accommodate human survival and welfare . . . needs while at
the same time respecting the integrity of other life-forms.

(Eckersley 1992: 34)

It is not intended, however, that the continuum is strictly linear, with each position
exclusive of other ideas. Some arguments within the human welfare ecology stream, for
example, may be more ecocentric than some offered within the generally more ecocentric
preservationist perspective. Similarly, the top row of Table 6.1 is illustrative of where
organisations might be positioned, but it is not definitive. Thus, environmental
organisations such as WWF may be represented in more than one stream. Nevertheless,
the continuum provides a useful framework for thinking about the way in which tourism
interacts with environmental politics and the politics of sustainability. It may be useful to
refer to Table 6.1 as each stream is explored below a little further.

Conserving tourism resources

Resource conservation is the least controversial stream of modern environmentalism,
although it is seen as anathema to the more radical ecocentric perspectives (Eckersley
1992). It includes the national parks movement that seeks both to conserve nature and make
it pay for itself, a compelling argument for environmental conservation for many cash-
strapped Third World governments. For example, such a perspective is likely to encourage
the costing of wildlife in terms of the potential income it can attract through safaris and
game watching, an economic justification for their retention and development. Within this
perspective, therefore, sustainability is conceived as 'sustainable development' and
involves sustaining the environment for human production (the creation of national parks)
and consumption (for the enrichment and enjoyment of tourists).

The notion of wilderness is also heavily implicated in this stream of environmentalism.
Wilderness is an excellent example of a socially constructed idea (Sarre 1995). To many
First World tourists it now represents ecological purity, an area free from human
interference and development and generally devoid of human habitation, and it is a
powerful notion that is repeatedly conjured up in tourist media such as travel brochures.
Of course, for local peoples living in these supposedly wilderness areas there is a
substantially different conception. As Deihl (1985) concludes of East Africa, '"wilderness"
was largely a creation of Western thought since most areas they called wild were in fact
used by native inhabitants' (Deihl 1985).

A number of commentators have been heavily critical of the wilderness concept for the
way in which it enshrines a division between humans and humanity on the one hand and
nature on the other, and for its undeniably ethnocentric connotations. It is capable of both
ignoring the ecological management undertaken by local indigenous communities and
assuming that First World conceptions of management are superior. In charting the costs
of wilderness preservation, John Vidal cites American philosopher Baird Callicott:
'Wilderness is a legacy of American puritanism. It played a crucial role in masking colonial
genocide and ethnic cleansing. It is a powerful conceptual tool of colonialism' (Vidal 2001:
1). It has also resulted in the displacement and exclusion of local people from areas
considered (more often than not from a First World perspective) worthy of protection.
The issue of displacement is covered in more detail in Chapter 8.

As Table 6.1 indicates, there are a number of policies and programmes aimed at meeting
the goals of resource conservation, some of which further underline the power invested in
First World interests. One of the most noted is the concept of debt-for-nature swaps.
Because few commentators envisage the solution to the Third World debt problem as being

Table 6.1 Tourism and the spectrum of environmentalism

	Resource conservation	Human welfare ecology	Preservationism	Ecocentrism
Organisations associated with this view	WWF, IUCN, UNEP, UNDP; Tour companies promoting 'green' holidays; Travel Adventure Society (US); Campaign for Environmentally Responsible Tourism (CERT); Conservation Foundation; Conservation International	Green Movement, Friends of the Earth; Individual citizens, consumers and householders concerned with the state of residential areas and the state of the planet; Tourism Concern	Coral Cay Conservation; Elefriends; Tusk Force; Born Free Foundation; Earthwatch	Greenpeace, Earth First, Wilderness Society
Place of origin and main ideas	USA, nineteenth century; Utilitarian – 'greatest good for the greatest number'; Wise use of natural resources	Industrialised Europe twentieth century; Enlightened self interest; Four laws of ecology: everything is connected to everything else; everything must go somewhere; nature knows best; there is no such thing as a free world; Pursuit of environmental quality	USA, nineteenth century; The reverence of nature; Aesthetic and spiritual appreciation of 'wildnerness'	North America, Australia and New Zealand; Nature and environment are of equal importance to humans
Views on resources/ conservation aims/ sustainability	Wilderness to be managed for the greater good of people; Regards non-human world in use value terms; A resource bank for industrial society to develop; Develop wilderness and prevent waste; Advocates 'sustainable development'; Sustaining natural resource base for human production and maximum economic yield	Concern with degradation of the environment; Concern for health, safety and amenity; Advocates 'sustainable development' – sustaining both natural resource base and sustaining biological support systems for human reproduction	Preserve nature from development; Defence of 'wild nature' for spiritual values to humans; 'sustainability' – preservationism at any cost	Nature and wilderness is of intrinsic worth and is not a resource

Table 6.1 *Continued*

	Resource conservation	Human welfare ecology	Preservationism	Ecocentrism
Policies and political programmes	Debt for Nature Swaps Conserve nature *for* development Biodiversity convention Global Environment Facility (GEF) Resource management National Parks and protected areas 'Costing' flora and fauna	Critical of unrestrained and inequitable economic growth Policies to counter pollution and for alternative technology and lifestyles Promotes pro-poor policy and sustainable livelihood approaches	Resistance to values of technological society Aim to create an alternative society National Parks and protected areas	Concerned to protect threatened populations, species, habitats and ecosystems *wherever* situated, irrespective of their use or value to humans
Views on application to tourism	Manage wilderness as a tourist attraction Cost 'natural attractions' as economic assets Natural Parks (as conservation areas) can pay their own way Would encourage high paying, well-heeled 'eco-tourists'	Concern with the environmental impact of tourism Heartland of the concerned, educated new middle classes wanting an ethically and environmentally enriching experience – alternative tourism as part of alternative lifestyles Pro-poor tourism initiatives	Selective and exclusionary Volunteer holidays – (research tours) Kutay – 'peace-corps type travellers' Archetypal 'ego-tourists' trekking in high places, rainforests Marine preservation	Likely to ignore tourism, which is concerned as part of the problem. But is full of academics and deep green ecologists having a fine time researching and analysing environ-mental issues (new tourists in their own right?)

Source: Adapted from Eckersley (1992); Open University (1995)

repayment, First World banks are willing to sell off the debts owed to them at a discounted rate. An environmental INGO with substantial funds pays off the debt, or a small part of it, to the First World bank that is the creditor. In return, the Third World government, unable to pay off the debt in dollars, pays it off internally in the local currency to a local NGO in partnership with or deliberately set up for the purpose by the First World INGO. This payment will then be used to protect a specified natural area. It is a relationship, as Adams (1990) argues, that greatly extends the notion of 'the one who pays the piper calls the tune':

> The principle of directing First World resources to Third World conservation has long been important in organisations such as the World Wide Fund for Nature . . . debt-for-nature swaps . . . arise from the exposure of First World Banks to Third World debt, and their willingness to sell off those debts at a discount to conservation organisations, who use them to bargain for expenditure on conservation in local currency.
>
> (Adams 1990: 200)

While debt-for-nature swaps do not necessarily implicate tourism, they certainly do indicate the lengths environmental organisations will go to in conserving certain Third World environments. Lewis (1990), for example, documents the 'bizarre agreement' in 1987 between the Bolivian government and US-based Conservation International in which US$650,000 of Bolivian debt was bought at a US$100,000 discounted rate on the condition that the Bolivian government pay US$250,000 in local currency to create a buffer zone around the Beni Biosphere Reserve. The buffer zone, however, was not a wilderness area, but one in which cattle rearing and lumbering were major activities whose promoters were not willing partners in the deal. Additionally, the money supplied by the Bolivian government to protect the reserve drained funds away from other environmental projects – the payment of US$250,000 was equivalent to the total annual budget for the national park system. In 1989, as Fred Pearce (1990) documents, the WWF paid US$2 million for a Costa Rican debt with a paper value four times greater. It then traded the debt for government bonds worth US$7 million to pay for the creation of the Guanacaste National Park.

Swaps, however, have covered only a tiny proportion of the Third World debt and are no longer widely considered to be a potential solution to the debt crisis. They divert the INGO's funds, energy and attentions away from tackling the structural causes of both the debt crisis and the conditions which in some circumstances force people and governments into exploiting and destroying their natural resources. They do not represent the flow of new money into a country but provide instead an illusory relief of the external debt. Moreover, they work against the local ownership and control of resources that are so critical to community tourism development by furthering the notion of private appropriation of land and resources (which in many cases in Third World countries used to be considered as common heritage), and by representing a transference of power over national resources to First World organisations.

The domination implied by debt-for-nature swaps is no less oppressive when it comes from a well-intentioned INGO than when it is imposed by the World Bank, IMF or a transnational company. But most of the criticism of debt conversion schemes appears to come from within the ranks of the INGOs themselves, which might suggest that such schemes are not promoted out of ignorance of their effects. Rather, they indicate that the international socio-environmental movement is being increasingly co-opted by the financial ethic of capital accumulation and by the supposed 'realities' of the global marketplace.

A second example of action-oriented programmes within the resource conservation stream is the commonly perceived global need to retain biodiversity. The term biodiversity has only relatively recently come into common usage which began with a draft report by the IUCN in 1987. David Pearce and Moran (1994) describe biodiversity in terms of genes, species and ecosystems. The significance of genetic diversity is normally related to seed production, control of which lies firmly with the transnational corporations (TNCs) which have tended to reduce genetic diversity and thereby make Third World farmers dependent on the TNCs. With regard to species diversity,

> We do not know the true number of species on earth, even to the nearest order of magnitude ... This lack of knowledge has considerable implications for the economics of biodiversity conservation, particularly in defining priorities for cost-effective conservation interventions.
>
> (D. W. Pearce and Moran 1994: 4)

Ecosystem diversity

> relates to the variety of habitats, biotic communities and ecological processes in the biosphere as well as the diversity within ecosystems ... No simple relationship exists between the diversity of an ecosystem and ecological processes such as productivity, hydrology, and soil generation.
>
> (D. W. Pearce and Moran 1994: 5)

Biodiversity is a term which is often used implicitly and sometimes explicitly to justify the designation of areas for some degree of protection from development and exploitation. It is important to note, however, that the notion of biodiversity and the Biodiversity Convention which came out of the 1992 Rio Summit are the subject of heated disagreement.

The main aim of the Biodiversity Convention is 'the conservation of biological diversity, sustainable use of its components, and fair and equitable sharing of benefits from the use of genetic resources' (Holmberg et al. 1993: 20). Although the spirit of the treaty is more in keeping with the sustaining and accessibility of genetic resources, it has clear resonance for tourism as an economic justification for 'conservation' of the 'natural environment'.

It is implicit in the Convention that biodiversity has an economic value. In order to determine this value, however, biodiversity change must be measured and its valuation is often related to the protection of areas against destruction by humans. This is especially so where the motive of national governments in designating areas for protection arises from its hopes of developing tourism (see Chapter 10), in which case potential tourist revenue can be estimated, on the assumption that the biodiversity that the tourists are going to see is retained. Achieving this valuation in a way that is agreeable to all, however, is fraught with difficulty. At its most fundamental, criticism of price valuation of natural features believes that: 'The most obvious wrong is taking something that is so clearly beyond value and reducing it to money terms ... It's a confusion of value with price, beauty with numbers, the sacred with the profane' (Meadows 1997). As E. F. Schumacher (1974: 38) warns: 'to undertake to measure the immeasurable is absurd and constitutes but an elaborate method of moving from preconceived notions to foregone conclusions'. Once an estimation of the monetary value has been made the problem of the distribution of costs and benefits then arises: who should pay how much to whom, for what, and what should the money paid be used for (United Nations Environment and Development UK 1993)?

This debate illustrates the way in which environmental questions and 'agreements' are mapped onto uneven and unequal development. Writing in Third World Resurgence on the Biodiversity Convention, the post-development critic, Vandana Shiva, argues that it started principally as an 'initiative of the North to "globalise" the control, management and ownership of biological diversity . . . so as to ensure free access to the biological resources which are needed as raw material for the biotechnology industry' (quoted in Holmberg et al. 1993: 22). In this sense, holiday environments can also be viewed as the raw materials of which Shiva writes.

The 1992 Rio Summit offered the First World the ideal mechanism to achieve this globalisation of the control and management that Shiva refers to: the Global Environment Facility (GEF). The GEF was set up in November 1990 by the World Bank, the UNDP and the United Nations Environment Programme (UNEP) to assist the so-called developing world in funding projects which either protect biodiversity against destructive development or promote development which does not destroy biodiversity. The GEF, however, is not a development agency:

> It operates via many development projects, but it modifies them so that the technologies used are cleaner than they otherwise would have been. Its purpose is not development as such, but the capture of global environmental value – the value that comes from reducing the 'global bads' of climate change, biodiversity loss and ozone layer depletion.
>
> (D. W. Pearce and Moran 1994: 132)

The 1992 Rio Summit allocated to the GEF the role of financial administrator for the Biodiversity and Climate Change Conventions which arose out of the conference. Again, Shiva (1993) is highly critical of this role and of the World Bank's part in it:

> The erosion of biodiversity is another area in which control has been shifted from the South to the North through its identification as a global problem . . . by treating biodiversity as a global resource, the World Bank emerges as its protector through GEF . . . and the North demands free access to the South's biodiversity through the proposed Biodiversity Convention. But biodiversity is a resource over which local communities and nations have sovereign rights. Globalization becomes a political means to erode these sovereign rights, and means to shift control over and access to biological resources from the gene-rich South to the gene-poor North. The 'global environment' thus emerges as the principal weapon to facilitate the North's worldwide access to natural resources and raw materials on the one hand, and on the other, to enforce a worldwide sharing of the environmental costs it has generated, while retaining a monopoly of benefits reaped from the destruction it has wreaked on biological resources. The North's slogan at UNCED [the 1992 Rio Summit] and the other global negotiation fora seems to be: 'What's yours is mine. What's mine is mine.'
>
> (Shiva 1993: 152)

In 1992 Oliver Tickell articulated the suspicion of much of the socio-environmental movement, that control of the funds by the World Bank could only lead to the imposition of a First World agenda on the allocation of those funds. He also pointed out the contradiction in the GEF's approach:

> The main qualification for receiving a GEF grant for preserving biodiversity is apparently to be running a World Bank project that threatens or destroys that biodiversity, like a giant dam, or a plan to develop logging in untouched forests. Marcus Colchester of the World Rainforest Movement estimates that 70 per cent

of GEF funds are tied to mainstream Bank projects, and are actually subsidising social and environmental destruction.

(Tickell 1992: 3)

Despite the misgivings of much of the socio-environmental movement and others, Fernandes (1994: 24) claims that 'Already the leading northern NGOs appear to have developed strong coordinating GEF links with the World Bank, UNDP and UNEP'. Chatterjee and Finger (1994: 155) claimed that the WWF is now the most consulted NGO on GEF projects. Fernandes' (1994) analysis of the GEF suggests that a large number of integrated sustainable tourism projects received funding under phase I of the GEF. He quotes Maria Soares of the Brazilian Institute for Economic and Social Analysis, who concludes that many GEF programmes, which are often linked to significant ecotourism and sustainable tourism components, represent 'mainstays for a (development) model that unceasingly reproduces conditions for the planet's deforestation, even while preaching its conservation' (Soares 1992: 48).

Fernandes also cites other fundamental flaws in the GEF mechanisms such as the stimulation of destructive competition between organisations working in the field of biodiversity, a failure to involve local communities, an over-dependence on international consultancies (see p. 114) and the creation of a number of 'paper parks' – a reference to the designation of parks by national governments which have few resources to provide management and protection systems on the ground (see Box 10.1). Moreover, the link between the World Bank and INGOs such as WWF, IUCN and Conservation International leads to these organisations adopting an approach more characteristic of the World Bank and advocating corporate schemes for a range of environmental programmes and projects (including tourism and ecotourism developments) which give 'total management control' to the 'private or NGO sector'. As a result they fail to even 'recognise the existence of village conservation movements opposing development projects ... much less to acknowledge their effectiveness in pressurising governments and corporations' (Lohmann 1991: 98–99, cited in Fernandes 1994: 26).

Conservation with a people dimension?

The human welfare ecology stream has the same general perspective of 'sustainable development' as the resource conservation stream but goes further in its concern for environmental quality. It is most clearly reflected in the 'green movement' and the individual concern with the state of the environment. It is this perspective that most readily captures the much-expressed concern over the environmental impacts of tourism and how diverse new tourism destinations such as the Himalayas, Zanzibar or Bali (as we discussed in Chapter 5) are reputedly being overrun and corrupted both culturally and ecologically.

It is also a stream that is reflective of the debates over the most appropriate way to travel and the most virtuous form of alternative tourism. As Morrison of the traveller's magazine, *Wanderlust*, suggests, rhetorically:

to make matters worse we now have to suffer that greatest hypocrisy of all – the lamentable cries of anguish from a latter-day explorer bemoaning the commercialisation of a part of the world that they were instrumental in opening up. It's a sadly familiar cycle. After the explorers come the travellers. After the travellers come the tourists. And after the tourists come those blasted explorers once again, this time heading up some conservation group to wail about the exploitation of a once virgin paradise.

(Morrison 1995: 68)

In a similar vein, Eckersley (1992: 36) suggests that the human welfare ecology stream appears to be a 'peculiarly late twentieth-century phenomenon' that can be attributed in large part to the 'emergence of "post material" values borne by the . . . New Middle Class'. It is an especially strong current of environmentalism in green politics in the First World, and tends to appeal to 'enlightened self interest', an individualistic concern centred upon 'our survival, for our children, for our future generations, for our health and amenity' (Eckersley 1992: 38). This is strongly reminiscent of the arguments put forward on the need to protect tourism resources from further development, from tourists discovering it, and so on. In other words, it appears to offer a thoroughly First World perspective on our needs and aspirations and why we should encourage sustaining natural environments for, among other purposes, our holidays. As Adams (1990: 200) concludes, conservation is 'dominated by global concentrations of wealth and power, and centralized decision-making. Despite the well-meaning rhetoric of environmentalists advocating sustainable development, it is not the Third World that stands to gain most.'

Preserving Third World tourism products?

The third stream of environmentalism, preservationism, can be contrasted to resource conservation which, Eckersley (1992) argues, aims for nature conservation to enable development. Preservationism, by contrast, seeks to preserve nature from development. Like the two previous streams, however, it tends towards anthropocentrism in that it recognises 'wild nature' and 'wilderness' areas as being of spiritual and aesthetic value to humans, on whose behalf it should be preserved.

There is an increasing number of Third World environments that are subject to pre-servation and to which 'discerning tourists' are attracted. The significance of authenticity in the discussion of tourism has already been emphasised in Chapter 3, where the importance of aestheticisation (the ways different facets of the Third World become an aesthetic experience) in forming a critique of new tourism was argued. The drive to preserve particularly natural environments, but also indigenous cultures, finds resonance in these concepts.

Beneath the push towards preservation is the perception that wilderness areas are rapidly being lost and the need to protect those areas remaining for the consumption, in part, of informed First World tourists. In this way preservationism tends towards exclusionism, for not only does it rely on the containment of visitor numbers to ensure the retention of a stable environment (the ecological carrying capacity) but also it requires fewer tourists to maintain the 'reverence' and 'spirituality' of 'wild' areas. Furthermore, in Chapter 5, it was suggested that it is exclusionary in another important respect in that it requires the ability to interpret and understand the environment (and the notion of sustainability); hence Brooks' (1990) reference to the prerequisite understanding of 'cultural values and ecological balance' in 'remote and unspoilt regions', and Brockelman and Dearden's (1990) view that only 'serious naturalists and Nature enthusiasts' should be considered for trekking in areas of Thailand.

Underlying these sentiments, it appears that preservationism tends towards a high degree of ecological and social selectivity. As Eckersley (1992: 40) suggests, 'wilderness appreciation has developed into a cult in search of sublime settings for "peak experiences", aesthetic delight, "tonics" for jaded Western souls'. In so doing it has singled out those places 'that are aesthetically appealing according to Western cultural mores' (Eckersley 1992: 40), including mountain ranges, ecologically rich rainforest areas and diverse marine environments. In other words, First World geographical imaginations are filled out with environments that are regarded as somehow sacred and worthy of our attention. Again, in parallel to the two environmental streams discussed earlier, it is First World citizens who are expressing their perceptions, their choices and exercising their will. It is a further illustration of the way in which uneven development is played out, as areas become targeted and coveted for their ecological content while other areas fall from our view and attention.

From his thoroughgoing review of the social dimensions of policies and initiatives for environmental protection, Colchester (1994) summarises four fundamental problems with what he terms the 'classical conservationist' approach (an approach that appears more akin to preservationism than resource conservation). First, humans come second in the conservationist's view of nature preservation and this has some serious implications for Third World communities affected by the push towards nature preservation (as Chapter 8 discusses). Second, and reflecting the discussion on the social construction and representation of wilderness, the generally holistic (or cosmovision) view of nature held by indigenous peoples, is radically different from the cultural notion of wilderness held by 'western conservationists'. Third, the view of indigenous peoples held by conservationists has been 'tinged with the same prejudices that confront indigenous peoples everywhere' (Colchester 1994: 56), and this is highlighted in Box 6.2 and Box 6.3, which focus on these views and associated attitudes. Again, there are overtones here of the ethnocentrism of the environmental debate, also known in other contexts as racism. And finally, conservationist interests have sought state legitimation for enforcing their view of conservation on local peoples.

Machlis and Tichnell (1985) argue that preservationist approaches to protected areas often result in what they refer to as 'militaristic defence' strategies or what Pleumarom (1994) terms 'militarization', arguing:

> Tourism, particularly in the context of 'sustainable development', often only serves to exacerbate the entrenched asymmetry of unequal relations in the world. Militarization is the most visible tendency of this process. For example, 'legitimate' violence in the name of resource control, such as 'shoot-to-kill' actions in Kenya's game parks . . . a tendency reinforced by the provision of para-military training for park rangers and anti-poaching equipment by international conservation groups.
>
> (Pleumarom 1994: 146)

As Pleumarom (1994) suggests, environmental INGOs have been implicated in this militarisation process. Box 6.4 illustrates this process with evidence of the WWF's alleged involvement in the hiring of mercenaries to protect the black rhinoceros population in Namibia.

One of the principal mechanisms for advancing preservationist policies has been through national parks and other categories of protected areas. National parks were first established in the United States in the nineteenth century and began a growth that is charted in Figure 6.3. Internationally, the categories most often recognised are those defined by the IUCN in 1994 and listed in Table 6.2. To the IUCN's categories should be added the biosphere reserves designated by UNESCO's Man and Biosphere Programme. Biosphere reserves are protected areas of environments internationally recognised for their value for the preservation of genetic diversity. Additionally, national governments may devise their own categories and add other standards of protection to those internationally recognised.

These categories of protection have a degree of legal standing, but, in Third World countries at least, this rarely translates consistently into appropriate security of protection on the ground. Even as early as 1992, Wells and Brandon (1992: 1) warned that many protected areas were suffering 'serious and increasing degradation as a result of large-scale

Box 6.2 He's destroying his own rainforest. To stop him, do you send in the army or an anthropologist?

This hurtful, insensitive slogan began a fundraising campaign by the World Wide Fund for Nature in 1994. It first appeared in the *Financial Times*, presumably for the benefit of WWF's growing band of corporate supporters, and set the tone for the advertisement that followed: glibly pro-nature and implicity anti-people.

Yet the WWF advertisement provides a useful warning, as it offers a perfect illustration of the dominant conservationist mindset. Unfortunately for indigenous peoples, most Western conservationists still cling to a romantic, Eurocentric conception of nature as an empty, unspoiled wilderness, separate from and uncontaminated by humanity . . . in which the indigenous inhabitants of these 'wildernesses' are at best an inconvenient disruption of the great romantic myth, at worst a menace to be repelled by barbed wire and guns . . .

Not surprisingly, the world's indigenous peoples are bemused and outraged by the behaviour of Western conservationists . . . They see the wilderness view of conservation as a new form of colonialism, with the same devastating results as military conquest or slavery. They also profoundly resent the idea that areas of land they have inhabited for generations – often for thousands of years – should be regarded as empty.

Like logging and mining, conservation has meant driving indigenous people from their lands by a mix of repression and fraud, but while miners and loggers are usually explicit about their intention to exploit, conservationists pacify public opinion with the language of environmental concern.

Conservationist rhetoric also makes politicians sound virtuous: call anything 'sustainable' these days and it is bound to win plaudits.

Sometimes, as a supposedly benign alternative to resettlement, people have been 'allowed to remain' on their land under certain strictly imposed conditions. Usually the intention of such schemes has been to encourage tourism by pandering to western ideas of 'noble savages'. The people concerned are thus dehumanised and presented as exotic species to be 'conserved' . . . as long as they play by the rules.

In South Africa the last group of 'Bushmen' (also known as San or Khoi-Khoi) were told that they could stay in the Kalahari Gensbok National Park provided they maintained a 'traditional' lifestyle and used 'traditional' methods of farming – in other words, as long as they continued to look like 'Bushmen'. Needless to say, this was not a success . . . The 'Bushmen' decided they wanted to have new housing, new clothes and hunting dogs . . . The wardens of the National Park concluded that 'their desirability as a tourist attraction is under serious doubt, as is the desirability of letting them stay for an indefinite period in the park. They have disqualified themselves . . .'

In practice the conservation movement has subjugated indigenous peoples to state or corporate control. It has violated their rights and, for the most part, failed in its own objective of environmental protection.

Source: Survival International (1995)

Box 6.3 Diamonds are forever: the Bushmen, the ecologist and the government of Botswana

The British colonial authorities set up the Central Kalahari Game Reserve (CKGR) in 1961 to protect its habitat and residents, and this policy was continued under the newly independent government five years later. Diamonds were discovered in the country in 1967 and now account for a substantial proportion of the country's export earnings. The government jointly owns Debswana, the company that controls and mines the diamonds, with the international diamond giant De Beers, but the Bushmen believe that they should receive a share of the profits if diamond mines are opened in the reserve.

In the 1980s, the government moved 2,500 Bushmen from the reserve in what they described as a 'voluntary relocation', but which was described as coercion by the Bushmen owing to the withdrawal of hunting permits and cutting of water supply. Since the mid-1990s, the government of Botswana has been 'squeezing' the Gana and Gwi Bushmen out of the CKGR for what looks suspiciously like the profit motive – specifically the search for diamonds, in which the reserve is thought to be rich. In 2002, 1,800 Bushmen were forcibly evicted from the reserve and those who resisted were beaten and tortured. Most now live in makeshift camps outside the reserve.

In 2004, the Bushmen began a court battle against their eviction by the government. The government gave two major reasons for its 'encouragement' of people to move out of the protected areas:

> Firstly, because their modern economic activities, be it hunting, arable and/or pastoral farming or some other commercial activity are inconsistent with the primary purpose of the Parks and Reserves, which is to conserve Botswana's unique wildlife heritage for its sustainable and beneficial way to all citizens of Botswana. Secondly, people have been encouraged to move out of game parks to give themselves and their children the benefit of development.
>
> (Personal communication)

In its attempts to remove the Bushmen, the government has used violence, threats, constitutional change, and the arguments for conservation. In September 2005, wildlife guards entered the CKGR and threatened at gunpoint the 200–250 Bushmen who had re-entered the reserve with re-eviction. At the same time, the court hearings of the Bushmen case against the government of Botswana developed an air of surrealism, as Survival International (2005b) explains:

> As special advisor to the President, Pilane [the government's lead attorney] may be hoping that his power surpasses that of the Botswana judiciary. He repeatedly refused to stand up when the judges were speaking to him. He was warned twice that unless he did so he would be cited for contempt. He still refused to stand up, leaving the court little alternative than to commit him to prison . . .
>
> This was the climax of Pilane's examination of the state's star 'expert witness', American ecologist and former employee of the Botswana government, Kathleen Alexander. Pilane was questioning her on material that the judges had already ruled should not be discussed . . . Alexander had already made no secret of her contempt for tribal peoples and hunter-gatherers. She wants them all out of any

protected areas in order to favour the animal inhabitants. Echoing remarks made by the president, she said that they had to evolve, claiming that 'culture' had nothing to do with ancestral land and they could practise it anywhere.

Among her asides was her comparison of inbreeding among animals 'to what happened to the ruling elite in England' . . . Although she claims to want everyone out of all protected areas, she herself lectures to American tourists – inside game reserves – who each pay US$10,000, plus tips, for their trip.

(Survival International 2005b)

Sources: Personal communication – letter to Martin Mowforth from E. S. Mpofu, Permanent Secretary of the Botswana Ministry of Foreign Affairs, 31 March 2004; Ryan Pearson (2004) 'Botswana Bushmen head to US to raise money', Associated Press, 25 August; Survival International (2005a) 'The Bushmen – their fight goes on', background information sheet, April; Survival International (2005b) Briefing paper, 1 September.

Box 6.4 WWF bankrolled rhino mercenaries

The World Wide Fund for Nature (WWF) has admitted that it provided funds to armed anti-poaching units in Namibia set up by a clandestine and proscribed operation run by a team of British mercenaries.

Documents obtained . . . reveal details of three payments that the WWF has previously denied making and which an independent inquiry failed to reveal.

The WWF's admission came a week after Scotland Yard's international and organised crime branch submitted a file on the secret programme, codenamed Operation Lock, to the Director of Public Prosecutions . . .

The WWF was anxious to dissociate itself from Operation Lock when it was uncovered last year. It had been set up in 1987 by Prince Bernhard [of the Netherlands, a former WWF president] . . . and Dr. John Hanks, then the WWF's Africa programme director, to infiltrate the illegal trade in rhinoceros horns and identify individuals and countries involved.

They were concerned that little was being done to stem the decline in numbers of black rhino from 100,000 in 1960 to just 4,000 in 1987.

Dr. Hanks had appointed KAS Enterprises Ltd. to run the operation. KAS's chairman was the late Sir David Stirling, founder of the Special Air Service. Many of its staff were former SAS members. It is thought they had little experience in rhino conservation.

In order to distance itself from the project, the WWF appointed Lord Benson . . . to conduct an independent inquiry. His report to the board of WWF International in Geneva is believed to have concluded that no WWF personnel other than Dr. Hanks knew of the scheme and no WWF money was involved.

However . . . a request for 109,400 Swiss francs (£43,760) was made to the WWF on 13 October 1989 by the Namibia Nature Foundation for 'urgent short-term' funding for anti-poaching units in Etosha and Damaraland. Dr. Hanks said last week that those units were probably set up by Operation Lock and would have been armed for self-defence . . .

Robert SanGeorge, director of communications . . . confirming the information received . . . said a total of Sfr. 157,160 (£62,864) was paid between Nov. 1989 and Feb. 1991. 'It would appear that we innocently and unwittingly funded anti-poaching units that were set up as part of Operation Lock,' said Mr. SanGeorge.

Asked how his investigation failed to identify payments made to anti-poaching units established by Operation Lock . . . details of which were contained in a file labelled 'Anti-Poaching Units' at the WWF headquarters in Geneva, Lord Benson refused to comment.

Source: Boggan and Williams (1991: 6)

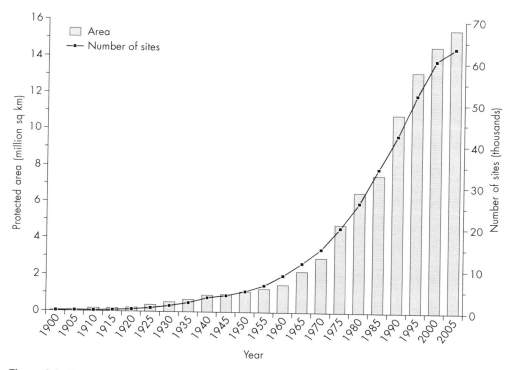

Figure 6.3 The global growth of protected areas

development projects, expanding agricultural frontiers, illegal hunting and logging, fuelwood collection and uncontrolled burning'.

In a manner not dissimilar to the Biodiversity Convention's promotion of debate and disagreement, the creation and protection of national parks, biosphere reserves and other protected areas have often been at the core of conflicts over land in Third World countries. Thrupp (1990), for instance, views tourism development (particularly its emphasis on wilderness protection through the creation of parks and reserves) as largely serving the interests of the privileged upper-middle classes, mainly First World tourists and scientists.

Box 6.5, which recounts the observations of Chris McIvor from a debate about conservation in Zimbabwe attended by both international environmental agencies and local representatives, illustrates a few of the important points concerning such conflicts and

Table 6.2 *Protected area categories*

Category	Type	Description
Ia	Strict Nature Reserve	Protected area managed mainly for science
Ib	Wilderness Area	Protected area managed mainly for wilderness protection
II	National Park	Protected area managed mainly for ecosystem protection and recreation
III	Natural Monument	Protected area managed mainly for conservation of specific natural features
IV	Habitat/Species Management Area	Protected area managed mainly for conservation through management intervention
V	Protected Landscape/ Seascape	Protected area managed mainly for landscape/ seascape conservation and recreation
VI	Managed Resource Protected Area	Protected area managed mainly for the sustainable use of natural ecosystems

Source: World Conservation Monitoring Centre website: www.unep-wcmc.org/wdpa/national.cfm
(accessed May 2007)

underlines the significance of conservation as a discourse and even as a guiding 'ethic' (see Chapter 4). First, there is a clear association of conservation with colonialism: reserved and protected areas were part of the system of securing raw materials (such as timber) for the imperial infrastructure (Colchester 1994). Second, there are sharply opposed views of the meaning and value of nature. As Colchester (1994: 5) observes, national parks and other protected areas have 'imposed elite visions of land use which have resulted in the alienation of common lands to the state'. This has also resulted in the displacement and resettlement of people to make way for park conservation – the case of the Maasai in East Africa stands out here, but other examples, as well as that of the Maasai, are illustrated and discussed in Chapter 8. Third, there is the question of who exactly gains from the construction of parks. It rarely seems to be the local people and, indeed, part of the answer appears to be found in the removal of local rights and a loss or denial of ownership. Instead it is the relatively rich First World consumer with leisure time and wealth enough to be a tourist, or even an ecotourist, who gains from the designation of national parks. The organisation Survival (previously Survival International) also rounds on conservation and its effects on tribal peoples: 'When national parks are set up to protect wildlife, tribal peoples are often the first casualties. They are thrown off their land. Denied the right to graze their herds or hunt for food they sink into poverty and despair' (Survival International, undated: 4). In this way, such approaches directly contradict sustainable livelihood approaches to development that seek to build upon the assets, capabilities and activities of poorer communities (a point to which we return in a discussion of pro-poor tourism in Chapter 11).

Box 6.5 Conservation and imagination: clash on environment

Despite the fact that both sets of contributors were to address the issue of 'conservation' one was left with the impression that they were speaking a very different language.

The Western environmentalists largely concentrated on the destructive impact of human activity. They lamented the disappearance of natural habitat, the loss of areas of wilderness ... Land should be set aside in national parks ... More resources should be spent on policing these areas, on preserving the wilderness that Africa can offer the rest of the world.

One ecologist ... warned Zimbabweans to beware the evils of development. Nature, he claimed, needs to be protected from economic exploitation so that society can enjoy the aesthetic and recreational benefits of an unspoiled countryside.

By contrast, Zimbabwean participants seemed to see no inherent contradiction between conservation and development ...

There was very little talk by local environmentalists of the recreational and aesthetic or the fact that the vast majority of visitors who frequent protected areas come from outside the continent. 'Conservation for us,' claimed one Zimbabwean, 'means the wise management of natural resources for economic use. It does not mean the absence of use at all costs.'

... arguments for a preservationist approach to the environment and the exclusion of human activities from protected areas are unlikely to find much support among African populations. The history of early conservation in Africa is indistinguishable from the history of colonialism and in particular the eviction of indigenous communities from land and resources that they once enjoyed.

In Africa ... where the majority of people still depend for their subsistence on agricultural production ... the land and its wild animals are not a source of aesthetic enjoyment but a resource to be managed so that people can survive.

As one Zimbabwean environmentalist concluded:

It is impossible to talk of wildlife preservation to a farmer whose fields have been raided by elephant and buffalo. Unless there is some tangible economic return to support his family, wildlife is a threat not an asset. When Western environmentalists talk to us about the aesthetic and recreational appeal of our landscape and the need to preserve our wild animals in their natural habitat, we wonder if they would continue to have such ideas if they shared our poverty.

Source: McIvor (1995: 35)

Ecocentrism?

The final stream of environmentalism is ecocentrism.[2] Unlike the other environmental streams, ecocentrism does not place humans above other life forms and instead regards nature and environment of intrinsic value and of equal importance. As Table 6.1 documents, this is a position adopted by the major environmental organisations such as Greenpeace and Earth First who take a holistic approach to nature. The similarities to preservationism in terms of programmes adopted are, however, detectable.

Ecocentric environmentalists are especially concerned to protect threatened species, habitats and ecosystems and 'strongly support the preservation of large tracts of wilderness as the best means of enabling the flourishing of a diverse non-human world' (Eckersley 1992: 46). In fact, many wealthy First World individuals and trusts have been buying up large tracts of land; as the UK *Guardian* newspaper headlines it: 'From Britain to Botswana, the Philippines to Patagonia, there is an explosion of individuals, charities, even billionaire financiers buying up vast areas of land in the name of protecting environments' (Vidal 2008: 6). A decade after the publication of the first edition of this book, the incidence of land purchases by individuals and INGOs (such as the World Land Trust and Conservation International) appears to be accelerating rather than waning in what some critics are referring to as 'a new wave of eco-colonialism' (Vidal 2008: 8). Vidal reports that:

> Tens of thousands of people have been evicted in order to establish wildlife parks and other protected areas throughout the developing world. Many people have been forbidden to hunt, cut trees, quarry stone, introduce new plants or in any way threaten the animals or the ecosystem. The land they have lived on for centuries is suddenly recast as an idyllic wildlife sanctuary, with no regard for the realities of the lives of those who live there.
>
> (Vidal 2008: 8)

Although organisations such as Greenpeace and Earth First have not really entered the debate over tourism, and indeed as Table 6.1 suggests, may wish to ignore tourism as an environmental problem, the relationship to new forms of tourism of the activities undertaken by organisations such as Earthwatch, The International Ecotourism Society, the World Land Trust and Coral Cay Conservation is clearly identifiable. 'These charitable or non-profit groups are allowed to collect money, employ police, build hotels and, in many cases, dictate how land inside the parks should be used, and even whether communities can live or hunt there. It may be good for conservation, but it can lead to hostility' (Vidal 2008: 9).

There are three overarching conclusions that can be made from this review of the different streams of environmentalism. First, when we talk of 'environmentalism', we are talking ostensibly about First World perceptions of environment, ecology and nature, and in particular the views formulated and advanced by organisations based in the First World and disproportionately represented in the new middle classes. As Table 6.1 suggests, these environmental streams have also permeated debates over contemporary tourism and the formulation of supposedly sustainable forms of travel. Second, the idea of the geographical imagination is a useful means of understanding how and why elements such as the 'environment' and 'conservation' are seen in radically different ways by different groups or interests and how certain interpretations are more powerful than others, and have the ability to impose themselves on others. This is especially important in the context of tourism, as the perceptions of local people and communities of their environments can be dramatically different from those of the environmental groups that seek to protect them and the tourists who visit them. Third, in a geographical sense, environmentalism more than any other political theory has drawn in both global and local dimensions. We are asked to think not only of our local environments but to consider the ways in which these environments are part of a global ecosystem (see, for instance, W. Sachs 1992a, 1992b). It is to the consequences of this that we turn next.

Environmentalism and power

In Chapter 3 it was argued that environmentalism and ecology are identified as one of the ways in which power is transmitted, and some critics invoke neocolonialism as a way of representing this and indicating the coexistence of a number of different forms of colonialism. It was suggested that thinking about environmentalism and sustainability, at least in part, as one such form of neocolonialism has produced new ways of formulating the contemporary debates on environmental issues. The ways in which this has been applied to the development of tourism were also suggested.

Box 6.6, concerning the impact of what Daltabuit and Pi-Sunyer (1990: 10) call 'new, "soft path"' tourism (and which they also refer to as a 'more intensified ecological tourism') on the indigenous Maya communities in the Yucatán region of southern Mexico, helps to illustrate these points. The authors assess the fortunes of local people in supplementing the now well-established mass tourism industry based upon Cancún with the emergence of these new forms of tourism. They refer to 'new correlations of power' and the effects of 'hegemonic' metropolitan cultures. Their work is also of importance as they refer to the 'environmental movement as a political force and powerful cultural construct' (Daltabuit and Pi-Sunyer 1990: 10). One case study is of Coba, a village situated close to the remains of a major Classic Maya metropolis in Quintana Roo, Mexico, outside the general orbit of tourism until recently and centre of what Daltabuit and Pi-Sunyer (1990) call the phenomenon of 'archaeotourism'. The story of Coba is a familiar one: the declaration of a national park, the expropriation of local lands, the building of a large and comfortable foreign-owned hotel complex (Villas Arqueológicas) and unilateral solutions imposed upon local people who are excluded from any processes of decision-making.

Box 6.6 Tourism development in Mexico

Daltabuit and Pi-Sunyer (1990) describe communities as

> experiencing intensified pressures to participate in the tourism industry as it undergoes a shift from highly localized resort development to a mode that increasingly stresses the marketing of the physical and human environment, including the archaeological remains of pre-Hispanic Maya civilization.
> (Daltabuit and Pi-Sunyer 1990: 9)

They ask if the penetration of tourism will 'result in improved material and social conditions for the local population'; whether these communities will 'be given an opportunity to play a significant role in determining their own future'; and, if the Maya can 'generate sufficient political power to protect their property rights and minimize the social and cultural costs of tourism'.

> the expansion of tourism in this part of Mexico . . . has been very much the work of national and international agencies and institutions, both public and private . . . Fundamentally, local people and local resources have been approached as elements to be 'managed' by planners engaged in macroeconomic strategies.
> (Daltabuit and Pi-Sunyer 1990: 9)

Daltabuit and Pi-Sunyer's work referred to pre-1990 conditions, but in the early 2000s Mexico's 'National Tourism Programme 2001–2006' confirmed this trend of macroeconomic management by planners.

As Daltabuit and Pi-Sunyer (1990) observe:

> in places like Quintana Roo – if history and experience are any guide – a 'managed' environment is bound to translate into an environment managed by outsiders chiefly to satisfy the needs of outsiders. It is a short step from external direction to what can be termed a process of *appropriation*, by which physical environment, and within it human societies and historical remains, become subtly redefined as global patrimony – universal property.
>
> (Daltabuit and Pi-Sunyer 1990: 10–11)

So it is interesting to note that by 2007, the total area of protected land in Mexico was 22 million hectares (11.2 per cent of the national territory), mostly consisting of biosphere reserves and 'areas for the protection of flora and fauna'.

These issues and questions were described by Daltabuit and Pi-Sunyer (1990) as 'fundamentally matters of power and its representation' in their comparison of such plans and visions with 'the realities of life experienced at the local level in Coba, a Maya village and major archaeological site'. As they remark, 'much that now makes Quintana Roo so appealing as an ecological treasure house and location of well-preserved Maya antiquities is attributable to environmentally benign patterns of subsistence and the . . . inaccessibility.' Yet they describe the new visitors as 'Affluent, educated, and dressed in fashionable expedition gear, the visitors who lodge at the villas clearly represent the up-market end of the business'.

As Mowforth et al. (2008) point out more recently, the Riviera Maya (a product of macroeconomic management by foreign as well as national planners) now covers a substantial swathe of Quintana Roo state, along the coastal zone south of Cancún.

> The Riviera Maya concept can in part be interpreted as a drive to sustain the vitality and market share of tourism in Quintana Roo in a competitive global tourism market by offering new products in the form of higher quality resort environments, accommodation and experiences. Yet in so doing, it has required environmental and social change and intrusion on a scale that seems far from notions of 'sustainability'.
>
> (Mowforth et al. 2008: 128)

Sources: Daltabuit and Pi-Sunyer (1990); Mowforth et al. (2008); see also Comisión Nacional de Áreas Protegidas (2005) 'Áreas Naturales Protegidas' www.conanp.gob.mx/anp; Secretaria de Turismo, Mexico (2001) 'Programa Nacional de Turismo 2001–2006'.

Doubts must also be raised over the nature of environmental concerns and how these are translated into practice. A global concern for the environment and the call to 'think globally, act locally', while lofty and perhaps laudable in practice, have a tendency to become a crusade (to 'think globally, impose locally') that is devoid of notions of social justice and a concern for local peoples' perceptions. As such it echoes top-down approaches to development. Take, for example, an invitation by the World Land Trust, issued in 2001, to purchase an acre of wilderness in Patagonia. On the reverse side of the invitation is an endorsement from Sir David Attenborough and an explanation that 'The WLT, together with its partners, have already helped save 300,000 acres of threatened tropical forest and other wilderness areas over the past decade'. For paying your £25 (approximately US$45–50) for an acre you will receive a personalised certificate and regular news updates. As was pointed out in Chapter 3, this example would appear to display the antithesis of local control.

In this context, Shiva (1993) sees the global debates around environmental concerns, such as those at the Rio and Rio + 10 summits, in a restrictive sense and as more akin to an application of eco-colonialism:

> Unlike the term suggests, the global . . . was not about . . . [the] life of all people, including the poor of the Third World, or the life of the planet . . . The 'global' of today reflects a modern version of the global reach of a handful of British merchant adventurers who, as the East Company, later, the British Empire raided and looted large areas of the world.
>
> (Shiva 1993: 150)

Drawing conclusions from their work on tourism developments in Mexico (Box 6.6), Daltabuit and Pi-Sunyer (1990) describe the contemporary version of this global reach:

> Environmental concerns clearly strike a deep chord in the educated Western public. Environmentalism often utilizes a powerful rhetoric that is virtually identical to that of nationalism. With it, however, comes an element of privileged discourse that harks back to the formulations that gave such authority to development theory, especially as it was applied to the Third World. In both cases, ideology legitimises interventionist policies. Problems are defined and solutions formulated not within the societies in question but by outside experts who are accorded extensive power and prestige.
>
> (Daltabuit and Pi-Sunyer 1990: 10)

Colchester (1994) reaches a similar conclusion, arguing that the fashion for linking the 'global' and the 'environment' has encouraged international and state interventions – and this quite clearly carries connotations of the exercise of international power: environmental globalisation.

Environmentalism and discourse

One way of representing the power reflected in environmentalism is through the notion of discourse (already discussed in Chapters 2 and 3). William Adams (2001: 3) argues that in adopting the terminology, or discourse, of sustainability, environmentalists have 'attempted to capture some of the vision and rhetoric of development debates'. But decontextualised from power and a more critical understanding offered through a political economy perspective, he argues that such usage has taken on a disturbing naivety and that sustainable development (as defined by Brundtland), although superficially attractive, remains a 'better slogan than it is a basis for theory' (Adams 2001: 5).

It is an emerging aspect of this discourse that the activity of tourism can be conveniently linked with terms such as 'environment', 'environmental management', 'biodiversity management' and 'ecosystem protection', to imply a notional sustainability of development. IUCN and WWF reports and pamphlets talk regularly of the protection of areas for and by the development of nature tourism. For example, the IUCN Central America Office's publication *Recursos* refers to the objectives for the IUCN's work in the Talamanca region of Costa Rica thus: 'The community will use the land to pursue and create conservation activities, scientific tourism, research and demonstration plots, with the aim of stimulating new development projects in the area' (IUCN 1989: 58).

And the WWF–UK's annual review for 1991 refers to 'great potential for tourism' exhibited by the Gashaka-Gumpti reserve in Nigeria which after years of decline still has bright prospects, especially for tourism. 'The Nigerian government has now declared Gashaka-Gumpti as a National Park and WWF has asked to assist in this work' (WWF–UK 1991: 15).

An added problem has been the intellectual weight afforded to the discourse of First World environmentalists. In Chapter 3 we referred to Sachs' arguments that the concept of the ecosystem as discourse has provided scientific credibility to the environmental debate and movement. In a similar vein, Yearley (1995) argues that such problems are posed by the manner in which expert knowledge and scientific diagnosis and reasoning are employed in environmental management, a problem that can be applied to the issue of biodiversity.

> Given the centrality of science, it is understandable that the discourse of science will affect the way that environmental problems are thought of . . . Unless there are powerful reasons to the contrary, scientists assume that natural processes are consistent throughout the natural world . . . the point is that science aspires to universally valid truths . . . a universalizing discourse . . . This orientation has left its stamp on the overall international discourse of environmental management.
>
> (Yearley 1995: 226)

It is interesting to reflect on Yearley's use of 'discourse' here. Clearly, he is signalling that 'science' and its use in addressing 'environmental problems' and 'environmental management' is a further way in which power is expressed. As he goes on to argue, doubts have been raised about the way in which the First World has analysed and interpreted 'global environmental questions'. A similar line of reasoning is equally applicable to the debates over both the environmental impact of tourism and tourism's role in addressing environmental degradation.

Cloning

There is a final route through which a First World conditioning of Third World environmental issues manifests itself – especially along the development and environmentalist channel, one which could be labelled 'cloning'. Reservations have been raised about the manner in which environmental INGOs reproduce themselves at the local level. It is, in one sense, a well-crafted technique of propagating self-like organisations and claiming that the concerns of some organisations are also reflected, as if by magic, in the Third World. The Third World is teeming with the 'sons and daughters' of development and environmental INGOs, many of which are created in a self-like image. In 1992 Jon Tinker, then President of the Panos Institute (UK), described this as the 'parrot syndrome'.

> A genuine global shift towards sustainable development in 1992? One barrier is the parrot syndrome: Northern environmentalists recruiting Southern NGOs to endorse Northern analyses and policies. Whales, the fur trade, global warming, Antarctica: the fact-studded, rhetoric-laden, morally impregnable and full-colour-lithographed NGO case often seems irresistible. Southern NGOs are too easily seduced into supporting it uncritically. They mirror Northern thoughts back to the North, and we call it a global network. But back home, the Southern NGOs begin to talk with strong Northern accents and rapidly become discredited. You can often detect the parrot syndrome when Latin American wildlife NGOs talk

about rainforests. Some of them have lost their domestic credibility, because they now sound like North American clones. The remedy? More self-restraint among environmentalists in the UK and elsewhere in the North: less eco-evangelism, less certainty that we are right, more listening. And greater efforts to help Southern NGOs prepare and disseminate their own analysis of development issues.

(Tinker 1992)

As Tinker (1992) implies, part of the answer must lie in acknowledging that First World agencies, organisations and institutions do not have all the right solutions, and it is necessary to move beyond the somewhat patronising viewpoint that we can 'help others from developing countries to help themselves' as Coral Cay Conservation put it (Expeditions brochure 1996). This cloning has its reflection in the field of tourism where there is a strong sense that the First World has the blueprints for successful environment-friendly tourism.

We return to the local dimensions of new tourism in Chapter 8; suffice it to note at this point the way in which the notion of a benevolent 'local community' has been so universally subsumed by First World organisations and commentators. So much so that it could even be argued that this is a further reflection and extension of commodification; Third World local communities have emerged as commodities utilised by the myriad INGOs that have emerged in the late twentieth century, to secure grants and aid, and on their behalf widespread fundraising is undertaken. As Korten (1996) concludes, while NGOs can clearly have an important role in building local economies and in advocacy for policies that strengthen local control, not all NGOs are created equal.

Conclusion

Environmentalism and environmental issues are central to the debates over contemporary tourism and the emergence of new, alternative types of activity that seek to escape the environmental harmfulness associated with mass tourism. Rather than cataloguing types of environmental degradation and the responses to it, this chapter has tried to demonstrate the scope of environmentalism and how this is reflected in the debates and types of action related to tourism and notions of sustainability and sustainable development.

Our discussion started from a recognition of broader contemporary change, and charted the linkages between environmentalism, the new middle classes and new forms of politics, especially as expressed through the socio-environmental movement. This is an important starting-point for comprehending the nature and content of the debate and the advocacy of new forms of tourism.

Although not strictly linear and exclusive, the environmental types identified by Eckersley (1992) help capture the continuum of environmentalism and environmental politics. Above all, this is a means of demonstrating that environmentalism itself is a highly contested arena, with socio-environmental organisations differing markedly over the most appropriate interpretations, policies and programmes, and especially how these relate to 'sustainable tourism'.

Central to the discussion is the manner in which these debates are transmitted through INGOs. This is important, for as later chapters discuss, while it is the most powerful countries together with the World Bank, IMF, other supranational agencies and the transnational tourism industry, that represent the centres of power and are often presented as the protagonists in discussions of Third World tourism, much less critical attention has been applied to the activities of the multinational INGOs and smaller NGOs. These organisations, often mobilised under the cloak of benevolence and respectability around issues of environment, have tended to escape criticism. However, the formulation of environmental politics in the First World and the drawing of other, Third World, places into a global environmental order are significant elements in the way in which power is reflected through new forms of tourism.

7 The industry: lies, damned lies and sustainability

Of all the players in the activity of tourism, those in the tourism industry itself have traditionally faced more blame than all others. They have been the target for some much-deserved criticism; and they have also been the easy scapegoat for many negative impacts of tourism developments.

This chapter considers the tourism industry's adaptations to the new forms of tourism and the different means by which it has made and is making these adaptations. In order to set the scene we first attempt to place the tourism industry in the context of the system of free trade that currently prevails in capitalist economies. This discussion is extended to include a consideration of the notion of fair trade in tourism and we ask if tourism as currently practised is fair. The scene-setting also necessarily includes a consideration of the size and structure of the industry which helps to explain some of the adaptations to the notion of sustainability that both the conventional mass and the new forms of the industry are making.

We recognise that in themselves the two descriptive labels, 'mass' and 'new', are problematic and that there is considerable blurring of the divide between them. There is great uncertainty about the relevant definitions, as Chapter 4 illustrated – what is considered as a 'new' form of tourism and as 'sustainable' varies from person to person, value to value, and within the industry. This blurring of the distinction between the two types is increasing as two processes unfold: first, the large companies offering conventional mass tours are increasingly diversifying their activities to offer exclusive tours with a difference; and second, as the growth in new and alternative tours continues, a number of specialist companies have started to emerge as dominant within this sector and, we would argue, to take on certain characteristics more commonly associated with the large-scale operators. And so we ask how sustainable are the operations of new forms of tourism.

Notwithstanding these difficulties of definition, the chapter develops the critique of new tourism outlined earlier in the book as it applies to the industry. The section 'Redefining sustainability' examines some of the ways in which both forms of tourism manipulate the notion of sustainability to their own ends, and the last section of the chapter briefly sketches some of the new personnel involved in new forms of the industry and some recently emerged features of the industry.

Trade and tourism

It is important to emphasise that the ideology of free trade as the most efficient means of operating and managing economies is synonymous with economic globalisation, and as we have argued, the spread of capitalist relations of production is pervasive and is witnessed in the rapid growth of tourism. Free trade also implies the need for global harmonisation between trading entities (sovereign nation states). More notably, free trade, economic growth and development are regarded as axiomatic in the hegemonic discourse of development; that is, the discourse that pervades the supranational and national agencies: the World Bank, IMF, United Nations, Organisation for Economic Cooperation and Development (OECD), EU and G8 governments.

Development, as Rist (1997) argues in his history of development, has a rich lineage. Discourses on development can be traced back to the so-called founding father of economics, Adam Smith, and the pervasive influence of *An Inquiry into the Nature and Causes of the Wealth of Nations* (first published in 1776). As Rist (1997) argues, the 'nature' of change and development is significant here in drawing parallels to natural progression and growth, the triumph of social evolutionism in which the superiority of the West was guaranteed and a general creed of development emerged whereby:

> there was general agreement on the three essential points: that progress has the same substance (or nature) as history; that all nations travel the same road; and that all do not advance at the same speed as Western society, which therefore has an indisputable 'lead' because of the greater size of its production, the dominant role that reason plays within it, and the scale of its scientific and technological discoveries.
>
> (Rist 1997: 40)

Herein lie the roots of a 'totalizing discourse' (Rist 1997: 39) of development as a single linear progression of economic growth and well-being, a model that, as discussed in Chapter 2, has been adopted by a number of theorists as a series of steps or stages (such as Rostow's (1960) *The Stages of Economic Growth*) to the 'promised land' – a developed state. As Chapter 4 demonstrated, similar conceptual models have been advanced in tourism studies. It is the universal acceptance that there can be no 'development' without economic growth, an equation that commands near universal respect and that needs to be questioned (Rist 1997).

And so it is open and deregulated markets – free trade – that appear to provide the cornerstone for 'development'. World Bank Senior Adviser William Easterly (2001) underscores the significance of free trade both to the World Bank and his analysis of development economics:

> Free trade allows economies to specialise in what they are best at doing, exporting those things and importing the things that they are not so good at producing. Interference with trade distorts prices so that inefficient producers will get subsidised. This distortion could affect growth because inefficient resource use lowers the rate of return to investing in the future.
>
> (Easterly 2001: 230)

Yet the myth of free trade itself manages to hide the uneven and unequal economic interdependencies manifest, for example, in trade protectionist measures adopted by First World governments. In this regard the hegemony of the USA is noteworthy, though equally the power vested in the G8 and key trading blocs (especially the NAFTA and the EU) should not be underestimated. Arthur MacEwan (2001) challenges these orthodoxies, arguing that neoliberalism is

> based on the ideology of free markets, and the practitioners of neoliberalism argue that free markets are the best means by which to support the interests of US businesses and, through US business, the interests of the US economy and US citizens. Yet guided by the fundamental principle that their goal is to promote

US business interests, US policy makers must continually violate free market principles.

(MacEwan 2001: 28)

MacEwan (2001: 30) argues that while neoliberalism fosters economic globalisation and the emergence of global markets, 'power always involves rivalries and those rivalries lead to inter- and intra-regional conflicts'.

Vandana Shiva (1999) is equally critical of the market liberalisation agenda pursued by the centralised and undemocratic World Trade Organisation (WTO, as opposed to the UNWTO) whose rules she argues 'are driven by the objectives of establishing corporate control over every dimension of our lives' and which need to give way to the 'rights of all species and the rights of all people . . . before the rights of corporations to make limitless profits through limitless destruction' (Shiva 1999: 8).

Fair and 'fairer' trade

The protests at the Seattle WTO Ministerial Conference in 1999, the protests at subsequent meetings in Prague (2000), at the 2001 G8 Summit in Genoa, in Cancún (2003) and in many other places and times less well publicised, have brought to a head the fundamental significance of trade, debt and development and the impact of economic globalisation; issues where emotions run deep and opinion is very sharply divided. The significance of trade and debt to global development is certainly not new, and the staggering accumulation of Third World debt, and the central role of the industrialised nations in fuelling this crisis, from the 1960s is well documented (George 1988; Payer 1991; see also Chapter 10). What is noteworthy is the way in which the impacts, opportunities and responses to globalisation have been steadily interpreted as 'economic' in nature.

There is little surprise that the myriad social movements protesting and advocating against globalisation have honed in on the impacts of trade liberalisation, unequal trading dependencies and the relationship between debt and poverty. Indeed, some suggest that the mounting campaign for fairer global trading rules is set to become the next global campaign mounted by a coalition of INGOs and assorted civil society interests (Denny 2002). Equally there are moves to consider fairer, more ethical and more sustainable forms of trade and corporate governance (Starkey and Welford 2001) and alternatives to globalisation through the growth of the localisation movement (Hines 2000).

In the 1990s the business sector slowly awoke to the sustainable development agenda and to the role of business in sustainability. In recent years there has been a steady growth in 'corporate social responsibility' (CSR) and 'triple bottom line reporting' within which the impact of business in the economic, environmental and social sectors are accounted for and audited (Elkington 2001), and the different forms of capital (economic, natural and social) have been defined. The degree of progress is questionable, however, and the fundamental inequalities in the structure of the global trading system remain intact. The questions remain to what degree is CSR an adequate response to global poverty and inequality, and to what extent is it a largely postmodern philanthropy centred upon the power of consumers and an incremental and partial voluntarist response in an inadequately regulated global environment (Tourism Concern 2002b). Or perhaps it is worse even than this and is, as the *New Internationalist* suggests in its December 2007 issue, nothing more than a confidence trick (New Internationalist 2007).

There is also scepticism from the business community, economists and observers. Former chief economist at the OECD, David Henderson, asserts that 'widespread adoption of CSR would undermine the foundation of the market economy' (Denny 2001: 27).

Commentators such as Klein (2001) contend that a large degree of the response has been to head off or respond to corporate 'public relations' disasters or alternatively to retain market share and customer loyalty in the context of heightened consumer awareness. Of course, when applied to the tourism sector where profit margins tend to be exceedingly tight, there are also issues of how far CSR can reasonably extend beyond relatively glib codes of conduct and tourist education exercises.

Codes of conduct, CSR and other techniques of regulating the industry are examined later in this chapter in the context of the debate around imposed regulation against self-regulation. As governments flirt and rut with globalisation they (and especially the powerful G8) have been negligent in accepting that 'business is in the business of business', at any cost so it seems, and have tacitly endorsed the worst excesses of economic globalisation on the premise that they are powerless to do anything. And worryingly protestors against inaction, unethical and unequal trade policy have been dismissed as naive fractional anarchists. The absence of ethical international leadership has been astounding. To many critics of the prevailing structures of power, one thing would appear clear: left to its own devices the western industrial and financial edifice is incapable of genuine self-regulation and the social and environmental consequences are ominous.

Box 7.1 illustrates the continuum of different approaches to trade; the table's presentation of material should not suggest a duality or polar opposites. Both approaches share a common objective of seeking to improve the incomes and working conditions of Third World producers and workers. However, as Traidcraft, one of the UK's leading fair trade organisations, argues, 'they differ in the groups of producers and workers they target, and in the methods used to achieve their objectives, as well as the underlying objectives of the organisations involved in them' (Traidcraft 2001b). Whereas ethical trade seeks the 'minimum' adherence to legally enforceable international labour standards, fair trade purports to pursue the deeper objective of working directly with marginalised Third World producer communities to eliminate poverty and assist 'development'. More specifically, fair trade approaches go beyond a number of common principles which include fixing a fair price, advancing payments and facilitating credit where necessary, establishing long-term relationships between producers and fair trade partners and customers and forming direct trading relationships within fair trade partnerships.

Fair trade and tourism

There is not a single definitive description of what fair trade is. The international movement to secure a fairer deal for Third World producers is most readily associated with the trade of bananas, coffee, chocolate and 'crafts', and the activities of niche trading groups such as Traidcraft (UK). But there have been recent attempts to extend and apply fair trade 'principles' to holidays.

Fair trade has worked most effectively to date with simple and tangible commodities. Because wages and the share of total profit are usually lowest at the source of production (the small growers, pickers and packers), fair trade seeks to achieve a realistic wage per production cost for the most economically marginalised and vulnerable in the production and logistic chain. It is a challenge to traditional economic theory and practice in that it seeks to set a price based on principles other than those of pure profit maximisation and on practices other than seeking the lowest cost of production in so-called efficient markets.

There are a number of initiatives that have some resonance of 'fairly traded tourism' most particularly through the 'awareness holidays' run directly or promoted by a growing number of non-governmental development organisations such as Oxfam and ActionAid. And there are elements of tourism-related activities, such as the production and sale of

Box 7.1 Fair, fairer, just and ethical tourism

Fair(er) trade	Ethical trade
Bottom-up people-focused development	Top-down development
Producer/community-driven – marginalised producers in the Third World	Consumer-driven – what the market will bear
Forward linkages (South focused) from community-based enterprises and small and medium enterprises – organisations with explicit aim of providing benefits to their workers	Backward linkages (North focused) from small and medium enterprises, multinational and transnational corporations
Social sustainability bottom line = fair prices and poverty elimination or reduction	Sustainanble development triple bottom-line accounting with retained markets and shareholder equity. Economic primacy dictates
Direct trading relationships	Mediated trading relationships
Moral responsibility to 'pro-poor' trading conditions and rules	Corporate social responsibility and market legitimation – codes of conduct
Price determines 'fair deal'; supported by access to credit and other forms of assistance	Enforcement of international standards: supplier codes; International Labour Organisation standards to monitor and verify
Fair trade and self-reliant trade	Managed trade
Fundamental economic change with new trade rules, global social justice and poverty elimination as key objectives	Manage democratic capitalist economies through corporate governance, and state and supranational agency regulatory frameworks and standards

Sources: Dunkley (1997); Ethical Trading Initiative (2001); Hines (2000); Traidcraft (2001a, 2001b)

local handicrafts, that provide the opportunity for elements of fair trade (Ashley et al. 2001). However, in one of the few attempts to provide a consolidated understanding of fair trade in tourism, Cleverdon and Kalisch (2000: 175) observe: 'Little or nothing is known about fair trade in services, let alone fair trade in the hospitality sector'. Involving primary research and including a dialogue with tourism providers from Third World countries, their work provides one of the few focused insights.

Cleverdon and Kalisch (2000) document a number of key differences between fair trade in services and primary commodities as a means of identifying the potential and shape of fair trade in tourism. Perhaps the most significant difference is that fair trading organisations are non-profit making, and those small independent tour operators who are practising some quasi-fair trade elements (such as direct trading relationships) also need to ensure commercial viability. Price is a significant element here, and the authors argue that the tourism industry depends on low prices that are flexible, and whereas, for instance, the price of products like coffee can be gauged in relation to a fixed world price, the same would be difficult if not impossible in the tourism industry. This begs the questions as to whether price should be the main determinant of fair trade tourism and whether there are other attributes such as the degree of local control or the distribution of benefits that may be more significant. There are also issues concerning the lack of collective organisation among Third World tourism producer communities, contrasted to the degree of collective organisation exhibited by primary commodity producers.

Other differences are perhaps less significant. Cleverdon and Kalisch (2000) argue that tourism is fundamentally different from traditional fairly traded commodities in that the 'product' – tourism – is intangible and invisible and is a multisectoral service activity incorporating diverse activities and sectors such as transport and agriculture. Of course, while the degree of complexity in tourism compared with fair trade commodities is beyond doubt, the production and distribution of commodities such as bananas also use a diverse range of services from the non-agricultural sector, not least the logistic chains necessary to get products from producer to consumer (Cook 1993). Although such products may be 'simple', the multiple service chains involved are not.

Cleverdon and Kalisch (2000) identify six variables as 'prerequisites for fair trade in tourism'. These are certainly not unique to the tourism sector, and must be understood within the context of a broader critique of development. As such they are reflective of the causes of unevenness and inequality that condition the activity of tourism. The prerequisites are 'access to capital, ownership of resources, distribution of benefits and control over representation of the destination in tourist-generating countries, and it needs to ensure transparency of tourism operations, including price and working conditions' (Cleverdon and Kalisch 2000: 178). In short, this brew of fair trade in attempting to 'provide a better deal for producers or, in the case of tourism, service providers in the South' (2000: 178) has reflections in new forms of tourism and is not dissimilar to pro-poor tourism initiatives (Ashley et al. 2001), which we address in Chapter 11.

In the context of economic globalisation and growing, not diminishing, inequalities, commentators have been quick to acknowledge the weaknesses and shortcomings of the fair trade movement. Small changes can undeniably make a significant positive impact to the lives of those involved. But the degree of change is likely to be minuscule in its impact on trading patterns and trends, at least for the foreseeable future. For example, while pro-poor tourism advocates claim their experiment is in its infancy, in terms of national poverty reduction pro-poor tourism has had little impact (Ashley et al. 2001: 41), and clearly tourism cannot be regarded as a panacea for rural development. (There is little or no literature on pro-poor urban tourism though it has begun to emerge in European cities involving immigrant communities.) Fair trade products also demonstrably involve niche markets. As Cleverdon and Kalisch (2000) observe, fair trade products constitute around 3 per cent of the commodity market, and although the number of fair trade products available rose from around 150 in 2003 to more than 2,500 in 2007 (Vidal 2007), there are no reasons to believe that fair trade tourism would comprise anything but a relatively small percentage of the total global tourism industry; though the percentage contribution in individual countries can vary considerably, a point to which we return in the case study of Costa Rica (Chapter 10).

This once again raises questions among commentators as to the relative merits of fair trade (small-scale exemplary holidays, for example) as opposed to corporate social responsibility and ethical trade in tourism (involving corporate codes of conduct, regulation of the industry and compliance with minimum standards), or ultimately to localisation and a reduction in western consumption (essentially travelling less). Although Hutnyk (1996) appears to sympathise with the good intentions of fair and alternative trade advocates, he concludes from his ethnographic study of alternative tourism in Kolkata (India):

Alternative travel, just as much as the alternative trade promoted by many organized aid groups, works as a reassuring front for continued extension of the logistics of the commodity system, even as it masquerades as a (liberal) project of cultural concern.

(Hutnyk 1996: 215)

From an ethical perspective there are question marks against the promotion of elite 'ethical' niche markets, consumption patterns and trading systems that may reinforce, rather than challenge, global inequality.

Size and structure of the industry

The effect of scale of operation: is size important?

An outline of the size and structure of the industry is important because, as well as being the world's largest single industry, it is also highly fragmented and diverse. It is composed of several different branches: tour operators; travel agents; accommodation providers; carriers; tourism associations (both NGOs and market-oriented associations); destination organisations (such as tourism chambers of commerce); consultancies; and formal and informal service providers. Each of these can be further subdivided. Figure 7.1 attempts to capture the significance of this scale and diversity.

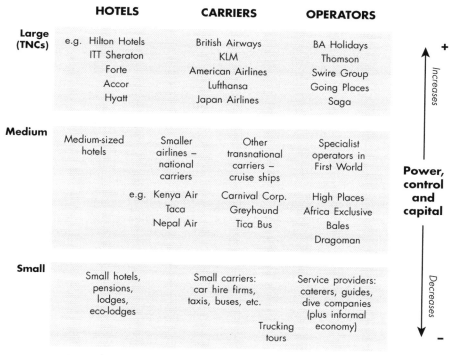

Note: The companies named are merely examples in the category: the list is not intended to be representative, still less all-inclusive

Figure 7.1 Elements of the tourism industry

This fragmentation is one of the factors which has led to minimal environmental regulation, while at the same time ensuring that plenty of examples of environmental 'good practice' can be found. These can be and are employed as marketing tools for the public image of the industry.

Given its size and structure, it is in fact tempting to ask, as Tourism Concern (1992) does,

> what exactly is 'the tourism industry'? Is there any such monolithic thing as the title implies? In fact, what we call tourism really embraces a vast and diverse range of activities, from large-scale mass or package tours to small-scale, individually-tailored holidays; from internal domestic visits to family or friends, to international or intercontinental journeys, to business trips and 'sun sand and sea' recreational breaks; from activity, sports, nature, health, 'green' or alternative holidays, to culture or adventure.
>
> (Tourism Concern 1992: 1)

We accept the fact of this diversity and range of activities within the industry and believe that the power held by different branches of the industry varies with size and scale. At its simplest level this means that transnational companies wield great power and influence over Third World governments while small-scale hoteliers have little clout, although at the local level this is not always true; transnationals and international industry organisations are able to influence the practices of the industry and the legislation of national governments, while local catering establishments cannot bring about changes in the industry even though they may adapt their own practices in response to external pressures; the northern tour operator has ready access to the clientele, while the Third World tour operator expends much energy attempting to capture the passing market of First World tourists. As stated here, this is greatly simplified, and of course there are small-scale companies and establishments which are highly significant within their local communities. But in general the scale of operation reflects the level of power and the ability to influence other relevant organisations and sectors in tourism such as government.

These differing levels of power are reflected in different ways of adapting to change. Companies vary considerably in the way they absorb the practices of new tourism or redefine sustainable practices so that they merge into existing practices with relative ease.

It may be argued that the greater the size and power of the company the stronger the profit maximisation motive – although this latter term can cover a number of business management strategies – and the more likely it is that the industry will use consultation, codes of conduct, internal management restructuring and other methods as public relations exercises to 'green' their image and as alternatives to introducing significant changes in practice. Such real changes would reduce the impact of their development and move towards sustainability, but would also run the danger of reducing profits. The imperative of sustaining profits has already been discussed in Chapter 2. Conversely, the weaker the profit motive, the more willing the operators of the company will be to change practices. This can be argued as a general rule, but it is clearly one that has many exceptions.

It is important to add that ethical issues are not necessarily uppermost in the minds of those who run small-scale, specialist tourism businesses. Any consideration of the effects of size, however, must obviously take in the power of the transnational corporations involved in tourism and their ability to determine or at least influence the nature and direction of developments in the industry.

Transnational companies (TNCs)

Lea (1988: 12) pointed out that 'the three main branches of the industry – hotels, airlines and tour companies – have become increasingly transnational in their operations in the 1970s and 1980s, to the point where these large enterprises dominate all others'. Hong (1985), Lea (1988) and Madeley (1996) have all claimed that the operations of TNCs dominate and control the development of tourism, and Kalisch (2001: 2) claims that 'Eighty per cent of the mass tourism market is dominated by TNCs'.

TNCs achieve this domination in a number of ways. For the hotel industry in Third World countries, for instance, Hong (1985) identifies five forms of TNC participation (see Box 7.2). Essentially, these forms of participation allow the TNC to exercise overall or substantial control of tourism activities through some form of contractual relationship while investing an agreed amount of capital in the development. This control of all or many aspects and activities of the industry is referred to as vertical integration within the industry. As Britton (1991) outlines, for instance, airlines illustrate this process with their

> resources to buy into hotels, tour operators and other transport operators to reap the benefits of vertical integration: and by forging alliances, set up computerised reservation systems which create captured international integrated (interindustry, for example, airlines, rental car, and accommodation) networks.
>
> (Britton 1991: 457)

When this relationship is extended across all three sectors of the industry, encompassing the carriers (most frequently airlines in the case of tourism to Third World countries), hotels and tour operators, it becomes evident that most of the finance paid into the industry by the tourist is controlled and retained by the TNC. A tourist travelling on holiday from Europe to a Third World country, for instance, spends the bulk of the cost of the holiday on the flight and accommodation. In general, the amount spent on consumables, and even durables, purchased locally will be small in comparison with the costs of flight and accommodation.

In many cases, the hotels, airlines and tour operators used are owned and operated by TNCs whose headquarters are in the industrialised world. The way in which these linkages affect tourism to Third World destinations is illustrated by Paul Gonsalves (1995):

> it is possible for a tourist to leave Japan on JAL [the Japanese airline], be transferred from a Third World airport by a Honda car to a Japanese-owned hotel, to be accompanied throughout his or her tour by a Japanese guide (from Japan, not just a local who speaks Japanese), eat at Japanese-owned restaurants, shop at a Japanese supermarket, and return by JAL to Tokyo, in order to tell his or her friends what a wonderful place the Third World is! . . . Given the complexity of the global economy, it is likely that the above story is a simplification of reality: for example, the car could have been a Mercedes, the hotel a Club Med, and the restaurant a McDonald's. What is inescapable, however, is that *ownership, control and therefore benefits, from Third World tourism, accrue mainly to the rich industrialised nations from where the tourists originate.*
>
> (Gonsalves 1995: 35–6, emphasis in original)

Gonsalves' examples are generalised, but they highlight the First World ownership and control of tourism activities. Clearly, then, only a relatively small proportion of the money spent by the tourist to a Third World country is spent in that country. This feature of the

---

Box 7.2 Five major forms of TNC participation in the Third World hotel industry

There are five forms of participation by TNCs in hotels in developing countries:

- *Ownership or equity investment*: Where the TNC owns a part or the whole of the share equity of the hotel. This is the most direct form of TNC control.
- *Management contracts*: 74% of TNC involvement in the Third World is through management contracts. According to the UN Centre on Transnational Corporations (UNCTC) [now defunct] report on *Transnational Corporations in International Tourism*, 'The management contract is the most far-reaching form of involvement of TNCs in developing countries'.
- *Hotel-leasing agreements*: In these agreements, the TNC pays the owner a proportion of the hotel's profits after deducting the expenses incurred in the operation of the hotel.
- *Franchise agreements*: These agreements allow the owner of a hotel to use the name, trademarks and services of the transnational hotel chain in return for a fee. The hotel is promoted as a member of the hotel chain and has access to the chain's worldwide sales, communications and reservations systems.
- *Technical service agreements*: These may relate to particular aspects of the establishment and management of a hotel. This may include market surveys, feasibility studies to determine the size, type and facilities required by the hotel, . . . personnel and staff training and financial or operational management planning and control systems and security.

This interlocking of interests and ownership within the various components of the tourist industry results in the domination and control of the industry by TNCs . . . Thus, 'the real beneficiaries are the rich industrialised tourist generating countries which control the entire industry – hotel cartels, airlines, tour operators and agencies' (UNCTC).

Source: Hong (1985: 18–21)

---

industry is known as 'leakage', which refers not only to the purchase of imported goods and services by tourists but also to the imports of goods and services by hotels and other tourism establishments or organisations and to the repatriation of profits by foreign owners of hotels and other services. Leakage of course is not a feature of the tourism industry exclusive to tourism to Third World countries.

The level of leakage is significant because it reflects the economic power held by TNCs in comparison with local communities and local governments. But estimates of the precise extent of leakage vary according to source, location and situation, and it should be noted that the calculation of leakage can be very complex, subject to problems of data collection and availability and based on arguable assumptions and definitions. Kalisch (2001: 2) reports on a leakage of 'approximately 69 per cent of the total expenditure of a mountaineering expedition' in Nepal; Madeley (1996: 18) reports on a leakage of 77 per cent for charter operations to Gambia; Pattullo (2005: 51–2) reports on a high level of leakages 'averaging from 50 to 70 per cent' in most of the Caribbean but on a figure of only 37 per cent for Jamaica, reflecting its more diversified economy and greater ability to block leakages; Becken (2004) reports that 'about 60 per cent of tourism foreign exchange leak out of Fiji'. In general, most estimates suggest that greater than 50 per cent of all tourist money paid either never reaches or leaks out of the Third World destination country. But the charge that leakage is a major problem in tourism is contested by some critics. Mitchell and Ashley (2007) have argued that leakage is not necessarily or always a significant factor and that it is not specific to the tourism industry. For example, the bulk of the profit from the mining industry is repatriated to parent companies in First World countries.

Given the hype around the development benefits of the tourism industry to Third World countries, ways of hiding this negative feature therefore have to be found, and some of these are illustrated below (see 'Redefining sustainability', pp. 191–214). Suffice it to say that the leakage has been well documented since the 1980s and is an accepted feature of tourism, especially of the operations of TNCs in Third World countries.

In some cases, the First World tourist who flies to a Third World destination (and more than 80 per cent of international tourists travel by air) will stay in a hotel owned by a TNC based in the First World, possibly even the same one which flew them out there in the first place. The quote above from Gonsalves (1995) illustrates this for the case of mass tourism.

Such tourists may not necessarily form part of the growing mass of new or alternative tourists who seek adventure, nature, wildlife, culture and authenticity. Increasingly, however, they are enticed on trips away from the safety of First World standards in the hotel enclave by the prospects of exotic flora and fauna in nearby national parks or by a riot of colour and culture in a local market. This is part of the process of diversification by the large tour operators. Increasingly such companies are using local contacts, organisations and companies or their own couriers or staff to offer variations on the standard, conventional theme of hotel accommodation, hotel food, hotel bar, and hotel swimming pool. Through this process, even the most unashamedly hedonistic tourists (who want nothing more than the swimming pool and the bar) can find themselves lured, even if only for a short time, into the category of new tourists. In this small way, even the conventional mass forms of tourism are being affected by the new, alternative and supposedly sustainable forms.

An important aspect of tourism TNCs and their attempts to address the issue of sustainability is their lack of accountability. Carothers (1993) quotes a former staff member of the United Nations Centre on Transnational Corporations (UNCTC, now defunct) who explains that the issues at stake were

> that multinationals fear even the semblance of public scrutiny. They shun any serious discussion of critical issues: their global market dominance, price fixing practices in small countries, wage cuts and job losses in the Third World, huge commercial debt repayments, and other 'negative' matters. A peak of absurdity was reached at the final preparations for the Earth Summit when there was heavy lobbying to remove the term 'transnational corporations' from the draft text of Agenda 21 . . . Third World debt, environmental destruction and the need for millions of new jobs cannot be addressed without understanding how multi-nationals operate. Governments and citizens need to know what transnationals are doing, within a legal framework that holds them accountable.
>
> (Carothers 1993: 15)

With this unequal distribution of power in mind, the UNCTC aimed to increase the TNCs' contribution to development, to provide technical assistance to Third World countries on foreign investment issues and to produce an international code of conduct for TNCs. The world of industry, commerce and finance, however, assisted by organisations such as the ultra-conservative Heritage Foundation and by US President Ronald Reagan, continually attacked the UNCTC. Despite the fact that the centre had achieved very little beyond establishing a few non-binding principles for foreign investment, the United Nations relegated the importance of its work in 1992 and finally disbanded it in 1993.

In July 1999, another example of corporate power against local control was exposed in Haines, Alaska, when Royal Caribbean Cruise Lines, one of the largest cruise lines, was convicted on many counts of bypassing purification systems, dumping oily bilge, illegal oil dumping, waste water discharge, forging documents and systematic obstruction of justice on numerous occasions in the Caribbean and Alaska. Much of the vocal opposition to this type of activity came from the town of Haines, which had earlier built a new pier to accommodate cruise ships. In December 1999 after the judgment, Haines was dropped from all the itineraries run by Royal Caribbean. Scott (2001) reports that many Haines citizens saw this as a blatant attempt at punishment and quotes one person as saying: 'Let all the other ports know that you can't just raise head taxes and tour taxes without paying a price.' Scott (2001) comments that 'the sudden unilateral removal of Haines from the cruise economy displays the power of multinational corporations to dictate the terms of business'.

With the disbanding of the UNCTC and with the recent push by the tourism industry for self-regulation (see 'Redefining sustainability', pp. 191–214), we now appear to be further than ever from the legal framework mentioned above.

Independent travel and independent tour operators

To tap into and cater for the growth in independent travel, specialist tour operators have increased in number and significance. Since 1976 in the United Kingdom some of these operators have been represented by the Association of Independent Tour Operators (AITO), which now has as its members 140 smaller specialist companies which annually carry over two million holidaymakers.

The AITO describes its members as 'owner-managed, specialising in particular destinations or types of holidays; . . . [they] strive to create overseas holidays with high levels of professionalism' (AITO website 2008). All its members are financially vetted and bonded in compliance with European regulations, but 'are also bound by AITO's own code of business practice', which includes provisions for the clear and accurate descriptions of holidays and the use of customer questionnaires for monitoring standards. The AITO's website states that 'Green tourism and responsible travel are key concerns for AITO and its members. Each potential member's sustainable tourism credentials are examined before they may join.'

The AITO first formally introduced its members to the notion of 'Responsible Tourism' in 1989, and set up a not-for-profit organisation called Green Flag International (GFI) to promote the idea. As the Green Globe initiative (see pp. 196–7) grew in importance in the early and mid-1990s, the AITO disbanded GFI and joined the Green Globe initiative instead. It found, however, that 'Green Globe, over the years, became too "corporate" in style for AITO', and in 2001 it established its own Responsible Tourism Guidelines. These cover general good practice in 'sales, marketing and pre-departure information'.

The AITO's implicit criticism of the corporate agenda associated with the Green Globe initiative is repeated elsewhere in its website where it claims that its own members are 'all independent of the vertically integrated groups within the industry' (AITO 2002). This distinction is important to the AITO because it allows its members to distance themselves from the widely perceived unsustainability associated with the practices of the large-scale mainstream tour operators. In turn, this distance gives them not only a status of specialists offering personal service but also allows them to claim, almost by default, environmental and ethical principles for their own.

There can be no doubt that the AITO values environmental sustainability, however it may be defined, highly. It lays considerable stress on its general environmental aim, its promotion of green tourism and its involvement in related debates. It promotes a code of conduct for tourists along with the Campaign for Real Travel Agents (CARTA), and it emphasises the importance of ethical considerations in the business practices of its members.

It is, however, an organisation designed to help its members market themselves and improve their performance (maximise their profits). In itself, this is fine, for it is clear and explicit and profits are not something to be avoided. But what we need to ask is whether this conflicts with its promotion of sustainability. One of its guidelines for tourists, for example, is 'Get your holiday off to a green start – if possible, travel to and from your airport by public transport' (AITO 1996b: 42). Despite the fact that the most polluting element of the holiday is likely to be the air journey, the code of conduct ignores this leg of the journey because it is essential to the holidays it is selling. Long-haul operations overwhelmingly involve air travel, an environmentally damaging form of transport. As Drukier (2001) notes: 'Hurtling across the world in metal cauldrons bubbling over with greenhouse gases looking for unspoiled vistas can hardly be defined as walking softly on the planet'. As Hall and Kinnaird explain:

> global travel to ecotourism destinations undertaken in fuel-hungry aeroplanes is in itself incompatible with ecological sentiments. As the very support upon which all life depends is under threat as a consequence of our Western lifestyles, it is acknowledged that patterns of consumption must shift away from fossil fuel burning and the use of non-renewable resources. The extolling of ecotourism development in faraway lands . . . may be thus viewed as paradoxical.
>
> (D. Hall and Kinnaird 1994: 111)

This paradox calls into question the claims of sustainability made by operators of new forms of tourism. Of course, the AITO is unlikely to suggest to potential customers that they should stay at home. But the example illustrates both where its real interests lie and the limits placed on its 'greenness' by the nature of its business, purpose and ethos. It should be noted that the AITO code of conduct now encourages its members to offset their carbon emissions where possible – an issue discussed further on pages 204–6.

We would suggest that the 'Independent' in the AITO is a relative term which cannot be defined without reference to the political and ideological contexts in which it is set. Likewise, the environmental sustainability it pursues is subject to constraints, not least of which is the fact that its definition varies with many factors, and should be interpreted only with reference to its political, economic, ecological and social contexts.

The new tourist, the independent traveller, the backpacker or the trekker, is one whose image, fashion and consciousness necessitate a clear, and sometimes explicit, acknowledgement that they seek sustainability in their holiday pursuits, with a minimum of negative impact. They may even be aware of the likely incidence of financial leakage from the destination country, and may well attempt to spend their cash in local shops rather than in the TNC-owned hotel parlours and kiosks. But for them too, the bulk of their holiday spending goes on the long-haul leg from home to Third World country. In terms of leakage, then, the new tourist is only marginally less associated with profit repatriation and TNC control of the industry than the conventional mass tourist, and their money is only marginally less likely to leak out of the destination than that of the hedonist mass tourist enclave.

Consider the following two types of tour:

- a Club 18–30 tour
- a trekking expedition to Nepal.

Assume that two average tourists – one seeking sun, sand, sea and sex, the other seeking something alternative, authentic and sustainable – live in the same town in England and earn similar amounts of money. Table 7.1 makes a qualitative assessment of the sustainability of each type of tour. The assessment is simple, intuitive and unscientific, and many of the categories of impacts depend upon definition. It uses simple qualitative scales for each type of impact in order to provoke debate and precisely because we do not wish to imply that there exists an indisputable definition of sustainability.

The conventional but superficial wisdom might tell us that the trekking expedition makes less of a negative impact on the environment than the conventional hedonistic tour. Table 7.1 gives cause to question this conventional wisdom and the general perception that trekking, and possibly other new forms of tourism, are sustainable. Oversimplified this comparison may be, but it suggests the need for further research to give a sounder footing for these highly generalised impact levels.

Table 7.1 *A qualitative assessment of some differences between a conventional mass tourist package and a typical trekking package*

Impacts often linked with sustainability	Club 18–30	Trek
Distance travelled to destination	1000–2000 km	7000–8000 km
Level of pollution associated with mode of travel (air)	high	high
Length of visit	1–2 weeks	3–4 weeks?
Cost of tour paid to operator in UK (£)	200–500	1500–3000
Daily money spent at destination (£)	medium–high	low–medium
Contact with local population	limited	limited
Number of jobs created in destination community	medium–high	low
Quality of jobs created in destination community	low	low
Secondary production and services created in destination community	medium	low
Social dislocation caused within destination community – dependency on tourism	possibly high – very dependent	limited – very dependent
Cultural impacts	limited	possibly high
Direct ecological damage at area of contact	high	high
Indirect ecological damage in surrounding areas (e.g. deforestation, changes in farming practices)	low/medium/high?	high
Carbon emissions	medium	high

Redefining sustainability

We would not argue that the profit motive of private companies in capitalist countries necessarily negates or dominates other motives. There are many examples from around the world of good environmental practice allied with profitability; there are examples of unquestionable altruism on the part of profit-maximising companies. Moreover, the very idea of publicising and promoting examples of good practice is eminently sensible. But the profit maximisation motive does have a tendency to subvert and subjugate other considerations, ethical and environmental. It is essential to keep this in mind in any analysis of the tourism industry.

The industry has responded to the growing importance of the notion and use of the term 'sustainability' in a range of ways which are examined in this section. The framework for the analysis here includes, where appropriate, companies which cater for conventional mass types of tourism to Third World destinations as well as those offering new types of holidays and tours for the new middle classes. This is because the focus in this section is on the term sustainability, and it is not just new forms of tourism to Third World countries which claim to be sustainable. The mainstream industry is also attempting to green its image, and this attempt is linked with the process of diversification which some large tour companies are undertaking. This section therefore looks at the techniques employed by both these forms of tourism, if indeed the two are so easily distinguishable and so mutually exclusive.

Many of the techniques of relevance here have been briefly discussed in Chapter 4. Broadly, those listed in Box 4.8 which are under some degree of control by the industry rather than other bodies fall into the following categories: advertising; industry regulation; corporate social responsibility; environmental auditing and EIAs; ecolabelling and certification; carbon offsetting and emissions trading; consultation and consultancies; codes of conduct; internal management strategies.

Advertising

For members of the tourism industry advertising is a means through which they can claim sustainability. Here we do not intend to analyse the discourse of tourism advertising – this has already been covered elsewhere, especially in Chapters 3, 4 and 5. We shall, however, look at some of the claims that the media used for the purpose of advertising and its effect on control over the industry.

Figures 3.2, 3.3 and 5.1 illustrate some of the language of new tourism, as do Boxes 4.4 and 5.4. That language may be different in vocabulary from the language of mass tourism, but it is almost as narrow in its range. Instead of words like pleasure, relaxation, carefree, resort and so on, the new tourism plays heavily on words such as conservation, ecology, responsible, environmental and so on. Pratap Rughani's description of tourist brochures as 'pleasure propaganda, selling escapism in a tone of juvenile orgasm, where "the natives smile welcomingly", everything is commodified, available and tremendous' (Rughani 1993: 12) seems to apply to new tourism as much as it does to conventional mass tourism.

The medium for advertising most commonly used by operators and exponents of the new tourism (apart from their own brochures and web pages) is that of magazines and journals. In particular, those magazines which will lend credibility to their claims of sustainability, environmental friendliness and cultural sensitivity are targeted as suitable vehicles for publicity: *New Internationalist, Resurgence, Green Magazine, New World* (newsletter of the United Nations Association, UK), the *Geographical Journal,*

Geographical Magazine, NACLA [North American Congress on Latin America] Report on the Americas, Wanderlust and the publications of the Sierra Club and the National Audubon Society. Various solidarity organisations such as the Cuba Solidarity Campaign and Nicaragua Solidarity Campaign also occasionally include with their newsletters advertisements or publicity fliers for tour operators which proclaim a solidarity association with their aims and objectives.

Associations with socio-environmental organisations allows operators or agents the chance to tap the conscience of the new middle-class tourists from the First World. Their advertisements offer the temptation of guilt-free, low-impact travel at the same time as providing a specific purpose to their tour, such as learning the language, researching the ecology, visiting specific development projects, promoting a solidarity twinning arrangement, joining a delegation on a specific study tour, or working in a brigade. In this way, the operators and agents are simply functioning in much the same way as the large-scale mass tourism operators who have identified their market population sector and then target their advertising at them through appropriate media channels. The market may be more specialised (niche) and fragmented and the operators may be smaller in scale, but the appeal to escapism (from the daily work routine) is similar, even if implicit rather than explicit, and the commodification of the tourism product (the wildlife, the national park, the scenery) is just as strong an appeal as mass tourism's appeal to the sun, sand, sea and sex (Selwyn 1993).

Many of the new operators deliberately align themselves with campaigning organisations which focus on environmental, social and ethical issues. Others produce their own newsletter or magazine, as distinct from their annual brochure, which feature articles on general tourism issues as well as news about their tours. It is an important indicator of their credibility to be able to show that they support some pertinent international organisation such as WWF or a local conservation organisation at the destinations they visit. Table 7.2 lists various features associated with a number of British-based new tour operators – new in the sense that they cater for the new middle-class desire for trekking, travelling or trucking to Third World destinations.

One important characteristic of their tours which nearly all the new tour operators report on is the size of their groups. Small group sizes indicate an awareness of the impact of tourist groups and thereby add to the environmental credentials of the operator. The practical considerations of managing a touring group may be a significant factor in group size, but it makes good public relations to show sensitivity by stipulating a maximum group size regardless of this consideration.

Some of the campaigning organisations themselves solicit the advertisements and attentions of the new tour operators. The National Audubon Society, for example, has a Marketing and Licensing Department which invites service providers and producers to use their name in a pamphlet entitled 'Looking for a New Niche . . .', in which it expounds upon 'the Selling Power of the Audubon Name'. As it states in the pamphlet, 'The NATIONAL AUDUBON SOCIETY has been licensing its name for fifteen years. Now more than ever before, our name is a powerful marketing tool for increasing sales, strengthening brand loyalty, and enhancing corporate image.' It would be difficult to find a clearer, more explicit confirmation that the new nature and conservation-associated forms of tourism are firmly rooted in capitalist accumulation than this. To question the sincerity of the ethical basis of this type of marketing and operation would not seem to be too surprising. As Wight (1994) says,

> There is no question that 'green' sells. Almost all terms prefixed with 'eco' will increase interest and sales. Thus, in the last few years there has been a proliferation of advertisements in the travel field with references such as ecotour,

Table 7.2 *Selected characteristics of 'new' and specialised tour operators*

Tour operator	Links with/supports	Specified group size
Abercrombie & Kent	Partner of Friends of Conservation; supports planting and caring for seedlings in Africa; offsets carbon emissions; supporter of Climate Change Challenge	None currently specified
Africa Exclusive	Supports 'Send a Cow'; Okavango Community Trust; Water pumps in Hwange Game Reserve (Zimbabwe)	None currently specified; offers individual, tailor-made tours
Bales	Partner of Friends of Conservation; business partner of WWF; the Travel Foundation; carbon offsets through Climate Care; supports local charities in destinations	Escorted groups between 12 and 16; offers tailor-made tours
Exodus: – Biking; Walking; Discovery	Partner of Friends of Conservation; carbon offsets with Tourism Industry Carbon Offset Service (TICOS); provides solar cookers to local areas in Nepal; helps many local projects in destinations	Maximum group size 16 or below
Explore	Travel Operators for Tigers Campaign; Sponsors a Sri Lankan elephant; Born Free Foundation; Galapagos Conservation Trust; carbon offsets with Climate Care	12–16 for cycling; 12–18 on walks/treks
High Places	AITO; Tourism Concern; International Porter Protection Group (IPPG)	Max. 12 or less
Himalayan Kingdoms	IPPG; AITO; Sir Edmund Hillary School at Khumjung; supports a school in Kathmandu; carbon offsets through TICOS; supports a range of other charities	Average group size in 2007 of 8; max. normally 12; max. on family treks up to 20
J. & C. Voyageurs	Tusk Trust; African Travel and Tourism Association (ATTA)	No maximum size policy, but offers small, individual tailored tours
Naturetrek	AITO; Environmental Investigation Agency; Butterfly Conservation; Birdlife International; Biscay Dolphin Research Programme; Bald Ibis Appeal Fund; carbon offsets through Climate Care; other conservation causes	'Small groups'
The Ultimate Travel Co.	Galapagos Conservation Trust; Lewa Wildlife Conservancy (Kenya); Tusk Trust; carbon offsets by donating to a small tree-planting scheme in southern India; many overseas suppliers signed up to a strict ethical code	'Groups seldom more than 15 to 20 passengers'

> ecotravel, ecovacation, ecologically sensitive adventures, eco(ad)ventures, eco-cruise, ecosafari, ecoexpedition and, of course, ecotourism.
>
> (Wight 1994: 41–2)

In their view of how (eco)tourist companies should set about 'marketing their product', Ryel and Grasse (1991) give an even clearer prescription of how the new, environmentally friendly companies should fit into the prevailing mode of capitalist accumulation. In an article which at times reads rather more like an instruction manual to the new ecotourism companies than a supposedly objective analysis of the marketing of ecotourism, they state:

> Nature travel companies should therefore invest in repeated advertising with their most productive advertising media . . . Advertising that complements editorial content also enhances the effectiveness of advertising . . . Ecotourism companies should keep abreast of upcoming editorial coverage in order to take advantage of special features that focus on their destinations.
>
> (Ryel and Grasse 1991: 173–4)

In-house brochures, newspaper supplements, magazines and journals are not the only media available to the new tour companies. Increasingly, advertisements for specific tours are appearing on travel, tourism and environment-related bulletin boards on the internet (as noted in Box 2.2 and Appendix 1). It is a particular feature of such advertisements that they appeal to the researcher or enquirer in the new middle-class tourist. Examples are given in Box 7.8 later in this chapter, where 'new academic tourism' is discussed. Major advantages of this medium are its cheapness and its audience, a population which contains a large body of potential and current ecotourists and ego-tourists (see Chapter 5).

Industry regulation

It was argued in Chapter 4 in 'The tools of sustainability in tourism' (pp. 109–18) that the issue of regulation of the industry can be represented as a struggle for control of the industry between different interest groups. These may be many and varied – as has already been noted, the industry itself is highly fragmented with many associations of hoteliers, travel agents, tour operators, caterers, transport companies and so on, at local, regional, national and international levels. Some of these groups undoubtedly have a role to play in promoting the attainment of ethical standards of practice, for the fragmentation is such that it would be impossible for all but the most bureaucratic of governments to regulate for all related practices and to enforce the legislation as well. Moreover, there is a clear and undisputed place for national and international legislation on a number of matters, such as airline safety and safety matters relating to other aspects of tourism. In most other areas, regulation for sustainability is a concept as contested as sustainability itself. And this contest leads to the ongoing debate around the issue of self-regulation. It is essentially between two camps: those who believe that the industry should pursue voluntary self-regulation on issues relating to sustainability; and those who believe that regulation should take the form of government imposed and enforced statutory legislation. It is not as simple as the descriptions of these two opposite camps would suggest – there are many combinations of company self-regulation, articles of association and government legislation which can be promoted – but the argument is often presented as a simple dichotomy.

Rebecca Hawkins and Victor Middleton of the World Travel and Tourism Environment Research Centre (WTTERC), which was set up by the WTTC in 1991, have pointed out that 'Despite concern about the environmental impacts of tourism . . . the industry overall has scarcely been affected by international regulation' (R. Hawkins and Middleton 1994: 104). In 1993, they identified six major categories of international environmental regulation and control which affect the tourism industry. We have adapted these in Box 7.3. Following their identification of these categories their general conclusion is that the effect of international regulation on the tourism industry is very limited; and this leads them to a defence, or even promotion, of industry self-regulation based on their belief that this limited effect is due to the inability of governments and international bodies to regulate. In this defence, their supposedly objective outline of the advantages of self-regulation (no disadvantages are listed) becomes part of the contest itself.

The areas of the industry affected most by this struggle are those which can be externalised in the industry's accounting procedures, where the industry has an impact upon the environment, society, culture and other factors which are not costed financially.

Box 7.3 Categories of international regulation and control of relevance to the travel and tourism industry

- Agreements which deal with the right to free time and a safe environment as an aspect of human rights – e.g. the Stockholm Declaration 1972; UN International Labour Organisation (ILO) labour rights and conditions.
- Agreements documenting the environmental impacts of travel and tourism – e.g. Manila Declaration 1980 – addressed regionally in international agreements such as the Alpine Convention and the Antarctic Treaty.
- General environmental policy dealing with specific pollution and ozone layer issues with implications for environmental practices of travel and tourism companies – e.g. Protocol for the Prevention of Marine Pollution of the Mediterranean Sea by Dumping from Ships and Aircraft (1978).
- Specific environmental policies relating to emissions into the atmosphere – e.g. International Civil Aviation Organisation's Rules (1986) limiting aircraft emissions of smoke, unburnt hydrocarbons, carbon dioxide and nitrogen oxides; Kyoto agreements, overtaken by the Bali Climate Change treaty agreement (2007); EU Emissions Trading Scheme (voluntary).
- Policies developed for specific areas, such as the Mediterranean Basin, the Great Lakes and Antarctica as a reaction to general environmental damage and in which tourism is or could be a major issue – e.g. the Barcelona Convention (1978), the Convention for the Protection and Development of the Marine Environment of the Wider Caribbean Region (1986); Antarctic treaties from 1959, 1972, 1991 and 1994.
- Regulatory and self-regulatory global policies developing as a result of the Rio (1992) and Johannesburg (2002) Summits, intended to respond to the new ethic for sustainable living and to ensure that growth remains within the Earth's capacity – e.g. Agenda 21; the Climate Convention; the Johannesburg Declaration 2002–2012, creating the UN Commission on Sustainable Development.
- Emerging self-regulation within travel and tourism corporations – usually implemented on a self-interest, voluntary basis, although it may have implications for membership of some organisations; ISO (environmental) and SA (labour) standards and awards.

Sources: various, including R. Hawkins and Middleton (1993: 165), and appropriate websites

It is self-evident that regulation of the activities of different branches of the industry constrains these activities. In the eyes of the industry and under the doctrine of the 'free market' (see Chapters 2 and 10), constraints on these activities are for the worse. They inhibit competition and consequent price reductions, they create 'unnecessary' bureaucracy, they cause delays, they may alienate those who work in the industry, and they stifle its performance and effectiveness.

But without these constraints, the industry is free to pursue profits with no regard for the external costs, the negative impacts on the environment, the culture, or the society. It is also free to use its voluntary attempts at self-regulation as a public relations exercise or marketing ploy. And without official, non-industry-based monitoring and inspection, it can deceive its consumers into believing that its operations are environmentally friendly or ethically sound.

Of course, it is not as simple a problem as stated by these two distinct and polarised camps. The issue may be more fairly and faithfully represented by a continuum of views between the two. But, as McKercher (1993: 11) points out, 'In a free market system, such a diverse and highly unregulated industry as tourism will likely continue to defy most efforts to limit its expansion'. The diversity which McKercher talks of refers to the highly fragmented nature of the industry already noted. He adds that 'Effective control measures can only occur through integrated programmes that incorporate federal, state and local legislation' (McKercher 1993: 11). Any attempt to regulate such an industry, however, would have to be clearly targeted unless it is to attract justified criticism from industry members.

At the same time, this fragmentation of the industry is one of the factors which permits operators within it a certain inconsistency. When government tries to regulate the operations of private companies, there are few industries in which it is allowed to do so without vociferous opposition from companies which vigorously uphold the benefits of voluntary self-regulation. When tourism companies are asked about the industry's responsibilities, on the other hand, their answers are somewhat at variance with their attitudes to regulation. As a Tourism Concern/WWF study of a number of companies showed: 'All operators stated that national governments had some responsibility, and nearly 60 per cent of operators said that governments had total responsibility. This view was echoed by travel agents, carriers and hotels' (Forsyth 1996: 31). This clearly points to the need for some form of authorised regulation rather than voluntary self-regulation.

The Green Globe scheme, launched in 1994 by the WTTC, is an international environmental management and awareness programme designed to encourage travel and tourism companies, whatever their size, sector or location, to make a commitment to continuous environmental improvement. It offers achievement awards, training, networking, publications, advisory and information services, branding and marketing assistance.

> Membership is granted on an annual basis to companies which declare their commitment to environmental improvement and show progress in achieving their stated objectives . . . it is expected that . . . Green Globe will be one of the principal ways by which the discerning tourist will choose a holiday company or destination in future.
>
> (United Nations Environment and Development UK 1994: 7)

The establishment of Green Globe by the WTTC was seen by the proponents of self-regulation as helpful because:

> it enabled tourism companies to seek advice on environmental matters from experts who also understood the travel trade . . . Furthermore, such progress had

been achieved, by and large, without the intervention of outside bodies. This demonstrated the commitment of the tourism industry to sustainable tourism.

(Forsyth 1996: 6)

As Brian Wheeller (1996: 15) points out: 'what they [the WTTC] do, they do very well, indeed excellently. The quality of their high profile publicity is what one would expect from such a professional body.' But as he also argues, the WTTC represents business interests which advocate the message 'no outside regulation, we can regulate ourselves', and who are acting, not altruistically, but only in the immediate interests of their members.

Noel Josephides, managing director of Sunvil Holidays and chairman of the AITO Trust, supports this view of Green Globe:

> the underlying reason for its launch is to prevent, by having in place a self-regulatory system, any government interference in the workings of the industry. There is no doubt that the large global players recognise the increasingly harmful impact the industry is having on the environment, which is now exciting considerable interest and anxiety among the media and inevitably the regulators. They also know that this unwelcome interest will interfere with the current freedom and market dominance they enjoy. If they have the Green Globe scheme in place before too many questions are asked, they will be able to hide behind the façade of self-regulation.

(Josephides 1994: 10)

To obviate such criticism, Green Globe 21 (as it is now known) established Green Globe Accreditation (GGA) which it describes as 'an independent organisation that has been given the contractual responsibility . . . for assessing and accrediting organisations seeking to offer Green Globe 21 assessment services'. GGA's supposed independence only means that it was established as a separate entity from Green Globe 21 itself. It does not represent true independence and it does not offer any form of governmental oversight of operations. Indeed, it was set up precisely to avoid interference from government.

Tourism Concern (1994a) has pointed to the contradiction between Green Globe's advocacy for companies of sound environmental principles through self-regulation and the WTTC's lobbying and promotion of the needs to expand travel infrastructure, liberalise policies to encourage growth in the industry, and remove physical, bureaucratic and fiscal barriers to travel.

It is perhaps too easy to interpret the WTTC's wholesale encouragement of self-regulation as promoted cynically, purely in self-interest and in pursuit of short-term profits. It could also be interpreted as a genuine attempt to help the industry adapt to what may become environmentally essential regulation. The former interpretation suggests that business is simply trying to avoid the inevitable; the second that it is prescient in trying to adapt to it.

Whether government legislation would really help to reduce the uneven and unequal nature of tourism development may be debatable (see Chapter 10). But self-regulation led by bodies such as the WTTC and the UNWTO, whose stated aims are the promotion of the tourism industry rather than its restraint, is likely to lead to policies which further the pursuit of profits in a business world where profit maximisation and capital accumulation form the logic of economic organisation.

Corporate social responsibility

Corporate social responsibility (CSR) is a specific application of the notion of environmental and social auditing to business practice, a technique which is increasingly being adopted by companies and promoted by NGOs such as Tearfund and Fair Trade in Tourism, an offshoot of the UK's Tourism Concern. The International Business Leaders Forum (IBLF) has also established a CSR Forum to promote the technique among its members.

Tearfund explains CSR in the following terms: 'Once there was just the financial bottom line. Now, companies recognise they must be accountable for the way they affect people, the community and the environment – the new triple bottom line' (Tearfund 2002: 10). It further suggests that the trend towards CSR is inexorable, although the New Economics Foundation (2000) reports that while one-third of FTSE companies issue environmental reports,[1] only 4 per cent produce fully verified reports. But given the fact that CSR is such a recent phenomenon, perhaps these figures can be taken as an indication of rising significance rather than as showing a lack of importance. Comparative figures for the end of the 2000 decade will hopefully be revealing in this sense.

As the term suggests, CSR is very much the preserve of corporations, those companies large enough and wealthy enough to fund departments which monitor and report on their environmental, social and community impacts. Despite this, the technique is strongly promoted by Fair Trade in Tourism Network (2002) which suggests that the technique of CSR emerged in the late 1990s out of NGO efforts to create 'a more equitable international trade system'. This consistent aim of many NGOs may be highly laudable and justifiable, but Bradshaw et al. (2002: 251–2) have summed up the general perception within civil society that: 'In the past, NGO lobbying has failed to change the structure of the world economy or the ideologies of its ruling institutions.'

Nevertheless, the possibility of changing management practice through CSR exists, and Fair Trade in Tourism suggests that:

> Experience in a range of business sectors shows that CSR and ethical trade practice can be put into action without jeopardising profit levels and share prices; indeed, they can actually enhance these. The tourism industry has an excellent opportunity to take this on board.
>
> (Fair Trade in Tourism Network 2002)

At various points in this book, and particularly in this chapter, we have already noted the possibility that exists for companies to promote such actions as CSR as public relations activities designed to enhance their image without necessitating a fundamental change in practice – a practice commonly referred to as 'greenwash'. To adopt such a perspective on all such efforts without investigative research would clearly be overly cynical and we would not recommend such an approach here. Enough cases of 'greenwash', however, have been highlighted to temper vigorous enthusiasm for CSR and similar techniques. Indeed, the *New Internationalist* of December 2007 highlighted many such cases, and in it Jess Worth declared:

> For many large companies, CR is primarily a strategy to divert attention away from the negative social and environmental impacts of their activities, and to continue operating without being forced by governments to change their core business practices.
>
> (Worth 2007: 6)

In January 2001 Tearfund published a report into the responsible business practices of 65 UK-based tour operators, and found that 'the tourism industry has made some good progress on environmental issues but lags behind other industries in terms of fulfilling its social and economic obligations' (Tearfund 2001: 25). It also found that 'only half of the companies questioned have responsible tourism policies, and many of these are so brief as to be virtually meaningless' (Tearfund 2001: 25).

Tearfund (2001: 25) also concludes that currently it is the smaller operators which offer a more ethical experience to holidaymakers, 'paying a higher proportion of profits to charity, offering more training to local operators and developing more local partnerships'. For instance, the AITO has developed a responsible tourism policy, whose implementation is likely to become a condition of membership in the future. While the AITO deals with the smaller and more specialised tour operators, the UNEP has also launched a Tour Operators Initiative (TOI) for Sustainable Tourism Development, which is open to any international tour operators. The IBLF also suggests that the industry is changing:

> In the mid-1990s, for example, its focus on issues of business ethics, social development and human rights was seen as rather radical and ill-focussed. Less than six years later, the same issues could be discussed comfortably in most Board rooms.
>
> (Corporate Social Responsibility Forum 2002)

Whether such discussions will bring about the change desired by so many NGOs and communities and by some company executive boards remains an open question for the future, although Worth (2007) is extremely doubtful:

> The fact that Corporate Responsibility is so popular today is an indication that big business is feeling the heat. But the solutions they advocate and pursue – voluntary unenforced codes of conduct, piecemeal eco-improvements, token philanthropic donations, endless rounds of meaningless "engagement" with "stakeholders" who might otherwise be publicly criticising them, dubious but lucrative techno-fixes – are actually dangerous diversions.
>
> (Worth 2007: 7)

Environmental auditing and environmental impact analyses (EIAs)

Two important techniques available to the industry for the purposes of adjusting or amending either its practices or its image are those of environmental auditing and EIAs. Goodall (1992) describes an environmental audit as

> a management tool providing a systematic, regular and objective evaluation of the environmental performance of the organisation, its plant, buildings, processes and products. In essence, environmental audit and EIA have the same goals and are complementary tools in the struggle to achieve sustainable tourism.
>
> (Goodall 1992: 62)

The principle of environmental auditing is of course laudable and the practice is worth further work and refinement, but we would argue that Goodall's description of the technique as 'objective' is rather naive. To be considered as objective, it would need to account for who is carrying out the audit or analysis, for whom and for what purpose. We would further argue that the broad context of the tourism operation (regional, national, international) needs to be considered. It is normally the case, however, that environmental audits are conducted on hotels, airlines, tour operators or other branches of the industry with the major point of reference being the efficiency, economic or otherwise, of the operation. Reference to the wider distributive effects of the operations are rarely made. Instead, they address issues such as the recycling of cans, bottles and other materials, the reduction of water use by, for instance, not changing towels every day, the installation of energy-efficient lighting systems, and reduction in the use of toxic chemicals. As Geoffrey Lipman, WTTC President, has said:

> Hotels are plunging in to reduce, re-use and recycle and [to adopt] sound management. Car rental companies in remote parts of the world are switching to lead-free petrol, and airlines are committing billions of dollars to quieter, less polluting and more fuel-efficient aircraft.
>
> (Lipman 1992)

These examples illustrate the narrow way in which the industry perceives efficiency: there is an underlying implication that such changes in practice should be made for the purpose of cost reduction; and where the environment is explicit as a consideration, it is so only in relation to the natural environment. Social, cultural, distributional effects may be more difficult to measure, but they could not be much further from consideration than they are at present.

It is the purpose of the technique of environmental auditing to improve the day-to-day environmental practices of the industry. The International Hotels Environment Initiative (IHEI – now known as the International Tourism Partnership, ITP) developed its Charter For Environmental Action in the International Hotel and Catering Industry in 1991 with this in mind. The stated aims of the Initiative are to:

- provide practical guidance for the industry on how to improve environmental performance and how this contributes to successful business operations;
- develop practical environmental manuals and guidelines;
- recommend systems for monitoring improvements in environmental performance and for environmental audits;
- encourage the observance of the highest possible standards of environmental management, not only directly within the industry but also with suppliers and local authorities;
- promote integration of training in environmental management among hotel and catering schools;
- collaborate with appropriate national and international organisations to ensure the widest possible awareness and observance of the initiative;
- exchange information widely and highlight examples of good practice in the industry.

(IHEI 1991)

Signatories to the Charter include the presidents and/or chairs of major transnational hotel companies such as Forte PLC, Hilton International, Holiday Inn Worldwide, Intercontinental Hotels Group and the ITT Sheraton Corporation. In 2004, the IHEI was replaced by the ITP in response to a perceived need for an 'integrated, industry wide approach to responsible business in tourism' (ITP 2007).

Despite the problems already outlined with these techniques, it is important to stress the need for industry to adapt itself along the lines of the Charter. But, given the list of signatories to the Charter, it might be worth questioning whether there is a hidden agenda behind such initiatives. It is difficult to believe that the companies listed are willing to adapt their procedures and practices to an extent which would bring about a change in operating motive and which would allow for the explicit and significant costing of socio-environmental factors. The power structures within the industry and their motives for operating are unlikely to be altered by techniques such as environmental auditing and it is unlikely that any redistribution of benefits will take place. Even for a hotel like the Crowne Plaza (whose group is a signatory of the Charter) in Managua, the capital city of Nicaragua, its environmental policy on water use appears rather diluted, to the point of meaninglessness, when one contrasts its regular draining of the clean water in its luxury swimming pool with the broken supply of poor-quality drinking water to many of the barrios surrounding it.

It will be noted, however, that the examples used here so far relate to the large-scale, luxury market – that end of the market which caters to the wealthy tourist and the expenses-paid business traveller. The new, still relatively wealthy middle-class traveller to the Third World, and the specialised operators who cater for them, are less likely to stay in such luxury accommodation. Their accommodation is less likely to be subject to environmental audits, despite the increase in recent years of (even Third World) government inspection of such facilities, largely because the small- to medium-scale accommodation companies do not have the capital resources or the internal management structure to conduct them.

The technique of environmental auditing, then, may more appropriately be seen as fine-tuning of the system, or at least one end of the system, or, worse, as a public relations exercise. And it is a fine-tuning exercise which allows the companies concerned to claim sustainability as theirs. To give these claims more substance, well-known environmentalists or environmental consultancies are brought into the development of such initiatives and programmes as an exercise in consultation (see 'Consultation and consultancies', pp. 206–8).

The credibility of claims to sustainability made by companies which are the subject of environmental audits or similar techniques is somewhat strengthened when the audit is conducted by an 'independent' organisation. In 1992, for example, the San José Audubon Society (SJAS), a Costa Rican chapter of the National Audubon Society (which now no longer exists), began to monitor the activities of tour operators in Costa Rica with the 'hope of satisfying the tourist's desire to know where to go for an ecotour, and be assured satisfaction with the product' (SJAS 1992). Together with Beatrice Blake, author of the guidebook *The New Key to Costa Rica*, the Rainforest Alliance and the Institute for Central American Studies, the SJAS researched and put together a list of Recommended Eco-tourism Companies that was updated and published monthly. To be included, companies had to comply with the SJAS's Code of Environmental Ethics for Ecotourism. Monitoring their compliance was achieved by observation, reports and questionnaires, and was carried out by representatives of Audubon, volunteers, community organisations and tourists.

As a conservation organisation, the SJAS may not have been truly independent, but its independence from at least the influences of the industry gave weight to the value of its audits of the companies it examined. The example illustrates that such an exercise can be fruitfully conducted and produce a meaningful and useful result.

With similar purpose, the Travel Centre of Henley (formerly known as Green Horizons Travel) used to be a travel agency set up in the UK to sell holidays with companies which meet criteria based on environmentally sound and sustainable principles – it has since metamorphosed into the D & G Travel Centre of Henley with rather different emphases in its operations. The company explained its modus operandi as offering:

> a range of holidays with operators who are taking action to increase the benefits
> for local communities in the destinations they visit and minimise the adverse

effects of their operations on the environment . . . we are the first retail travel agency to offer such a service . . . Our recommended operators have been selected following an examination of the companies' brochures and research based mainly on the response to a questionnaire designed to assess each one's environmental responsibility.

(Green Horizons Travel 1995)

Criteria for the assessment included the company's stated environmental policy, the existence of some form of monitoring of impacts, its support for local businesses, organisations and the community and the nature and quality of the information supplied to customers before and during their holidays.

These examples of environmental auditing from Costa Rica and the UK are minor in coverage and significance, but they highlight the kind of exercise that can be carried out, even at the local level, with some effect. But by their former existence, they also exposed the lack of national and international coordination of the monitoring of tourism companies, a factor which will be crucial if environmental auditing of tourism companies is to become both useful for the customer and helpful in moving the industry towards increasing sustainability. The fact that neither of these organisations now exist is not encouraging.

A further technique related to EIAs and environmental auditing is that of ecological footprinting, a technique that can be adapted for use by tour operators to gauge a 'holiday footprint' for the tours they run. Essentially, this technique measures the impact of a given holiday in terms of the resources it uses relative to a notional annual allowance per person of resources available for use. It has already been described in Chapter 4 as one of the techniques of measuring sustainability indicators.

Ecolabelling and certification

As a technique of product review and assessment, environmental auditing has given rise to the certification of specific products as having satisfied certain environmental and/or social criteria of production. Early examples of products certified as having been produced under environmentally friendly conditions that awarded a just economic return to the producer were coffee and chocolate produced organically for the NGO Oxfam and which were labelled with the Oxfam logo. Such certification of products was also extended to the provision of services, and in the decade of the 1990s the application of service certification to the tourism industry gained momentum.

In essence, an ecolabel is a seal of approval awarded to a service whose provision satisfies a number of conditions. It should enable both producers and consumers to identify service providers whose practices do not harm the environment and society in which they occur. In the field of tourism, ecolabels are intended primarily to indicate that the hotel or tour operator acts responsibly towards the environment and the society in which it is located or that all aspects of the tour are arranged on an ethical basis taking due consideration of the local people, the local environment and the labour rights of employees.

Labelling schemes generally offer use of a logo to those service providers which meet their criteria for membership of the scheme. Ecolabelling and eco-certification refer to the level of responsibility with which the tour company or hotel carries out its operations.

Given the intention of ecolabelling and certification, it would seem self-evident that such schemes should be widely recognised and limited in number. Worldwide, however, the 1990s saw the proliferation of over a hundred such schemes, which clearly present the tourist with a bewildering array of labels and logos, and renders it impossible to distinguish the relative merits of each scheme without a degree of research which is not a normally accepted element of the tourism experience. Not surprisingly, then, as pointed out in an independent report commissioned by WWF–UK (2000a), the confusion caused by this proliferation 'has, in turn, led to a lack of consumer demand for certified holidays' (WWF–UK 2000b). As the same report adds, 'Furthermore, less than one per cent of businesses have joined up to these schemes. The failure to establish clear brand recognition could seriously undermine the potential of certification to bring about sustainable tourism' (WWF–UK 2000b). Rosselson (2001) has also bemoaned the confusion clouding certification schemes: 'there is currently no comprehensive symbol that is universally recognised by tourists as a guarantee that a tour company is both socially and environmentally responsible'.

In 1997, UNEP conducted a survey of tourism and environment ministries, national tourism boards, tourism industry associations, environmental NGOs and educational and research centres around the world in order to identify as many existing or planned tourism ecolabel schemes as possible at a global level. From the replies to 400 questionnaires distributed, 28 were selected for detailed analysis which is reported in UNEP (1998).

The UNEP report acknowledges the confusion caused by unnecessary duplication. It also recognises the need to develop measures of the effectiveness of tourism ecolabels, remarking that most schemes are too young for their effectiveness to be gauged. In common with other reports by many UN agencies and international finance institutions (IFIs), however, the UNEP report sees ecolabelling as 'part of the worldwide movement by industry towards self-regulation – a movement which is demonstrating how responsible industry can be towards environmental issues and how voluntary action can stimulate or even replace formal legislation' (UNEP 1998: 5). In this sense, it repeats, frequently, one of the mantras of capitalist globalisation, that of self-regulation, regardless of the results of the survey. In other words, UNEP acts as just one more agency which uncritically promotes voluntary self-regulation seemingly unaware of the discourse and debate surrounding the use of such terms and the effects of the application of such policies.

Ecolabelling is potentially useful as a tool for assisting both producers and consumers, especially where it is used for a specific interest, focus or theme, but its current practice suggests that it may be used more as an image-greenwashing or as an ideological support rather than a genuine guide for tourists. To overcome such suspicions, myriad schemes need to be reduced and rationalised; a system of coordination of schemes needs to be established; an effective mechanism for monitoring schemes is essential; the award of logos should be made for compliance with specified criteria rather than merely a commitment to improve; and certification should cover all aspects of sustainability (environmental, social, cultural and economic) rather than only environmental performance.

Work on the rationalisation and coordination of schemes began in 2001 when the Rainforest Alliance, a global organisation which works in environmental certification, hosted a feasibility project on a proposed global accreditation body to be called the Sustainable Tourism Stewardship Council, which would set international standards for the certification of sustainable tourism organisations. The project was supported by a comprehensive range of organisations including the UNWTO, the WWF, TIES and UNEP among many others. The feasibility project lasted for 18 months and resulted in a number of reports. One solid outcome has been the creation of the Sustainable Tourism Certification Network of the Americas, a regional network of all certification programmes as well as NGOs, academic institutions and other interested parties in the Americas (Sustainable Tourism Stewardship Council website 2008). Whether it will meet the conditions set out above or whether it will simply serve to reinforce the use of certification schemes as an image-greenwash or as an excuse to produce further justification for self-regulation remains to be seen.

Carbon offsetting and emissions trading

Since the late 1980s there has been a growing awareness of the problem of global warming and of the part played in this phenomenon by greenhouse gases, the most significant of which is carbon dioxide, CO_2. As the amount of CO_2 in the atmosphere increases, so the atmosphere heats up. CO_2 in the atmosphere is increasing largely because of the emissions created by human activity and because of the loss of vegetative material (especially forests) to absorb the CO_2. The science behind this conclusion was confirmed in 2007 by a report of the Intergovernmental Panel on Climate Change (IPCC 2007).

From the point of view of the tourism industry, global warming is likely to have both positive and negative effects. The negative effects are alarming, especially for specific destinations. Small island states, for example, are particularly at risk from inundation and loss of coral fringes as a result of sea level rises and coral bleaching. As some holiday zones such as parts of the Mediterranean become uncomfortable as a result of increasing average summer temperatures and an increasing incidence of malaria and other diseases, other holiday destinations may become substitute destinations. The effects of such substitution may be offset, however, by a more positive environmental impact of global warming: namely, the generally warmer summer weather conditions in higher latitudes as well as increases in the real cost of air travel which may persuade Europeans to take holidays nearer to home rather than traveling to exotic locations.

Recently, then, carbon has become an item of relatively common discussion. The idea of measuring an individual's carbon budget may be very recent, but thanks to this discourse, popular understanding of the notion should not present a great obstacle to its use. A range of websites can currently be accessed to assess the carbon emissions made by specified journeys and activities. Several of these are given in Appendix 2 of travel- and tourism-related and carbon budget calculation websites.

Ultimately, the questions raised by such websites is whether, having expanded our horizons through travel and tourism, we should now rein them in so that we may reduce the adverse effects of carbon dioxide emissions on the climate. If science and society tell us that we must indeed do so, then will we all be allocated a carbon quota (a kind of personal carbon budget) in the future, which we will not be allowed to exceed (unless we have a licence to do so by purchasing another person's unfulfilled quota)? Such an idea may seem a little far-fetched, but in July 2006 the then UK Government Environment Minister (David Miliband) revealed his ministry's intentions to examine schemes for implementing and managing tradable personal carbon allowances.[2]

On a scale slightly greater than the personal, Climate Care is a company formed in 1998 'to tackle climate change by reducing greenhouse gases in the atmosphere. 'We do this by offsetting – making CO_2 reductions on behalf of individuals and companies' (Climate Care 2007). The company provides funding for projects that reduce CO_2 in the atmosphere, for example, through forest restoration and energy efficiency schemes. The aim of such funding is that it offsets flight emissions. The funding comes from companies which donate to Climate Care a nominal amount for each client who uses services, such as flights, which add greenhouse gases to the atmosphere. A range of transport and tour companies support Climate Care in this way.

A similar service is offered by the UK-based CarbonNeutral Company (formerly Future Forests) which helps organisations

> to measure and reduce the CO_2 from some or all of their operation, and then 'offset' unavoidable emissions. When emissions are reduced to net zero, the organisation, product or service can carry the CarbonNeutral quality stamp, our registered trademark.
>
> (CarbonNeutral Company 2007)

There is no compulsion on companies to support Climate Care and so it would seem churlish to dispute the motives of the companies that do so. The crux of the argument, however, lies not with the motives of the companies which support Climate Care, but with the claim of Climate Care that their funding genuinely offsets the emissions made in the course of air travel. At the same time as its popularity has grown, the exercise of carbon offsetting has increasingly become of dubious value. Tree-planting schemes in particular are notoriously unreliable and subject to fire, land takeover and change of use. The *New Internationalist* exposed several carbon offset reforestation schemes which lasted for only short periods of time and managed to offset only a small fraction of the CO_2 emissions caused by the journeys of those who contributed towards the schemes (New Internationalist 2006). Criticism of such schemes appears to be mounting (especially among environmental organisations) as their ineffectiveness at combating climate change becomes clearer. The mainstream media, however, is slow to adapt to the critique and still regularly offers advice and possibilities to travellers to offset their flights in articles on 'How to be a responsible tourist' or similar.[3]

Of course it is possible to estimate the tonnage of greenhouse gases absorbed by newly planted forests during their lifetime or to calculate the amount of energy saved through the use of renewable sources of energy – assuming that this same amount would therefore not be produced by other emissions-producing sources. But do these absorptions or savings really offset the emissions produced by the clients of tour companies? Critics of carbon offset schemes point out that a reduction in emissions can only be achieved by reducing the number of flights sold and that the offsetting approach is being used solely as a gesture to the problem – a kind of greenwash. Ultimately they ask if it is possible to reconcile a tourism industry which encourages people to fly as frequently and as far as possible with attempts to reduce carbon emissions and global warming.

It is a mark of the recent rise to significance and to public perception of the notion of climate change that in 2005 the UK and European governments began to implement the European Union's Emissions Trading Scheme. The UK's Department of Environment Food and Rural Affairs (2005) explains that

> participating companies are allocated allowances, each allowance representing a tonne of the relevant emission, in this case carbon dioxide equivalent. Emissions trading allows companies to emit in excess of their allocation of allowances by purchasing allowances from the market. Similarly, a company that emits less than its allocation of allowances can sell its surplus allowances. In contrast to regulation which imposes emission limit values on particular facilities, emissions trading gives companies the flexibility to meet emission reduction targets according to their own strategy; for example by reducing emissions on site or by buying allowances from other companies who have excess allowances.
>
> (DEFRA 2005)

British Airways has joined the UK Emissions Trading Scheme. The company

> does not accept that the right way to limit emissions is to discourage flying – by punitive taxes or constraints on industry growth. . . . if applied to air transport, [this] would lead to extremely negative social and economic effects for the European economy.
>
> (British Airways 2007)

However, the company no longer points out, as it used to, that international aviation emissions are not included in the agreements, and therefore that the scheme only includes its domestic services and UK ground energy sources.

Even if emissions trading were to be extended to international travel, it may be argued that such schemes effectively sanction the continued emission of pollutants into the atmosphere. The perpetrators are given the opportunity to avoid taking action to reduce their own emissions; their 'purchase' of the clean record of others as a counterbalance to their own dirty record cleanses their conscience for emitting so many pollutants themselves. But to 'offset' their own high emissions in this way does not imply that they are *reducing* them, which is widely perceived as the only solution to global warming. Whether these arguments against emissions trading are justifiable is open for debate; certainly proponents of trading would argue strongly that this is a major new technique which allows the world to begin to combat global warming.

These arguments are not raised here in order to cast doubt on the motives and meaning of the companies concerned; rather, they are presented in order to illustrate different facets of the debate. Moreover, they raise again some of the points made in other debates involving the travel and tourism industry, especially those concerning corporate social responsibility, the use of codes of conduct, voluntary self-regulation, ecolabelling and certification.

Consultation and consultancies

Consultation is another technique which can be used by the industry either to assist in efforts to achieve sustainability or as a public relations exercise designed to demonstrate 'genuine' motives while at the same time making few if any significant changes to established practices and procedures.[4] A number of techniques of consultation were listed and briefly discussed in the section 'The tools of sustainability in tourism' (Chapter 4). One of these, the Delphi technique, will be illustrated and discussed in Chapter 8. Here, however, the term 'consultation' is taken as having a number of distinct directions which are shown in Figure 7.2. The three major types of consultation identified here are not exclusive, and the illustration in Figure 7.2 is partial and much simplified. The governmental level in the consultation procedure has been deliberately omitted partly in order to maintain its simplicity and partly to sharpen the focus on the industry – environmental consultancy companies operating on behalf of the tourism industry can be seen as part of the industry itself.

Linking the words 'consultation' and 'consultancy' is significant here, for environmental consultancies are sometimes used by the industry as a vehicle of consultation with local people who may be affected by the industry's tourism proposals. This relationship between the industry and its environmental consultants is important in setting the type and quality of consultation which takes place, and therefore whether it can be seen as consultation which moves towards a presumed definition of sustainability or as a consultation exercise designed to avoid a move towards sustainability.

Often the consultation which takes place is one which cuts out the local community. The industry prefers to consult with environmental consultancies, and in some cases those companies large enough to do so may establish environmental departments within their own management structure (see 'Internal management strategies', pp. 210–13).

The kinds of consultancies relevant to this field include not just those dealing directly with travel and tourism (such as Green Globe, discussed earlier in this chapter) and technical and engineering firms which carry out EIAs on tourism development projects, but also those whose remit is more general. These latter include Forum for the Future, which was established in 1996 by three well-known British environmentalists, Paul Ekins, Sara Parkin and Jonathon Porritt, with a mission 'to accelerate the building of a sustainable way of life' (Forum for the Future 2001). The Forum's aim of promoting

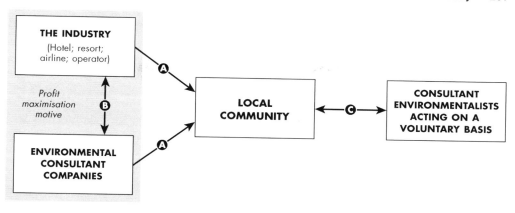

ⓐ Downward local consultation From the company to local people who will be affected by the tourist development. Can come from either the private consultancy firm operating on behalf of other bodies or directly from the tourism company.

ⓑ Co-optive consultation Where a tourism company enlists the advice or association of a person or organisation with ethical/environmental credentials to assist the company.

ⓒ Voluntary joint consultation Where a community enlists the help of an independent, often unpaid, environmental consultant or organisation.

Figure 7.2 Consultancies and forms of consultation

best practice in sustainability gives it a reason to approach industry itself without waiting to be asked.

Forum for the Future 'does not see itself as a mass membership organisation', choosing to work instead 'with around 60 core and project staff' (Forum for the Future 2002), all of whom are well known and who can give the organisation kudos and access to different arms of the world of business and finance, the captains of industry. Its 'corporate partners', which include two major international companies involved in the tourism industry, Center Parcs and Taj International Hotels, support the Forum, financially, and in return gain access to leading environmentalists, such as the three directors. In this way, they can green their corporate image.

The Way Ahead (a radical green publication by a faction of the British Green Party) cynically describes one of the Forum's early initiatives, a conference on 'Business as Partner in Social Development' in India in March 1996:

> The venue was a five star establishment in New Delhi. The accommodation rates at the Taj Palace Hotel are astronomical, but do include free 'airport transfer and daily cocktails'. Clearly those 'partners' to be 'socially developed' were not expected to be involved.
>
> (Way Ahead 1996: 4)

This form of consultation is obviously severely restricted in its reach. It is designed for the upper echelons of business decision-makers and seems to accept that, at least at this stage in the development of the notion of sustainability, other voices can be ignored. Despite its stated aim of delivering 'strong communities and practical answers to poverty and disempowerment' (Forum for the Future 2002), it firmly maintains the decision-making at its present level. It is a two-way co-optive consultation, mutually beneficial to both partners. In return for implementing some of the environmental best practices promoted and publicised by Forum for the Future as well as supporting it financially, business is able to improve its environmental credentials and to promote the belief that it is addressing the problems of sustainable development in what The Way Ahead calls an 'inherently unsustainable' system.

Of course it is not one of the Forum's stated goals to give green credentials to industry simply by association, and the Forum does not aim to set itself up as a consultation body between the developers and the developed. It aims to inform, influence and inspire these 'decision-makers and . . . key agents of change to meet the challenge of achieving genuine sustainability'. This is a daunting objective, but no doubt they will accumulate many examples of the effectiveness of their influence and these will have merit in their own right. But to achieve a degree of sustainability (or commonly agreed good practice) in a few cases, to have a catalogue of best practice examples, is precisely what the industry would want in order to justify self-regulation, to work against government legislation, and to continue doing business as usual without the systemic, structural change that is required to prevent continued concentration of wealth and power in the hands of those who already take the decisions. As a Green Party commentator remarked, 'Perhaps industry sees these consultancies as the chance to wear a clean conscience as well as a good suit?' (personal communication 1996).

Tim Forsyth's work for Tourism Concern and WWF surveyed eight consultancies whose total or partial role was to advise tour operators on the establishment of tourism abroad.

> Half of consultants questioned believed there was a commercial advantage to be gained by being environmentally aware before industry standards and market expectations make it compulsory . . . Two also wrote 'sympathy leaflets' or brochures for companies. These were partly to educate tourists about destinations, but also to add value to services offered.
>
> (Forsyth 1996: 25)

Again, the association between environmental credentials and the profit motive, using the vehicle of consultancies, could not be clearer.

Other forms of consultation can and do take place, as Figure 7.2 shows. Moreover, there are examples of healthy and generally beneficial consultation exercises which are more widely participatory and which can genuinely claim to move development practice towards a greater degree of sustainability. We believe that in any analysis of consultation exercises in tourism developments, it is important first to consider the different forms that consultation may take and second to examine the exercise's inclusivity/exclusivity, its direction of consultation (that is, who initiates it) and its beneficiaries.

Codes of conduct

It was remarked in Chapter 4 that codes of conduct publicised by tour companies, or others, can be misused as exercises in public relations. They can be seen as attempts to jump on the 'green' bandwagon for the purposes of establishing environmental credentials. This 'spin' can be used to show an environmentally responsible attitude to the supposedly discerning consumer or to head off potential antagonism and bad publicity from critical environmentalists.

Naomi Klein describes corporate codes of conduct as the 'most controversial by-product of brand-based activism', suggesting 'it's difficult to get swept up in the starry-eyed idealism of it all' (Klein 2001: 430) drafted as they are by public relations outfits wooing consumers or steadying the corporate ship in the aftermath of public scandals. As Klein (2001: 434) continues, the 'proliferation of voluntary codes of conduct and ethical business initiatives is a haphazard and piecemeal mess of crisis management'. Her conclusion is that self-regulation provides 'unprecedented power . . . the power to draft their own privatised legal systems, to investigate and police themselves, as quasi nation states' (Klein 2001: 437).

Codes of conduct, many of which are self-designed by the user and few if any of which are monitored by independent bodies, are the epitome of self-regulation, and we have already suggested that self-regulation is at best questionable and at worst meaningless – tantamount to asking the metaphorical bull to tread carefully in the china shop. So what meaning and purpose can we find for them?

The WTTERC has established a database of codes of conduct relevant to the travel and tourism industry, such has been the growth in their number especially in the international sphere. Mason and Mowforth (1995) suggest that most codes offer no measurable criteria and conform to no widely accepted set of standards. A good many of these have been devised by tour operators merely to give some palliative to the pressure exerted by the environmental lobby and in anticipation of potential criticism – the 'crisis management' of which Klein (2001) writes. They may also serve as attractors to potentially discerning and critical customers who are seeking a relatively ethical holiday, allowing them still to travel to faraway and exotic places, have a holiday and return environmentally enriched. In these senses, the code of conduct becomes more a marketing ploy than a set of standards by which to guide a company's behaviour and practices.

A good example of the ineffectiveness of a code of conduct is the Thomson Holiday Code which used to be produced by Thomson Tour Operations. This used to appear on the inside back cover of all their brochures, most of which are large and replete with colourful photographic images that suggest unlimited hedonism. The blandness of the advice given in the code of conduct and its positioning within the brochures suggest that it was included more as a public relations exercise than as a committed stance on environmental protection. Perhaps because this had become so clear, Thomson withdrew its Holiday Code from its brochures after 2000.

In March 2000, the UNWTO, along with UNEP and UNESCO, launched the Tour Operator Initiative (TOI) for Sustainable Tourism Development, to which many UK tour operators have so far committed themselves. The initiative is designed to promote protection of the environment (in its widest sense), cooperation with local communities, respect for local cultures and compliance with local laws and customs. Such an initiative may help to overcome the banality of earlier codes such as Thomson's and to inject a little rigour into the application of codes of conduct; but they can still be used as a means of marketing, lending false moral and ethical high ground to those whose principal and overriding aim is to make a financial profit.

WWF's *Beyond the Green Horizon* (prepared by Tourism Concern 1992) contains a chapter on 'Marketing tourism responsibly', which includes a list of recommendations for the tourism industry to follow. This list serves effectively as a code of conduct for the industry's marketing practices. It also points out that: 'A European Commission directive on package tours stipulates that brochures "may not make misleading claims, but must provide clear and precise information . . ."[5] on such practical aspects as price, transport, accommodation and visa requirements' (Tourism Concern 1992). Although it should be recognised that some detail may have to be sacrificed for communicability, it is also necessary to acknowledge that codes of conduct can clearly be used to make the kind of misleading claims referred to by this directive.

A suitable system of reference for codes of conduct, made easily accessible to tourists and others who wish to verify the claims made by any member of the industry, would help to improve the credibility that can be placed in codes of conduct. But any such system would need to take account of who initiates them, for whom they are intended, and who monitors their use if indeed any assessment is made of their effectiveness. To this end some system of environmental auditing of codes of conduct would be helpful. The UNWTO or WTTC are best placed and best resourced to achieve this, but as has already been established at several points in this book, these industry organisations represent the interests of their members and companies which provide their funding.

Where joint efforts are made to devise codes of conduct, the balance of power on the coordinating body often favours representatives of the industry. 'Interested parties' tend to move the aims of such efforts towards the externalisation of as many costs as possible and away from genuine environmental, social and ethical considerations. As *The Ethical Consumer* magazine suggests about ecolabelling schemes: 'It seems obvious to us that for any such scheme to appear, let alone be, impartial, representatives from industry are the last people you want on the board' (Ethical Consumer 1994: 7).

We believe that, for the tourism industry and its codes of conduct, this may overstate the case. There is a clear argument for industry representation in order to move away from simple exhortation towards incorporation of business practicalities (but see '"Reality" and "practicality" in achieving sustainability', pp. 213–14). Moreover, the active involvement of the industry is important, for without it, no codes are likely to be effective. Despite which, we believe that the problem of vested interests, as described above, also applies to the design, operation and monitoring of codes of conduct in the tourism industry. As with other techniques of sustainability, they enable the industry to redefine the concept for its own purposes so that changes to operating practices need only be minimal.

Internal management strategies

The internal management systems and strategies of tourism and other companies may be manipulated in order to allow for the environmental and social costs of operations and projects. These costs are those which have customarily been externalised by industry, that is, costs which do not have to be borne by the company but which are borne by the wider community – the costs of infrastructure construction, for instance. Many companies large enough to afford the creation of a new department to deal with these externalities have restructured themselves to include departments, divisions or branches which, supposedly, account explicitly for environmental matters arising from their operations.

One such company is British Airways (BA) whose social and environmental strategy is characterised in Box 7.4. In British Airways' (2008) own words, 'Responsible management of social and environmental issues is central to the long-term success of British Airways'. In making available so much information from its environmental audits, there is no doubt that the company has exposed itself to public scrutiny. In order to produce its reports and large amounts of data, BA created an environment branch which meets regularly with directors of other branches, offers awards for environmental achievement, produces relevant publications, and requests external consultants to undertake research when appropriate. It now produces an impressive annual 'Social and Environmental Report' which covers its management strategies, its relations with employees, customers and communities, and its environmental policies relating to aircraft noise, emissions, fuel efficiency, waste, congestion and tourism (British Airways 2001), although the last report available on its website is for 2004.

Box 7.4 British Airways' environmental strategy

Since 1989, British Airways has:

- appointed a Director of Safety, Security and Environment and a Head of Environment
- undertaken a series of reviews and audits of its main airport locations and their impacts on the local environment in terms of noise, gaseous emissions, congestion, waste and tourism
- sponsored annual Tourism for Tomorrow Awards
- sponsored the WTTERC
- conducted an environmental audit of its related tour operator company, British Airways Holidays
- initiated a Light Green programme which allows employees to suggest ways of reducing the environmental impacts of the company
- introduced a Company Travel Plan which, among other things, subsidises free public transport use by its employees
- produced an annual *Environment Report* which makes public a considerable amount of data about its operations
- established an internal Corporate Responsibility Board
- adopted a Code of Business Conduct and Ethics, applicable to all employees
- supported the Climate Care company
- joined the UK Emissions Trading Scheme.

This is its internal management strategy for the environment, and in 2001–02 the direct costs of its environment branch (including consultancy work) amounted to £1.5–2 million. The higher end of this range represents just under 0.53 per cent of the company's operating profit of £380 million in 2000–02, although it should be added that these are only the direct costs of the branch created specifically to cover environmental issues. Doubtless other departments undertake work which can be categorised as 'environmental'; doubtless also, environmental factors have influence in many technical decisions concerning such things as engine noise or emission controls.

But the question which has to be asked is whether this internal environmental management strategy is genuine and effective or whether it is an attempt to display a corporate conscience for the sake of its public image. Indeed, its website describes its attempts to improve its management of social and environmental issues as strengthening 'British Airways' corporate reputation' and acting 'as a "corporate conscience"' (British Airways 2008).

It is important, then, to ask whether the creation of BA's environment branch has had or is likely to have a significant effect on the company's operations. Is it possible that changes that have been attributed to the influence of its environment branch would have occurred in any case as a result of changing regulations and changing public awareness? And has the growing importance of the public image of airlines played a significant role in the company's perception of the need to create such a branch? Beyond these questions, the real test is whether the creation of such a department really affects the ethos and operations of the company. In the case of BA, the answer to this will depend heavily upon one's viewpoint and employment, especially if the latter is with BA or a BA subsidiary such as BAH (British Airways Holidays). It is worth noting, however, that as regards the practice of carbon emissions offsetting, which it offers to its customers through the Climate Care company, it is at the same time 'vigorously promoting massive expansion of British airports, ramping up its inter-city commuter flight services and has now launched a budget airline to popular short-haul holiday destinations' (Reddy 2006). A definitive answer is not offered here; but it is stressed that it is important to ask the questions in any analysis of the tourism industry. It is also important to recognise that such management restructuring exercises offer the potential for deception as well as genuine change.

There is no clearer illustration of this type of use of internal management restructuring as an attempt to claim false environmental credentials than the World Bank's actions through 1987 and 1988. The World Bank in itself is not a private enterprise or a trans-national corporation of course, but in essence it operates on behalf of such companies and actually leads rather than follows the objectives which guide them. It is therefore used here, briefly, to illustrate a point that is relevant to the rest of the industry.

At the start of 1987 the World Bank employed 2,700 staff above secretarial level, of whom only three were trained ecologists. Their task was to monitor the environmental and social implications and impacts of over 250 new projects each year. In response to mounting awareness and criticism from many quarters throughout the 1980s, the Bank created a new environment department which employed some forty people. By the end of 1987 it began to assess environmental threats in the thirty most vulnerable developing nations, and to design initiatives against desertification and deforestation. Particularly through its then President, Barber Conable, it also began to express a commitment to environmental and social criteria. At the time, the Bank's Acting Director of its Environment Department, J. Warford, spelled out this new commitment:

> World Bank staff have clear instructions to carefully consider the environ-mental consequences of projects, and, if necessary, amend or reject them on environmental grounds. The Bank has recently announced plans to substantially increase its staff devoted to this work, and will recruit additional ecologists, biologists, anthropologists and other expertise as necessary to ensure that our projects and policies are consistent with environmental objectives.
>
> (Personal communication 1987)

Despite its admission of past mistakes and failure and despite its newly stated commitment to reform, the Bank remained wedded to the kind of large-scale projects and major investments in energy, infrastructure and industrialisation which had previously caused so much environmental damage. Its commitment to local residents and indigenous people has taken the form largely of resettlement programmes and cash compensation. Its commitment to more conspicuous environmental issues is also highly dubious – in 1995 a billion-dollar World Bank package to help prop up private Mexican banks was partly funded by cutting previously approved Bank loans for the environment and other 'not such high' priority projects (Chatterjee 1995). As Ian Linden, then General Secretary of the Catholic Institute for International Relations, made clear:

> Unfortunately this intention of amendment did not extend . . . from the diakonic to the evangelistic. Poverty programmes were merely clamped on to the main body of doctrine. The World Bank would try to help the strangers, widows and orphans, but the content of their crusade, the Great Doctrine of structural adjustment, remained the same.
>
> (Linden 1993: 3)

It is often pointed out that awareness of an issue is the necessary stage before action. In this context this means that company directors and presidents will talk about the environment before changing their practices. It may take longer than the save-the-world groups would like, but the change will occur if given time. This argument might be based more on wishful thinking than on evidence, for the industry, more than any other branch of tourism, will change rapidly once it perceives some factor or phenomenon that will bring advantage to it.

A reorientation of the aims, ethos and operations of the tour industry implies the need for structural rather than cosmetic change in its management operations. Unless such structural change is evident, then there must be serious doubt about the intentions behind the creation of new management divisions or departments, especially where they are given severely limited budgets and are marginalised within organisational structures. Such moves as the creation of a new department and the design of a new mission statement may be necessary steps in the process of change but are not in themselves indicative of a significant shift in approach. Indeed, they may simply reflect the industry's wish to redefine sustainability in order to allow its operations to continue as usual.

With this in mind, the obvious suggestion is the centralisation of some form of rating or ecolabelling system, independently designed and applied, to which tourists and other users of the industry (such as suppliers) could refer. The problem with such a system is that it suggests some form of regulation, and, as already noted earlier in this section, most of the industry is fixed determinedly against regulation imposed by others. As Jonathan Croall (1996: 5) says, 'Sadly, such a scheme is likely to be firmly resisted by the industry'. This only serves to further throw into question the motives of industry-based moves to alter its management structures. The industry will gain credibility for its internal changes only if it subjects them to external criteria and assessment. And even then, if it is done under the guise of a consultancy, as seen earlier, there is a chance that the consultancy will act as an integral part of the industry with its currently prevailing ethos of profit maximisation.

'Reality' and 'practicality' in achieving sustainability

The tourism business community is much the same as other sectors of business in its invocation of 'business realities' in order to justify or excuse its resistance to change and to external influences. 'Commercial practicalities', 'the real world', 'the need to keep the competitive edge' and similar phrases are used to argue against regulation from government and interference from environmentalists and conservation, labour and human rights organisations.

Recently, the arrival on the scene of many companies which claim to operate a form of sustainable, environmentally friendly and ethical tourism has begun to challenge this resistance. The alliances made by these companies with environmental and conservation organisations, the stipulation of maximum group size for tours, the promotion of codes of conduct, attempts at consultation, and other activities designed to display environmental and ethical credentials have shown to companies which typified the no-change attitude that a degree of change is possible. But what needs to be asked is how far profit maximisation has compromised the changes? Or has it simply served to preserve the existing social, economic and political structures in which the problems of tourism identified by many authors since the 1980s are inherent?

We have tried here to question whether the techniques used by the industry to convey an atmosphere of change and to claim sustainability are cosmetic and superficial regardless of whether they originate from the conventional mass tour companies or the new and specialist tour companies. Are these techniques used effectively to bolster respect for the prevailing mode of production and service and all the problems that this creates? Or do they genuinely address the problems in such a way that the system no longer creates them? Have consultancies such as Forum for the Future, formed by environmentalists who are often considered to be, or at least portrayed as, radical, simply lent their good name to a variety of companies in return for minor, marginal changes in practice and a healthy consultancy fee? How valid is their justification that the real world is run by businesses whose mode of operation dominates, and that therefore we must work with and inside them in order to bring about change? Is this the only practical strategy of action to pursue? And if so, how effective is it likely to be in altering the basis of the system of capital accumulation, which we have suggested throughout is the origin of the problems caused by both mass and new forms of tourism development?

The excuse of 'the real world' or 'business practicalities' can be used as a justification for doing nothing or making the least change possible. These 'realities', however they are defined, can be used to persuade those in search of change to work with the industry in order, jointly, to examine and implement the ways in which change can be made, sustainability sought, and impacts reduced. It is precisely this 'request that environmental groups change campaigning "from exhortation to a discussion of practicalities and tools"' (Forsyth 1996: 22) which is used in an attempt to rein in the demands of the environmental lobby, to persuade it that cosmetic change is adequate, and then to subvert it by using it (the environmental lobby) as a public relations ploy.

It can be contended that both new tour operators, in their attempt to present a distinct environmental image, and conventional tour operators, in their attempt to take on environmental credentials, are in fact simply carrying on business as usual but with selected 'add-on' features and marginal changes to established practices. As one of BAH's area managers described his company's policy on social and environmental issues, 'It is basically a PR exercise in many respects because obviously there is a growing awareness of green issues' (personal communication, 19 September 1996). In an examination of environmental valuation, Colin Price (1993) summarises both the way in which the industry usurps the notion of sustainability and the logic it uses to justify this:

> The mood of the 1990s has made it mandatory for public, corporate and private bodies to embrace the sustainable development idea. The let-out clauses of weak and metaphorical sustainability ensure that this need not be financially burdensome. Sustainability objectives will be adopted by the politically astute, while continued application of discounting underwrites a 'business as usual' practice.
>
> (Price 1993: 142)

What Price (1993) calls discounting refers to the ignoring of externalities such as environmental and social costs. This captures our own analysis of what is happening in the industry. But regardless of the contentious nature of this view, it is crucial that the above questions are asked in order to test the veracity and sincerity of the industry's stated policies.

New personnel and new features of the new tourism industry

The following pages give thumbnail sketches of a variety of personnel and features involved in the provision of new forms of tourism. Although they are based on real characters, they are sketchy and stereotypical, despite which they are illustrative of many of the features and phenomena of new tour companies and new tourism that have been discussed in this chapter.

The new tour operators and their teams

Because the scale of operation and the size of company of new tour operators is generally smaller than those of conventional mass tourism – there are exceptions – it is possible to identify individual exponents and employees of the new tourism business, whereas the latter are characterised by large departments. Indeed, many of the new operators which offer independent or small group travelling, trucking and trekking tours are closely associated with one or two individual characters or with a family. High Places, for instance, is directed by Mary and Bob Lancaster, who head a team of about twelve tour leaders, all of whom are identified with a photograph and potted biography in their brochures. Similar potted biographies are given by many of the trekking and trucking companies such as Explore, Truck Africa, Naturetrek and Guerba Expeditions. Some biographies for tour leaders were given in Box 5.3.

David Sayers Travel, specialists in botanical and garden travel with several Third World destinations on their books, is a small company (now part of Coromandel) which plays strongly on its personal approach, with its early brochures inviting potential clients to visit its director, David Sayers, and his partner at their Lincolnshire home. Bales, an older and well established tour company catering for the less adventurous, older and more wealthy clients, also exploits the family basis of the company to convey its personable and approachable style.

Leadership characteristics played upon by some companies which offer expedition adventures, including overland trucking, mountain climbing, biking and rafting tours, are the youthfulness, vigour, success and machismo of their guides or leaders. 'Dave has a Guiness World Record for his 10,000 km trans-Africa ride' (Exodus Biking Adventures). 'He also descended the infamous Rhondu gorges on the Indus River in northern Pakistan, featured in the television documentary "The Taming of the Lion"' (Ultimate Descents). 'He has made a number of films of Chris Bonington's expeditions and has been on eleven trips to the Himalayas' (Himalayan Kingdoms). One 'lives close to the Cairngorms in Scotland where he and his team run the Avalanche Patrols each winter' (High Places). 'Chris has worked as a mechanic in many areas of the world, including a stint in the Yukon as a gold mine engineer!' (Truck Africa). 'A veteran of two Everest expeditions and many other Himalayan climbs' (Guerba Himalaya). 'A skilled photographer, Chris has captured an enormous variety of African wildlife on film' (Naturetrek).

Clearly, lifestyle is an important feature of the directors, leaders and partners of the new tour operators, and this is as much so in the USA as in the UK. Box 7.5 highlights a description of the country of Honduras aimed at selling a lifestyle for wealthy retired people from the USA, and it clearly reflects the characteristics of neocolonialism discussed in Chapter 3. It is interesting to contrast the notion of arranging tours which have the purpose of selling land and property in Honduras and emphasising its cheapness to potential US purchasers and investors with the fact that this country is characterised by chronic poverty, malnutrition and landlessness.[6] Such juxtapositions are symptomatic of the paradox that underlines so much tourism and further emphasise the relative power and powerlessness that are involved.

The new environmental managers

Reference has already been made to the way in which large tourism companies are restructuring their management systems to include environment departments. Along with the creation of such new departments come new posts. Thomson Tour Operations, for instance, whose environment department was noted in the last section, employ

---

Box 7.5 Offshore real estate: Third World for sale

Honduras is a country for dreamers

With 400 miles of unspoiled pristine coastline and construction costs as low as $25 per square foot, Honduras is your best buy in the Caribbean.

Honduras is a country for dreamers . . . where Caribbean breezes lift fresh ocean air through your windows . . . the sunlight glints off the water at odd angles . . . and the sound of waves lapping at the shore is just hypnotic enough to spirit away the dissonance of the 21st century.

All year round, temperatures hover in the 80s – with cooler, forested areas inland. Jaguars, armadillos, toucans and herons complete this scene of a wild, untamed paradise. The Bay Islands – and the coral reef surrounding them, which is second in size only to the Great Barrier Reef – have been a well-kept secret of divers.

Not only is Honduras a place of great natural beauty, but property prices are reasonable, the cost of living is low, crime is almost non-existent, and health care is top-notch and inexpensive. In addition, it's home to some of the best-preserved Mayan ruins in the world, and, with the second largest barrier reef on the planet just off its coast, offers some of the best scuba diving in the world.

Tax incentives make residency and investment very attractive

If you choose to retire in Honduras, you'll find that the Honduran government won't tax your US Social Security and pension payments. In fact, as a foreign resident, you'll be entitled to tax breaks. And the government encourages foreign investment in the tourism industry – a promising sector that has been largely neglected until now – with attractive tax incentives. If you are involved with a government-approved tourism project – and that could be anything from a restaurant to a hotel to a souvenir shop – you'll pay no income tax on your profits for 20 years. Even if you make 10 million dollars, you will not pay one penny in tax to the Honduran government.

Source: Kathleen Peddicord, *International Living* section of the *Offshore Real Estate & Investment Quarterly*, www.escapeartist.com website (accessed 29 May 2007)

(Kathleen Peddicord is the publisher of *International Living*, a 25-year-old business that publishes several free e-letters, a monthly print newsletter, and a growing line of books and reports, all detailing the best places in the world for Americans to live, travel and invest.)

---

an environment manager, a post held in 1995 by Paul Thornton, after earlier employment with the Disney Corporation and ten years with Thomson Travel (Commonwealth Institute Symposium, 'Managing Tourism: Education and Regulation for Sustainability', 16 November 1995: biographical notes of speakers).

Reference has already been made to Thomson's tourist code of conduct and of the code's palliative nature and lack of significance. A few of the statements made by the company's environment manager during a conference at the Commonwealth Institute in 1995 served to underline the company's lack of commitment to environmental sustainability and its use of the terms 'environment' and 'sustainability' to demonstrate publicly a supposedly ethical lining to their policies. In response to questions regarding the responsibility to educate about the impacts of tourism, Thornton explained that they were there to sell holidays and that the question of 'commercial realities' meant that there was no culture within the tour operating business to consider the impact of tourist developments on the environment. Thornton suggested that tourists should have as many baths as they want and because they have paid for it they should be able to 'abuse [the environment] if they wish'.

Andy Kershaw is BA's Manager of Environmental Affairs. Jess Worth describes his explanations of the company's environmental policies as

> the most blatant example of a company using the language of Corporate Responsibility to avoid serious consequences . . . He bemoaned the fact that the aviation industry is 'under siege' from environmentalists and the media. This really isn't fair, as according to him flying is actually responsible for a very small proportion of global emissions and BA are being 'castigated for providing a public service'. He backed up his case with a series of outdated and misleading statistics. When challenged that aviation has a much larger impact on climate change than he'd allowed, and is the fastest-growing source of emissions, he didn't deny it . . . BA, Kershaw claims, is working towards 'sustainable aviation'.
>
> (Worth 2007: 6)

There is of course an argument expounded by free marketeers that sustainability can only be achieved if market forces are given the freedom to operate. But even free marketeers acknowledge that environmental, social, cultural and other ethical factors are externalised or ignored by the market. The logic of the argument then assumes that environmentally unsustainable practices allowed by the market can only be rectified after the fact; that is, after they have produced the particular form of pollution associated with that factor. But sustainable development, as far as it can be defined, is something that happens before or during the fact rather than after it.

It has to be added, however, that other (types of) environment managers exist.

The new service providers

Not only are tour operators, airlines, hotel chains and travel agencies attempting to adapt their practices, publicity, language and clientele to the new tourism, so too are the smaller-scale service providers, the hoteliers, lodge owners and managers, restaurateurs, minibus companies, guides, and even craft salesmen and saleswomen. Take the example of the Illusig family from Italy. The family is illustrative of a new wave of First Worlders seeking to try out their luck and their entrepreneurial skills on a new 'frontier' – a kind of globalising gentrification. In Box 7.6, Jacopo Illusig describes and explains the reasons for his family's move to set up a pizzeria in Puerto Viejo de Talamanca on the Caribbean coast of Costa Rica.

For the Illusig family, the move was opportunistic; they sought to escape their disenchantment with life back in Italy. In other cases, the move is linked with altruistic motives, such as a desire to save the tropical rainforest. Also in Costa Rica, for instance, Amos Bien, an American ecologist, 'decided that the best way to convince people not to cut down rainforests is to demonstrate that conservation through tourism and ecologically sound management is the best use of land from an economic standpoint' (Blake and Becher 1991: 171). The result is Rara Avis, a 1,500-acre private forest reserve, established in 1983, close to the Braulio Carrillo National Park. But Bien has since been joined by many of his fellow Americans in Costa Rica, not all with the same sincerity of motive as his. Indeed, the North American influence is in strong evidence in most of the tourist locations within

> ## Box 7.6 The Illusig family in Costa Rica
>
> 'In Italy, we worked so that we could earn, eat, dress ourselves and pay taxes. We had to dream about going on holiday. Then he [my father] went to Spain to open a pizzeria. We were in Spain for one year. It was a different kind of tourism, one of drunks, fiestas, wild holiday living.'
>
> 'Then some people told us that Costa Rica was very pretty and very cheap and that it was possible to buy a small restaurant easily. He returned to Italy and sold the house before leaving for Costa Rica. He was here for a couple of months and then he called us and told us to come. I arrived first, in April 1994, and then the rest of the family.'
>
> 'I find it enchanting here; we live differently; it's calmer and quieter; there are no fiestas, or if there are, they are calm fiestas. We earn less than we did in Italy, but there's more time here – to go to the beach, to rest, to read a book, to live. There's less money, but there are fewer costs. It's a different life – you work four or five hours a day and say that you have worked hard; in Italy, you work 12 hours a day and say that you haven't worked very hard.'
>
> *Have you observed a growth in the number of tourists since your arrival?*
>
> 'From what I have seen during my year here, there has been a big increase. Every month, there seems to be a new place opening, a new lodge, a new restaurant – always more services, followed by more tourism, and so it goes on. I believe that more services, more infrastructure will bring more tourists – it's going to grow. It's not always good competition.'
>
> Source: Author's transcripts (1995)

Costa Rica's remaining natural ecosystems, especially in the business of providing the services that the tourists require.

This is not to say that all the service providers for the new tourism are from the First World. A majority come from within the national boundaries, but even in many such cases the seeping influence of the First World and its business values and practices have an effect through the mass media, conferences, and general contact with more and more First World tourists and operators. Gradually, as more contact at conferences and trade fairs takes place and as more linkages between the companies in the First World and the service providers in the Third World are made, the balance between what the supplier can offer and what the visitor wants tips towards the latter. The service providers purchase more and different items and expand their range of services. And slowly the prevailing and dominant value system shifts to that of the consumer from that of the indigenous caterer.

The new financiers

The Overseas Private Investment Corporation (OPIC) is a US-based organisation that was established in 1971 as an agency of the US government to link US investors with investment opportunities overseas. Annually it transfers slightly more money from the US private sector to the Third World private sector than the United States Agency for

International Development (USAID) manages with public money to public organisations in the Third World. One of the organisation's special interest industries is that of ecotourism, and Box 7.7 illustrates the importance it attaches to ecological sustainability. It is interesting to note, however, that its understanding of sustainability does not stretch much beyond the ecological, despite its recognition of the existence of indigenous cultures.

In order to qualify for an OPIC investment scheme, which would help those requiring investment, any potential Third World developer must submit at least 25 per cent of the shareholding stock to the US investor. Effectively, this condition surrenders considerable control and power over the development to the investor.

> There is no requirement that the foreign enterprise be wholly owned or controlled by US investors. However, in the case of a project with foreign ownership, only the portion of the investment made by the US investor is insurable by OPIC.
>
> (OPIC 1995: 3)

OPIC conducts environmental impact analyses for any schemes which apply for its assistance. Despite the implication in Box 7.7 that the World Bank guidelines are not sufficient for its ecotourism standards, under the heading of 'Environmental Standards' it declares:

> In determining whether a project will pose an unreasonable or major environmental, health or safety hazard, or will result in significant degradation of national parks or similar protected areas, OPIC relies on guidelines and standards adopted by international organisations such as the World Bank.
>
> (OPIC 2004: 16)

Box 7.7 OPIC's ecotourism standards

OPIC-supported ecotourism projects seek to balance profitability with ecological sustainability and respect for indigenous cultures.

Ecotourism is a means of enabling tourist dollars to flow into local communities in developing countries while simultaneously conserving ecosystems and wildlife through responsible travel that preserves cultures and natural environments.

Tourism in natural areas can generate significant adverse impacts beyond those normally associated with large-scale tourism in commercial areas. The World Bank Guidelines on Tourism and Hotels is designed for tourism in a conventional setting and does not address the specialised impacts of tourism in natural ecosystems.

As proposed in guidelines issued by Conservation International and the Ecotourism Society, all [OPIC-supported] ecotourism projects should address the following issues: (1) A comprehensive plan to protect ecological integrity and enhance community participation. (2) Local community capacity building that provides necessary skills for ecotourism development, while ensuring that this development merges with traditional practices. (3) The primary revenue source of the project must be directly linked to the conservation effort. As a result, OPIC-supported ecotourism projects can be a profitable conservation and community development model.

Source: OPIC (2004: 21)

The irony of this will be clear to all who have followed the constant, at times bitter, and damning criticisms of World Bank environmental impact assessments and its environmental and social policies by a range of environmental and campaigning organisations.

It is also of great assistance to potential overseas investors to have the services of a local investment organisation such as the Foundation for Investment and Development of Exports (FIDE) in Honduras. FIDE has focused on the creation of conditions that would attract investment under the Export Processing Zone (EPZ) concept, in which the investing company is exempt from paying local taxes. The Honduran EPZ Law of 1987 allowed privately owned and operated industrial parks and free trade zones to be established anywhere in the country, and the free-trade zone concept was subsequently extended to include tourism developments. FIDE offers its overseas clients who wish to invest in the country a service which includes individually tailored itineraries, accompanied site visits, meetings with government, banking and international officials, contacts with possible partners and suppliers, appointments with lawyers and relevant consultants, the identification of buildings and sites for rental or purchase and assistance with immigration, housing, health and schools. Overseas investment is clearly a two-way process. It can be attracted as well as pushed and to this end FIDE has offices both in Tegucigalpa, the capital of Honduras, and in Florida.

'New' academic tours

Private companies have been involved in providing travel services to groups of school and college students for as long as school journeys and fieldwork trips have been operating. But in the 1990s, three types of change in this sector of the market were witnessed. The first change was that Third World destinations were increasingly considered suitable for this type of tour; second, the industry began to exploit the possibilities for studies of the natural world and issues of environment, society and development in Third World settings; and third, operators instigated the tours themselves, without waiting for schools and colleges to organise them first.

Not only are companies now actively seeking school and college groups (see Box 7.8) but also they initiate 'academic tours', attracting First World professionals and tapping into the rise in environmental awareness. Nature, Third World culture, difference and otherness are clearly being sold in the examples given in Box 7.8. As Richard Cahill, formerly of Eco-Tours de Panamá (see Box 7.8), says, 'there is a great market for student travel' (personal communication, 1995), and there are similarities between such trips and the gap year phenomenon discussed in Chapter 5. Eco-Tours de Panamá no longer exists, but many 'new' tourism companies are beginning to exploit this 'great market'. In its

---

Box 7.8 Academic tours

Eco-Tours de Panamá

Richard Cahill, former Sales Manager of Eco-Tours de Panamá S.A., explains:

> We have been working on new products and one of them is conducting a 'Rainforest workshop' for universities. The idea is to conduct field studies for

students interested in learning about tropical forests and their inhabitants. We feel there is a great market for student travel that receives credit [for academic qualifications]. The program would be led by a professor and we would give the logistical help and guide expertise.

[The students] can come from the United States . . . and the idea is to bring them down here where they would make a specific study . . . At the end of the two weeks they would receive . . . a credit or certification, depending on what the authorities in the United States require.

(Cahill in interview, 1995)

Operation Crossroads Africa

AFRICA: Internships   Work Projects   Travel   Study   Camps

Crossroads participants do not go to Africa to impose their own Western values, but to seek comprehension of African values; and they are challenged to adjust to local ways of doing many things. You will 'experience Africa from the inside out', as one participant described her experience. Students generally arrange with their schools to receive credit (typically 5 to 10 units) for their summer experience.

(Email communication 1996; Crossroads website [www.igc.org/oca/] 2002)

Amazon, education, research, ecotourism, ecology, preservation

14 Day Excursion in the Brazilian Rainforest

Purpose: To experience, understand and explore conservation, research, education and ecotourism opportunities in the Brazilian Amazon.

The trip will be guided by Hilton P. da Silva, MS, MD, a native from the Amazon Basin currently finishing his PhD in Medical Anthropology/Public Health at the Ohio State University, and an expert in Amazon conservation and biodiversity. Other scholars and scientists from Brazil will join us along the trip.

Places to be visited include:

– The Ferreira Penna Research Station
– A Native Caboclo settlement
– Belem
– Mosqueiro, a fresh water beach
– Scenic airplane trip over Marajo Island
– and Brasilia

Since this is a not-for-profit activity the trip's price is only $2,890.00 (DBO) per person for the entire fourteen day period.

(Email communication, 10 June 1996)

magazine *Horizons*, for instance, in 2008 Explore advertised 'School Adventures One Big Learning Curve' in eleven destination areas.

Beyond the advertisements featured in Box 7.8, new organisations, which cannot be classified either as educational institutes or as tour operators, have begun to exploit this sector of the market. Ecopaz, for instance, is the Centre for Ecology, Peace and Justice Studies and is based in Teresópolis, Brazil. They advertise a two-week travel and study seminar as 'an adventure of learning and sharing in Brazil' at a cost of $1,195. Fred Morris of Ecopaz, however, followed up an earlier and similar notice on the internet with the admission that 'we at Ecopaz took some heavy hits . . . for having posted information about our Ecopaz Travel/Study Seminars, as we were accused of being "hotshot operators" trying to exploit a new hot item' (personal communication, 1995). Excerpts from his justification for the posted information or advertisement (depending on your viewpoint) are given below, and are presented here simply as an illustration of a feature of the new forms of tourism to Third World destinations:

> An experience in travel can create a larger awareness of the nature of the problems facing the worldwide environment and begin to stimulate a sense of solidarity with all peoples and the planet . . . For nine years, during the 80s, I did 'political tourism' in Central America, taking North Americans to Costa Rica, Nicaragua and Honduras to enable persons to see for themselves the realities of that part of the hemisphere, realities quite different from what was being presented in the US media . . . More than 600 persons participated in [these] and . . . returned home with a new sense of solidarity with the people of that region, as well as a new understanding of the political realities . . . While efforts at persuading tour operators to be more sensitive toward the environment and the peoples they are visiting are all important, it is more important that increasing numbers of tourists have a chance, as part of their travel experience, to come to a real understanding of the social, economic, political, environmental realities of the places they are visiting so that they can become part of the solution to the environmental problems.
>
> (Email communication, 1995)

The word 'realities' echoes the discussion on reality and practicality earlier in this chapter. As we have argued elsewhere, reality is, so to speak, in the eye of the beholder, and is open to social construction and interpretation, especially with regard to intangible, highly politicised notions such as sustainability, environment, development and even tourism.

Conclusion

This chapter has examined tourism in the context of international trade and has contrasted the concepts of free trade and fair trade as they relate to the tourism industry. The ethical basis of fair trade in tourism was considered along with its links to sustainability. The question was raised of whether the tourism industry under conditions of either free trade or fair trade is likely to challenge or reinforce the unevenness and inequality of global development.

The chapter also examined ways in which different branches of the tourism industry have adapted their operations to absorb the notion of sustainability. The techniques used in claiming sustainability of operations vary partly according to the size of operation, ranging from the vertically integrated TNCs to the small-scale lodges and service providers. Clearly, sustainability is no longer the exclusive claim of new forms of tourism. Techniques such as advertising, regulation, environmental auditing, ecolabelling, consultation, codes of conduct, carbon offsetting and management strategies are employed to 'green' the image of large- and small-scale operators and to sell the products of both mass and new forms of tourism.

A crucial question to be considered with all these techniques concerns the extent to which they promote genuine change in practices or cosmetic change which serves as good publicity but which makes little effective difference. It has been suggested here that this will depend, at least in part, on the motive behind the operations. Where the profit maximisation motive externalises all other possible motives and factors, sustainability will most likely be redefined to fit in with a business-as-usual approach. At the same time, the automatic assumption made by new forms of tourism that their operations are environmentally friendly and sustainable has also been brought into question.

The chapter has also briefly described and questioned the roles of a number of recent features of and employment possibilities associated with the tourism industry, especially those linked to the drive for sustainability and new forms of tourism to the Third World.

8 'Hosts' and destinations: for what we are about to receive . . .

In 1963, Katherine Whitehorn in the UK *Guardian* wrote: 'The only unspoilt village is the one no outsider has ever visited, not even you.' While this is extreme in its denial of the dynamic element and benefits of social integration and acculturation, it makes the point about the effect of visitors and tourists on local communities. Over forty-five years on, there is a vast body of work which demonstrates that local communities in Third World countries reap few benefits from tourism because they have little control over the ways in which the industry is developed, they cannot match the financial resources available to external investors and their views are rarely heard. This chapter focuses on local communities which receive tourists and examines their levels of power, control and ownership of tourism; Chapter 11 will further consider the effectiveness of so-called 'pro-poor' tourism interventions.

In the chapter title the word 'hosts' is in inverted commas. This draws attention to the implication that there is a willingness on the part of those who receive guests and possibly even an assumption that they have a degree of control over tourism developments in their community. As already discussed and as is well documented elsewhere, it is not often the case that local people derive benefit sufficient to outweigh the disbenefits of their community receiving tourists. Mike Davis (2002) describes the collision of the different worlds of the visitor and the visited thus:

> the gentrification of wild places . . . is always a theft of tradition, an uprooting of community. Eventually all the world's ruggedly beautiful landscapes of toil and struggle seem destined to be repackaged as 'heritage', wrenched from unemployed locals and sold off to scenery-loving burghers fleeing the cities.
>
> (Davis 2002: 116)

Chapter 3, particularly, illustrated these uneven and unequal relationships of power within local communities.

The terms 'destination community' and 'visited population' are used interchangeably rather than the word 'host', but 'hosts' is used in the title because, as will be seen in this chapter, there are examples of communities managing to take a degree of control of, and to exercise power over, the developments of tourism in their localities. The word 'destinations' is used in the chapter title because all too often the local communities visited by tourists are viewed precisely as that – places, to be collected, as if the people who live there are either irrelevant or at best incidental to the place. Alternatively, where 'experiencing the local culture' is considered to be important as part of the tourist experience, then the local people may be considered as objects or commodities. In no instances is this more so than in the case of organised tours to visit tribal peoples, a feature of the industry which is also discussed in this chapter. Pratt's notion of transculturation and the ways in which cultural domination is transmitted from one group (the tourists) to another (the visited population) are discussed.

Also throughout the chapter the term 'local community' is used somewhat loosely, reflecting its common usage which fails to acknowledge that the term is often contested, between different groups within the community, for instance, and is often assumed to represent a homogeneous population. As we shall see, this assumption is questionable and fails to acknowledge the heterogeneity and different interest groups within what are commonly referred to as 'local communities'.

The body of this chapter examines the different levels at which local communities participate in tourism and the levels of ownership and control that they, and others, hold over the resources of the tourism industry. The relationships of power between local populations and the tourists, the governments, the industry, the NGOs and the supranational institutions produce effects which reflect and promote the unequal development of visited populations and these other players in the activities of tourism. The differences in the approaches taken in pursuit of community control and government control are also outlined. Throughout the chapter, we also examine the demands made upon the visited populations that they be 'authentic'.

Local participation in decision-making

The two words, 'local' and 'participation', are regularly used together to emphasise the need to include and involve local people; and it is this juxtaposition of the two words which implies, paradoxically, that it is local people who have so often been left out of the planning, decision-making and operation of tourist schemes. At various points in the 'age of development' (see Figure 2.4), however, participation and people-focused approaches have become axiomatic with development.[1]

As discussed in Chapter 4, one of the criteria often agreed as essential to the conditions of sustainability and development in any 'new' tourism scheme is the participation of local people. For the most part there has been an overwhelming benevolence towards the process of participation and a once marginal activity has become mainstreamed in the work of many INGOs, multilateral (World Bank 1996) and bilateral agencies. Indeed, the 1990s was the decade of participatory development. As Henkel and Stirrat (2001: 168) argue, 'It is now difficult to find a development project that does not . . . claim to adopt a "participatory" approach involving "bottom-up" planning, acknowledging the importance of "indigenous" knowledge and claiming to "empower" local people'. As Jules Pretty (1995) points out:

> In recent years, there have been an increasing number of comparative studies of development projects showing that 'participation' is one of the critical components of success . . . As a result, the terms 'people participation' and 'popular participation' are now part of the normal language of many development agencies, including non-governmental organisations, government departments and banks. It is such a fashion that almost everyone says that participation is part of their work.
>
> (Pretty 1995: 4)

Moreover, Survival International (1996) has noted that 'it has become fashionable for conservationists to talk about "consulting" local people . . . This looks good on paper, but [is] hardly an adequate substitute for land ownership rights and self-determination.'

Through the evolution and development of Local Agenda 21, participation has become part of the apparatus of development, an inseparable process. The association of participation with 'empowerment' and 'sustainability' and the multi-beneficial direct and indirect impacts identified as arising from it have tended to place it on a pedestal.

Participation is not, however, without its critics (Cooke and Kothari 2001; Desai 1995; Rahnema 1992). Cooke and Kothari (2001), for example, refer to participation as the 'new tyranny', a critical attack on much development practice some would argue, but also a critique which seeks to expose and understand the sanctity in which participation is held, and the manner in which there are at times 'evangelical promises of salvation' (Henkel and Stirrat 2001: 172); and there is a whiff of spiritualism in participatory practices that mirrors the discourses of travel. As a powerful discourse, participation, it is argued, must be subjected to a critique and must be alive to the possibility that participation has 'the potential for an unjustified exercise of power' (Cooke and Kothari 2001: 4). Thus, phrases such as 'targeting local people' and 'eliciting community-based participation' (Brandon 1993: 136), and sentiments such as 'environmentally sustainable development . . . rests on gaining local support for the project' (Drake 1991: 132), and 'projects must provide direct benefits to local peoples' (Epler Wood 1991: 204) come from the perspective of the project planner, usually from the First World, as are all these examples. The planners are often associated with a major INGO (such as WWF, Conservation International and TIES as in these cases), supranational institutions such as the World Bank (as in two of these cases) or external consultancies, and all seek their own form of sustainability through their appropriate projects.

It is not so much the good intentions or ethical and theoretical value that lie behind participation that are open to question, but rather, the often uncritical manner in which participation is conceptualised and practised that has drawn increasing attention. Commentators have pointed to the manner in which participatory exercises have been conducted and the way in which it has been subsumed into contemporary developmental practice – codified and 'manualised' as part of a technical activity. Neatly linking participation with the significance of power structures, Easterly (2006) highlights one of the failings of the IMF and World Bank approaches to participation (the contemporary developmental practice):

> The participation chapter of the IMF and World Bank's PRSP [Poverty Reduction Strategy Paper] Sourcebook advocates consultation of the poor 'stakeholders'. The Sourcebook does not address how aid deals with tyranny and political conflict – if 'stakeholders' disagree with a dictator, as seems likely, who does the IMF and World Bank listen to? What if stakeholders disagree with one another? It is hard to think of how the IMF and World Bank are in a position to do anything constructive to referee political conflict and opposition to tyranny.
>
> (Easterly 2006: 144)

Cleaver (1999, 2001) argues that a new faith in participation arises from three key tenets: that participation is inherently good, that good techniques can ensure success, and that considerations of structures of power (and politics) should be avoided. Before turning more directly to its application to tourism, we consider these points in turn.

As noted in Chapter 3, commentators have already referenced the 'spiritual whiff of righteousness' in elements of the development discourse and participation is no different in this respect. Participation has been regarded as an inherently positive force for change and development. Henkel and Stirrat (2001: 177), for example, refer to Robert Chambers' 'theology of development' requiring practitioners to undergo an experience 'akin to that St Paul underwent on the road to Damascus' (Henkel and Stirrat 2001: 177), with Cleaver

(2001: 37) positing participation as an 'act of faith in development'. But far from the exercise of a value-free (perhaps 'ecumenical') approach Henkel and Stirrat (2001) suggest that what the 'new orthodoxy boldly calls "empowerment"' has special resonance in what Michel Foucault (1980) calls 'subjection', where the technical framework, approach and means of participation in participatory rural appraisal (PRA – see pp. 233–5) is preordained and fixed. Ultimately, critics argue, this form of participation drives participants to see and represent their world within the context of the PRA 'expert's' vision. The process and its failings are the subject of Oren Ginzburg's and Survival International's cartoon booklet (*There You Go!*), from which Figure 8.1 is shown here to illustrate the dissonance between the expert's plans and the local indigenous group's approach to them. Or perhaps, local people are simply pragmatic and are able to offload local knowledge into predetermined structures, but with the view to realising opportunities and resources from external programmes.

Second, and leading on from the inherently positive nature of participation, there has been an overwhelming belief that problems exist only in terms of the methods and techniques employed. But a number of critics have questioned the underlying methodology – that is, the philosophy of methods. Kothari (2001) argues that there are a number of tropes in participatory discourse expressed as dualisms that favour the South, the local community and participation over the North, the global community, government, and non-participation; the former are imbued with morality, the latter immorality. There is an underlying assumption that participation is a trip-switch to development. But as Cleaver (2001: 47) questions, 'Are we in danger of swinging from one untenable position (we know best) to an equally untenable and damaging one (they know best)?'

We started with Participatory Community Development...   *but they did not fully participate.*

Figure 8.1 Participatory Community Development – There You Go!

Reproduced by kind permission of Oren Ginzburg and Survival International from Oren Ginzburg (2006) *There You Go!* Published by Survival International.

There is also an interesting parallel to be drawn from the application of Goffman's (1997) writings on the presentation of self in everyday life, to the staged authenticity thesis in tourism and the application of participation processes through techniques such as PRA. Kothari (2001: 148) argues that PRA represents an act with participants performing distinct, 'contrived' roles and practitioners or facilitators acting as 'stage managers or directors who guide, and attempt to delimit' the performance of participants. In this way only partial or distorted representations of everyday lives are offered up or participants provide the information they believe is required to secure support and manipulate interpretations to serve their interests (Mosse 2001); as Cooke (2001: 111) suggests, 'participatory processes may lead a group to say what it is they think you and everyone else want to hear, rather than what they truly believe'.

A further point of criticism is the degree to which participation manages to challenge the traditional top-down and mechanistic approach to development practice. As Mosse (2001: 24) argues, rather than 'local knowledge' structuring and modifying develop- ment projects, formulaic project frameworks (such as widely used logical frameworks) relying on participatory planning techniques, in effect structure and articulate the local knowledge. As such there is evidence that development practice is increasingly influenced by western managerialist thinking, especially human resource management (Taylor 2001).

Third, and as Cleaver (2001: 53) forcefully argues, an emphasis on perfecting method has inevitably resulted in a belief in problem solving through participation 'rather than problematization, critical engagement and class'; and this belief in problem solving through participation fails to acknowledge the structures of power, both within so-called 'communities' and between these communities and outsiders conducting participatory exercises. This neutralisation of power structures and political priorities is also a noted limitation of the sustainable livelihood approach (Beall 2002; Carney 1998; Carney et al. 1999; Devas 2002) to development, of which participation is a critical element. This aspect is especially significant within the overall context of this book, with its emphasis on uneven and unequal structures and relations of power.

The problems start with the notion of a community as the '"natural" social entity' (Cleaver 2001: 44) and identifiable reality, and the manner in which the heterogeneity and unequal access to power is assumed away. There exists a further assumption that members of a community are willing and able to participate equally. This has been an enduring debate and problem within community development studies. The emphasis on solidarity in communities together with a closed and bounded conceptualisation of place, culture and community (S. Hall 1995; Massey 1991, 1993, 1995a) leads to the relegation of con- flict and exclusion in communities, and a failure to understand social and power structures that greatly influence the conduct and outcome of participatory processes.

As suggested above, however, a consideration of relationships of power and the discourse of participatory development also necessitates 'an investigation of the motives and ideology of the "experts" who advocate such an approach' (Hailey 2001: 98), for as Hailey argues, 'There is a suspicion that those "experts" who advocate participatory approaches to development appear to sit on some moral high ground and as such are immune to criticism' (Hailey 2001: 97). Most critically some commentators have argued that participatory discourse and practices must be understood within the broader context. Attempts 'to obscure the relations of power and influence between elite interest and less powerful groups such as the "beneficiaries" of development projects in local communities in developing countries' (Taylor 2001: 122) are indicators that this broader context is being ignored. In Taylor's view, participation is simply not working, because it has been promoted by the powerful, and is largely cosmetic, but most ominously because 'it is used as a "hegemonic" device to secure compliance to, and control by, existing power structures' (Taylor 2001: 137). As such then, participation simultaneously veils and legitimises existing structures of power.

While participation is a fundamental means of interaction and 'development', it is certainly not a panacea and does not automatically or necessarily lead to a change in the underlying structures of power. There are many well-documented examples of the relative lack of power held by local people in tourism developments in their locality – Brandon (1993: 135) cites over fifty schemes, 'many of [which] had initiated nature tourism activities, but few of the benefits went to local people'. (See also Johnston (1990), Wells and Brandon (1992), West and Brechin (1991) and a multitude of such examples in the publications of Tourism Concern and the TIM-Team.) This exclusion of local people from involvement and decision-making in the operation and benefits of tourism can be seen in some of the examples cited in this chapter (see Box 8.1, for example, on p. 231) and elsewhere in this book.

Pretty's typology of participation

The principle of local participation may be easy to promote; the practice is more complex, and clearly participation may be implemented in a number of different ways. Pretty has identified and described different types of participation as shown in Table 8.1, which offers a critique of each type.

Local circumstances, the unequal distribution of power between local and other interest groups, and differing interpretations of the term 'participation' are reflected in Pretty's

Table 8.1 *Pretty's typology of participation*

Typology	Characteristics of each type
1 Passive participation	People participate by being told what has been decided or has already happened. Information being shared belongs only to external professionals.
2 Participation by consultation	People participate by being consulted or by answering questions. Process does not concede any share in decision-making, and professionals are under no obligation to take on board people's views.
3 Bought participation	People participate in return for food, cash or other material incentives. Local people have no stake in prolonging technologies or practices when the incentives end.
4 Functional participation	Participation is seen by external agencies as a means to achieve their goals, especially reduced costs. People participate by forming groups to meet predetermined objectives.
5 Interactive participation	People participate in joint analysis, development of action plans and formation or strengthening of local groups or institutions. Learning methodologies used to seek multiple perspectives and groups determine how available resources are used.
6 Self-mobilisation and connectedness	People participate by taking initiatives independently of external institutions to change systems. They develop contacts with external institutions for resources and technical advice they need, but retain control over resource use.

Sources: Pretty and Hine (1999), adapted from Pretty (1995)

typology of participation, which is just as applicable to the idea of 'partnerships', another mantra of the current phase of development, as it is to the idea of participation. Pretty's typology is especially helpful in developing an understanding of the factors which affect the development of tourism schemes in local communities, and the case studies illustrated in this chapter are evaluated with reference to this typology.

The six types of participation range from *passive participation*, in which virtually all the power and control over the development or proposal lie with people or groups outside the local community, to *self-mobilisation*, in which the power and control over all aspects of the development rest squarely with the local community. The latter type does not rule out the involvement of external bodies or assistants or consultants, but they are present only as enablers rather than as directors and controllers of the development. The range of types allows for differing degrees of external involvement and local control, and reflects the power relationships between them. For local people, involvement in the decision-making process is a feature of only the *interactive participation* and *self-mobilisation* types, while in the *functional participation* type most of the major decisions have been made before they are taken to the local community. The only forms of local participation that are likely to break the existing patterns of power and unequal development are those which originate from within the local communities themselves. This chapter provides a few such examples, but even these illustrate the fact that local circumstances always manage to complicate the best of intentions.

It would be easy here to make the prescriptive assumption that the greater the degree of local participation, the better (by whatever definition) the project. There are those, however, who might disagree with this assumption, especially, but not exclusively, those who represent a vested interest in a particular development project – the development agencies, governments, supranational institutions, or operators, for instance. In these cases, some of the lesser types of participation might be considered preferable. It is precisely this point which emphasises the importance of the power relationships involved in any (tourist) development project, and the fact that Pretty's typology reflects this underlines its value.

At this point, it is worth contrasting a number of examples of local participation in tourism developments in order to illustrate the manifestations and effects of different levels of involvement. We have attempted simply to describe the situations of each case study in the appropriate box and in the text to relate it to Pretty's typology, which allows us to make a consideration of the power vested in each interest group and their relation to the local community.

Box 8.1 includes excerpts from an article by Phil Gunson on the large-scale Mundo Maya (Maya World) project which covers five southern Mexican states plus Guatemala, Belize, El Salvador and Honduras. It is described in Mundo Maya publicity material as 'a ground-breaking tourism and regional development initiative . . . [which] seeks to improve the lot of area inhabitants with low-impact projects which give visitors the opportunity to explore the area'. In 1991 the project initially received US$1 million from the European Commission to promote three kinds of tourism in each country: cultural tourism, coastal tourism and eco/adventure tourism. The project promotes infrastructural improvements, new hotel construction, archaeological projects and extensive international marketing through glossy brochures, in-flight magazines and travel trade shows.

As Box 8.1 makes clear, there appears to be little or no attempt to involve local communities in decision making. As the editor of *Tourism Link* (Belize Tourism Industry Association 1992: 4) explained, 'full decision making powers for all Mundo Maya affairs lie in the hands of only five persons – basically the top public sector tourism officials of each country'. The fact that this statement came as part of an article of complaint by private sector representatives about public sector control of the project underlines the irrelevance

Box 8.1 The curse of the tourism industry

Marketing men put curse of tourism industry on Mayas

Order a prawn cocktail in a hotel in Chetumal, south-east Mexico, and it will probably come smothered in 'Mayan sauce'. A trivial example, but one that shows how the tourist industry, helped by Latin American governments, is turning a great pre-Columbian civilisation and its present-day descendants into a marketing concept.

But critics, including Mayan organisations, claim that archaeological sites and indian villages face being turned into a giant theme park, and that the millions of indigenous inhabitants have no part in decision-making. 'The bottom line is that they are just exploiting the resources of our people', says Greg Cho'c of the Kekchi council of Belize. 'Mayan people are not involved and cannot influence the project.'

The aims of the Mayan World scheme . . . include improving the quality of life of local residents, protecting the environment, and safeguarding historical and cultural heritage. But the Mexican government's own archaeological and cultural institute, the INAH, is sceptical. 'They have no awareness of what ecology is', says the director of the local INAH office, Adriana Velásquez. 'If they put up a palm-thatched hut they think it's "ecological".'

She cites the once-unspoilt Xcaret ruins, which have been turned into a park for day-trippers from the up-market resort of Cancún. The entrance fee is about £13, out of the reach of local people . . .

Rolando Pérez, a Quiché Maya, is one of about 30,000 Guatemalan refugees living in south-east Mexico . . . Mr. Pérez . . . believes white and mixed-race people want to eliminate the indians. 'They see us as an obstacle to development', he says. 'They just want to build big hotels for the tourists. They're the ones that benefit, not us.'

Local initiatives, such as a village guesthouse scheme started by Mayan villagers in Belize, have been ignored, says Stewart Krohn, managing director of Channel 5 television in Belize . . . 'If you go to a meeting of the Mundo Maya you won't find a Maya there, except maybe serving dinner', Mr. Krohn says. 'The Mayan people are just being used as low-cost labour. If I was a Maya, I'd put sugar in their gas tank.'

Source: Gunson (1996)

of local communities in this contest for power. According to Pretty's typology, this example might be classified as *passive participation*.

Box 8.2 provides an account of a small-scale tourism scheme in a community of mostly Salvadoran exiles in Costa Rica. Although there has been a degree of external assistance in this case, the idea for the scheme arose from within the community itself and all the tourists' activities are under the direct control of the community. Moreover, it is one of the advantages of this type of tourist scheme that money for services rendered goes direct to those who render them without being 'creamed off' or cut down to a minimum by middlemen and agents. Although the community received considerable assistance in its early years, its tourism venture has been largely unassisted, which classifies this scheme as *self-mobilisation* in terms of Pretty's typology.

Box 8.2 Finca Sonador, Costa Rica

The Longo Maï movement, based in France, aims to give war refugees a positive and productive home and work environment rather than a temporary and transient camp. In 1978 the movement helped a group of Nicaraguan refugees fleeing the dictator Somoza's terror to form a small community in southern Costa Rica. After Somoza's overthrow in 1979 they returned to Nicaragua but were soon replaced by Salvadoran refugees fleeing the same type of state terror.

The community is now called Longo Maï – although some call it Finca Sonador – and has a population of around 400, mostly Salvadorans with a few Costa Rican campesino families, refugees from poverty and landlessness in their own country. The Finca is a relatively self-sufficient agricultural village which produces coffee and a few other products for sale beyond the village. In low income months their traditional survival agriculture is based on corn, beans, pumpkins, rice and yucca (casava). Since 1992, the Finca has diversified its economic activities by attracting visitors. About 40 families are willing to accommodate visitors and have the space to do so. The *Comité de Turismo* ensures that all associated families receive visitors at times and that prices are uniform.

Prices for meals and accommodation (in 2007) vary between 8 and 12 US dollars depending on the length of stay. Most visitors are students doing projects and staying for weeks and often months. By the standard of the northern professionals, prices are ridiculously low. Other activities offered in the village include horse-riding, guided tours, involvement in farming and general inclusion in communal and family life. Project work, donations (particularly for scholarships) and a small 'tourism tax' to co-finance communal projects through the *Comité de Turismo* ensure that the tourist money is distributed further afield within the community. Advertising for the scheme is largely by word of mouth, although a publicity sheet is posted in the Quaker lodging house in San José, and the community also has a website: www.sonador.info/

The small scale of the scheme is an essential feature both for the tourist and the host. For the visitor it is important that the experience has its air of exclusivity in the sense that this is not the usual tourist experience. For the host, it earns a little extra income with little extra cost and does not disrupt the community's or the family's way of life. In 2004 Finca Sonador won the German To Do! Prize for socially responsible tourism.

Sources:Cristoph Burckardt (2007) Personal communication; Finca Sonador (2000) 'Volunteer Work in Longo Maï and UNAPROA, Costa Rica', Finca Sonador information sheet; Mowforth, M. (1996) 'Co-operativo Longo Maï, Costa Rica', Newsletter of the Environmental Network for Central America, 19: 6–7; Pérez, J. and Richmond, L. (2003) 'Ecotourism keeps coffee farmers afloat in Costa Rica', Newsletter of the Environmental Network for Central America, 34: 9–10.

These two examples come from Central America, but later examples in this chapter are from Africa and the Himalayas, as well as Latin America. You may also find it useful to refer other case studies presented elsewhere in this book to Pretty's typology.

Participatory appraisal and inquiry techniques

In their efforts to involve local populations in the planning, decision-making and operating of tourist schemes, planners and academics have developed a range of techniques. Some of these techniques were listed in Box 4.8 and briefly mentioned in 'The tools of sustainability in tourism' (pp. 109–18). As was pointed out, techniques which allow for consultation and participation are still young in their development and suffer various shortcomings. It is debatable whether any of the relatively sophisticated techniques that have become available recently are able to improve on the traditional and well-used technique of the meeting. Local communities the world over traditionally use both formal and informal meetings to debate the courses of development and issues which may affect them. Of course meetings are not always all-inclusive; for example, women and children are excluded from many but by no means all such meetings.

More sophisticated survey techniques, public attitude surveys, stated preference techniques and contingent valuation methods all suffer the disadvantage of being conducted, administered, promoted and publicised by persons outside the local community affected by a tourism development. They are tools used by professionals administering the surveys on local communities, who by definition do not therefore enjoy control over it. Both inputs and results are often open to dispute. In terms of Pretty's typology such techniques may help to improve the level of participation, but they are unlikely to attain a high level unless they focus on the degree of decision-making devolved to the local community as well as its active involvement in the operation of the scheme. There is little doubt, however, that with their systemic and structured learning processes they can increase the likelihood of sustained success of schemes. A number of the examples given in this chapter illustrate this.

These techniques represent recent attempts to involve local people in research, policy appraisal and implementation themselves. But in reviewing the origins of participatory appraisal, Chambers (1994a, 1994b, 1994c) notes that rural appraisal techniques can be traced to the late 1970s and early 1980s as a reaction to the 'biased perceptions derived from rural development tourism (the brief rural visit by the urban based professional) and the many defects and high costs of large-scale questionnaire surveys' (Chambers 1994b: 1253).

Accompanying the general fashion for 'local participation' discussed above, recent years have seen the development of a trend for participatory approaches to enquiry and research. Participatory action research (PAR), participatory research methodology (PRM), participatory rural appraisal (PRA), rapid appraisal (RA), rapid assessment procedures (RAP), rapid assessment techniques (RAT), rapid ethnographic assessment (REA), rapid rural appraisal (RRA), participatory learning and action (PLA) and a bewildering array of other acronyms and initials have entered into use. Although they are often formally stated to involve many steps in the process, essentially they all follow the three-step procedure of participatory enquiry, collective analysis and action in the locality. (For a more detailed outline of the general procedure, see the work of the International Institute for Environment and Development (IIED).)

The principle of local participation underlies all these techniques, but they involve differing degrees of participation often in different stages of the appraisal, and the same technique may be interpreted and implemented in different ways by different people.

A specific example is given in Box 8.3, which includes extracts from a 1996 advertisement (on the internet) for members of a team to conduct a PRA in Ecuador. While the request for support or assistance in this case came from the community, through the Ecuadorean NGO EcoCiencia, to Jon Kohl at Yale University, there is an implication that it is the 'outside tecnicos' and professionals who, out of largesse, will solve the problems of the local community. The leader has a clear preference for the establishment of an environmental interpretation centre, and it is uncertain whether the idea for this arose from within the community and/or is what the local people had in mind. Nevertheless, it is likely that this example would be classified as *functional* or *interactive participation* by Pretty's typology. A little more precision in the classification may be possible with more information regarding the number and kind of major decisions which were made before the process began to involve the local community.

While the techniques of local appraisal are well intentioned by those who lead and conduct them, the critical questions concerning the balance of power are who leads them

---

Box 8.3 Recruiting a PRA team for Ecuador

This summer I will be working with an Ecuadorean NGO called EcoCiencia and will be leading a Participatory Rural Appraisal (PRA) in a small peasant community located in Pululahua Geobotanical Reserve, a 30 minute bus ride north of Quito.

EcoCiencia has held some successful campamentos with kids and parents but now sees potential and has interest in establishing a more permanent environmental education program with complete local participation. They suggested possibly a small Andean zoo.

I'm taking the zoo suggestion one step further by doing a PRA and thus figuring out with the residents what are their biggest problems and what are suggested solutions, some of which might be addressed by an interpretation centre of some kind . . .

For those who are unfamiliar with PRA methodology, in short, it's a workshop of 2–3 weeks on site where a team of outside tecnicos works with a local community team to investigate the site, discuss problems and solutions, and present them to the entire community who then discuss and prioritise them. A document results, done in part by the people and it is designed to promote autogestion on the part of the locals to use action plans and start their own development process.

In my case I will also be using the PRA results as a feasibility study for the establishment of an environmental interpretation centre which would provide economic, educational, and any number of other benefits.

PRA is an impressive methodology catching on in Latin America (while it is mainstay in India and Asia) which really puts into practice the principle of local participation that many talk about and few do.

I am currently recruiting tecnicos for the interdisciplinary team . . . Requirements are proficiency in Spanish, some experience working with rural or local people, disciplinary skills that complement the yet unformed team, and a desire to work hard during the short stay. The offer is open to all students and professionals. I welcome recommendations of other people, especially latinos, as well.

Source: Jon Kohl, Yale School of Forestry and Environmental Studies, in ENVIRONMENT IN LATIN AMERICA NETWORK internet bulletin board (21 March 1996)

and to what ends. In general they are led, or at least significantly advised, by First World professionals, and the idea that a group of outsiders visiting for a short period of time can appreciate, let alone solve, the problems experienced by local communities is rather pretentious and patronising, and suggestive of neocolonialist attitudes. It is no doubt exciting, and a little 'ramboesque', for the First World professional to be whisked off to help a community somewhere in Latin America, Oceania, South East Asia or Africa, and will certainly add kudos to their curriculum vitae. But such approaches may not be appropriate for addressing the structural and long-term problems of community development. This is not to say that collaboration between First World professionals and local communities is not possible or desirable. But a crucial element in such collaboration might be to redress existing imbalances of power so that the outcome of the exercise represents the interests of local people rather than the interests and values of the 'outside tecnicos'. To this end, Heeks (1999) suggests that there are specifically several questions to be asked where participation is being considered: What are the political and cultural contexts? Who wants to introduce participation? And why? Who is participation sought from? Do they want to, and can they, participate? (Heeks 1999: 6).

Other participatory techniques

Another related technique which attempts to involve the notion of participation in the making of decisions is the Delphi technique, which is used to set threshold values or critical levels or standards of specific aspects of a development (such as pollution levels or maximum visitor numbers) or to identify positive and/or negative impacts of a development. It is a judgemental technique involving the subjective assessments of those who take part, although it is often seen 'as a means of collecting expert or informed opinion and of working towards consensus between experts on a given issue' (H. Green and Hunter 1992: 37).

The technique uses responses of the participants to an initial questionnaire about the issue under study. The second stage compiles these responses and informs participants of the results (that is, the total responses). The third stage repeats the first but participants have the benefit of knowing all other responses. The stages can be repeated numerous times if a consensus is not close enough.

Although meetings between its participants can take place as part of the process, one of the advantages of the technique is that it provides anonymity, or at least separation, for each individual participant, thereby reducing peer pressure in the formation of opinions and permitting more honest responses. Meetings and/or collection and dissemination of responses form an important part of the attempt to reach a consensus from all individual responses.

The example given in Box 8.4 illustrates the use of the technique. It highlights the general limitations on the depth of participation attained by the technique and the consequent suspicion with which its results may be treated by local people. Despite this, there is no doubt that such techniques may be useful in the field of tourism planning, as the example cited in Box 8.4 illustrates. It is not possible to be precise in a classification of this scheme according to Pretty's typology, largely because of the limited amount of information available. One important issue concerns the adequacy of the representation of the local village population by the mayor – some groups may feel inadequately represented. On present information, however, its classification could range from Pretty's *passive participation* to *functional participation*.

Disadvantages of the Delphi technique include its method of selection of participants, the possibility of the dominant influence of particular personalities, its design by

Box 8.4 An illustration of the Delphi technique

A landowner intends to lease an area of land close to the rim of an active volcano to a building consortium which has plans to develop the site for a hotel and tourist observation post. The hotel will incorporate a restaurant which will be open to non-residents. The site is in a National Park area and is adjacent to a small wildlife reserve within the Park. Other than the crater itself, which has little vegetation apart from a few mosses, the area outside the crater rim is covered by cloud forest and has all the rich and varied wildlife and plant life associated with that vegetation type.

At present, there is only a rough track, just suitable for four-wheel drive vehicles. The scheme will necessitate the construction of a surfaced road. The Park authorities are seeking agreement from all interested parties about several factors concerning the development. These are:

- the width of the road
- the capacity of the hotel
- the height of the hotel
- the numbers of tourists allowed into the reserve
- the training and management of the tourist guides
- a possible minibus system from a village 3 km away to the observation post
- the possible need for a car park next to the observation post.

The people to be consulted are:

- the managing director of the building consortium
- the landowner
- the National Park director
- the mayor of the nearby village
- a biologist who works at the research station in the wildlife refuge
- a vulcanologist (who works at the same place)
- the director of a tour company interested in running tours to the volcano
- a conservationist from a leading environmental organisation concerned particularly with tourism.

All these people are to be asked to answer the following questions as part of a Delphi process of approaching a consensus:

1 What should be the maximum height of the hotel?
2 How should sewage from the hotel be dealt with?
3 Should there be a maximum capacity fixed for the hotel?
4 What should be the maximum width of the road?
5 Who should pay for the road?
6 The hotel will have a car park for its residents' cars. Should a car park be constructed next to the observation post for other members of the public? Or should the public be encouraged to use a minibus service? (It is possible to prevent cars, other than those of the hotel users, from using the road.)
7 As very few people visit the crater at present, there has been no need to restrict the numbers of people on the trails in the wildlife refuge. But many more visitors are expected once the hotel is built. What should be the maximum carrying capacity of a trail in the wildlife refuge?
8 Guides will be needed in the refuge. Who should train them?

professional planners rather than those most affected by the plans, and the arbitrariness of the selection of its values. The selection of participants is normally made either by the professional planners or by the interested party who wishes to see the proposal go ahead, and is most unlikely to be made by those affected adversely by the plans. The anonymity of participants does not necessarily preclude the inordinate influence of a dominant personality over the outcome of the technique as a result of the power relationships between the participants. As with any subjective assessment technique, it is feasible that the same group could produce a different outcome at a different time and a different group could produce a different outcome at the same time. Moreover, it has the potential to be used as a means of ensuring that control stays with the 'experts' and out of the hands of the local people. (For further discussion of the Delphi technique in the context of indicators for sustainable tourism, see G. Miller 2001.)

Chapter 4 introduced the notion of carrying capacity for tourism destinations and illustrated its calculation in Box 4.6. Although Box 4.6 calculated the physical, real and effective carrying capacities of a trail in the Guayabo National Monument in Costa Rica, each of these measures refers to some tangible and physical factor pertaining to the area under study. John Clark (1990), however, outlines the need to add a further dimension to these calculations by incorporating the social carrying capacity to measure the level at which tourist activity becomes a cause of social unrest and/or tourist discomfort. Clark (1990) expands the number of factors under consideration and the number of output measures or results, and makes it clear that he does not see carrying capacity purely as a negative idea which results in restriction, but relates it instead to the notion of sustainable development. Despite acknowledging difficulties in measurement and definition, he does however assume that a scientifically rational balance can be reached between all these factors, resulting in an 'objective' measure.

Watson and Kopachevsky (1996), however, argue that the result of carrying capacity measurements will always depend on the context of the situation being measured and that this context will vary not only with the physical and social environments, but also with the values of those asking the questions and establishing the conditions for measurement: 'carrying capacities cannot be determined in the absence of value judgements that specify the type of experience a given area is attempting to provide . . . the establishment of target levels is fundamentally an exercise in human value judgement' (Watson and Kopachevsky 1996: 175). They identify different types of carrying capacity (see Box 8.5) and are adamant in their belief that values 'influence all phases and elements of social research' and 'play a critical role in the choice and application of science' (Watson and Kopachevsky 1996: 177). Referring to the work of Thomas Kuhn (1962), they state that 'conceptual frameworks and paradigms rise and fall . . . as much on political grounds as on scientific ones' (Watson and Kopachevsky 1996: 177). In other words, human judgement will always be required in assessing appropriate threshold levels for a given activity, in this case tourism.

It is also clear that carrying capacities may vary with time, a point made by the Belize Centre for Environmental Studies (BCES 1994: 1): 'the physical carrying capacity of a road may decrease at night when visibility is less, or the environmental carrying capacity of Half Moon Caye, in terms of visitor numbers, may decrease when the boobies are nesting'. The way in which visitor carrying capacity has had to be recalculated over time as visitor numbers have increased to the Galápagos Islands also illustrates the point about calculation changes with both time, perceptions and values. Mowforth et al. (2008: 126) chart the rise in the 'sustainable' capacity of the Galápagos Islands from 12,000 tourists per annum in 1978 to a limit of 50,000 per annum soon after that – a limit that was exceeded by the actual number of visitors which reached 125,000 in 2005.

This growing realisation, that the setting of limits is a normative process which cannot be divorced from the objectives of the exercise or from the values of those who set them,

Box 8.5 Types of tourist carrying capacity

Ecological-environmental capacity

The level of tourist development or recreational activity beyond which the environment as previously experienced is degraded or compromised.

Physical-facility capacity

The level of tourist development or recreational activity beyond which facilities are 'saturated'; or physical deterioration of the environment occurs through overuse by tourists or inadequate infrastructural network.

Social-perceptual capacity

The level reached when local residents of an area no longer want tourists because they are destroying the environment, damaging the local culture or crowding them out of local activities.

Economic carrying capacity

The ability to absorb tourist functions without squeezing out desirable activities. Assumes that any limit to capacity can be overcome, even if at a cost – ecological, social, cultural or even political.

Psychological capacity

This is exceeded when tourists are no longer comfortable in the destination area, for reasons that can include perceived negative attitudes of the locals, crowding of the area (traffic jams) or deterioration of the physical environment.

Source: Adapted from Watson and Kopachevsky (1996)

(For an illustration of the practical implementation of the concept of carrying capacity, see Box 4.6.)

has led to enquiry into the techniques of setting limits of acceptable change (LACs), which are to some extent developing out of the work on carrying capacity. One essential element that has been built into the development of LACs has been the involvement of different interest groups in the technique, on the grounds that the setting of limits based on value judgements would be more acceptable to users if they were involved in setting them. As Roger Sidaway (1994: 1) says, 'In LAC, the entire process involves the interest groups from the outset'.

Sidaway (1994: 3) identifies the features which distinguish the LAC approach from carrying capacity and other management planning systems as 'its attempts to identify measurable aspects of quality, to monitor whether environmental quality is maintained and the degree of interest group involvement throughout the process'. The first two of these features show a similarity with another recent development in planning systems, that of sustainability indicators; and in its degree of interest group participation it resembles the Delphi technique discussed earlier.

Despite the recognition of LACs that the universal standards implied by carrying capacity calculations are not compatible with temporal and spatial variations and that the definition of standards of quality varies with time, space, interest group and value, the technique still attempts to define maximum or optimum thresholds applicable to a given situation. But as Munt (1992: 213) asserts about carrying capacity, LACs also tend to provide 'both the ecological and social justification for forms of exclusionism – an objective carefully nurtured by a growing corps of eco-missionaries and ego-travellers in Third World countries'.

Government control and community control

In tourist-speak, suitable destinations are just as likely to be countries as they are to be specific small-scale resorts, towns or settlements. In fact a browse through the brochures of new forms of tourism shows that most are organised by country or even by groups of countries rather than by resort or community. But the countries that tour operators speak of are nation states and are run by governments which often represent different interests and have different priorities from those of local communities. The Malaysian government, for instance, promoted 1990 as 'Visit Malaysia Year' with advertisements featuring images of indigenous peoples in colourful traditional dress while at the same time imprisoning members of these same groups for protesting against the logging of their land. Burma is a similar case (see pp. 330–3), and there exist many other examples of government marketing policies being at odds with the same government's treatment of indigenous and other communities.

This raises the question of what is a community. Gujit and Shah (1998), cited in Cooke and Kothari (2001), argue that:

> simplistic understandings of 'communities' see them as homogeneous, static and harmonious units within which people share common interests and needs. This articulation of the notion of 'community', they argue, conceals power relations within 'communities' and further masks biases in interests and needs based on, for example, age, class, caste, ethnicity, religion and gender.
>
> (Cooke and Kothari 2001: 6)

Although a community can be defined by scale, sector, interest, level of power and by numerous other features which express its diversity and heterogeneity, it is taken here as an amorphous term over which there is considerable debate. For the purpose of discussion, community is not regarded here as a homogeneous construct; rather, it is seen as something locational within which there are divisions of differing degrees of contrast according to many criteria. The formation and influence of local elites as a result of these divisions are discussed later. The definition of the community, then, differs according to the case study under question, and where divisions between sectors or groups within the community are significant these are pointed out and discussed if necessary and appropriate.

Gerardo Budowski (1995) has posed 'the possibility of a mass ecotourism towards natural areas' and asks whether this will affect the development of ecotourism. He further asks whether the development of 'megahotels' will operate in opposition to the small rural and rustic hotels noted for the low impact of the tourism they promote. It is this last question which highlights the different courses of action available to national governments. On the one hand, they can promote relatively small-scale, locally owned, community-based tourism facilities (small hotels, pensions, restaurants and other facilities which form an integral part of the community in which they are located). On the other hand, they can attract transnational investment into the country in the form of large-scale, luxury hotels with all associated facilities integral to the hotel.

The former approach to tourism development offers more chance that the economic benefits of the exchange will remain in the hands of local people, although the government is still able to extract indirect revenue from taxes and from tourist contributions such as park entry fees, transport tariffs, banking charges and other enterprises which the tourists make use of and which the government either taxes or manages. The latter form of tourism development concentrates the economic benefits (after the TNC has taken its lion's share) with the government, although it would be contested by the supporters of this approach that the employment deriving from such schemes would also ensure some trickle-down of benefits to local workforces.

Of course there are more than two approaches available to national governments and local communities. There are numerous ways in which revenue can be shared, both indirectly (through taxes) and directly (through allocation of a proportion of profits or takings – see the case of the Maasai, pp 253–7), and there are combinations of these.

The examples of Belize and Costa Rica have been cited in a number of places throughout this book, partly because they have both built international reputations as destinations for new forms of tourism. They have also both publicised themselves as pursuing community-based tourism development. In 1994, Henry Young, then Belizean Minister of Tourism and the Environment, for instance, delivered a speech entitled 'Community-based tourism development in Belize: government policies and plans in support of community initiatives' in which he outlined plans 'to direct tourist dollars to . . . flow into and stay in local communities' (Young 1994: 21). In Chapter 10 we give further details of how the Belizean government has used the language and rhetoric of sustainable tourism and its stated promotion of community-based tourism to further its international reputation and attractiveness.

It is noteworthy that the quotation above from Henry Young is found in a booklet funded by USAID and jointly published by the Ministry of Tourism and Environment and the Belize Enterprise for Sustained Technology (BEST). The parts played by BEST, USAID and the Belizean government in supposedly promoting community-based tourism development but actually undermining it in the region of Toledo are outlined below.

In 1991 the Toledo Ecotourism Association (TEA) initiated the construction of guest-houses in six indigenous villages (of the Kekchi, Mayan and Garifuna indigenous groups) in the district of Toledo in the south of Belize. Each guesthouse sleeps eight visitors, and is built in traditional style, using local materials. As extras, concrete floors, water tanks, screened windows and private shared bathroom facilities are also included. The outside and inside of the guesthouse in the village of Laguna are shown in Figures 8.2 and 8.3. In early 1996 there was a temporary lapse in activity in all but two of the villages because of the need to upgrade the facilities, but there are now thirteen villages involved in the scheme.

From the start the scheme encountered a number of difficulties. Local businesses, hotels and lodging houses in Punta Gorda, the district capital, opposed the scheme in the belief that it would take clients away from them. Cement for the construction was not easily obtained. Money and support from the government of Belize was slow in coming. Rivalries and local political squabbles between villages within the TEA hampered the smooth operation of the programme.

Figure 8.2 The TEA guesthouse in Laguna, Toledo district, south Belize
Source: Martin Mowforth

Figure 8.3 Inside the Laguna guesthouse
Source: Martin Mowforth

Time has helped overcome some of the local political squabbles and the government's suspicion of a scheme over which it did not have control. Recently, however, another problem has presented itself in the form of competition in a number of the villages. In a misguided attempt to improve an environmentally and socially tarnished image in the region, USAID began to promote the development of village-based and community-controlled tourism on condition that these developments also promote the benefits of competition. Through the organisation BEST, the Agency funded the construction of a new guesthouse at Laguna village. The BEST guesthouse is in competition for tourists with the TEA guesthouse.

In practice such developments are likely to cause rifts and rivalries between and within families in these villages, and this case was no different in that respect. The villages are small and, despite individual effort, enterprise and family-based cultural development, many of the practices, customs and norms of village life have depended strongly on cooperation and community action rather than on the spirit of competition.

Additionally, the BEST promotion works directly against the aims of the TEA. The need to coordinate and plan the programme and to rotate visitors to different villages to ensure a fair distribution of the benefits has been at the heart of the efforts of the TEA. In one move, the USAID created untold difficulties for the TEA and all the villagers who take part in its programme. As if not to be outdone, the United Kingdom's Overseas Development Administration (as it was then) also funded the construction of a new guesthouse in a village which already has a TEA guesthouse. The suitability of this type of promotion may be considered to be flawed. In the words of Chet Schmidt, an adviser to the TEA:

> These agencies showed what I consider to be no cultural sensitivity at all, not an overall idea of planning, nor an holistic view of the whole development process here . . . All the ugly things that happen with uncontrolled tourism begin with this kind of thing – and it was done by foreign aid.
>
> (Chet Schmidt, personal communication, October 1995)

Box 8.6 presents the case from the point of view of one of the villagers. The USAID and BEST efforts in this case would appear to be highly selective in their participation and somewhat arrogant in their willingness to override already existing community structures. On Pretty's typology their exercise could well classify as *manipulative participation*.

A further example from Central America also helps to illustrate some of the problems and limitations of the processes of participation and development. In the Cosigüina Peninsula of Nicaragua a local NGO has devised and initiated an ecotourism development programme whose ultimate aim is the improvement of the quality of life of the residents of the region. Box 8.7 outlines the programme and gives the background to a number of issues arising from the links and contradictions between new forms of tourism and what is generally understood as development.

To Mowforth, who carried out the feasibility study outlined in Box 8.7, there is no doubt that the scheme will benefit a small minority of people in the Cosigüina Peninsula of Nicaragua. But it is difficult to envisage the alleviation of the poverty suffered by the majority of the area's population through the trickle-down effect. Five hundred years of resource extraction for the sake of satisfying the consumption patterns of other societies have not yet managed to reach them. Indeed, it is debatable whether in relative terms the area's level of poverty is any lower now than it was five hundred years ago. Certainly neoliberalism imposes market and financial disadvantages on the campesino sector which it never experienced before the latest round of globalised development. Throughout the

Box 8.6 A letter from Belize

Rainforest S.O.S.

It has been very difficult for us to write what is happening here, probably because it is hard for us to understand . . .

With organised groups in 13 communities we have been told the TEA [Toledo Ecotourism Association] is the largest indigenous ecotourism conservation association in the western hemisphere . . .

The unfortunate fact is that even with this overwhelming evidence of the TEA's ability to help the local people to unite, to plan, operate, control and directly benefit from ecotourism in their areas, the major sources of funding in the country for conservation, sustainable national resource management and ecotourism development have refused to assist the TEA . . .

More difficult for us to understand is the fact that they are, on the other hand, helping to organise and are funding other independent individual groups to be involved with tourism in direct competition to the TEA groups . . .

These competing groups have divided our people and seriously threatened to weaken and destroy the unity the TEA has painstakingly developed over the years. This unity represents the only realistic hope and chance the rural and urban people of Toledo have to continue to control and benefit from responsible ecotourism.

We have been told that this is happening because there are other national and foreign interests who are not from our villages who want to take advantage of the opportunities for tourism and other developments . . .

The indigenous and other poor people know what this is all about. Before the coming of the TEA, 95 per cent of all tourism in Toledo was controlled by a small group of foreigners and wealthy Belizeans . . .

Many of them still resent and resist what they consider to be a loss of their exclusive political and economic powers. Those who oppose the real empowerment of the . . . people of Toledo are falling back on the old colonial system of divide and conquer. They have influenced the major funding agencies to use the money . . . earmarked to strengthen local community-based organisations for ecotourism and conservation development, to weaken the most successful association established for this purpose – the TEA.

. . . what about the major ecotourism organisations in the world and our government leaders who have endorsed this programme? It appears that they have been intimidated by the wealthy people . . .

Our only hope now lies with individual citizens abroad, who will write to the large conservation and ecotourism organisations to which they contribute . . . They could demand to know why the TEA has been consistently denied this assistance, and to write to the Belizean Prime Minister and Minister of Tourism and the Environment to encourage them to continue to stand up for and encourage aid for the TEA.

Source: Tourism Concern (1994b)

Box 8.7 SELVA's ecotourism development programme

In 1998 the Methodist Relief and Development Fund (MRDF) financed a feasibility study of an ecotourism development programme for ten small communities in the Cosigüina Peninsula of Nicaragua – see map. The programme was devised by SELVA,* a local environmental organisation.

The study concluded that, despite the problems of this rural area – the lack of potable water, the lack of a sewage system out-side the only town of El Viejo, the lack of electricity in most of the area's 130 villages, the poor road system, lack of social infrastructure and the widespread and evident poverty of the area – there were also significant potential attractors to backpackers, adventurers and nature tourists (Mowforth 1998). These include the Cosigüina Volcano, accessible by foot or horse, thermal waters, artesanal fishing, boat trips into the Gulf of Fonseca, a still relatively diverse wildlife despite defor-estation, and completely unused beaches and cliff shorelines. It also concluded that the development of tourist facilities should progress slowly in order to integrate local people into the programme as genuine beneficiaries and to avoid the creation of divisions between those involved in the programme and those not involved.

The Cosigüina Peninsula is an area not normally visited by tourists, other than Nicaraguans (mostly from the local area) during the Christmas and Easter holidays. There are no facilities in the area to cater for foreign tourists.

Two years later the MRDF funded a two year programme for the construction of the necessary facilities on a small area of land owned by SELVA in the town of El Viejo, which would serve as the point of entry into the peninsula for any future visitors. The facilities now include a visitor centre offering accommodation for up to 16 people, an adjacent dining room and bar with all the appropriate equipment, and a rather rustic toilet and shower block. All these facilities are built in the local rancho style with palm-thatched roofs and cane walls lined with tule matting. The photograph shows the visitor rancho with the San Cristobal Volcano behind.

In 2008, several years after the opening of the facility, it still lacks a significant number of visitors, possibly in part because it has not developed a link with any tour operator who can guarantee a constant supply of tourists.

* SELVA: Somos Ecologistas en Lucha por la Vida y el Ambiente/We are ecologists struggling for life and the environment.

most recent decade of IMF structural adjustment policies imposed on Nicaragua, UN sources have consistently reported poverty rates at well over 50 per cent of the population, those living in extreme poverty at over 40 per cent, and the country as a whole as being one of the three poorest in the hemisphere (UNDP Human Development Report, annual). Nearly 40 per cent of the population has no access to clean drinking water. All these statistics are worse in the rural areas such as the Cosigüina Peninsula. SELVA's tourism programme offers a genuine chance to improve their quality of life to a small number of people in the Cosigüina Peninsula of Nicaragua.[2] For all that, its proponents will admit that there is little chance of the programme, if successful, affecting any generalised indicators of poverty, economic well-being or sustainability in the region.

The list of beneficiaries from the successful conduct of the programme is not particularly long. It includes the members of the local NGO, SELVA, who are responsible for its operations and who will benefit financially through salaries, those few people who are employed by SELVA in the establishment and the day-to-day functioning of the activity (service providers such as boat operators), and of course the visitors. If the programme achieves its long-term goal of involving ten villages in the peninsula as well as SELVA's own centre in the town of El Viejo, then a few persons in each village will also benefit. Doubtless, there will also be a multiplier effect on the economic activities carried out by others in the villages, such as farmers who might hire out horses and storeholders whose stores may be patronised by the visitors. Even counting the whole population of each village in the total number of beneficiaries does not bring the proportion of the region's population beneficially affected by the programme up to a significant level. The effect may be significant in individual terms or to a family, but as with many small-scale NGO-inspired schemes, it gives no significant widespread response to the problem of poverty, even on the scale of a small region such as the Cosigüina Peninsula with a total population of about 80,000. Yet poverty alleviation is one of its principal stated aims.

In more than one of the selected villages, SELVA personnel have experienced difficulty in explaining to their village contacts that the benefits of the scheme and involvement in it should extend beyond the immediate family and friends of the SELVA contact. In one particular village, the local *cacique* or village head assumed that all persons involved should be members of his family and all others should be excluded. In another village, SELVA risked losing goodwill and interest by alienating a local village head who did not take kindly to the idea that as many villagers as possible should be involved. If the programme begins to show some signs of success, SELVA also risk converting themselves into a local elite, whom others will approach with deference. This is not their aim and to their credit it can be said that they are aware of the problem.

Another group of potential beneficiaries exists: that is, the national tour operators. Currently, they show only the vaguest signs of interest in SELVA's programme, but if SELVA show any degree of success in their efforts to attract visitors to the area, then it has already been suggested by one of the most prominent Nicaraguan ecotour operators that they would be interested in considering a deal with SELVA to provide ecotourists to use SELVA's facilities. In such an eventuality, the new beneficiaries will be both the operators and SELVA, with a little more trickle-down to those already benefiting in this way.

SELVA now faces the following issues and problems: How to attract visitors? How to promote the centre's use as a springboard for visiting other points of interest and the ten villages in the peninsula? How to involve the local people in the centre's use? (Local people were employed in its construction.) How and whether to turn themselves into a small-scale tourism micro-enterprise? How to ensure that the economic benefits of the activity can be used to improve the general quality of life in the area and to alleviate poverty?

The slow development of the programme is not accidental. SELVA wishes to maintain control over its programme and to be in a position to monitor progress and to spot problems. In this way, the worst aspects of cultural intrusion and colonialism by the visiting culture can be avoided and changes can be incorporated gradually and sensitively. In the sense that SELVA is a local NGO, clearly the scheme may be said to be in local hands. As a NGO, however, it does not have the power of a local municipality nor the finances associated with some national and international NGOs. In truth, the slow development of the programme has to some extent been due to the lack of finance, which is another factor which might qualify the project for funding under a First World governmental pro-poor tourism initiative (see Chapter 11).

This case study illustrates the difficulties facing local communities attempting to enter the western capitalist economy by diversifying their sources of income, a strategy conventionally considered as wise, and in this case widely considered necessary because of the failure of traditional agricultural production to provide an adequate living. Paradoxically, earlier advice from IFIs for areas such as this concerned the concentration of production on those crops for which the area has a natural advantage – following the neoliberal theoretical plank of comparative advantage. This still remains a part of the dogma we hear from IFIs despite its abject failure to meet the needs of both the local producers and local consumers. It is tempting to ask whether the advice was always meant to reflect the desires of the First World rather than the needs of the Third World, and further to ask if the advice to develop tourism reflects the same.

Because of the roles of TNCs and supranational institutions, the case of the Costa Rican government's promotion of both community-based, sustainable tourism at one moment and large-scale tourist condominia at another is presented later (see Chapter 10). It is no less relevant to the discussion here, however, and the gulf between the two policies is clearly demonstrated by the case. The arguments of those supporting community control of the development of tourism are aired regularly in the pages and letters columns of the Costa Rican daily and weekly newspapers, and with tourism being the single most important foreign exchange earner in Costa Rica, the debate is high on the political agenda and in the forefront of general discussion. Despite the international renown it has achieved through its earlier promotion of small-scale, community-based ecotourism projects, however, the 1990s saw the Costa Rican government's approval of a number of large-scale, foreign-owned tourism development projects.

In the same region, the policy divergence between small-scale, community-based tourism and large-scale, mass tourism is emphasised for Honduras by Ron Mader:

> There are two strategies Honduras is pursuing simultaneously – though from different quarters. A World Bank consultant and officials from the Institute of Tourism tout the development of a handful of luxurious five-star hotels near various 'ecotourism' destinations – Cusuco, Tela, La Tigra, Celaque and Copán . . . Another approach promotes community-based efforts. USAID has sponsored the creation of an ecotourism association – APROECOH – which has trained dozens of Hondurans. This type of grassroots, community development should be more successful (in my view).
>
> (Ron Mader, email communication 1996)

These examples highlight a number of points about this divergence. First, it is obvious that both governments and communities face a range of options and courses of action between the two extremes of government-inspired mega-projects and community-inspired local projects. While it is important that ideas for and control of tourism developments should come from within the community, it is also important that local communities can make use of and benefit from the assistance of national government resources to help establish and coordinate their ideas and schemes. This is especially so where management of protected areas is pertinent to the scheme and where the advice of specialist professionals may be helpful. It can also be crucial in enabling communities to gain access to the tourists themselves. Indeed, channels of communication and information should be an intrinsic part of the tourist system so that local communities can seek assistance when they deem that they require it.

Second, the examples discussed here draw attention to the division between the rhetoric of national politicians and their actions. Governments which in public espouse the language of the sustainable and ethical high ground of local community tourism development may be subject to external pressures (from supranational lending agencies, for example) which dictate a policy of economic liberalisation and foreign exchange maximisation. Such pressures and circumstances are more likely to lead them into the development of megaprojects which at best ignore and at worst trample on local communities, regardless of their rhetoric and stated aims.

Third, local communities may lack the base of resources, skills and finances required without assistance from a higher-tier authority such as provincial or central government. Hence, a partnership arrangement may often be more suitable than a community attempting to do everything entirely from within its own human, physical and financial resources. Partnerships do not necessarily have to include government departments or ministries, and indeed may be best advised to exclude them if they are perceived by the local community to be corrupt or likely to pervert the aims of the development. Industry or the academic community can also be involved in partnerships, especially where the tour operator's access to tourists is required or where survey or research work is required as part of the process of tourism development. Where the resources required are financial, however, then the assistance of governmental or international bodies may be necessary. But it needs to be stressed that partnerships are subject to the same types of manipulation and power-brokering as is the idea of participation.

A number of aspects of partnerships are illustrated in Box 8.8, which outlines one community's development of the CAMPFIRE scheme (Communal Areas Management Programme For Indigenous Resources) in Zimbabwe. The CAMPFIRE initiative was designed to help rural communities manage their wildlife and natural resources for the benefit of the community as a whole, and treats wildlife as a resource rather than a threat and a problem. The CAMPFIRE scheme is heavily cited in tourism's academic literature of the 1990s, and more general details can be found in B. Child (1996), G. Child (1996), IIED (1994), McIvor (1994), and Murphree (1996). In all the CAMPFIRE examples and in others where partnerships may be more complex, there is a clear need for ultimate control to rest with the local community.

The generally acknowledged success of the CAMPFIRE scheme in Zimbabwe in the 1990s has now fallen victim to postcolonial political circumstances in the country. Major problems began in 1999 with severe fuel shortages which

> made it difficult for tour operators to guarantee that tourists could travel to and from different attractions. The situation worsened in 2000 when longstanding protests over the distribution of land spilled over into violence. Thousands of Zimbabweans invaded and occupied the land of over one thousand (mainly white-owned) farms and some exclusive tourist resorts. TV pictures and newspaper reports of murders, brutal beatings and burned out farm buildings were broadcast around the world. Some governments issued travel warnings, and holiday insurance companies suspended travel cover to tourists traveling to Zimbabwe . . . In 2000, visitor numbers fell by 60% . . . and 66 local tour operators closed

---

Box 8.8 A photo-tourism partnership in the CAMPFIRE scheme

The safari hunting contract provided the initial impetus for the programme, and still produces the bulk of the income. However, further schemes for generating revenue are now being developed. In 1993 a contract was set up with a photo-tourism safari operator, to run walking safaris from a tented camp. From the beginning, the local people were more involved in setting this up than they had been with the hunting contract, which was done purely at Council level. Still the business belongs to the operator, so the risks and responsibility lie with him – the joint venture is purely a financial one.

In 1994 a further step was taken in devolution, when one community began building their own small tourism camp. Limited funding was obtained from an outside donor, so that the local WWC [Ward Wildlife Committee] had to budget and plan the project properly. They have been the decision-making body at every stage . . . When it is completed it will be staffed and managed by the WWC . . .

The money-making potential is not large, since the project is very modest, aimed at the budget traveller or weekenders from the city, but it has been, and will continue to be, an enormous learning experience for the people involved, as well as being a source of pride within the community. Having worked through this project, I can see the need for a centralised training facility to serve all such ventures . . . [and] provide a forum for sharing ideas and experiences, so that each District can learn from the activities of others. This would require outside funding to set up and run.

The connection between the hunting, photo-tourism and small scale projects is community participation and responsibility.

Source: Bird (1995)

---

shop by the end of the year . . . For the communities that rely on income from tourism, the effects have been devastating. [Between 2000 and 2002] up to 10,000 Zimbabweans lost their jobs as hotels and tour operators cut their workforces.

(Global Eye website 2007)

Despite these local difficulties, community benefit tourism initiatives based on similar lines to those of the CAMPFIRE scheme, however, are still very much a part of the southern African region. Simpson (2008) discusses a number of such schemes which perhaps fail to live up to the early promise of the CAMPFIRE programme, but which nonetheless concentrate benefits and participation on the local communities involved.

Fourth, despite the need for partnership arrangements with outside bodies, there is also a need to balance the advice of the 'experts' or professionals with the advice of the community, where the term community is as all-inclusive as practice allows. Experts and professionals are as subject to values and the sway of competing interests as are local people. They may have greater knowledge of a particular field, they may have greater funds of money, they may have greater access to particular market sectors, but they all represent interests of some kind, whether it be market forces, a political interest, a requirement for more research funding, or an ego; and it is important to balance the nature of their advice and expertise against the local interest.

Local elites

Notwithstanding the potential problems created by outside bodies, Krippendorf (1987: 54–5) noted that: 'Some locals do, unquestionably, make a nice profit out of tourism, but they are usually a very small minority belonging to the propertied classes. It goes without saying that they are staunch advocates of a further development of tourism.' It should be acknowledged, then, that the local community itself is not immune to the divisions which may come from within its own number and which may expose either the influence of a dominant local elite or the need to balance the demands and wishes of different sectors of the community. Again, an analysis of the distribution of power, in this case within the local community, is essential for an understanding of the dynamics and effects of tourism developments.

The example of the Kuna indians of Panama (Box 8.9) illustrates this point. In the main group of the San Blas Archipelago of islands on which the Kuna live (off the Caribbean coast of Panama) there are only three hotels, one of which, the Hotel San Blas, is shown in Figure 8.4. It is generally perceived by the Kuna, for a number of reasons, as being in their own interests to keep it this way in order to prevent too great a number of tourists from visiting their islands and in their attempts to prevent foreign involvement in their islands' tourism industry (Bennett 1997). But many of the Kuna manage to derive financial benefit from the tourists by selling their appliqué cloths (*molas*) (see Box 8.9). The greatest profits from tourism, however, undoubtedly accrue to the owners of the hotels and their families. Doubtless also these families gain respect from this position and in turn derive more than average influence in the development of their islands. In the case of the Kuna, this is not an influence which is generally begrudged on account of the general benefit derived from tourism and the decision-making systems in the community. Regardless of how this power and influence is used, however, the case serves to illustrate how local elites may be formed.

The case of the TEA discussed earlier illustrates the potential dangers of tourism benefits being divided among only a few. As the case makes clear, one of the essential objectives of the TEA is to distribute tourism benefits widely between and within the associated villages. In villages of up to 500 or 600 people, however, it is not possible for all families to participate in the activity of catering for tourists – not that all families would wish to do so. But inevitably, there is space for the rise of favouritism within the allocation of tourists to households for meals, for instance. In the TEA villages this has occurred in the past, but it is to the credit of the structure and operation of the TEA system of tourist allocation that the problem has thus far always been detected and corrected, in one case with the suspension of one family's involvement in the scheme. In the case of the TEA, though, the competing scheme promoted by USAID and BEST is likely to create divisions within the village which could lead to the formation of a local elite, especially as this competing scheme has access to far more funding opportunities than does the TEA.

Finally, it is worth pointing out that elitism may not relate solely to the financial aspect of tourism. Local district councils may develop an elitism of influence and decision-making without necessarily benefiting financially from it. In such cases a social distance and a communication gap may develop between the decision-makers and those they represent. Some district councils in the frequently acclaimed CAMPFIRE scheme in Zimbabwe during the late 1980s (see Box 8.8), for example, have been described by McIvor (1994: 33) as 'almost as remote as the central government in the minds of the people'. In such circumstances, representatives of local communities may take decisions on behalf of interests other than those of the people they represent. This is alluded to by O'Riordan (1978: 153), who states that 'participation on a mass scale is an idealistic dream. In a

Box 8.9 Cruising round the Choco and Kuna

To visit the Choco tribe in the Darién jungle, we stepped off the ship into *cayucos* at 4 am and made the first two hours of our journey in these dug-out canoes in total darkness. We were amazed to see at first light the intricate palisades of mangrove on either side . . .

We powered upriver for another two hours in a convoy of eight vessels to . . . their little riparian village, about 40 houses on stilts built along a broad avenue that features centrally a basketball pitch . . . Beyond are some clearings of light cultivation, but mainly they are hunters of meat, though their small stocky build suggests that over the centuries protein has been hard to come by.

They were waiting for us all along their riverside avenue, behind platforms and logs spread with the goods they had fashioned for sale, and were being continually reinforced by others who'd been alerted by messages in Coca-Cola bottles dropped from a plane by a dynamic American Mr. Fixit who has lived in Panamá for 40 years.

The men make music; the women sell. They are the advance guard of a 'nation' of about 6,000 people. They carve beautifully in rosewood, imaginative little ornaments and earrings from ivory nut. The women make the dyes and the baskets. They used to make them large, but they've learned that tourists can only handle small ones. Their goods sold on merit too – 'Who would imagine that I'd get up at 3 am to do my Christmas shopping in the jungle?' said my schoolteacher friend – and I estimate that we spent $5,000. Our cruise director takes along a float of $3,000 to bankroll those who run out.

We left more like $10,000 with the Kuna on Acuatupu in the San Blas Islands. The Kuna is a larger nation, about 50,000, with a much greater exposure to tourists. They are not as good-looking as the Chocos, their features are sharper, but they're more together commercially . . . for the past 25 years they've had the rights and control for domestic purposes of the archipelago.

They don't carve, but they make brilliant 'molas' (a word for clothes that now means specifically their appliqué designs on squares which people have been known to buy for $10 and sell for $60 at Nieman Marcus) . . . The garment of one was overprinted, '*500 años de resistencia indígena . . .*'

And they make a dead set at the photographers, offering not only their own images but carefully contrived little tableaux. For instance, a little girl with an umbrella sitting on a bench, smoking a pipe and affecting to launder a brightly coloured shirt . . .

As the Kuna see the trade flagging, they rapidly pack up and jump into dug-outs, pulling plastic covers over themselves and their goods for the choppy journey . . . to a secondary outlet, the large cruise ship *Radisson Diamond*. I remarked to our ship's official photographer that we'd just been looking at what the anthropologists call 'staged authenticity'. He was shocked by this comment, and replied that that was surely an oxymoron. 'That's what they like about it', I said.

Source: Hamilton (1995)

Figure 8.4 The Hotel San Blas, Panama
Source: Martin Mowforth

representative democracy, it is impractical and unnecessary; in a political culture with a tradition of elitism, it is out of the question.' In the prevailing economic and political system, arguably, nothing else should be expected.

Displacement and resettlement

Of all the problems experienced by local communities facing tourism development schemes, the most harrowing involve accounts of people being displaced. Such events normally reflect the distribution of power around the activity of tourism and highlight the powerlessness of many local communities. And it seems to be rare that displacement and subsequent resettlement of displaced people result in a more even and equal development.

In the literature most case studies of displacement and resettlement illustrate a deteriorating situation for those displaced. This has been especially well documented in cases where the development promotes mass forms of tourism, as in Guatemala, where three hundred campesino families were evicted in June 1996 from land they claimed belonged to the state – police burned down their homes and arrested several of them – to make way for a Spanish businessman's plans to build a tourist complex (Flynn 1996: 4). But the UK Tourism Concern's (2002b) *In Focus* magazine makes clear that displacement, often by violently enforced eviction, is also a feature of a surprising number of ecotourism projects in many parts of the world. The example of the forced relocation of the Padaung communities in Burma for the development of tourist complexes and the creation of tourist attractions – a kind of mass ecotourism or eco-voyeurism – is given in Chapter 10; and details of the displacement of groups in eastern Africa are documented in this section.

Equally conflictive situations between the tourism industry and local communities are likely to arise over the demands for and control over water resources. Figure 8.5 outlines

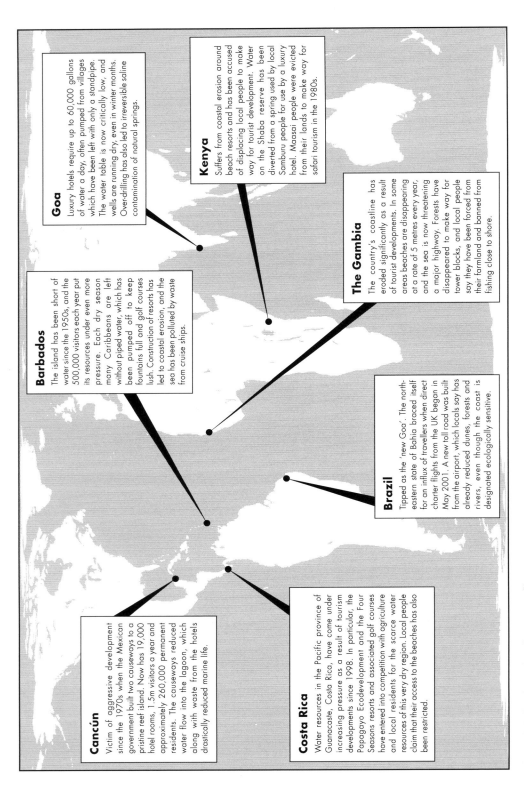

Figure 8.5 Tourism-related water conflicts

Goa

Luxury hotels require up to 60,000 gallons of water a day, often pumped from villages which have been left with only a standpipe. The water table is now critically low, and wells are running dry, even in winter months. Over-drilling has also led to irreversible saline contamination of natural springs.

Kenya

Suffers from coastal erosion around beach resorts and has been accused of displacing local people to make way for tourist development. Water on the Shaba reserve has been diverted from a spring used by local Samburu people for use by a luxury hotel. Maasai people were evicted from their lands to make way for safari tourism in the 1980s.

Barbados

The island has been short of water since the 1950s, and the 500,000 visitors each year put its resources under even more pressure. Each dry season many Caribbeans are left without piped water, which has been pumped off to keep fountains full and golf courses lush. Construction of resorts has led to coastal erosion, and the sea has been polluted by waste from cruise ships.

The Gambia

The country's coastline has eroded significantly as a result of tourist developments. In some areas beaches are disappearing at a rate of 5 metres every year, and the sea is now threatening a major highway. Forests have disappeared to make way for tower blocks, and local people say they have been forced from their farmland and banned from fishing close to shore.

Brazil

Tipped as the 'new Goa'. The north-eastern state of Bahia braced itself for an influx of travellers when direct charter flights from the UK began in May 2001. A new toll road was built from the airport, which locals say has already reduced dunes, forests and rivers, even though the coast is designated ecologically sensitive.

Cancún

Victim of aggressive development since the 1970s when the Mexican government built two causeways to a pristine reef island. Now has 19,000 hotel rooms, 1.5m visitors a year and approximately 260,000 permanent residents. The causeways reduced water flow into the lagoon, which along with waste from the hotels drastically reduced marine life.

Costa Rica

Water resources in the Pacific province of Guanacaste, Costa Rica, have come under increasing pressure as a result of tourism developments since 1998. In particular, the Papagayo Ecodevelopment and the Four Seasons resorts and associated golf courses have entered into competition with agriculture and local residents for the scarce water resources of this very dry region. Local people claim that their access to the beaches has also been restricted.

Sources: Addley, E. (2001) 'Tourists' water demands bleed resorts dry', *Guardian*, 12 May; Institute for Central American Studies (2007) 'Costa Rica: As development rages, municipalities react', San José, *Mesoamerica*, April; Personal communications

a few such cases but those listed here are merely the tip of the iceberg of a problem which promises to get worse in the coming years. Leo Hickman (2007) highlights the problem in an article headed 'Sun, sand and slavery': 'Many places – from Cancún to Costa Rica, Thailand to Ibiza – say that water supply has become one of their biggest concerns now that they rely so heavily on tourism.'

Around Third World countries, the list of stories of displacement and resettlement is endless and is charted regularly in the newsletters and publications of tourism campaigning groups, such as Tourism Concern, and human rights groups. One would assume, however, that the supposed ethical base of new forms of tourism would avoid such pitfalls. Unfortunately, much evidence appears to contradict this assumption, and displacement and resettlement have become frequent outcomes of policies aimed at conservation and protection. Two examples are cited here, both from Africa, together with others elsewhere in the book, in order to illustrate the ways in which the goals of tourist money, conservation and 'sustainable' development policies may be linked together in the dispossession of local communities and indigenous groups from their land. Neither example here, however, is entirely pessimistic and both include an element or two of a positive nature for local communities. We will return to displacement in Chapter 11 when discussing the aftermath of the 2004 Asian tsunami.

The Maasai in Kenya and Tanzania

Box 8.10 introduces the first case study of displacement, the Maasai in Kenya and Tanzania. It makes clear that the displacement was not a single move, a one-off event; rather, it has been a prolonged and systematic persecution of the Maasai by the Tanzanian and Kenyan authorities, although it should be stressed that these authorities were guided in their actions by First World conservationists and operated not only in their own interests but also in those of developers in the tourism industry.

It is clear from Box 8.10 that after the Second World War, all the policies of exclusion and resettlement in this case have been pursued in the name of conservation, especially the conservation of wildlife. The mechanism for doing this has been the creation of national parks and wildlife reserves, and the impetus for creating them came largely from First World conservationists and scientists who suspected that pastoralism was responsible for environmental degradation and decreases in wildlife numbers. George Monbiot (1995) accuses some scientists of having

> maintained that local people have always been a threat to wildlife: that they hunt the game with destructive methods and over-graze the land. These arguments have been well-rehearsed among conservationists, and are known to many of the tourists visiting Kenya.
>
> (Monbiot 1995: 11)

Monbiot argues that the Maasai's activities did not threaten the wildlife and that the work of earlier scientists was clouded by 'colonial disdain' and 'genuine misunderstandings about savannah biology'. He summarises:

> What is incontestable is that a fantastic abundance of wild game continued to exist alongside the herds of the Maasai and other nomads up to and beyond the arrival of the British in East Africa. It was indeed because the Maasai had not destroyed the populations of game that the Europeans wanted to conserve their lands.
>
> (Monbiot 1995: 11)

Box 8.10 Displacement of the Maasai

Early 1900s

European hunters eliminated some wildlife species and decimated others which had survived over 2,000 years of contact with Africans and their livestock.

Second World War

Hundreds of thousands of wild animals killed to feed British troops.

1959

Serengeti National Park in northern Tanzania divided into the Ngorongoro Conservation Area (NCA) and smaller Serengeti National Park. The Maasai who lived in the latter (and had done so for over 200 years) were moved into the NCA so that the Serengeti would become a game park with no human interference. In the NCA, 'should there be any conflict between these interests, those of the latter [the game animals] must take precedence' (Governor of Tanganyika, 27 August 1959).

1960

Establishment of the Maasai Mara Game Reserve in Kenya, adjoining the Serengeti in Tanzania, further restricted the movements of the Maasai.

1974

Maasai forced to evacuate the two crater areas within the NCA on the grounds that their presence was detrimental to the wildlife and landscape.

1975

All cultivation within the NCA prohibited.

1976

Maasai prohibited from entering the Olduvai Gorge on the grounds that their 'livestock were detrimental to the archaeological value of the site' (Olerokonga 1992: 6).

1980

Collection of resin, a source of cash for the Maasai, stopped. Burning grasses in the highland areas also restricted.

1987

Anti-cultivation operation mounted by the authorities against the Maasai, who were farming small plots: 666 people arrested; nine jailed for six months; 549 fined (Olerokonga 1992: 6).

1994

Allegations of torture, false imprisonment, theft and corruption by the Kenya Wildlife Service against the Maasai (Corroboration by Paul Ntiati, an African Wildlife Foundation representative, cited in Monbiot 1994: 93).

2007

Groups of Maasai seek restitution for land taken from them by the British in the twentieth century when Kenya was British East Africa and when 'thousands of people were forcibly moved into reserves, resulting in a massive rise in diseases, drought and famine' (Hughes 2007: 45).

Nowadays, it is difficult to escape the realisation that wildlife and the pastoral activities of the Maasai have managed to coexist in the region for many centuries, and that the landscape (at least until recently) was the product of their grazing and burning practices. To Deihl (1985: 37), it seems 'ironic . . . that one of the first steps in establishing a national park is to rid the region of its original caretakers'.

As the Maasai have been excluded, so the tourists have been allowed access. Despite the exclusion of the Maasai from the crater areas within the Ngorongoro Conservation Area (NCA), for example, there are several camp sites for tourists on the crater floor, although casual camping is no longer allowed. Tracks and roads have been created to allow tourist vehicles easy access to wildlife, thereby destroying natural vegetation. Although the Maasai have been excluded from the Olduvai Gorge, many tourists enter it every day, some even removing stones as souvenirs (Olerokonga 1992: 7). Tourists also gain access in hot air balloons, gliding low over herds of wildlife, stampeding and disorienting them (Olindo 1991: 37), although there is no recognition of this problem in Abercrombie & Kent's 2005 brochure in which they advertise 'Hot Air Ballooning: To drift over the herding game of Africa's great savannahs is one of life's unforgettable experiences. Celebrate your flight with a glass of bubbly and breakfast in the bush.'

In 1991, Perez Olindo of the African Wildlife Foundation in Kenya and formerly director of the Kenya Wildlife Department outlined several of the plans of the Kenya Wildlife Service (KWS) which were aimed at addressing the tourism and conservation problems in the game reserves and national parks. These included road construction, a ban on the development of new tourist accommodation and on casual camping, minimum flight levels for balloons, and the promotion of ecological sensitivity in the tourists. There is no mention there of the guardians of the original environment, the Maasai, neither in terms of their land and grazing rights nor in terms of an acknowledgement that tourism and conservation have been largely responsible for their dispossession and displacement.

Although seen by some as the best disciplined conservation management force in Africa, Monbiot (1994: 121) describes the KWS as a 'para-military sustainable tourism-con-servation organisation', and Fernandes (1994: 11) says it 'has implemented sustainable market driven model growth strategies which deserve to be condemned'. Moreover, it is clear that much of what is done in the name of conservation is actually done to protect the profits from tourism: 'Several times I was told by (tourism) conservation officials that the Maasai had to be kept out because the tourists did not want to see them there' (Monbiot 1994: 119). Olerokonga (1992) makes it clear that the tourist and the Maasai are currently alienated from each other.

Fernandes (1994) maintains that such strategies owe much to the Brundtland-inspired sustainable development approaches. It is the action plans and agendas that have emerged from this generalised approach which Adams (1990: 67) claims are 'firmly anchored with the existing economic paradigms of the industrialised North. This might be called the approach of "green growth".' The implication is clear: the power exercised here derives from the First World, is expressed by conservationists, acted upon by a powerful local elite, benefits First World tourists, and serves to increase the inequality of development by, first, preventing the local communities from conducting their traditional ways of life and, second, excluding them from the benefits of the activity of wildlife tourism. As Olerokonga (1992) explains:

> Since the mid-eighties many stand along the main road to Serengeti and the Ngorongoro Crater waiting for tourists and hoping that they will pay to take pictures of them . . . For both, their only interest is to profit as much as possible from each other – the tourists by taking pictures of the Maasai, the Maasai by getting money from the tourist. They don't see each other as dignified human beings.
>
> (Olerokonga 1992: 7)

In the first half of the 1990s, then, conservation in East Africa was still largely a matter of separating land from its traditional human inhabitants. New approaches are currently being explored, however, with the aim of attaining some redress of dignity and power. In 1996, for example, a small group of the Maasai opened the Kimana Community Wildlife Sanctuary covering over 6,500 acres in Kenya, having negotiated a deal with a British tour operator to construct a luxury lodge, from which a proportion of the tourist payments (approximately US$12 out of each US$80–100 paid per tourist per night) are made to them. Having calculated their takings before the deal was signed, they developed plans for a school and clinic for their community. In November 1996, the British Guild of Travel Writers recognised the significance of this deal by the Kimana community with one of their annual awards. By the year 2002, the school and clinic had not materialised and the benefits had not been as widely distributed as originally intended, which detracted from the significance of the scheme. Nevertheless, this was still an important step for one group of the Maasai – in the past such deals directed benefits into the hands of a single individual or a single family rather than a whole group. Other examples, perhaps more successful than the Kimana Community scheme, exist. The Eselenkei Community Wildlife Sanctuary in Kenya, for example, is owned by a Maasai community in which the financial benefits of the wildlife tours in the sanctuary are reported to be widely distributed and appreciated throughout the community (David Lovatt Smith, personal communication, June 2002).

For the Maasai, these cases may not make good their displacement, nor retrieve their former lifestyles, nor compensate them for all the losses and betrayals they have suffered in the past. But now for some groups of the Maasai at least such schemes offer a hope of improvement, even if they can hardly be said to threaten the dominant model of development which has been at least in part responsible for the displacement. Such mild 'success' as this highlights a deep division among the Maasai over the wisdom of involvement in these kinds of tourism projects. The views of the Kimana community proponents of the sanctuary and lodge contrast sharply with the views expressed in the following statement from one of the Maasai:

> We know there is money to be made from tourism. We already have tourists staying on our lands in tented camps. And, yes, they bring us an income. We don't need the Kenya Wildlife Service to tell us that. But you can tell Dr Leakey [director of KWS until 1994] this.[3] We don't want to be dependent on these tourists. We are Maasai and we want to herd cattle. If we stopped keeping cattle and depended on tourists, we would be ruined when the tourists stopped coming.
>
> (Cited in Monbiot 1994: 98)

Mountain gorilla conservation in Uganda, Rwanda and Zaire

The case of the international conservation agencies' attempts to save the remaining mountain gorillas from extinction in Uganda, Rwanda and Zaire illustrates an example of wildlife conservation that has had some success, although the onset of genocidal inter-ethnic strife in the region in the early 1990s has led to uncertainty about the long-term prospects for mountain gorillas. The case is included here because publicity about the success of the wildlife conservation and tourism revenue attraction measures has overshadowed the human displacement which these measures caused.

Between 1960 and 1973 evidence showed that the mountain gorilla population in the area of the Virunga volcanoes had declined from 450 to 260, mainly through poaching (Weber 1993). The gorillas were under pressure not just from poachers but also from local farmers who required more land to provide food for a growing population. As a result of this pressure, the Mountain Gorilla Project (MGP) was established in Rwanda in the late 1970s. The project had the linked goals of promoting ecologically sensitive tourism, improving park security and spreading conservation awareness. The ecologically sensitive tourism essentially took the form of gorilla watching similar to that described in Uganda's Mgahinga National Park by Melinda Ham:

> Holding their breath, the four tourists crouch in the grass, their long-lens cameras at the ready. Less than five metres away, their prey gently breaks off a piece of bamboo as thick as a human arm, tears off strips with his teeth and chews the soft green core. The silver-backed gorilla stands proud. Beside him a female grooms her neighbour while a baby gorilla somersaults playfully beside her. They take little notice of the human presence . . . The tourists have spent nearly two hours climbing up through the dense, muddy bamboo forest, accompanied by two trackers and a Uganda National Parks (UNP) ranger. The trackers carry machetes to cut a narrow path through the forest. The ranger carries an AK47 assault rifle in case of a chance encounter with armed poachers. Under the strict rules governing gorilla-tracking tourism in Uganda the tourists can spend only one hour with the gorillas once the trackers find them.
>
> (Ham 1995: 1–3)

For Rwanda's Parc National des Volcans, Lindberg and Huber (1993) show that the revenue from tourist fees to see the gorillas rose from US$7,000 in 1976 to US$1 million in 1989, a sum which far outweighed the cost of running the park (approximately US$200,000 in 1989). Demand was high and only 24 visitors per day were permitted entry. This allowed the fixing of a high individual fee – almost US$200 per person for a one-hour visit which, as Sherman and Dixon (1991) suggest, was around the highest charged anywhere in the world and may have been near the upper limit that visitors were willing to pay. Moreover, it has been estimated that an extra US$3–5 million annually was paid into the national economy by foreign tourists (Weber 1993).

The high fees have drawn charges of economic elitism, but the scheme was widely deemed to be successful. From the late 1970s to 1989, the gorilla population in this area rose from 260 to 320, which was due, according to William Weber (1993), a specialist in primate conservation at the New York Zoological Society, to tourism revenues paying for more forest guards and in turn reducing gorilla poaching. Since 1994, when genocide in Rwanda took up to a million lives, the popularity of gorilla watching declined from almost 7,000 people per annum in 1993 to just over 2,000 in 2001 (Vick 2002).

In Uganda a similar success in terms of wildlife conservation was achieved only after the displacement of over 1,300 people. Despite the success in Rwanda, it was clear that very considerable threats – civil unrest, poaching and loss of habitat to cultivation – still loomed over the primate population. Accordingly, in 1990 several international conservation agencies led by WWF formed the International Gorilla Conservation Programme (IGCP) in cooperation with the Ugandan, Rwandan and Zairean governments. Under the Global Environment Facility (see pp. 160–1) the World Bank provided US$4 million to fund gorilla conservation projects in Bwindi and Mgahinga National Parks in Uganda.

Uganda has only begun to realise and exploit the potential tourist revenue from its gorillas since the creation of the IGCP. The Mgahinga National Park covers only 33 square kilometres in the Ugandan part of the Virunga Mountains and shares the gorillas with Rwanda and Zaire. Ten square kilometres of this bamboo forest had been cleared illegally by over 1,300 peasant farmers, who planted peas, potatoes and wheat on terraced plots. Apart from the food grown by these families, the land was also used to provide timber and bamboo for construction, and plants, barks and fibres for medicinal purposes. They also used the forest for their beehives and their water supply came from streams in the reserve. Every year the gorillas were pushed further up the slopes of the volcanoes and across the border into Zaire and Rwanda. Melinda Ham explains the official response to these actions:

> Originally Mgahinga ... was only a forest reserve. But in 1992, the IGCP encouraged efforts to turn it into a protected National Park ... In mid-1992 the UNP [Uganda National Parks] evicted the peasants from their illegal plots. However, after a year of bitter hostility between National Park authorities and the community, the villagers were compensated with funds provided by the US Agency for International Development ... Didas Mutabazi, one of Mgahinga's six park rangers, says: 'Even though the villagers have received money, they are still very angry about losing their land' ... Only two farmers were offered employment in the new national park as trackers. Both refuse to talk about the loss of their land.
>
> (Ham 1995: 1–3)

Melinda Ham (1995: 3) also points out that Mgahinga was already 'being eyed by major tourism companies for investment'.

Such displacement, resettlement and later compensation certainly compromises the success of the conservation measures. Ham describes the Mgahinga case as an 'ideological battle between international and local conservationists' and cites Jaap Schoorl, a Uganda National Parks adviser seconded from the international aid agency CARE, as saying 'We have learned that if you don't get the cooperation of the local people you can forget about gorilla conservation altogether. People have to see the benefits of having a national park on their doorstep' (Ham 1995: 1).

After the eviction in 1993 the Ugandan government established an advisory committee for the region, which included representatives from the local community.

> The committee decided that the whole community could still have access to the park, to collect water, gather plants or place their beehives in the forest. But access would be limited and subject to detailed agreement between all groups. The committee also agreed that the community would receive ten per cent of the revenue generated from the park entry fees paid by tourists.
>
> (Ham 1995: 3)

This example shows up the potential conflicts between wildlife conservation and human survival strategies and serves to highlight the importance of tourist revenue in these conflicts. Tourism and conservation can be the cause of displacement and are factors which serve to increase the unevenness and inequality of development. But potentially they also offer a partial solution to some of the problems which they created in the first place.

The Ugandan example further highlights the gap between two opposing conservationist views widely discussed in Chapter 6 – *parquismo*, as Lorenzo Cardenal (1991) calls it, and an integrated approach. The former is the policy of excluding humans from the area to be protected and conserved (the *parque*), and the latter refers to the integration of human activity with flora and fauna conservation. In the case of the Mgahinga National Park, the difference is emphasised by the words of Jaap Schoorl above. Jean Carriere of CEDLA (Centro de Estudios y Documentación Latinoamericano) also highlights the former approach in his description of US-influenced environmental institutions as tending 'to see environmental protection in isolation from the social context, and [would] soon convert Costa Rica's forests into fenced-off green museums surrounded by starving peasant families' (Carriere 1991: 198).

Colchester (1994) has pointed out that conservation NGOs have generally derived their funding from the establishment and have attempted to use the power of the state to impose their visions and goals. In search of funding for their conservation programmes, international conservation organisations tend to form alliances with supranational organisations such as the World Bank, First World governments, or even transnational corporations. In so doing they find their programmes becoming geared to the kind of economic justifications – hence, the need to raise tourist revenue – which fit with the dominant economic paradigm promoted by these organisations. With such justifications, and with the moral high ground of ecological sustainability on the side of the conservationists, it is relatively easy to treat local people and communities as inconvenient and to confirm the need to displace them.

Visitor and host attitudes

Tourism is widely touted as a beneficial cultural exchange for all parties involved, as having contributed to the general well-being of peoples around the world, as having stimulated economic development and as 'promoting better understanding between races, religions and human beings worldwide' (UNWTO 1999). With such descriptors one might be forgiven for believing that the relationships between tourists and local people in the visited destination areas are always rosy. Many friendships are indeed made between the visitors and visited, but the relationships between visitors and hosts give rise to a range of other outcomes. In the following pages we discuss three possible outcomes of the relationships created by tourism to Third World destinations.

Transculturation

One of the most protracted criticisms of tourism in the Third World has concerned its impacts on indigenous cultures. Critics have consistently rounded upon cultural 'bastard-isation', 'trinketisation', the destruction of indigenous cultures, and so on. Tourism, they contend, is a process of acculturation through which Third World cultures are assimilated into materialistic First World lifestyles.

In the first part of this book, it was suggested that advocacy of the need to protect cultures finds strong resonance in colonialism and romanticism of the past, an approach that has the potential for institutional racism that celebrates primitiveness. As Robins (1991: 25) describes, in a process of unequal cultural encounter, 'foreign' populations 'have been compelled to be the subjects and subalterns of western empire, while, no less significantly, the west has come face-to-face with the "alien" and "exotic" culture of its "Other"'.

Similarly, in discussing travel writing, Pratt (1992) refers to 'contact zones', 'social spaces where disparate cultures meet, clash, and grapple with each other, often in highly asymmetrical relations of domination and subordination – like colonialism, slavery, or their aftermaths as they are lived out across the globe today' (Pratt 1992: 4). Contemporary tourism, and particularly new tourism in the Third World, is staged in these so-called contact zones, which serve to emphasise that tourism is experienced in sharply differentiated ways by visitor and host. Pratt continues:

> A 'contact' perspective emphasises how subjects are constituted in and by their relations to each other. It treats the relations among colonisers and colonised, or travellers and 'travelees', not in terms of separateness or apartheid, but in terms of co-presence, interaction, interlocking understandings and practices, often within radically asymmetrical relations of power.
>
> (Pratt 1992: 7)

However, this asymmetry of power between host and guest tells only half the story. First, this has been encapsulated in the ideas of Edward Said (1991) in *Orientalism* (Chapter 3) – and the images portrayed in the collage in Figure 3.2. More often than not, it is charities, social movements and tourists that talk about the rights, cultural practices and uniqueness of Third World cultures, as if these people do not have a voice (which of course many do not) and are unable to represent their own views. Hence, Said's satirical 'quotation' of Marx: 'They cannot represent themselves; they must be represented.'

Second, through a process termed transculturation, Pratt (1992) attempts to encapsulate the way in which marginalised or subordinated groups select and invent from materials transmitted to them by dominant 'metropolitan' cultures. Stuart Hall (1995) refers to this as a 'cultural strategy' which operates between previously sharply differentiated cultures which are forced to interact. It is this process of change that those engaged in the promotion and undertaking of new tourism find difficult to accept. It is a feeling that we are somehow being cheated of 'authentic experiences', that this is no longer the real thing. This search for authenticity lies at the heart of much new tourism activity, as noted in Chapter 3, where it was suggested that authenticity might be understood as a part of the desire for (cultural) sustainability. It is an aspect of new tourism that is sharply reflected in trekking, an activity about solitude and distance from other tourists, but also about contact with 'real' cultures.

Trekking is the visiting of off-the-beaten-track locations and involves walking, often but not always in organised parties accompanied by a number of porters. The names of some of the small independent tour operators (High Places, Himalayan Kingdoms) which organise trekking tours testify to the rugged mountainous areas often visited by trekking parties and individual trekkers, but gentler highland areas such as Thailand, Kenya and Tanzania are also favourite trekking destinations. The phenomenal growth of trekking in South-East Asia, Latin America and Africa (Brockelman and Dearden 1990; and as testified by the brochures of the 'new', specialised tour operators) underlines its importance in new middle-class travel. In Nepal the number of trekking permits issued rose from eight in 1966 to approximately 13,000 in 1976, 47,000 in 1987, 61,000 in 1988 and 80,000 in 1996. In the late 1990s the Annapurna area attracted over 50,000 trekkers annually, approximately one and a half times the area's year-round population; the Khumbu (Everest) region, home to around 4,000 people, received 15,000 trekkers in 1996 (Nepal 2000; Tourism Concern, undated).

The environmental effects of trekking in the Himalayas are documented in many papers and articles. Nepal's forest area is believed to be decreasing at a rate of 3 per cent per year with higher rates in lowland areas and heavily trekked routes in the hills. One hectare of cleared forest loses 30–75 tons of soil annually. Also regularly cited is the problem of litter, with a special topic of interest appearing to be the ugly and unhygienic streamers of used toilet paper, which are the exclusive contribution of western trekkers.

In Nepal trekking also provides an estimated 24,000 full-time jobs with as many as 70,000 people employed as porters on a freelance basis. 'Porters are poor people, and the majority work without proper insurance, without proper clothing and for very low wages. They get next to nothing if they are injured or disabled while working, and they receive no proper training' (Agha Iqrar Haroon, President of the Ecotourism Society Pakistan, quoted in Tourism Concern 2002a).

As Tourism Concern make clear, the cultural impacts of trekking are almost impossible to quantify, but Box 8.11 gives anecdotal evidence of some of the subtle cultural impacts of trekkers upon local populations. On the basis of his observations of trekking in Thailand, Erik Cohen (1974, 1989) modified the idea of staged authenticity identified by MacCannell (1973, 1976) into the idea of communicative staging (see Box 3.7). Central to this debate has been the way in which local populations adapt to tourism. The last example cited in Box 8.11 particularly reflects the notion of transculturation in its description of those manifestations of First World influence which have been 'selected' for use by the younger Sherpa men.

One advantage of the concept of transculturation is that it allows us to explore possibilities that lie beyond the often repeated, even slavish, charge that tourism distorts, disrupts and bastardises Third World cultures. In some instances this is, of course, exactly what happens, especially where the power of the tourism industry is intense, as described in the next few pages. Indeed, these problems are also common features of trekking, as illustrated in the UK Tourism Concern's (2002a) campaign entitled 'Trekking wrongs: porters' rights'. But it is a view that debars us from considering how the visited actually adapt and borrow from cultural practices and in turn modify their own cultural practices or ways of making a living, even in circumstances where their power is differentially distributed.

Tribal peoples and zooification

Survival International, an INGO which supports indigenous groups, has in recent years adopted the term 'tribal peoples', reflecting their representation of people who live by tribal

Box 8.11 The cultural effects of trekking

We wait on a path in the Hinku Valley as another weather-battered group creaks towards us from Mera, one of Nepal's 20,000 feet trekking mountains. Their strained, peeling faces contrast with their porters' clear complexions and bored expressions. One scabby Lancastrian in hi-tech gear gasps through wind-cracked lips, 'It's amazing. But now I'm shattered; I'm emotionally and physically drained.'

Twenty over-burdened, under-clad porters rush past, anxious that nothing should interrupt their journey home to the comfort of a wood fire. It is just as well they hurried away, as they might not have cared for the parting words of one of our number . . . 'They have to learn,' he said 'They can't keep chopping down trees.'

His remarks epitomised one of the chief ambiguities in what has come to be known as 'sustainable tourism'.

(Gordon-Walker 1993)

After he [Sir Edmund Hillary] had climbed and succeeded, he was so grateful to the Sherpas because without them he couldn't have done it. He wanted to do something . . . and decided that what these people needed were schools. After the school had been built, as elsewhere in Nepal governmental authorities took over. They wanted to use the schools to homogenise the country's diverse ethnic communities. So they're using the Hillary schools to teach the Sherpa children Hindu ways and Nepali, and English and Mathematics as well, so that they can serve the tourists and bring tourist currency to the country.

In these schools they teach nothing about the Sherpa-Tibetan culture – and nothing about their own 1,200 year old written language, which is classical Tibetan, of course. Most of the Sherpa children growing up in Kathmandu . . . do not learn a word of Sherpa. If you get off the aeroplane at Lukla . . . you are met by a whole group of youngsters who speak 'Hillary School English' to serve you. Over the years they have been completely incorporated into the tourist economy. Most of the younger inhabitants of this area don't know how to run their farm any more.

(Kvalöy 1993)

Younger Sherpa men, of all Nepalis, are the most 'hip'. They dress in expensive jeans, tracksuits, baseball caps. They have Walkmans, tend to speak good English and smoke designer cigarettes. The women, however, have on the whole kept up the traditional – and hard – way of life . . .

Western influence, which followed expeditions to Everest . . . has greatly affected the Sherpas' way of life. Schools, hospitals and clinics, postal services, air transport and radio communication changed a semi-nomadic life to one much more dependent on tourism, trekking and expeditions.

(Klatchko 1991)

norms, customs and practices rather than those of mainstream society. The term indigenous groups includes tribal peoples, while the term tribal peoples excludes members of indigenous groups who live by the norms and practices of mainstream society. In the tourism industry not all host communities are tribal peoples but tribal peoples who have had contact with the 'civilisation' of the First World are potential host communities. This chapter has already cited a number of examples of tribal peoples and their interaction with the tourist industry: the Maya, the Choco and Kuna, the Sherpas, and the Maasai. This section highlights one particular point about their experience as 'hosts' in the tourism industry: namely, their treatment as objects to be viewed, a process which 'forces Indigenous Peoples to become showcases and "human museum exhibits"' (World Council of Churches 2002) and which might be called the 'zooification' of tribal peoples.

The nineteenth-century and earlier Christian fundamentalist and explorer's view of tribal peoples as savages whose souls needed saving may not be as pervasive a public viewpoint now as it used to be in the early part of the twentieth century. But it helped to shape the common perception of tribal peoples as 'noble savages'. The characteristic of nobility owes something to the development of this early First World view by conservationists and environmentalists in their interpretation of wilderness as inclusive of the peoples who are indigenous to it.

> [T]hese images [of godless, natural, wild and blameless savages] are retained to this day and lie behind conservationist policies of 'enforced primitivism', whereby indigenous people are accommodated in protected areas so long as they conform to the stereotype and do not adopt modern practices.
>
> (Colchester 1994: 3)

The buildup of area protection policies and an associated conservation ethic (see Chapter 4) have clearly been important in the promotion of the common perception of tribal peoples as natural and wild. It is this perception which, in some cases, dominates the relationship and exchange between tourists and tribal peoples and which confirms and strengthens the already prevailing prejudices and can, at times, lead to the process of zooification.

Visiting tribal peoples' settlements is an activity especially associated with new forms of tourism. Such visits are advertised as small group tours, implying low impact; tribal communities are described as 'almost untouched' or 'totally unchanged', implying an authentic experience; conditions are referred to as 'primitive', implying an experience with a difference, thereby conferring status on the tourist; and some tours to tribal settlements mention the possibilities for artefact purchases, implying that the exchange will assist in the development of the settlements and peoples visited.

In general, 'small group tours' is an accurate description, but these can hardly be claimed to be low in impact. Even the lone traveller can have an insidious and disruptive effect on local culture, especially if the host community has had little contact with mainstream society. Description of the supposedly primitive nature of tribal peoples by tour operators is used to emphasise the 'otherness' of the experience. Again, examples have been cited where this cultural authenticity is often falsely enacted, such as the divesting of T-shirts solely for the sake of the visiting tourists (see the first example given in Box 8.12). The economic exchange involved can of course be significant, as the Choco and Kuna cases illustrate (Box 8.9). Equally, however, it can contribute to the 'trinketisation' of a culture and can lead to conditions of near-slave labour, as in the case of the Yagua in Peru where traders or middlemen take all the profit (see Box 8.12).

The authenticity of the tourist experience is called into question by many of the examples cited in this chapter. Even in those cases where control of the visit is in the hands of the tribal people, the nature of their ceremonies or products is altered for the sake of the visiting

Box 8.12 Tourists and tribal peoples

The scene is a Bushman camp in a remote part of Botswana. In the distance a plume of dust shows the arrival of a jeep. The people drop whatever they are doing, quickly pull off their T-shirts, trousers and cotton dresses, and begin to dance.

In the Himalayas, fields lie uncultivated. The men who once farmed them have become porters to climbing expeditions.

In the Peruvian Amazon, women of the Yagua tribe make bags, hammocks and jewelry for the tourists. They work hastily because they are being paid almost nothing. The traders make all the profits. Decoration is crude, the finish careless – 'Well,' the women say, 'these people don't know any better.'

In Tuareg camps around Tamanrasset in Algeria, the tents of the drought-stricken refugees are normally covered with plastic sheets – only when the tourists arrive are the old coverings of animal hide brought out.

After an Amazon trip, Mick Jagger told an interviewer: 'What the tour guides do is take all the Indians' clothes off and put their little skirts on them, hand them a spear which they hand back at the end. The Indians dance a little around you.'

A Survival member writes to us in disgust: 'In the Mulu National Park (Sarawak) I realised how the Penan are being treated like animals in a zoo. Almost every tourist group that visits the park is taken there to walk around and look at the Penan "way of life". A longhouse is now in the process of being built for them, which as you know is not the way the Penan live. The atmosphere is one of despair.'

Another Survival member describes how her party were taken to stay in a village of the Meo tribe in the hills of Thailand. She paints a picture of the tragic gap in understanding that remained between the hill people and their well-intentioned visitors, who were quite unprepared for the 'primitive' conditions they found. A young German, looking bewildered, asked where the toilet was. Our guide laughed, and exclaimed 'Everywhere!' As the initial exhilaration of our adventure gave way to fatigue, relations with the villagers became strained and our interaction with them was reduced to stares and picture taking. Occasionally, as an inadequate response to our own guilt and to demonstrate our gratitude we made them gifts of whatever we had. A soap box or a second hand toothbrush.

Source: Survival International (1991)

tourists. The Kuna, for instance, carry out rehearsals of dances and songs on the night before a cruise ship is due to arrive. As Box 8.9 outlines, the Kuna are adept at staging authenticity. Many other cases, a few of which are given in Box 8.12, show that the exchange is much more unequal. One example actually refers to the treatment of the Penan (in Sarawak) as being 'like animals in a zoo', but most of the other examples also hint at this kind of treatment. Knowledge of the fact that the tourist experience is staged often fails to deter the tourist from wanting to experience it, a reflection perhaps of the new tourist's need to collect cultural capital (see Chapter 5).

The zooification process involves turning tribal peoples into one of the 'sights' of a rainforest expedition or a trek, giving rise to Rigoberta Menchú's statement that 'our costumes are considered beautiful, but it's as if the person wearing it didn't exist' (quoted in Survival International 1995).[4] As a 1995 Survival Background Sheet says, 'All too often tour operators treat tribal peoples as exotic objects to be enjoyed as part of the scenery.' But the following example, from the Green Travel internet bulletin board (see Appendix 1), shows that this attitude is just as likely to come from professionals such as archaeologists and anthropologists as it is from tour operators.

> There is a group of archaeologists and anthropologists who are travelling the jungles of Mexico, Belize, Guatemala and Honduras totally self-contained on X-country bikes. They are uploading journals and graphics via a satellite linkage of their adventures for all to see on a World Wide web site . . . The purpose of the expedition is to explore remote and nearly inaccessible Mayan pre-Columbian ruins and to study the present day Mayas in their native habitat. These travellers are all young people . . . and their adventures make . . . interesting reading for anyone.
>
> (Email communication, 1995)

The process of zooifying tribal peoples leads inevitably to a position of powerlessness for them as well as a complete loss of human dignity. The key to avoiding such situations is control of and participation in the tourism activity, which do not necessarily mean simply a greater share of the financial profits. As Pretty's typology (Table 8.1) suggests, it also implies control over all the conditions of the tourism development. Again, as was noted in the discussion on local participation, one of the most important elements in the success of a tourism scheme is that the idea and impetus for it should come from within the community itself.

It is of course too simplistic to demand a single-minded, blanket policy of total control to the tribal groups involved in any tourist development. There are dangers, as Colchester (1994: 57) points out, in making 'an assumption that once an area is under indigenous ownership and control the problem is solved . . . This is patently not the case.' Notwithstanding these dangers, it may be argued that the community has to own and control the development if it is to avoid the pitfalls associated with external control.

Doxey's levels of host irritation

Broadening out the analysis of relationships between hosts and visitors to include all local communities from the Third World rather than just tribal peoples, it is a helpful starting-point to use Doxey's index of irritation, first put forward in 1975. This is often referred to as Doxey's Irridex and is illustrated in Table 8.2.

The Irridex is a causal model of the effects of tourism developments on the social relationships between visitors and the visited. Beginning with a state of very little tourism development and only the occasional passing visitor, the model's four stages describe different states of tourism development and the ways in which tourists and local people perceive each other in these stages. Its final stage is that of antagonism in which the stresses and tensions between the visitors and visited, resulting from high levels of development for the tourists, are at a peak and are likely to lead to a deterioration in the reputation of the destination.

Clearly, this is a highly generalised model, and its sequence and relevance will be subject to a wide variety of factors which differ with time and space. Its original application was in a First World, mass tourism context, but it is feasible that the relationships between Third World communities and the new tourists who visit them will follow a similar sequence to that of the relationships in First World resorts. (Tourist motivations may be somewhat different, but tourism effects are not likely to be dissimilar.)

Table 8.2 *Doxey's levels of host irritation extended*

Doxey's Irridex	Social relationships	Power relationships
EUPHORIA	Initial phase of development; visitors and investors welcome	Little planning or formalised control; greater potential for control by local individuals and groups in this phase
APATHY	Visitors taken for granted; contacts between residents and outsiders more formal (commercial)	Planning concerned mostly with marketing; tourism industry association begins to assert its interest
ANNOYANCE	Saturation points approached; residents have misgivings about tourist industry	Planners attempt to control by increasing infrastructure rather than limiting growth; local protest groups begin to assert an interest
ANTAGONISM	Irritations openly expressed; visitors seen as cause of all problems	Planning is remedial but promotion is increased to offset deteriorating reputation of destination; power struggle between interest groups may force compromise

Source: Adapted from Doxey (1975, 1976)

Community control of the developments from the outset may go some way towards breaking what may appear from the Irridex to be the inevitability of the sequence of worsening social relationships. The Irridex relates the type of social relationship (euphoria, apathy, annoyance, antagonism) directly to the level of development of tourist facilities and infrastructure. The last two stages indicate that a level of change to local lifestyles above what is considered acceptable by local people has been reached, and especially in the final stage has been surpassed. This may come about as a result of dimensional changes, such as overcrowding (in which case planning and visitor management techniques may be able to provide solutions) or structural changes (such as the outside influence of foreign investors or national politicians pursuing goals different from those of the local community). The latter cause especially implies that local control of development may act as a solution, and it is interesting to speculate on the association between Doxey's levels of irritation and the degree of local control. This association may be a pointer for worthwhile future research, a starting-point for which we have provided by extending the Irridex in Table 8.2 to include speculation on the power relationships implied by the level of irritation as well as the social relationships.

It should be stressed that the Irridex is offered here only as a loose framework for considering the relationships between visitor and visited. It should be noted that its applicability will be compromised by circumstance. It should also be noted that Doxey's is not the only attempt to characterise the different stages and features of these social relationships. Butler (1975) and Murphy (1983), for instance, offer more detailed models and descriptors of social interaction that make explicit allowance for a number of variable factors. Both acknowledge that communities can adjust their lifestyles in order to overcome stresses caused by uneasy social relationships between visitors and visited. But perhaps rather than looking for remedial action to counter the inequalities and unevenness of tourism developments, it might be more suitable for local communities to control developments from the start, as has been emphasised throughout this chapter.

Conclusion

The relationships of power, one of the book's key themes, are central to a consideration of the role of local communities in tourism. In this chapter these relationships have formed a background to a consideration of local participation in tourism developments. Although widely seen as a wholly benevolent feature of tourism planning that is associated with the notion of sustainable development, there is evidence to suggest that participation is becoming something of a buzzword, a sine qua non for development project proposals, to such an extent that its role may even be seen as tyrannical, in itself a determining force for project proposals. Our examination of issues of participation has used Pretty's six-point typology of participation to describe differing degrees of involvement and control by local people over (tourism) developments in their communities. The case studies presented in this chapter have been related to this typology and it is suggested that reference to the scale of participation is something that could profitably happen at the outset of all tourism development schemes. Likewise, techniques for assessing the degree of local participation in schemes need themselves to be subjected to a consideration of who is doing the assessing and for what purpose.

A note of caution needs to be sounded, however, for the general assumption that the greater the degree of local control and participation the greater the scheme's supposed sustainability and the wider the distribution of benefits within the community does not always hold true. First, as has already been well established, sustainability differs according to the interests of those who are defining it, and the interests of the local community will not necessarily coincide with those of others; nor is it likely that the interests of the local community will be the same for all within the community. Second, local power relationships within the community can be as factional as those which include players on a broader stage such as national governments, INGOs and supranational institutions. Thus, the emergence of local elites is as likely to produce inequalities within the community, just as these other players produce disparities of benefits at a different level.

At its worst, tourism and the conservation measures which have been used to support it have been responsible for the displacement and resettlement of local communities. The case studies from eastern Africa used to illustrate these phenomena are disturbing, but it is noteworthy that they are not entirely negative examples. A number of other case studies highlight the important point that only where the impetus for tourism development comes from within the community is the prevailing inequality of development likely to be challenged.

The notion of transculturation, the process by which local communities adapt themselves to the cultural mores and habits of those with whom they interact, was used to demonstrate that this interaction is not purely a case of the imposition of one set of cultural values upon another, as it is often represented in tourism analyses. The term 'zooification' was introduced, however, to illustrate that in the cases of some tribal peoples this dehumanisation of local peoples does indeed take place, leaving them no power or dignity. Less extreme cases of First World new tourist–Third World local community interaction may be appropriately analysed according to Doxey's Irridex, which can be extended to describe the relationships of power as well as the social relationships between the visitors and the visited.

9 Cities and tourism: guess who's coming to town?

It is now commonplace for research, policy papers and institutional strategies on urbanisation in the developing world to commence with quantifying the rapid rate of city growth, the type and scale of problems caused by urbanisation, and to reiterate that globally we are now living in an urban world (United Nations Centre for Human Settlements (UNCHS) 2001). Yet as a number of commentators have critically observed, the scale of this urban challenge and the problems and opportunities presented by the expansion of cities is not matched by a commensurate interest in cities within the general field of development nor among the external supporting agencies comprised of multilateral and bilateral development lenders and the wider development community (Satterthwaite 2001).

The fact is that we live in an increasingly urbanised world, and at least in our lifetime this is likely to accelerate rather than reverse. In 2006 a report by the United Nation's 'city agency', UN-HABITAT, confirms that the global urban transition is only at mid-point with projections showing that over the next 25 years the world's urban population is set to increase to 4.9 billion people by 2030, roughly 60 per cent of the world's total population (United Nations Department of Economic and Social Affairs 2006). Moreover, the most significant growth is projected to occur in less developed regions with sustained and rapid increases culminating in 3.9 billion urban dwellers in these regions by 2030 (United Nations Department of Economic and Social Affairs 2006). And it is the nature of this growth that is of great significance, with a rapid increase in the number of the poor, the majority of whom are likely to be concentrated into city slums (UN-HABITAT 2003) – see Figure 9.1.

The centrality that cities and towns perform in the movement and accommodation of tourists, and the potential economic impact that urban-based tourism can contribute, is beyond doubt. For example, Bangkok's accelerated growth (physically and economically) has been spearheaded by rapid expansion in the service and financial sector and by its equally rapid growth as a tourist gateway to Thailand and the broader region. Similarly as Colantonio and Potter (2006) observe, tourism became central to Cuba's re-entry into the global economy in the 1990s and 'urban tourism in Havana has been pivotal' in this political and economic process. In the years 1995 to 1998 total revenue from tourism in Havana was a little over US$1410 million (48 per cent of the national total tourism income), with 54 per cent of all visitors to Cuba visiting Havana in 1999. In this chapter we consider if, and how, new tourism has a role to play in this urban challenge and its ability to provide a 'development dividend'. However, we approach this task by setting our discussion in the deep-seated bias against Third World cities in much development thinking and new tourism literature and practice more broadly.

Initially however, just as Chapter 2 contrasted the potential benefits of new tourism in the Third World with migrants' remittances sent home from the First World, it is useful to set our discussion within a broader context and make wider connections for three main reasons. First, as we have argued throughout, tourism illustrates – par exemplar – broader

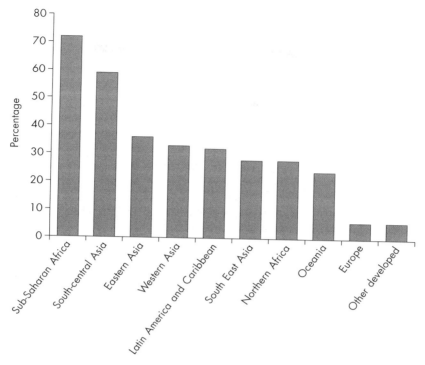

Figure 9.1 Slum dwellers as a percentage of urban population by region, 2001

Source: United Nations Human Settlements Programme (2003) *The Challenge of Slums: Global Report on Human Settlements 2003.* UN-Habitat, London: Earthscan.

unequal and uneven geographies of development. The very concept of a Third and First World is messy and rife with contradictions. Deprived areas with high incidences of poverty are not somehow confined to distant worlds but are as visible in First World cities often characterised by successive waves of in-migration from former colonies. Reflecting on the new townscapes resulting from immigration, Koptiuch (1991), for example, refers to the 'third worlding' by 'exotic others' of US cities in the late 1980s. Much the same trends have been witnessed in European cities. Somewhat ironically it is these very places, ranging from East London's Brick Lane (or 'Banglatown') to Toronto's Parkdale district and Rotterdam's southern inner city that have increasingly attracted new migrants and 'new tourists' eager to experience the other side of cities: the proverbial 'urban jungle'. As Biles (2001: 15) suggests, 'Intrepid holiday-makers are generally found in the wilds of Africa or the deserts of Arabia, not the industrialised heartland of Rotterdam'. Interestingly, while there has been voluminous research on the growth and impact of urban tourism in general, Hunning and Novy (2006) commenting on tourism in New York and Berlin draw attention to the precious little attention on tourism in city areas 'beyond the beaten path' – the very form of tourism that is a 'consequence of a broader trend towards a more individualised and differentiated mode of travelling' (Hunning and Novy 2006: 2).

Second, and linking us back to Chapter 5, First World cities have undergone intense economic restructuring from industrial and manufacturing bases to service sector-oriented growth. Not only have the economic and physical landscapes of these cities dramatically changed, but also their social and cultural landscapes have markedly changed too, marked in part by the swelling ranks of the new middle classes. Cultural vibrancy, diversity and an ethnic cutting edge have become as important, if not more important, to the life and prosperity of First World cities than more traditional markers of efficient managerial approach to urbanisation. This gentrification, underscored by the accumulation of cultural capital (Zukin 1982), has become the seedbed of the consumers of new tourism.

Both the above points emphasise that the process and outcomes of development are not confined to the conventional Third World, but are processes that are fought out daily in cities globally. And akin to the charges of new tourism as an expression of contemporary colonialism and uneven and unequal development, some commentators have been compelled to label gentrification as the 'new urban colonialism' and draw attention to the centrifugal spread of gentrification from the cities of the USA, Europe and Australasia, 'into new countries and cities of the global "south"' (R. Atkinson and Bridge 2005: 2). As we suggest in this chapter, there are some uncanny connections between the growth of the new middle classes, gentrification and new tourism as constitutive of neoliberal growth (see e.g. N. Smith 2002). In short therefore, the issues of heritage and gentrification, and of the prospects of pro-poor urban tourism, discussed in this chapter are as applicable to First as to Third World cities (and are a much needed area of research and critical discussion).

Third, and returning to our principal focus in this book, we must ask why it is that viewed from the 'West' there is a certain nobility, sympathy, even romanticism, of the rural poor and poverty, but a neglect – close to despite – for the poor that crowd into the ramshackled slums that cling (sometimes precariously) to the peripheries of Third World cities. A strange question perhaps in a book on tourism, but there has always been a tendency in the alternative and responsible travel world to accentuate the possibilities and potentials of culture-rich rural repositories and to downgrade or worse demonise (supposedly culturally deficient) cities in the travellers' tales and travelogues we discussed in Chapters 3 and 5. The sense of hopelessness encountered by the iconic travel writer Paul Theroux is emblematic of the experience of gazing as travellers on this seemingly chaotic heart of darkness:

> These huts, in a horrific slum outside San Salvador, are the worst I saw in Latin America. Rural poverty is bad, but there is hope in a pumpkin field, or the sight of chickens, or a field of cattle which, even if they are not owned by the people in the huts, offer opportunities to the hungry cattle rustler. But this slum outside Guatemala City, a derangement of feeble huts made out of paper and tin, was as hopeless as any I had ever seen in my life. The people who lived here, I found out, were those who had been made homeless in the last earthquake – refugees who had been here for two years and would probably stay until they died, or until the government dispersed them, and set fire to the shacks, so that tourists would not be upset by this dismal sight. The huts were made out of waste lumber and tree branches, cardboard and bits of plastic, rags, car doors and palm fronds, metal signboards that had been abstracted from poles, and grass woven into chicken-wire. And the slum, which remained in view for twenty minutes – miles of it – smouldered; near each house was a small cooking fire, with a blackened tin can simmering on it. Children rise early in the tropics; this seemed to be an entire slum of children, very dirty ones, with their noses running, waving at the train from curtains of yellow fog.
>
> (Theroux 1979: 129)

Is there an alternative way of seeing and travelling in cities that charts a way between this unmitigating hopelessness on the one hand, and the colonial heritage and slick city centre bars on the other? And are there alternative ways of interpreting and reading many of the city landscapes experienced by travellers, for as David Harvey (2006: 13) contends: 'Reading a book . . . will likely affect how we experience that place when we travel there even if we experience considerable cognitive dissonance between expectations generated by the written word and how it actually feels upon the ground.' In a counter-intuitive perspective for example, Mumtaz (2001) suggests:

> slums are not only inevitable, they are a mark of success of a city . . . Insisting on a 'city without slums', especially when no alternative housing has been developed, can mean even more hardship for the very group that is so essential to urban development: the rural migrant.
>
> (Mumtaz 2001)

Mumtaz (2001) concludes, 'Just as slums and slum dwellers need cities to survive, so do cities need slums to thrive'. Similarly, urban community specialist Arif Hasan (2003) of the Urban Resource Centre (Karachi, Pakistan) argues that Asia needs more slums, not less; but slums that are secure, safe, in the right place and provide the potential for evolving into 'non-slums' within a decade or so. To unpack the potential of cities and the role of new tourism in this we need to start with exorcising the demons and consider why cities are generally vilified within the field of development.

Urbanisation as the antithesis of development

If, as advocates suggest, new and more responsible forms of tourism are in part an attempt to get into those spaces of development marginalised by mainstream mass tourism, then it is indeed an attempt to support those who have least (if any) economic opportunity. Like development more generally however, tourism suffers from the same blind spot. Twenty years past, cities were barely on the radar screen of international development agencies. Nowadays they are there, but the signal is weak. For the purposes of tourism many Third World cities are perceived as dirty, unhealthy, violent, noisy, alienating – and are the foundation for hair-curling travellers' tales (see Chapter 3). Having journeyed through Latin America in the memorable travel account *The Old Patagonian Express*, Paul Theroux's assessment of many of the Latin American city landscapes is not encouraging: 'I had no strong desire to see Mexico City again', says Theroux, 'It is, supremely, a place for getting lost in, a smog-plagued metropolis of mammoth proportions.' Theroux is no less gracious of cities further south in Central America. Guatemala City is brutal and on its back where 'east of the capital, on the other side of the tracks . . . desolation lies . . . For a full hour as the train moves there is nothing but stone-age horror of litter huts' (Theroux 1979: 161).

Cities present undeniable problems that work to repel, rather than attract, tourists. Crime and insecurity, to name just one set of problems, are significant barriers to both development and the expansion of tourism. Latin America, for example, saw a dramatic increase in crime and violence in the last quarter of the twentieth century, and is acknowledged as a serious social and economic problem especially in urban areas where crime is at an unprecedented level (Carroll 2007; Environment & Urbanization 2004; Koonings and Kruijt 2007; World Bank 2003). The UN reported in 2007 that the rapidly expanding metroplitan areas of Caracas, Mexico City, Rio de Janeiro and São Paulo account for more than half of all violent crimes in their respective countries (UN-HABITAT 2007). As the World Bank concludes, 'rapid urbanization, persistent poverty and inequality, political violence, the more organized nature of crime, and the emergence of illegal drug use and drug trafficking' are commonly referenced as major causes for such increases (World Bank 2003: 7).

The root causes of violence appear to be found in political, economic and social factors (Winton 2004). For Latin America, many countries have experienced protracted periods of political violence, civil wars, unrest and repressive authoritarian regimes. As Winton (2004) suggests, there are 'major repercussions of sustained state repression' and many newly democratic states are yet to reform the judicial system and state policing with the result that there has been 'no systematic dismantling of past institutional structures of terror and oppression' (Winton 2004: 167, 168). But such violence is not only confined to so-called post-conflict transitional states. Duncan and Woolcock (2003) note that poor urban communities in Kingston (Jamaica) are impacted by strict political clientelism arising from the two dominant national parties. Most significantly, as Caroline Moser (2004: 6) reviews, the 'sheer scale of violence in the poor areas or slums means that, in many contexts, it has become "routinized" or "normalized" into the functional reality of daily life'. As might be expected, the economic impacts of crime, including the tourism industry, are equally marked. For example, Guerrero (1999) estimated that the financial costs of murders in Latin America ran above US$27,000 million each year and that 14 per cent of the region's GDP was lost to violence.

If the demonisation of cities were merely the reserve of 'opinion-formers' (such as Theroux above), arguably it would be a matter of less concern. But there is a more deep-seated prejudice against cities that runs to the heart of the meaning of development and underlies the politics of tourism in cities. As the UK-based charity WaterAid pondered,

> the vilification of cities and their marginalisation within development . . . is not just reserved to critical academic debate, it has a clear resonance in the practice of development too . . . cities have struggled to get on the agendas of multilateral, bilateral and the burgeoning international non-government sector, and the few INGOs that have addressed urbanisation have had to initially dismantle the 'demonisation' of cities.
>
> (Black 1994)

Many reasons have been offered for why cities have faired so weakly in development, not least that despite the accelerating growth in urban poverty it 'gets ignored because it happens slowly, inexorably' (McLean 2006). One of the most persistent problems has been the dominance of the 'rural' in development and the casting of cities as a significant part of the development problem by encouraging people to migrate from the land to urban labour markets (United Nations Population Fund (UNFPA) 2007). Indeed, for some countries, the challenge has been interpreted as the need to manufacture legislative restrictions on the movement of poor rural migrants to urban areas. As Angotti (1995: 16) argues, favoured urban strategies throughout Latin America 'often target cities themselves as the problem, and seek to stop urban growth instead of improving the urban – and rural – quality of life'. This 'ruralisation' of development as a core strategy asserted its prominence in the influential and popular environmental 'shockers' of the late 1960s and 1970s. Schumacher's (1974) *Small is Beautiful*, for example, rounded on the 'footlooseness' which resulted in swollen cities sucking the vitality out of rural areas:

> let me take the case of Peru. The capital city, Lima . . . had a population of 175,000 in the early 1920s, just fifty years ago. Its population is now approaching three million. The once beautiful Spanish city is now infested by slums, surrounded by misery-belts that are crawling up the Andes. But this is not all. People are arriving from rural areas at the rate of a thousand a day – and nobody knows what to do with them . . . Nobody knows how to stop the drift.
>
> (Schumacher 1974: 64)

In addition, UN-HABITAT (2003) has forcefully argued that Structural Adjustment Programmes (discussed in Chapter 10) were deliberately anti-urban in nature and designed to reverse any bias towards cities that had previously existed in government investments or welfare policies. It is not surprising therefore that there is a sizeable deficit in recording and promoting the potential of cities. As the UK submission to the Fourth World Urban Forum in 2006 reflects: 'people need positive and progressive examples in order to develop visions and make demands of their leaders, but virtually all of the current press coverage and commentary on urban growth and change is negative, alarmist and doom-laden' (Hague et al. 2006: 86).

Deborah Eade of the international development charity Oxfam argues that the reluctance to engage in urbanisation issues is further exacerbated by the international non-government sector's virtual silence. If cities make the headlines it tends to be as a result of 'natural' disasters, terrorism or insurrection. As significant as this may be, this pays lip-service to the lives of hundreds of millions of city dwellers each day. Eade's (2002) conclusion is:

> The prevailing attitude is either that cities are a problem in and of themselves and shouldn't be encouraged, or that their residents enjoy better facilities and so are less 'needy' than their rural counterparts, or that the challenges posed by rapid urbanisation are simply too big, too expensive, and too complicated to handle. A glance through the grants lists and literature of some of the best-known international NGOs suggests that . . . if they get involved at all, most find it easier to deal with the specific problems of specific population groups in the towns and cities of the South – street-children and sex-workers topping the list – rather than getting involved in the messier processes of urban management . . . Ironically, the largest human settlements in which many NGOs take a more holistic approach to the planning and management of basic services are refugee camps – usually cramped and often squalid settlements that earn their description as 'rural slums'.
>
> (Eade 2002: xi)

In summary, there are compelling reasons for looking at a critical discussion of cities and new tourism in development. As Chapter 2 suggested there is a significant heritage to development theory and critical political economies of tourism (including the potential of alternative tourism) which argue that it is the industrial, metropolitan core of the so-called 'developed' world that maps, translates and dictates the development process, and that cities play a significant and (for some) increasingly powerful role within globalisation. Development has cultivated an anti-urban bias that leads us to the conclusion that development is, or should be, everything that the 'metropolitan world', and by inference urbanisation, is not. Development has been interpreted therefore as ostensibly about addressing rural change, supporting rural livelihoods and stemming urban migration. Where cities have been invoked, and reflecting back to the evolution of development theory in Chapter 2, it is within the envelope of their economic potential. It is to this that we turn next.

Cities as economic machines

Located in Mexico's easternmost state (Quintana Roo) on the Yucatan Peninsula, Cancún is most readily characterised by its major hotel zone built on a coastal spit by the Government of Mexico's National Trust Fund for Tourism Development (Fondo Nacional de Fomento al Turismo). Joel Simon (1997) charts how Cancún transformed from a fishing village of just 800 people into a seaside city attracting a transient tourist population of one and a half million annually. Cancún was an experiment to create a tourist city in paradise; a snub to the urban decay and 'third world chaos' inflicting Acapulco and that had resulted in tourists being chased away to competing destinations. With the end to Cuba's lucrative tourism industry following the 1959 Revolution, and with the US embargo against travel to Fidel Castro's Cuba, the East Coast market was open to new localities. An hour's flying time from Havana and less than two hours from Miami, Cancún was Mexico's response to soaking up this demand for a new (eastern seaboard) playground. But Cancún is typical of the problems that emerge and is the outcome of a 'dramatic and inequitable process of local urban development that can be traced back to national policies of the 1970s and which created new economic growth poles' (Aguilar and Garcia de Fuentes 2007: 244). The strip of mass tourist hotels is in stark contrast to the thousands who live without adequate water and sanitation systems. Like all land and urban development, the stakes are high and the outcome of such experiments sadly predictable in their balance of power. As Simon (1997) concludes:

> megaprojects like Cancún may have made money for the federal government, they have also spawned a large-scale, capital intensive style of development geared toward quick returns. They have institutionalized land speculation and fomented a tourist economy in which capital is highly concentrated, developers are used to thinking big, and power is concentrated in the hands of the federal bureaucracy.
>
> (J. Simon 1997: 195)

While conceived well before the full force of neoliberalism blew in the direction of Latin America and the Caribbean, Cancún is a physical manifestation of the 'lost decade for development' (see Chapter 2). Cities provided a spatial logic for the economic fundamentalism of the Reagan–Thatcher axis premised on a belief in the power and application of free market principles and trickle-down economic growth strategies. Trickle-down urban regeneration was relentlessly pursued in western cities and economic structural adjustment policies in the Third World had marked impacts upon cities through the drive to implement the neoliberal policies of the Washington Consensus (see Chapter 2). The principle was straightforward – privatise, liberalise, deregulate – and allow the 'magic of the market' to dictate and manage urban growth and development. The result is significantly changed spatial patterns and cities that construct and maintain a viable and attractive tourist product, in part by pushing the 'problems', such as the marginalised and unsightly urban dwellers cramming into urban areas, into the so-called peri-urban fringe (or slums). By the close of the decade the pattern of Latin American urban development assisted by the policy consensus was characterised by 'high levels of spatial segregation; uneven distribution of population densities, infrastructure and services; large areas of under-utilized space and facilities; and a marked and escalating deterioration in environmental and social living conditions' (Burgess et al. 1997: 118) – the very conditions that actively discourage tourism and tourists.

Although the fundamentalism of economic growth is now widely challenged, and the primacy of economics is tempered by the need to demonstrably improve the lot of the (urban) poor, the pervasiveness of the primacy of economic growth as a *sine qua non* remains. There is a critical problem in defusing the correlation that cities and urbanisation are primarily (and in some cases exclusively) an economic phenomenon in even the most pro-poor, anti-poverty, inspired tracts. As Mitlin (2002) concludes it is an approach which belies that city life is much more than an economic phenomenon or relationship.

The World Bank's position is of special interest as the largest provider of urban development assistance and for the influence of its policy on other lenders and national government interventions. The World Bank's (2000a) *Urban and Local Government Strategy* refocused multilateral attention back onto cities in the face of rapid urbanisation with the central strategic aim of promoting sustainable human settlements as defined by four characteristics: liveability, competitiveness, good governance and management, and bankability. Although the strategy emphasises and is preconditioned by the significance of more highly integrated and holistic interventions than previously, and by comprehensive development frameworks for the 'urban arena' (World Bank 2000a: 8), it is nonetheless steeped in the discourse and primacy of economic health and competitiveness. Indeed, the focus on cities is in part driven by the agglomeration economies yielded by geographical proximity not as a social or environmental entity but as an 'urban economic area that represents an integral market', and which conditions the 'prospects for economic development' (World Bank 2000a: 15). As the strategic objectives of competitiveness and bankability suggest, the strategy necessitates adopting a range of economic measures and incentives including a commercial approach to many urban services and functions, 'market-friendly land use planning' and 'buoyant, broad-based growth of employment, incomes and investment' while 'keeping social concerns in view' (World Bank 2000a: 12, 19, 9). The translation of such an approach to a form of urban development that serves the interests of tourism as an economic sector is predictable. The World Bank concludes that tourism is

> becoming important in many cities, both for end-visits and as transit points in the transport system . . . Tourism, however, requires a well-ordered city, secure, clean, and healthy; that is to say, the quality of life in the city is fundamental to its capacity to earn income from tourism.
>
> (World Bank 2000a: xx)

If the likes of San Salvador and Guatemala City

> were hosed down, all the shacks cleared and the people rehoused in tidy bungalows, the buildings painted, the stray dogs collared and fed, the children shoed, the refuse picked up in the parks, the soldiers pensioned off . . . and all the political prisoners released, those cities would, I think, begin to look a little like San José.
>
> (Theroux 1979: 161)

An exceptional city, he concludes.

There are however few examples of cities in the Third World that match the World Bank's benchmarks. Cities often lack a technocratic 'order', often struggle to cope with mounting problems of waste management and invariably find human health compromised. They are very often unsafe. Ironically perhaps, as our earlier discussion implied (see Chapter 3), it is these very characteristics that attract new tourists. As some authors argue, these so-called 'middle-city urban cultures' give full expression to the informality, flow and tapestry of urban life that are critical to the development process (Samuels 2005).

It is not the importance of economic development to the life of cities that is of questionable value; clearly it is important. Cities are largely founded on cash-based economies and urban residents therefore need work – both formal and informal. Rather it is the primacy, in some cases exclusively, of economic growth in cities that has resulted in a biased and uneven approach to development, together with the obsessive push to make cities 'world-class' magnets of international investment. Some commentators have drawn attention to the discernible trend among governments globally, regardless of political orientation, to adopt 'gentrification as a form of urban regeneration policy broadly connected with an entrepreuneurial style of governance . . . and a focus on the middle classes as the new saviour of the city' (R. Atkinson and Bridge 2005: 2). As Arif Hasan suggests,

> Local governments are obsessed by making cities 'beautiful' to visitors and investors. This means building flyovers and elevated expressways as opposed to traffic management and planning; high-rise apartments as opposed to upgraded settlements; malls as opposed to traditional markets (which are being removed); removing poverty from the centre of the city to the periphery to improve the image of the city so as to promote FDI [foreign direct investment]; catering to tourism rather than supporting local commerce; seeking the support of the international corporate sector (developers, banks, suppliers of technologies and the IFIs) for all of the above.
>
> (Arif 2007)

It is therefore a narrow and potentially socially explosive approach, and a reminder that within and beyond an urban context the 'dimensions of development and poverty alleviation obviously include more than economics and economic growth' (Berwari and Mutter 2005: 1).

From our discussion so far therefore, it appears that cities are caught between two stools. On the one hand cities are cast as anti-developmental. On the other hand cities are cast primarily as economic development machines capable of catalysing national growth: as Special Advisor to UN Secretary General, Jeffrey Sachs (2005: 36), concludes, 'Modern economic growth is accompanied first and foremost by urbanization'. But is there an alternative way of seeing cities and development, and of gauging the potential of new forms of tourism to one of the greatest challenges of our time? As Beall and Fox (2007: 20) forcefully remind us, the 'time is long overdue for those concerned with issues of poverty, inequality, and social exclusion to devote more energy, attention, and resources to engaging with issues of urban development'. The following sections consider whether the foundations for a pro-poor new urban tourism are in place.

Recycling places: heritage and the urban poor

The notion of 'cultural heritage' has been traditionally linked with ancient monuments and archaeology, and in the context of the discussion above more readily associated with rural resources than the current fabric of globalisation and urbanisation. But, as we will argue here, 'culture' is by no means the preserve of rural communities, and the Swedish International Development Cooperation Agency (Sida) boldly suggest that a 'sensible and sensitive "heritage tourism" could ideally be seen as the urban equivalent to "eco-tourism"' (Sida 2004: 25). The presence of iconic colonial architecture in the cities of the Americas (Rojas 2002) and parts of Africa and Asia (B. Shaw 2006) is undeniably a powerful draw card for new tourism attracted by both the historic value of such places, but more significantly by the aesthetics and ambience that is developed around it. As Eduardo Rojas, urban development specialist at the Inter-American Development Bank, records, for example, there are many cities in Latin America that 'are blessed with a rich legacy of buildings, public spaces, and urban structure' (Rojas 2002). In combination, pre-Columbian, colonial (both Spanish and Portuguese) and postcolonial industrial (late nineteenth-century) architecture provide for the requisites of urban-based heritage tourism in this region.

As Swedish development specialists Tannerfeldt and Ljung (2006: 106) suggest, tourism is growing quickly in historic cities and 'spectacular sites are obviously magnets that attract capital and business. In many places cultural heritage is the most important single asset, although its potential may not have been realized.' It is an asset of such potential that Michael Cohen (2004) goes so far as to refer to it as 'financial heritage' and one that should be taken seriously by the development and aid industries: 'The role of local culture, and cultural heritage, in the debate on cities is important,' Cohen argues, and if

> we understand past investment in infrastructure, museums, public space, and other facilities as part of a wider definition of urban cultural heritage, we need to reconsider how the patrimonio can be valued and utilized as an economic, cultural, and social resource as well. This is far beyond the common argument about tourism, but it involves a serious examination of the flow of benefits that urban areas can receive from earlier investments.

> (M. Cohen 2004: 8)

The fortune of these historic urban areas has changed dramatically since the late 1970s, especially accelerated by the introduction of the United Nations Educational, Scientific and Cultural Organisation (UNESCO) World Heritage List in 1972. Figure 9.2 provides a brief illustration of a few of these potential new tourist 'honey pots'. The World Heritage List includes 830 properties forming part of the global cultural and natural heritage, which the World Heritage Committee considers as having outstanding universal value. These include 644 cultural, 162 natural and 24 mixed properties in 138 states and an increasing proportion of sites listed which are city centres and historic towns (approximately a quarter of all listings). Listing remains important, and a goal of many historic centres, not least because cultural tourism is becoming an increasingly significant component of the global tourism industry and a package of the new tourism industry in particular. Figure 9.3 provides an image of one such city that appears in the listing, Luang Prabang (Laos). Described by UNESCO as a unique and outstanding example of the fusion of traditional architecture and Lao urban structures with nineteenth- and twentieth-century European colonial architecture, the Lonely Planet concludes that 'the city's mix of gleaming temple roofs, crumbling French architecture and multiethnic inhabitants tends to enthral the most jaded travellers' (Lonely Planet, *Laos* 2005).

The presence of these historic tourism resources is not, however, without its own problems and tensions. The pressure of urban expansion and change, together with the socio-economic 'poverty' of many city centres, undermines the conservation efforts of these richly textured urban landscapes, and the promotion of tourism as a potential (and sustainable) mechanism in support of rehabilitation and renewal does not always live up to its billing. Equally, there are protracted issues in the balance between 'conserving' and 'preserving', and the need for these areas to continue to work within the existing grain of urban life and economic activity (often characterised by thriving informal economies) rather than catalysing gentrification, transforming these areas into the equivalent of urban theme parks, and effectively cleansing such areas of the vulnerable urban poor. There are therefore considerable challenges to using 'Heritage as a tool for poverty alleviation' (Sida 2004), when the economic 'boosterism' offered by urban-based tourism undercuts channelling tourism activities and revenue to pro-poor tourism strategies and initiatives that support the presence and livelihoods of the urban poor. Sylvio Mutal (2005), an international consultant for the organisation World Historic Cities, addresses this relationship between urban 'heritage' and urban tourism, and the 'inherent conflict between preservation and renewal and the importance of striking a balance between the two' (Sida 2005a: 25):

> in some areas of the world, some projects concerning historic city development may well have created new problems for local populations e.g. excessive stress

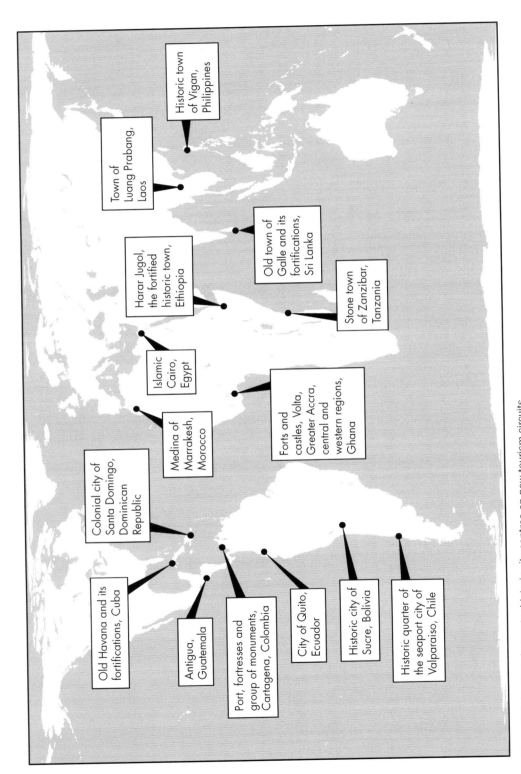

Figure 9.2 Heritage hotspots: historic city centres on new tourism circuits

Figure 9.3 Luang Prabang

Source: Ian Munt

on "tourism". Tourism and other income generating focus need to be put together. Tourism cannot be the magic key for investment. Our challenge is to reverse negative tendencies and create a symbiotic relationship between urban development and heritage preservation for *the improvement of social economic conditions of residents of all walks of life*. As one of the aspects of historic city life is urban poverty and social exclusion, there is a trend to improve the quality of life of inhabitants of cities through historic city programmes in ways that at the same time improve the urban environmental context and preserve and enhance cultural values, conserving adequately the built urban cultural heritage.

(Mutal 2005: 24, emphasis in original)

What is beyond doubt is the intensity of urban problems to which new forms of tourism could be applied and the potential candidate areas which are many. As Mike Davis (2006: 32) observes, 'Whatever their former splendour, most of Guatemala City's palomares, Rio's avenidas, Buenos Aires' and Santiago's conventillos, Quito's quintas, and Old Havana's cuarterias are now dangerously dilapidated and massively overcrowded.' And there are, of course, examples of the successful marriage of heritage and tourism. Gutiérrez (2001) draws from observations of Quito and Lima, reporting that 'strategies for functional regeneration and recycling, the generation of diversifying activities, the encouragement of cultural tourism, and other means of improving the quality of life of the people have made outstanding contributions to the resurgence of historic centres.'

The arrival of trendy cafés and bars and upmarket craft emporiums in cities as diverse as Quito, Antigua and Luang Prabang may however come at a price for the urban poor. As urban commentator Charles Landry (2006) comments on Havana, the flow of old classic cars and the music excite, but on the down side you are aware of the clash between tourists and poor locals. The latter are tied into an oppressive relationship with the tourist; their relaxed laid-back lifestyle contrasts with the need to hassle and compete for tourists.

(Landry 2006: 121)

The same tensions and contradictions are experienced by these new urban tourism resources as is experienced by new tourism in rural areas in that 'tourism has a tendency to turn cities into museums, often compromising authenticity and either expelling the inhabitants or turning them into exotic exhibition objects for the tourists – as if they were in a zoo' (Hemer 2005: 7). Acknowledging the complex pressures of contemporary cities and the occupancy of the urban poor, the Director of the UNESCO World Heritage Centre, Francesco Bandarin, reportedly concedes: '200 cities = 200 headaches' (Hemer 2005).

In one of the few systematic studies of Third World urban heritage and tourism, in this case of the historical centres of nine Latin American cities, Scarpaci's (2005: 95) primary research concurs that heritage tourism is 'not conflict free' and that 'defining authenticity, heritage, and historic periods are difficult tasks' that run the risk of cultivating living museums of indistinguishable and ubiquitous urban cultural landscapes. Clearly, as the results of focus group discussions testify, it is a task that directly affects the lives of local residents as the pressures exerted by gentrification take hold. Of discussions held in Habana Vieja and Cuenca, Scarpaci (2005) reports that

historic district residents hold strong feelings about public authorities, investors, tourists/tourism and the future of their neighbourhood . . . Generally, residents of all three historic districts were not optimistic about the future, had negative things to say about local authorities, and felt that tourism will bring more harm than good.

(Scarpaci 2005: 141)

If a form of gentrification creep is detectable through new tourism in 'natural' ecologically rich areas, it tends to be institutionalised and economically sanctioned in some historic cities, requiring the full participation of property owners and the private sector. In common with physically 'run-down' urban neighbourhoods globally, the variously termed renewal, restoration and regeneration strategies tend to come at a high price, both economically and socially. Even the Inter-American Development Bank, at the vanguard of neoliberal financed change in Latin America and the Caribbean, signals the downside:

while gentrification benefits municipalities and landowners, it tends to expel low-income families and less profitable economic activities from the area. The poor lose access to cheap housing and to the economic and social opportunities offered by a downtown location.

(Rojas 1999: 17)

The need for carefully crafted public policy capable of offsetting the drawbacks in historic restoration is clear (see e.g. Box 9.1), for no matter which city one chooses, the undertow of restoration and rehabilitation tends to be characterised by the same tendencies and tensions: spiralling land prices and land speculation, prices that are too high for the urban poor and clearances or 'relocations' towards the mile upon mile of unremitting misery that Theroux laments. As Hemer (2005: 6) asserts, there are World Heritage listings that have 'saved the cultural heritage at the cost of aggravated social exclusion'. Gentrification is therefore regarded increasingly as an unavoidable by-product of urban conservation

Box 9.1 Restoring Old Havana

Old Havana has an irregular grid of narrow streets and small city blocks, with buildings sharing party walls and inner courtyards: a coherent urban fabric with dominant squares and churches. As the city expanded during the 1700s, it developed typical calzadas: wide streets with tall porticoed pedestrian corridors opening into stores and dwellings above.

About half of the residents of tenements with high ceilings have built barbacoas – makeshift mezzanines or loft-like structures that create an extra floor. They are often unsafe, poorly ventilated and their bricked up windows deform building facades. Moreover, barbacoas add considerable weight to load-bearing walls, already weakened by leaks, often leading to partial or complete building collapses. Another source of extra residential space, as well as extra building weight, are casetas en azoteas – literally, 'shacks on roofs' – which are usually wooden structures built on top of multi-household buildings. The Cuban regime's encouragement of development away from Havana has indirectly helped to shield Old Havana from some overuse; nevertheless, most slums are still concentrated in the inner-city municipalities of Old Havana (Habana Vieja) and Centro Habana. The result of density, additions and poor maintenance is regular building collapse.

The restoration of Old Havana and San Isidro started after Havana became a World Historic site in 1982. In 1993, Havana's Historian's Office was granted the right to run its own profit-making companies in the real estate, building, retail and tourism fields, and to plough back part of its earnings into restoring the historic district. In addition, it could devote a portion of its own resources to financing community facilities and social programmes for local residents and to repair and rehabilitate dwellings, even in non-historic areas. Most residents remain in the area, and gentrification has been avoided, to some extent, since housing for local residents is included in the upper floors of restored buildings. Some, however, are displaced to apartments built and financed by the Historian's Office where some residents welcome the more spacious, well-equipped new dwellings, while others find commuting extremely difficult . . . Local economic development also takes place; some residents have received training and jobs as skilled construction workers for the restoration process, others have received incentives to produce crafts for sale to tourists, or obtained other employment in the tourist industry.

Source: UN-HABITAT (2003: 86)

Photo: Martin Mowforth

efforts (Sida 2005a), and resignation that 'comprehensive restoration might result in gentrification – that the rents go up so that only the well-off can afford to live there' (Tannerfeldt and Ljung 2006: 108).

The inconvenient fact is that while the 'poor' are a potential asset to new tourism focused in rural areas (keeping authenticity high and costs low) they are a potential liability in cities (supposedly forcing down aesthetics and the new urban chic, and forcing up perceived urban insecurity). In the former they are geographically enriching and iconic, in the latter they are misplaced, unsightly and unwanted.

There is a further consideration in the transformation of cities and city attractions that are the embodiment of colonial and trading relationships that resulted in and sustained colonial powers, and to which new tourism should be alive. As Nick Robbins of Henderson Global Investors' Socially Responsible Investment Team argues, one of the 'tasks of history is to rescue the memory of those cast aside by the powerful, to seek justice across the centuries' (Robbins 2002: 87). Kolkata (formerly Calcutta), for example, is described by Philip Davies of English Heritage as 'one of the great cities of the world . . . the neglected jewel in the crown of India's heritage' (Davies 2000: 12) and is regarded as a 'unique architectural heritage' and counterpoint to the 'negative images which prevailed later – Kipling's City of Dreadful Night exacerbated by popular images of Mother Teresa's City of Joy' (Davies 2000: 14). It is little wonder therefore that Kolkata, the most significant node of power and control in the British Empire – the London of the East – and trading post for the East India Company, has attracted increasing new tourist interest (beyond those fractions of new tourism described by Hutnyk as travellers-cum-volunteers in Chapter 3) in the undisputable beauty of its architectural legacy. In the context of the corporate power wielded by the private joint-stock East India Company in the eighteenth and nineteenth centuries, for Robbins (2002: 80) there is a paramount need to redress a 'corporate amnesia and begin the process of remembrance and reparation'. Robbins presents a penetrating critique of one of the most powerful 'commercial dinosaurs that once straddled the globe' that at the height of its operations 'ruled over one-fifth of the world's people, generated a revenue greater than the whole of Britain and commanded a private army a quarter of a million strong' (Robbins 2002: 79). But as he argues, there is nothing to mark 'its power and its crimes' and 'the remorseless logic of its eternal search for profit, whether through trade, through taxation or through war' (Robbins 2002: 79, 87). As Jeffrey Sachs (2005: 171) concludes, it is the history of 'greed-driven private armies running roughshod over a great civilization'. Seeing Kolkata's colonial legacy as part of the fabric of remembrance, as well as architectural gems in their own right, is one potential route for mapping this power through tourism, and of new tourism puncturing the skin and aestheticisation of urban destinations.

Similar discussions may be applied to the slave economies of the Atlantic (also known as the 'triangular trade'), often regarded as one of the first systems of globalisation, and one in which the wealth and power of the East India Company was intermittently bound. The slave trade was the biggest deportation in history and a determining factor in the world economy of the eighteenth century. Millions of Africans were torn from their homes, deported to the American continent and sold as slaves. The year 2007 marked the two hundredth anniversary of the abolition of slavery in the Parliament of the United Kingdom (through the Abolition of Slave Trade Act, 1807). There has been considerable debate and argument in the representation and interpretation of the role of the abolitionist movement, together with the more complex issues of reparation and restitution resulting from the personal effects of the slave trade. Less consideration has been given to the role and interpretation of those towns and cities, such as Elmina Castle (Ghana), the first permanent slave trading post built by the Portuguese in 1482 (the dungeons of which are popular with visitors), that acted as points of departure, transhipment and arrival of slaves (largely from Africa). These sites are becoming increasingly significant in reconstructing history, and for African Americans in particular, for retracing genealogy through a particular brand of new tourism focused on slavery.

Whether the conservation of this grisly heritage is yet more evidence of the aestheticisation of powerlessness (as discussed in Chapter 2) – an unquenchable thirst for voyeuristic experiences – or by contrast a powerful representation and monument to inhumanity and a significant part of reconciliation and truth, is a matter for debate. Sida, for example, has promoted cultural heritage as a potential mechanism for longer-term processes of reconciliation and 'healing'. In Bagamoyo, Tanzania's oldest town on the coast opposite Zanzibar, slaves were sold until the end of the 1800s and historical preservation has been considered an effective and appropriate method of reflecting the town's multicultural and terrifying history. (Discussions are being held on including Bagamoyo on UNESCO's World Heritage List.) Arguably a large part of the answer lies not in the physical presence of such sites, but, as with so many activities through tourism, with the sensitivity in interpretation and the receptiveness of tourists themselves.

Our discussion so far suggests the potential of new forms of tourism are tempered by both the struggle of the urban poor to remain in their homes once city areas become earmarked for cultural development, and the interpretation of these places that reflect their heritage and histories. In the next section we take these challenges further. As the urban commentator Charles Landry (2000) notes, while tourism feeds off 'culture', most tourism focuses on a narrow conception of 'culture' including architectural heritage, museums, galleries and theatre, rather than the rich cultural distinctiveness of individual cities. This begs the practical and ethical question: Can new tourism expose the distinctive culture of the urban poor and the peripheral areas in which most of them live?

Pro-poor city tourism?

Initially of course we need to ask a more fundamental question as to whether new forms of tourism should be considered as a vehicle for promoting poverty-related initiatives and supporting the lives of the city poor. In short, should the development assistance offered to Third World cities demand a pro-poor orientation? As Goran Tannerfeldt, former senior urban development adviser to Sida, provocatively asks:

> why should we preserve the cultural heritage and why should we support this in development co-operation? Will it contribute to the eradication of poverty? Will it improve gender conditions, promote human rights and democracy or any of the other objectives for development co-operation? My answer is: So what? Maybe it is something that is justified on its own merits and not in relation to other objectives.
>
> (Sida 2004: 12–13)

Reflecting on Tannerfeldt's challenge, the reality is that there are few systematic studies on either the impact of tourism on Third World cities or the potential of tourism in reducing poverty (including the role of cultural heritage). This is therefore a major area for innovation and application in research. Where such urban studies do exist they tend to either conflate correlation and causality (see e.g. Colantonio and Potter 2006), or they are inconclusive and superficial in their recommendations. The main findings of a regional workshop bringing together mayors and officials from the Asia-Pacific Region, for example, 'was the need for a significant paradigm shift in the way tourism development occurs' (Jamieson 2002: 7). Tourism consultant Walter Jamieson reported the need for 'tourism officials [to] shift from a situation where tourism arrivals are the primary indicator of tourism success to one concerned with a sustainable approach which improves conditions for the poor through tourism development' (Jamieson 2002: 7). A laudable goal perhaps, but how this is achieved in practice, and in detail, is far less clear.

In response to Tannerfeldt, and putting the potential contribution of urban tourism to poverty reduction aside, there should be agreement that new tourism must avoid exacerbating the condition of the urban poor: the right to reside in the city and the availability of adequate shelter (a 'home'), services (water, sanitation, electricity, waste collection) and security (from crime and eviction). And this returns us to one of the underlying tensions of tourism in cities. For the urban poor inhabiting the expanding cities and towns of the Third World, especially those crowded into the chaotic informal settlements on the outskirts and traditionally less desirable parts of the city, home is often far from safe, secure and settled. Slums have traditionally been an anathema to tourism. The urban poor sit uneasily with the heritage management discussed above and are literally 'out-of-place' in the more conventional forms of tourism (that demand an airport–city transfer that is free from the eyesores of urbanisation and are reminiscent of the World Bank's well-ordered, secure, clean and healthy city referred to earlier in this chapter). It is these pressures and perceived opportunities that work towards eviction and displacement. The vulnerable poor, despite the ingenuity of adapting to the harshness of cash-based urban economies, have few rights and little security:

> The idea of *remoção* (removal) is nothing new to Rio de Janeiro, where nearly 20% of the population, a million people, now live in about 750 slums. During the 19th century, town planners forced thousands from the slums of central Rio in a bid to turn the city into a 'tropical Paris' . . . Now in a campaign by sections of the Brazilian media, the question of favela removal has been thrust back on the political agenda . . . Sectors of the tourist industry also back removal. With Rio hosting the Pan American Games in 2007, several partial removals are being planned . . . I'm absolutely in favour of removal,' the president of the Brazilian Association of Travel Agents, Carlos Alberto Ferreira, recently told Globo newspaper, arguing that without the favelas blighting the landscape, tourism levels would rise, the profits of which could be channelled into fighting poverty.
>
> (Phillips 2005: 11)

Whether new forms of urban tourism can redress this imbalance of power is a moot point. Nevertheless, the potential of cities for reducing poverty is increasingly advocated, and the role of city-based tourism as a new form of urban 'ecotourism' is highlighted as a potential vehicle. The Asian Development Bank (ADB), for example, reports:

> Properly planned and managed, urban tourism can be a significant tool for pro-poor urban development. It is labor-intensive and a recognized job creator. It does not require high academic skills. Vocational and basic skills, which the poor can easily acquire, are sufficient.
>
> (ADB 2006a)

The ADB sees its role as a significant one, and as

> well positioned to promote pro-poor tourism as a key ingredient of urban development and renewal. Its policies of sound environmental management, community participation, and decentralization – coupled with its support for broad-based urban development – are conducive to catalyzing pro-poor tourism

as part of the Asia and Pacific region's effort to reduce urban poverty, promote social equity, and enhance heritage management. It is a challenging task – but achievable.

(ADB 2006a)

It is a task, however, set within a vociferous neoliberal advocacy of the significance of property ownership to the development process. The view is promoted by the Peruvian international celebrity economist, Hernando de Soto (reviewed in Chapter 2), who advises that lack of formal ownership is increasing the vulnerability of the urban poor to removal. The insecurity of communities as tourism grows and expands has emerged from the critical appraisal of new ecotourism initiatives in rural areas and the evidence of eviction and displacement of local people – see Chapter 8. Such critical analysis needs to be rapidly applied to cities too. Evictions (and the perceived threat of evictions) have always been a burden of the urban poor, but tourism-related development and infrastructure projects, large international events (such as the Olympics) and urban renewal and so-called 'beautification' initiatives all have a tendency to increase the vulnerability of communities judged to be 'in the way' (Centre On Housing Rights and Evictions (COHRE) 2006). Some proponents argue that at the very least, new tourism can help acknowledge and promote the cultural vibrancy, resourcefulness and organisation of the urban poor, as an integral part of the urban social fabric. But the question of how this can be achieved has to be addressed, and it is to this matter that the final section turns.

Slum tourism: aestheticising the poor or taking control?

Characterised by a lack of so-called basic services (such as water and sanitation and adequate collection of waste), substandard housing, unhealthy and often hazardous living conditions and insecure residential status (subject to the regular threat of being evicted), urban slums may share similar attributes familiar to those rural areas popular for new tourism.[1] While the sheer overcrowding and density may, on the surface, make cities a less attractive tourism proposition, we have argued throughout that tourism has a voracious appetite for consuming new experiences and places. Given the arguments developed in Chapter 3, it should be of little surprise therefore that the omnipresence of poverty makes the other side of cities an increasingly popular niche form of new tourism (known as 'slum tourism' and increasingly referred to as 'reality' tourism).

In previous sections we considered how there is a tendency to use the vehicle of urban cultural heritage to aestheticise cities, rather than reconstructing their sometimes painful histories. In this section we turn to a different form of interpretation through tourism, the representation of slums and the urban poor. Consider first the attempt of a travel operator in Kenya's capital, Nairobi, to make slums a bona fide object of the tourist gaze.

In a world full of visitor attractions (actual and potential), tourism has an uncanny incidence of making fashionable stars of some: over-hyped, over-analysed and over-quoted. Kenya's largest slum, Kibera, on the outskirts of Nairobi (and the destination of the package excursions referred to in Box 9.2) is one of the more unlikely to experience such focus with a conveyor belt of short- and longer term visitors eager to observe and record the reality of slum dwelling. But as Neuwirth (2006: 67) commences his exploration of Kibera, it is a paradoxical example of a ready-wrapped new tourism super-product: 'One glimpse is enough. You have discovered the famous misery of the Third World. A sea of homes made from earth and sticks rising from primeval mud-puddle streets.' Covered by daily national newspapers across the world, Reuter's correspondent Andrew Cawthorne provides a flavour of the public controversy that such reality tourism has given rise to (see Box 9.3).

Box 9.2 Kenya Slum Tours

Reality tourism in Kenya – escorted tours of the slums of Kenya to sample first-hand the difficulties faced by the poor in Kenya's Urban Cities and Rural areas.

Kenya has its fair share of the world's poorest people. 5.4 million Kenyans live in informal settlements. A large number of Kenya's poor, living on less than a dollar a day, stay in the urban centers such as Nairobi and Kisumu . . . These pro-poor tourism activities take Victoria Safaris' clients to these rarely visited regions and enables interaction with the local people even as the guests experience first-hand the problems that these urban people face in their day-to-day living.

Victoria Safaris has come up with this new noble idea of Kenya Slum Tourism as a means of creating awareness of the plight of the poor in Kenya to both foreign and domestic tourists with an intention of wiping out the slums in Africa and Kenya in particular as a long-term measure by using tourism business, the highest Kenya government revenue earner; reducing poverty by engaging the poor to participate more effectively in tourism development in Kenya and at the same time receiving an increase in the net benefits from tourism as a short-time measure. As the aims of pro-poor tourism range from increasing local employment to involving local people in the decision-making process, Victoria Safaris has hired and is continuing to recruit its local staff for the Slum Tour programmes among the inhabitants of the slums areas where it performs the Slum Tours.

These include the tour van drivers from the affected slums, the slum tour guides among which are the community leaders who understand the slum community locations better and the Slum Community policing security teams. All these personnel live within the Slums where the escorted tours are performed.

Source: Victoria tours http://www.victoriasafaris.com/kenyatours/propoor.htm (accessed May 2007)

Box 9.3 'Slum tourism' stirs controversy in Kenya

It's the *de rigeur* stop-off for caring foreign dignitaries. It reached a worldwide audience as a backdrop to the British blockbuster *The Constant Gardener*. Any journalist wanting a quick Africa poverty story can find it there in half an hour. And now at least one travel agency offers tours round Kenya's Kibera slum, one of Africa's largest. 'People are getting tired of the Maasai Mara and wildlife. No one is enlightening us about other issues. So I've come up with a new thing – slum tours,' enthused James Asudi, general manager of Kenyan-based Victoria Safaris. But not everyone in Kenya is waxing so lyrical about the trail of one-day visitors treading the rubbish-strewn paths, sampling the sewage smell, and photographing the tin-roof shacks that house 800,000 of the nation's poorest in a Nairobi valley. Indeed, the recent well-meaning visit of UN Secretary-General Ban Ki-moon . . . drew a stern editorial from Kenya's leading newspaper. 'What is this fascination with Kibera among people who do not know what real poverty means?' asked the *Daily Nation*. 'More to the point, how do Kenyans themselves feel about this back-handed compliment as the custodians of backwardness, filth, misery and absolute deprivation?'

Answer: Not a lot, at least according to an informal, random survey by this correspondent in Kibera itself. While all recognise the potential for good from such attention, plus the pressure it puts on the government and others to help slum-dwellers, most said tangible benefits so far were few, while the embarrassment factor was growing every day. 'They see us like puppets, they want to come and take pictures, have a little walk, tell their friends they've been to the worst slum in Africa,' said car-wash worker David Kabala. 'But nothing changes for us. If someone comes, let him do something for us. Or if they really want to know how we think and feel, come and spend a night, or walk round when it's pouring with rain here and the paths are like rivers.'

Even groups working day-in, day-out in Kibera – and dependent on foreign funding – are getting weary. Salim Mohamed, project director for the Carolina for Kibera charity, said the stream of high-profile visits to the 3 km-long corridor was raising expectations among residents which, when not quickly fulfilled, fuelled frustration with the appalling living conditions. Visits by tourists, which reached a crescendo during the recent anti-capitalist World Social Forum in Nairobi, were testing the local hospitality culture to the limit, he added.

Echoing a constant complaint by Africans of Western media, office administrator Christine Ochieng, 20, said the image of unmitigated misery in Kibera was not fair to her community. 'I can see how visiting the largest slum in Africa is very attractive to people, but there are so many untold stories here,' she said, rattling off ideas she and friends would like to include in a local magazine they want to start. 'But people just want to talk about poverty, poverty, poverty all the time,' she said.

Victoria Safaris' manager Asudi, from the same Luo tribe which constitutes the majority of Kibera residents, insists the tour he offers of Kibera and other slums in Nairobi and Kisumu in west Kenya, are beneficial to locals. They raise awareness, and he hands his tourists back a percentage of their payment to donate to a cause they have seen on their walkabout, he says, such as a health or school project. His publicity, however, has ruffled feathers. 'After lunch, proceed to the Korokocho slum where you will be amazed with the number of roaming children,' reads a typical paragraph.

Nairobi's chattering-classes are not amused. 'Kibera is the rave spot in Kenya,' wrote one columnist sarcastically. 'For where else can one see it all in one simple stop? The AIDS victims dying slowly on a cold, cardboard bed. The breastless teenager . . . Plastic-eating goats fighting small children . . . and – ah yes – the famous "shit-rolls-downhill-flying-toilets". It is unbeatable.'

Government spokesman Alfred Mutua has led a campaign to promote the bright side of Kenya and clean up its cities. He shakes his head when asked about the Kibera phenomenon. 'It is very sad that when dignitaries come here, the first place they run to is Kibera, the residents are getting tired of people coming and giving lip-service,' he told Reuters. 'Kibera is the "in" place everybody wants to be associated with, whether they are doing anything about it or not. . . . People look at others who are poor and destitute and get a "feel good" attitude about themselves, that they are above that.'

Source: Andrew Cawthorne (2007) '"Slum tourism" stirs controversy in Kenya', Reuters, Nairobi, 9 February

Of course, tours of 'slums' have existed for some time, originally associated with the tours of townships in post-apartheid South Africa and in the volunteerism associated with cities such as Kolkata (of which Hutnyk wrote brilliantly in 1996). But from modest beginnings, the desire to experience the reality of slum life and for slum dwellers and their associations to raise modest finance from these activities, is on the increase. Critically, Amelia Gentleman (2006) asks whether a new travel experience (offered to visitors to experience the harsh lives of Delhi's street children) is, in her words, a 'worthy initiative' or 'voyeuristic "poorism"'. 'For anyone weary of Mughal tombs and Lutyens architecture,' Gentleman commences, 'a new tourist attraction is on offer for visitors to the Indian capital: a tour of the living conditions endured by the 2,000 or so street children who live in and around Delhi's main railway stations.' For just £2.50 (or 200 rupees) both western and Indian participants take a two-hour guided tour of 'Delhi's railway underworld' by former street children, we are told, with the proceeds filtering back to a charity focused on rehabilitating street children. 'The trip is designed as an awareness-raising venture and organisers deny that this is the latest manifestation of "poorism" – voyeuristic tourism, where rich foreigners come and gape at the lives of impoverished inhabitants of developing countries,' Gentleman (2006) reports.

The degree to which this is a pragmatic response to the lives of the poor mediated by local charities that 'speak on their behalf', or a voyeuristic and staged consumption of the Dickensian underworld of Third World cities that tacitly endorses rather than challenges complex relationships of power, is a matter of debate. Much like rural-based tourism, however, the barometer should remain the level of access to, ownership and control of resources and the degree of power vested in poorer communities as determinants of pro-poor development potential. But what exactly can we learn from these mediated experiences that are no longer than an average feature film? Gentleman (2006) concludes:

> By the end of the walk, the group is beginning to feel overwhelmed by the smells of hot tar, urine and train oil. Have they found it interesting, Javed asks? One person admits to feeling a little disappointed that they weren't able to see more children in action – picking up bottles, moving around in gangs. 'It's not like we want to peer at them in the zoo, like animals, but the point of the tour is to experience their lives,' she says. Javed says he will take the suggestion on board for future tours.
>
> (Gentleman 2006)

As we have argued, there is scant academic literature which addresses and details the dynamics of new and supposedly responsible forms of tourism in urban areas, and the problems and prospects that arise from them. Deborah Dwek's (2004) study of new tourism in Rio de Janeiro's largest *favela*, Rocinha, is therefore of special interest. Many *favelas* (the terms used to describe slums in Brazil) have their origins in overnight migrations of large numbers of people from rural areas to the cities, having been dispossessed of their land or made jobless, and whose flimsy encampments, constructed from materials of plastic, tin sheeting and even cardboard, are reminiscent of Theroux's earlier descriptions. Rio has around six hundred *favelas* and Rocinha, like all others, is an illegal community (or as de Soto (2001) would argue, a sea of dead capital, devoid of formal property titles) created over years of successive land invasions. Established initially without any services, if these slums escape clearance by the authorities and become established, house improvements appear, businesses take root, water pipes and electricity lines are 'tapped' and a degree of formalisation emerges. Indeed, like all urban areas, scratch beneath the surface and there are myriad processes and systems that tend to challenge our perceptions of what slums are. As investigative journalist Robert Neuwirth (2006: 31) remarks on arriving in Rocinha, 'I still had the idea that squatter communities had to be primitive. But Rocinha was like nothing I had ever imagined.' As Dwek records, most of today's *favelas* do not deserve the label of 'slum', and many *favelados* would object to its use to describe their residential area.[2] Indeed, as Neuwirth (2006) suggests:

> Rochina has been such a commercial success that residents have coined a new word to describe the process they see unfolding in their neighbourhood: *asfaltização* (asphaltization). It is the squatter city version of gentrification. It refers to businesses from outside the favela – from the asphalt city, the legal city – invading illegal turf.
>
> (Neuwirth 2006: 43)

As Box 9.4 describes, it is no surprise perhaps that the government plans to capitalise on this upward trend with the introduction of further tourist infrastructure.

Despite *asfaltização*, it is also clear that these are areas which in popular perception have become associated with poverty, deprivation, crime and violence, and the extraordinary

---

Box 9.4 Tourist guesthouses in Rio's shantytowns

The Brazilian government has announced multimillion pound plans to build tourist guesthouses – pousadas – in one of the most notoriously violent corners of Rio de Janeiro.

President Luiz Inacio Lula da Silva revealed the plans . . . as part of a development project which also includes the construction of roads, creches, hospitals and a convention centre in Rocinha, Rio's largest shantytown . . .

The pousadas . . . are expected to be located in Laboriaux, one of the highest sections of the favela. The area boasts spectacular views over Rio's undulating landscape but is also known for shoot-outs between drug traffickers and police and is located near clandestine cemeteries used by traffickers to dispose of their enemies.

Yesterday the mood on the Ladeira do Laboriaux – a steep incline that leads into the shantytown – was buoyant. 'Gringos?' said Cristiane Felix de Lima, sitting in her husband's bar, Seven Lives, in the hilltop favela. 'It will be good for business.'

Ricardo Gouveia, an architect and human rights activist, said the new attempts to combine 'land rights, urban redevelopment and social projects' represented a significant step towards improving living conditions in Rio's biggest shantytown, home to over 100,000 impoverished Brazilians . . .

Few deny that developing the sprawling shantytown represents a huge challenge. 'To sort out Laboriaux and Rocinha, you have to look at all the problems,' said community leader Paulo Sergio Gomes as he stood in a shack constructed out of branches and abandoned wardrobe doors. Beside him three naked children – aged five, six and seven – played in the dust. Their mother, who works 15 hour days, had abandoned them at home . . . 'Most of the houses around here are even worse than that one,' Mr Gomes added, as a teenager hurtled past on a motorbike, a pistol tucked in his belt. 'We need more than half a dozen pousadas to fix things.'

Source: Extracts from Tom Phillips (2007) 'Brazil to build tourist guesthouses in the heart of Rio's shantytowns', *Guardian*, 20 January

rise to prominence of cocaine and its introduction into the *favelas* at the end of the 1970s and early 1980s led to the establishment of drug gangs and organised crime networks which still dominate the image of the *favelas* held by outsiders. This image leads to a mixture of prejudice and fear about the *favelas* and *favelados*, which is felt at least as much by Brazilians as it is by foreigners. As Dwek (2004) remarks:

> I found that this attitude which revealed both prejudice and fear was representative of the attitude of many middle-class Brazilians who, while cohabiting and interacting with *favelados* on a daily basis, confined this contact to a hierarchical relationship with maids, porters or others within the service industries. Few had been inside a *favela* . . . associating them immediately with drugs and danger. Most thought it was still too dangerous to go into Rocinha, a tribute in part to the powerful job the media has done in promoting this view.
>
> (Dwek 2004: 24)

It is paradoxical that the *favelas* are also so strongly associated with a number of positive aspects of Brazil's image, especially the samba and carnivals, and it is this mixture of images – fear and danger, yet excitement and sensation – which makes them an attraction to visitors. 'While Brazil uses popular culture to successfully market her image on the world stage, prejudice against *favelados* within Brazil runs deep' (Dwek 2004: 24). Most visitors actually aim to experience a sense of threat and danger, whether real or imagined.

The *favela* tour is a recent feature of the Brazilian tourism industry, set up to exploit this 'sexy' image, as Dwek calls it, and what has become known as '*favela* chic': an 'external fascination of the West with the favela has always been about projecting outsiders' perceptions and values on to this world' (Dwek 2004: 11). Dwek's research, carried out in 2004, is representative of this relatively new line of inquiry, although officially the *favela* of Rocinha in Rio de Janeiro has been receiving tourists since 1992. It is commonly perceived that *favela* tours are run by residents of the *favela* and are thus an indication of local entrepreneurial spirit rising out of the hardship of life. Dwek's (2004) analysis suggests that most of the tours are actually managed by outsiders and surprisingly few residents act as guides. Additionally there are question marks over the supposed localised economic impacts. A number of artists were recorded as having developed a degree of dependence on the tours for selling their work with their economic livelihoods rendered precarious if tours were cancelled. The small-scale trickle-down of economic benefits to a few individuals appears not to have contributed returns to the community as a whole. Despite such shortcomings she concludes that most residents viewed the tours in an extremely positive light, and there is little doubt that some residents gain financially from the interchange.

What research such as this does provide is an insight to the underlying tensions and contradictions of supposedly new and responsible forms of tourism. Inevitably 'there is still a stratum of tourists who find the idea of these tours ethically dubious', and a voyeuristic and sometimes uncomfortable expedition. 'I was witness to this safari-style viewing on my Jeep Tour with an extended family of American tourists complete with cameras,' she recalls:

> After entering a house of one of the local families in Parque da Cidade favela a 'typical scene' was pointed out to us of a family relaxing on their balcony below and I felt our role as 'gazer' was made uncomfortably clear as we were given the impression of viewing the natives in their natural habitat, almost as if we should not be making too much noise in case we disturbed them!
>
> (Dwek 2004: 21)

Dwek records a 'superior and patronising attitude' which was

> displayed by some of the tourists who were frequently surprised at how friendly, well-behaved and receptive the favelados were, having expected them to be more hostile as well as noting how they seemed to treat each other very well, even sharing their food among themselves. It was as if they were another species of people who belonged to a world unlike anything they knew.
>
> (Dwek 2004: 25–6)

The ethical doubts might also reflect the concern of some critics regarding the search for the 'authentic other' by these postmodern tourists. The *favela* tour is the chance for many of the middle-class tourists and backpackers (identified by Dwek as the major groups of visitors using the tours which are on offer to the *favelas*) to get closer to the 'real' Rio. 'Today's tourist, coming predominantly from the post-industrialised and post-modern western world, travels in search of the antithesis to capitalism and modernity' (Dwek 2004: 22), and it is this search that leads them into a *favela* tour. A number of tour companies have made this search for the 'other' and for the 'real' Rio a major focus of their tours, while others have given a more objective portrayal of the areas. Of course, the very existence of tourism in the *favelas* alters the nature of the interchange and experience for both the visitor and the visited, bringing into question the authenticity of this experience. As suggested in the discussion in Chapter 3, the fetishistic nature of tourism and the aestheticisation of the experience – of living on the 'edge' – also have a tendency to embrace and celebrate the risky side of tours:

> if the police ever stop him while he is guiding a tour it only adds to the excitement for the tourists. . . . They also clearly liked the 'danger' element – which Fantozzi played up to for the thrill-seeking tourist – emphasising the 'underworld' elements of the favela such as drug-related graffiti and the traffickers themselves – some of whom the tourists were convinced were following them.
>
> (Dwek 2004: 21)

Despite the observed problems, Dwek concludes that such tours do have their place in dispelling prejudices about the *favelas*, though clearly 'development' in the *favelas* of Rio will inevitably require much more than just tourism. What this study does underline is that local people and communities must be at the heart of these developments and that initiatives which spring from the *favelados* themselves will produce the greatest economic returns for the communities. As the *favelado* Jorge Ricardo, a financial director cum disc jockey, tells Neuwirth (2006: 65), 'Brazil has passed from Third World to Second. And the reality of Brazil is Rocinha. We are a community that serves as a model for other communities. We are the future.' But as for controlling and channelling *favela* tourism for the greater community good, Dwek's analysis suggests there is some way to go. While it was observed that *favela* residents have their own ideas about how to exploit tourism, they 'do not appear to have a voice with which to express them or any power to effect change' (Dwek 2004: 22).

Conclusion

Although cities have been traditionally associated with tourism, there is relatively little written on the impacts of tourism on cities in the Third World. This chapter has sought to demonstrate why such a focus is necessary and to emphasise the need for an assessment of whether new forms of tourism can support broader pro-poor development efforts. This developmental focus is of special significance within a context where cities remain largely seen as 'economies' (well-ordered and oiled machines) rather than 'societies', where their economic and labour market potential is given primacy over the social development potential (arguments that reflect back to the discussion in Chapter 2), and where the rapid growth of cities more generally has resulted in swelling the ranks of the urban poor. It was also discussed how a deeply held anti-urban bias has characterised the evolution of development as an idea and practice, a bias that is equally reflected in the analysis of Third World tourism. While the environmental and cultural impacts of tourism in rural, coastal and natural areas are well documented and critiqued, city-based tourism receives less attention and for the most part the coverage is limited to the size, growth and economic impact and potential of tourism.

The chapter took a closer look at two forms of new urban tourism. The first, cultural heritage, has a long lineage in the cities of Europe (in particular) but has been less prevalent in Third World cities. For some cities, the presence of highly textured cultural cityscapes represents a rich resource that is acknowledged globally (through UNESCO's World Heritage Listing, for example). As the chapter discussed, however, it is less clear if such tourism can avoid the displacement and marginalisation of communities reminiscent of more traditional forms of tourism. Equally, the chapter signalled that there are considerations of how cultural assets are represented and how histories are retold and interpreted, an especially important process in the context of conquest and colonialism.

The second form of new urban tourism, slum or reality tourism, comes into direct contact with a city's poor. On the surface this may present a more effective foundation for channelling the economic benefits of tourism into activities and initiatives that support poorer communities. But as with all forms of tourism it was argued that the ownership, control and level of influence of such tourism activity are critical to shaping its development potential. Equally there are significant ethical questions over a form of tourism that has the potential to be voyeuristic and to aestheticise urban poverty and insecurity, and increase rather than challenge inequality and powerlessness.

Given the heritage to development theory and Third World tourism studies and the rapidly increasing interest in the potential of new (and supposedly) more responsible forms of tourism as a tool for development, it is clear why research has honed in on community-based rural tourism. This chapter has signalled why efforts should be immediately applied to understanding the role of new tourism in cities (from policy and practice, to sociological and anthropological analysis) as one of the least researched and most open tourism research agendas.

⑩ Governance, governments and tourism: selling the Third World

The central governments of nation states together with the global multilateral institutions are the last of the key players in this analysis of tourism. It is governments that have a pivotal role and possess the potential power to control, plan and direct the growth and development of tourism. And it is largely through governments that tourism-related international investments and loans and overseas aid is agreed and channelled. As such, it is widely agreed that for tourism to assist in development, a favourable national policy environment is required.

It has already been argued that tourism has come to represent a considerable attraction to many Third World governments. It has been widely promoted both within the Third World and by First World 'experts' as a means of economic diversification and an important mechanism in producing foreign exchange. There can be little surprise that in Third World countries whose economies are characterised by indebtedness and by primary industries (such as agriculture and mining) adversely affected by world market prices, tourism has come to represent something of a panacea.

In this chapter we take global governance as a further way of exploring the politics of Third World tourism development. By governance, we mean the web of institutions and agencies that are central players in the political environment, and the focus of our discussion is on national governments, bilateral development agencies and the supranational institutions (such as the World Bank, IMF and United Nations). As already argued, the consideration of politics has been a poor partner in tourism analysis. The assessment of government activity has tended to focus on strategies, policies and programmes of national-level planning. While such analysis is an essential part of assessing the approaches of Third World governments to tourism, it tends to minimise broader consideration of the inherently political nature of tourism development. In other words, the pressures that governments face from external influences such as lending policies, First World foreign policy or the activities of international NGOs tend to be bypassed; similarly internal forces may also be overlooked. In short, the politics of tourism is another way of exploring unequal and uneven development.

In the first half of this chapter, politics is established as a critical factor in Third World tourism analysis. Most importantly, the need to acknowledge that tourism is not a neutral factor that can be assessed as just another government economic activity is established. Tourism is used for a variety of political purposes and there is a wide range of external influences. The structural adjustment policies of supranational institutions such as the World Bank and IMF, for instance, are considered along with the policies promoted by the World Trade Organisation (WTO),[1] World Tourism Organisation (UNWTO) and the World Travel and Tourism Council (WTTC). This provides the opportunity to reflect on the political globalisation outlined in Chapter 2 and in particular to consider the nature of sustainability promoted by such institutions.

The second half of the chapter uses a number of case studies to illustrate the development of new forms of tourism and the ways in which these are intimately related to sustainability. In particular, we analyse sustainability and sustainable tourism as part of a political discourse. Throughout this chapter, the ways in which politics is reflected through the power jigsaw (Figure 3.1) are considered.

The politics of tourism

We are constantly reminded by the media of the hazards of travel in the Third World and by implication the critical nature of the politics of tourism. At its most dramatic are the newsworthy events, the killing or kidnapping of western tourists, often by political factions eager to further their causes. For these factions, tourism becomes a means to an end: access to the system of global communication. To First World populations, such events provide a succession of countries – Algeria, Cambodia, Colombia, Egypt, India, Indonesia, Morocco, Peru – which, we are told, are plagued by fundamentalist groupings and where terrorism is rife. The most dramatic of these events was the 11 September 2001 terrorist attacks on the USA. This event clearly had an immediate impact on the perception of travel security of First World tourists. The international airline and hotel industries were hit hard, with the bankruptcies of airline companies of significant size (Sabena and Swissair) and hotels all over the world reporting dramatic falls in visitors and revenue in September, October and November 2001 over the same period a year before. Lay-offs of workers involved in the tourism industry became commonplace in these same months.

Although this was an incomparable event, evidence from it and other traumatic events confirms that the effects on international tourism may not be long-lasting. In November 1997, 58 foreign tourists were killed by terrorists at the Temple of Hatshapsut at Luxor in Egypt. Tourism to this part of Egypt, and to most of the rest of the country, dried up overnight. The livelihoods of a majority of the population of the town of Luxor came to an abrupt end. The effect was also generalised to some extent to other countries in the Middle East. In 1998, international tourist arrivals to Egypt fell by 14 per cent, and tourism receipts by 45 per cent. As Figure 10.1 shows, however, by 1999 international tourist arrivals were back to a level that might have been expected had the yearly rate of increase up to 1997 continued without the terrorist interruption. We discuss further the significance of terrorism to tourism in Chapter 11.

Western government presentations of supposedly dangerous countries have been one of the principal ways in which our geographical imaginations are filled out, and show how tourists and the tourism industry decide where, and where not, to visit and invest; for example, in Chapters 3 and 5 it was suggested that 'dangerous' regions may actually act as a bonus for some adventurous 'new tourists'.

As Richter (1994) demonstrates, the USA has dozens of countries for which DON'T GO warnings are provided, of which over 80 per cent are in the Third World. But such warnings, Richter (1994) continues, appear to be based as much on ideological reasons as on potential security risks to US tourists. Countries to which the USA is more kindly disposed, such as Israel, Brazil, Mexico and Egypt, appear to require much higher levels of danger to tourists before travel advisories are made against them. For example, up until 1994, although tourists had been signalled as potential targets for attack by Islamic groups in Egypt, this large-scale receiver of US foreign aid had not been mentioned in State Department advisories (Richter 1994). Similarly, despite the seeds of a civil war in southern Mexico, this country too has escaped travel advisories: a reflection, some would argue, of Mexico's 'integration' into NAFTA.

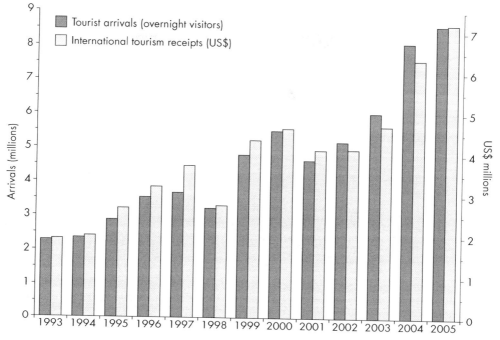

Figure 10.1 Tourist arrivals and receipts, Egypt, 1993–2005

Source: WTO (2007) *Compendium of Tourist Statistics, Data 2001–2005,* Madrid: UNWTO (and earlier editions)

At the other extreme, countries to which the USA considers itself ideologically opposed, such as Cuba and North Korea, remain off-limits to US tourists. The notion that *tourism is politics* is forcefully expressed by Conrad Hilton (of Hilton Hotels), who is reported as saying 'each of our hotels is a little America' and 'we are doing our bit to spread world peace, and to fight socialism' (quoted in Crick 1989: 325). Conversely, tourism is used by some Third World governments to gloss over glaring social inequalities and in some cases, as discussed later in this chapter, the systematic abuse of human rights.

Of course, for many Third World countries, political stability is one of the principal keys to securing a steady stream of First World tourists. The case of Gambia shows how profoundly political instability, or just the First World's judgement of such instability, can affect a country's tourism fortunes. Gambia experienced a period of economic restructuring with two major IMF-instructed economic reform programmes (1985 Economic Reform Programme and the 1990 Programme for Sustained Development) which attempted to diversify the economy. As C. Thompson et al. (1995: 573) report, 'Both programmes emphasise the important role played by tourism in the development process'. Tourism made a substantial contribution to the Gambian economy, estimated to be 12 per cent of GDP by the end of the 1980s and became the principal source of foreign exchange.

In 1994, Captain Yahyah Jammeh seized power in a coup d'état and established the Armed Forces Provisional Ruling Council (AFPRC). A failed counter-coup in November 1994 resulted in the British Foreign Office issuing travel advice not to visit Gambia. As a result of this advice, air charter arrivals (December to March, the peak of high season) dropped from 45,733 in 1993/4 to 8,363 in 1994/5. There was an estimated 60 per cent reduction in the contribution of restaurants and hotels to the GDP and an estimated direct job loss of around 10,000. The indirect or knock-on effects are estimated to be at least double this impact. Overall, tourism contracted by 60 per cent and the economy shrank by 6.2 per cent. The Gambian tourism industry all but collapsed (Evans et al. 1994).

Despite the lifting of the formal travel advice for the 1995/6 season, the Economist Intelligence Unit identified what appeared to be further pressure applied by the British Foreign Office: 'the UK appears determined to maintain pressure for an early return to full democracy and civil freedoms, and this position is likely to influence other trading and aid partners' (Economist Intelligence Unit 1995: 17). 'Tourism, accounting for around 12 per cent of GDP, has yet to pick up after the most recent season (November 1994–May 1995), ruined by European governments' advice to their nationals to stay away' (Economist Intelligence Unit 1995: 17).

The crucial significance of First World governments' travel advisories was shown again in early 2008 by the turmoil after Kenya's disputed presidential elections. The unrest decimated Kenya's tourism industry, with tours and charter flights cancelled and the US, British and other European governments initially issuing travel warnings for the whole country. By February 2008, less than six weeks after the first unrest, Britain and the USA amended their travel advice, 'warning citizens to avoid only certain areas rather than the entire country' (Rice 2008). Kenyan hoteliers had little else other than this change of travel advisory in which to place their hope.

Teye (1986, 1988) has argued that the perception of political instability arising from the liberation wars and frequency of coups d'état after decolonisation in sub-Saharan Africa has seriously stunted tourism development in some states. The perception of political instability is not the only likely deterrent to new tourists. After Hurricane Mitch hit Honduras, Nicaragua and El Salvador in October 1998, the tourist money on which many Central Americans depended dried up, and the Honduran Institute of Tourism began to promote a type of 'disaster tourism', attempting to attract wealthy North Americans to witness the devastation caused by the natural disaster and to assist in the recovery programme. More details of this programme are given in Chapter 11.

Clearly, then, it is not only the actual occurrence of political and civil instability that is of critical importance, but rather the way that it is perceived, constructed and represented in the First World.

These are perhaps dramatic examples, but they help to emphasise the inherently political nature of tourism development and the way in which power is transmitted through tourism. Countries promoting tourism (both mass and new) can be influenced by a range of factors, from the decisions of First World institutions and tourists, to the way in which particular Third World destinations are perceived in the First World. Consequently, Third World governments are also eager to control the way in which their tourism image is projected. We must therefore ask whether the messages we receive truly reflect what is happening in reality.

Assessing the politics of tourism

Although the size and scale of the global tourist industry would seem to suggest that tourism is a hot political issue, so far we have argued that the politics of tourism has remained relatively underexplored. In terms of assessing the activities of Third World governments, the majority of analyses have been policy studies focusing upon national tourism policies, including both individual case studies and comparative policy analyses between countries (see, for example, Richter and Richter 1985; World Tourism Organisation 1994). A good deal of this work is also focused upon the identification of relevant and appropriate methodologies for planning and implementing tourism policy (for example, Gunn 1994). Clearly such discussion is of interest and use, and has a wide appeal to practitioners.

Its weaknesses, however, lie in both its often prescriptive nature and its formulation from within the First World and, in many cases, the failure to set policy and action in a broader and more critical framework which acknowledges that there are competing interests. In Chapter 3 it was argued that the political economy approach is one of the few attempts to illustrate the way in which control of tourism development is held primarily in the First World. As Britton (1982) argues: 'The World Tourism Organisation, International Monetary Fund, United Nations, World Bank and UNESCO, among others, set the parameters of tourism planning, promotion, identification of tourism products, investment and infrastructure construction policies often in conjunction with metropolitan tourism companies' (Britton 1982: 339).

Of course, it is not just external influences that affect tourism policy and development, but internal factors as well. This includes the conflicts between governmental and non-governmental interests, and conflicting interests and priorities within governments themselves. The latter are particularly prevalent where responsibilities over tourism resources overlap more than one government department. In Uganda, for example, there were three separate government departments with responsibilities for national parks, the focus of Uganda's tourism. Box 10.1 illustrates a similar problem of different outlooks adopted towards its national parks by different government departments in Guatemala during the 1990s. Within Guatemala this problem appears to have been resolved during the 2000s, but clearly, it is necessary to avoid the crude suggestion that government policy and action is an undifferentiated whole.

A non-critical approach also tends to assume that there are rational policy decisions that will lead to the surmounting of potential problems. As C. Michael Hall and John Jenkins (1995: 65) argue, while 'rationality' represents an influential approach to policy analysis, it is also 'extremely misleading as it fails to recognise the inherently political nature of public policy'. Take, for example, the approach of Edward Inskeep, a tourism planner who has worked with the UNWTO, World Bank and UNDP in many Third World countries: '[the] planning of tourism is necessary not only for scientific purposes and to conserve the environment for the benefit of residents, but also for the protection of long-term investments in tourism infrastructure, attractions, facilities, services, and marketing programs' (Inskeep 1987: 119).

What is important here is to question who benefits from planning and policy formulation and who controls these decision-making processes. If anything the 'benefit to residents' is a poor partner of the need to protect 'long-term investments', but the latter is couched in terms of conserving the environment for local people. The suspicion that environmental conservation is not really about a concern for local people is confirmed: 'Increasingly, tourists are demanding that their environments be high-quality and pollution-free as well as inherently interesting, and some tourists will change travel patterns if environmental quality expectations are not met' (Inskeep 1987: 119).

Ultimately then, we must ask who controls government policy and planning and what forces drive the processes of its formulation and implementation. Sustainability is a critical consideration here, for Agenda 21 and later initiatives designate national governments as having the lead responsibility for the process. The overall aim of developing a sustainable tourism programme has been set out by the WTTC, which defined Agenda 21 as the process 'to establish systems and procedures to incorporate sustainable development considerations at the core of the decision-making process and to identify actions necessary to bring sustainable tourism development into being' (WTTC 1995: 38). We must also ask in whose interests such actions are undertaken and explore how the uneven and unequal nature of development is reflected through this.

Box 10.1 Guatemala's protected areas

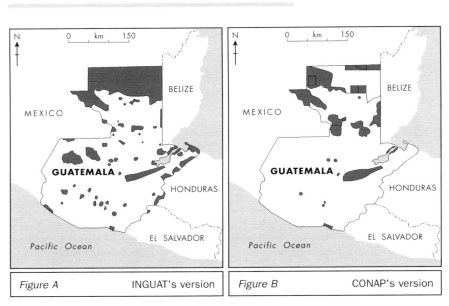

| Figure A | INGUAT's version | Figure B | CONAP's version |

Figures A and B show the 1994 protected areas of Guatemala – according to two different arms of the Guatemalan government. Figure A gives the version according to INGUAT, the Guatemalan Institute of Tourism; Figure B according to CONAP, the National Council of Protected Areas, which is charged with overseeing the protection of these areas.

The differences between the two were technically due, in the main, to INGUAT's inclusion of all the areas whose status as a protected area was only proposed rather than actual. But these differences also reflect the roles and outlooks of the two departments. INGUAT is in the business of enhancing the country's image, of making it look attractive to potential visitors; so it covered as much of the national area as possible with the status of a protected area, with the greenwash of pristine natural areas. CONAP, on the other hand, is more realistic; it was aware that it did not have the resources necessary to oversee those areas which were already designated as protected, still less all those which were proposed.

The notion of 'interests' is a useful one in focusing attention on which groups, organisations and institutions have influence and power in decision-making. As Table 10.1 implies, such influences can vary from the structural adjustment policies of the World Bank and IMF to the campaigns against visiting certain countries (due, for example, to the infringement of human rights within that country – see the case study of Burma outlined later in this chapter). Our discussion of Gambia (above) helps to identify different ways in which power is expressed through political processes. It is illustrative of the typical problems faced by Third World countries in their efforts to develop tourism. In this case, the expansion of the sector has been encouraged by the internationally inspired structural reform programmes which have clearly pushed Gambia along a particular development path. But First World countries also retain the potential to control the tourism industry

Table 10.1 *Tourism interest groups*

Scale	Industry groups	Non-industry groups	Single interest groups
International	UN World Tourism Organisation, WTTC, World Bank, regional development banks, World Trade Organisation	Environmental and social organisations, e.g. IUCN, WWF, ECTWT, FOE, Third World Tourism European Network (TEN) The International Ecotourism Society, Friends of Conservation	Occasional environmental or social issues, often location specific, e.g. End Child Prostitution, Pornography and Trafficking (ECPAT), Global Anti-Golf Movement (GAGM), Burma Action Group (BAG)
National	National tourism industry associations (e.g. Belize Tourism Industry Association – BTIA), trade unions, national professional, and trade associations	Environmental and consumer organisations, e.g. National Trust, Tourism Concern, Wilderness Society, Sierra Club, Audubon Society, ANCON (Panama)	Single-issue environmental groups, e.g. those opposing airport development, Surfers Against Sewage
Local	Chambers of commerce, regional tourism business associations, local area promotion partnerships	Local government; rate-payers' and residents' associations, e.g. Talamanca Association for Ecotourism and Conservation	Groups opposed to tourist development in a specific location, e.g. anti-resort development groups

Source: adapted from C.M. Hall and Jenkins (1995)

(especially when much of the industry caters for mass packaged charters, as in the case of Gambia).

Tourism as politics

Chapter 3 showed how political economy critiques have revealed the shortcomings of Third World tourism and how tourism is an additional element in uneven and unequal development. However, this has not prevented countries which pursue socialist strategies and ideologies from dabbling in tourism, and in some cases arguing for forms of tourism that are not subordinating and are responsive to the issues of power and control exercised by the First World so heavily criticised in the political economy framework. As Crick (1989) argues, in some countries this has resulted in seemingly glaring contradictions to the anti-colonial ideologies adopted by newly independent countries. Indeed, as Chapter 3 showed, some critics suggest that tourism may help to maintain systems of neocolonial power and control.

Tanzania is a frequently cited example where an emphasis on tourism development in the 1970s ran contrary to the anti-colonial socialism adopted by Nyerere in the wake of independence and the policy of self-reliance pursued by the government (as discussed in Chapter 2). The Tanzanian intellectual, Shivji, was critical of the separation of politics from economics: 'The justification for tourism in terms of it being "economically good" though it may have adverse social, cultural and political effects, completely fails to appreciate the integrated nature of the system of underdevelopment' (Shivji 1973, quoted in Crick 1989: 321).

Other independent countries pursuing socialist ideals have also explored the development of tourism. In Cuba, for example, as much effort has been put into widening the appeal of Cuba as radical chic, with solidarity and study tours, as has been devoted to stemming the foreign exchange leakage characteristic of other destinations in the region. In order to reduce the leakage of foreign exchange, Cuba has established a tourism industry that seeks, if uneasily, to separate tourism spending from the national economy, and forces tourists to buy goods and services in foreign currency (particularly the Euro) or in convertible pesos which are still tied to the US dollar, even though the US dollar is now one of the foreign currencies no longer accepted in the country.

Following a successful coup d'état in 1979 in Grenada, a regional neighbour of Cuba, Maurice Bishop, head of the People's Revolutionary Government (PRG), acknowledged the shortcomings of mass tourism as it existed in the Caribbean and started to identify the basis of an alternative type of tourism. Outlining the parallels of tourism with colonialism which we discussed earlier, the socialist Prime Minister argued:

> in the early days and even now, most tourists are white. This clear association of 'whiteness' and 'privilege' is a major problem for Caribbean people just emerging out of a racist colonial history where we had been so carefully taught the superiority of things white and inferiority of things black.
>
> (Bishop 1983: 69)

Bishop elaborated his ideas further at a Caribbean regional conference on the impact of tourism, where he identified the need to replace 'old tourism' with what he termed 'new tourism'. Along with Jamaica under the socialist leadership of Prime Minister Michael Manley, Bishop attempted to chart the development of a tourism designed to escape the problems that had been associated with 'old' tourism, bound up with 'colonial and imperialist connotations'.

> It was foreign-owned and controlled, unrelated to the needs and development of the Caribbean people, and it brought with it a number of distinct socio-cultural and environmental hazards such as the race question and undesirable social and economic patterns such as drug abuse and prostitution.
>
> (Bishop 1983: 71)

Bishop's vision was for a 'new' tourism to escape these characteristics. It was to involve all people (as both 'guests' and tourists); it would seek to create linkages between the different sectors of the economy; and above all it was conceived as a tool for development. However, the construction of Grenada's airport, a key part of the country's attempt to attract tourism (see Box 10.2), encapsulates and symbolises the political nature of tourism development. Bishop was murdered in October 1983 and with him died this vision of new tourism. Of course, it is questionable whether such a vision was realisable given the context of uneven and unequal development in which Third World tourism operates.

A rather different aspect of the inherently political nature of tourism is the way in which the USA sought to destabilise both Grenada and Jamaica, under Bishop and Manley respectively, through negative publicity and foreign policy statements that were hostile to both countries (C. Thomas 1988). In the case of Grenada this is particularly poignant. Grenada quickly gained an extraordinary international reputation, with Fidel Castro

Box 10.2 Grenada's tourism development

The absence of an international airport was a clear weakness in Grenada's tourism plans and a brake on its development, a shortcoming that Maurice Bishop readily acknowledged in describing the airport as 'the gateway to our future' (Marcus and Taber 1983). The majority of capital (US$60 million) and technical assistance for the airport's construction came from Grenada's ideological neighbour, Cuba.

President Reagan, however, claimed that the construction involved a new naval base, a new air base, storage bases and barracks for troops, and training grounds. 'And, of course, one can believe that they are all there to export nutmeg . . . It is not nutmeg that is at stake in Central America and the Caribbean, it is the United States' national security' (President Reagan, March 1983, quoted in J. Ferguson 1990).

As James Ferguson points out, despite the US development agency's (USAID) hostility to the People's Revolutionary Government policies and projects, it was deeply ironic that USAID were financing and administering the completion of the international airport, especially given the initial announcement that the USA had no plans to assist a project that was never intended to promote the development of Grenada's tourism.

After the murder of Bishop and the US invasion of the island, post-revolutionary Grenada provided US technocrats an opportunity for a structurally adjusted blueprint – a 'USAID inspired restructuring and free market deregulation' (J. Ferguson 1990: 114). Grenada was subsequently a laboratory and showcase of US ideological superiority. Much USAID/NNP [New National Party] policy and finance was targeted on environmental and infrastructural improvements designed to attract foreign investment and tourism. As Pattullo (2005: 16) states of Grenada, 'mass tourism became entrenched by the 1980s'. Indeed tourism has been one of the principal factors in the US-approved development strategy for the Caribbean. In many other places too, tourism is alluded to as a peace industry, and a way of allowing the expansion of capitalist economies in a stable and peaceful environment. 'The overwhelming majority of USAID funds have . . . gone towards dismantling the state sector, encouraging private enterprise and wooing foreign capital – with negligible success' (J. Ferguson 1990: 39; see also McAfee 1991). As Ferguson argues, the failure of the model is all the more significant given the symbolic importance placed on Grenada's development by the US government.

Source: Information and some extracts from J. Ferguson (1990)

assessing that in conjunction with Nicaragua and Cuba they were representative of 'three giants rising to defend their rights to independence, sovereignty and justice on the very threshold of imperialism' (S. Clark 1983, quoted in C. Thomas 1988). As Thomas concludes from the radical agendas pursued in these countries and the reworking of tourism that they advocated, 'they confirmed that any attempt at radical social reorganisation in the region would be met by insistent efforts on the part of the US to destabilise the process (or worse), especially if the proposed reorganisation involved new options for foreign policy and external relations' (C. Thomas 1988: 246).

Globalisation and the politics of external influences

Supranational institutions

Supranational organisations exist in many fields of activity. Those described as socio-environmental organisations were discussed in Chapter 6. While these may in certain instances influence the policies pursued by national governments, this section deals with those organisations dedicated to direct and deliberate involvement in the formulation of national economic policy through policy regulation and/or money-lending activities. The three most powerful and dominant of these are the International Bank for Reconstruction and Development (known as the World Bank), the IMF and the World Trade Organisation, although regional finance and development organisations, such as the European Bank for Reconstruction and Development (EBRD), the Inter-American Development Bank (IDB), USAID and increasingly other bilateral aid programmes also wield very considerable power over the policies pursued by Third World countries. Given the significance of the World Bank and IMF, a brief background to their formation and role is given in Box 10.3.

Box 10.3 The World Bank and IMF

The International Bank for Reconstruction and Development (the World Bank) and IMF were born at the Bretton Woods conference in the USA in 1944. Their purpose was to provide an order to the global economy to prevent the collapse of international trade and the development of isolationist economic policies, which it was believed had led to the Depression of the 1930s and the rise of fascism. The conference rejected the proposals of John Maynard Keynes to establish a world reserve currency administered by a central bank. Instead, it 'opted for a system based on the free movement of capital and goods with the US dollar as the international currency' (New Internationalist 1994: 14).

Initially, a more specific purpose was to provide funds to assist in the rebuilding of war-torn Europe. The European countries objected to the severe loan conditions, so the World Bank worked through the Marshall Plan which provided US finance to rebuild Europe mainly through grants rather than loans. Third World countries did not receive the same treatment. Instead, they were pressured to keep their economies completely open to foreign goods and capital; and in exchange for financial assistance, countries were expected to adopt structural adjustment policies (see Box 10.4).

Partly as a result of the debt crisis of the 1970s and 1980s, both institutions now exercise enormous leverage over the policies pursued by most governments which require financial assistance. 'That influence is enhanced by donor insistence upon an IMF and World Bank seal of approval as a condition for aid and debt relief' (Oxfam 1995: 8).

As Duncan Green (1995: 34) points out, the two organisations are not democratically run: 'decisions at the IMF and World Bank are taken on the basis of "one dollar, one vote", guaranteeing the dominance of both by the US government'. In 1995, for instance, the USA had 17.18 per cent of the total vote at the World Bank while the total vote for 23 of the poorest African countries together amounted to 1.58 per cent of the total.

The economic system promoted by these institutions is based on the free movement of capital and goods with the US dollar as the international currency. In such a system, tourism can be seen as much the same as any other cash crop, and pressure exerted by the development banks upon the government of a country to increase its export earnings by boosting production of export crops is as applicable to tourism as it is to sugar, coffee, cotton, bauxite, or other products. Foreign exchange earnings can be boosted by attracting more international tourists and the foreign exchange which they bring with them.

The debt crisis of the 1980s and the First World's reaction to it increased the power of the IMF and the World Bank. Spurred on by the prevailing ideology of President Reagan and Prime Minister Thatcher, the two institutions of the World Bank and IMF embarked on a mission to structurally adjust Third World economies so that they could meet debt obligations and fall into line with the right-wing economic orthodoxy sweeping the globe.

The origins of the debt crisis can be traced back to the huge profit made by members of the Organisation of Petroleum Exporting Countries (OPEC) in the early 1970s. These profits were surplus to requirements within the OPEC countries and much of the money was therefore banked and invested in the commercial banks of the First World. Awash with petro-dollars, the banks sought to make the most profitable use of this money. One of the ways in which they recycled it was by financing development projects in Third World countries, including tourist resort schemes.

In the rush for 'development', many of the projects financed were ill-considered and poorly planned. Some of the money was spirited away by Third World elites with access to power brokers and decision-makers, and more was paid direct to First World companies taking part in the projects. A proportion of it therefore found its way back into the same First World banks that lent it in the first place.

In the late 1970s and the first half of the 1980s, interest rates on these loans rose sharply, an event that had not been anticipated by the lenders or borrowers in the early 1970s. Many borrowing countries found themselves unable to pay back even the interest on the loans, quite apart from the principal loan itself. First World bankers and politicians became concerned 'that the sheer volume of unpayable loans would undermine the world financial system. They turned to the World Bank and the IMF, who were to restructure Third World economies so they could meet their debt obligations' (New Internationalist 1994: 15); hence, the rise of structural adjustment programmes (SAPs), which in 2000 were renamed poverty reduction strategy papers (PRSPs), and which are produced by national governments themselves. It should be pointed out, however, that they are produced with substantial input from IMF and World Bank officials and the two organisations retain the power of veto over PRSPs. Critics such as the World Development Movement (WDM) believe that PRSPs are merely SAPs containing language about poverty reduction.

From the point of view of our analysis, what is crucial here is that SAPs/PRSPs imposed by the IMF and World Bank on Third World governments effectively force those governments to pursue specific policies not of their own design, although it was precisely in anticipation of such criticism that their means of production and title (to PRSPs) were changed. These policies make living conditions for the population harsher and more difficult while supposedly enabling the government to increase its foreign exchange earnings and thereby pay off its debt. The kinds of policies associated with SAPs/PRSPs are listed in Box 10.4. It is worth noting, however, that most Third World countries already had little effective control over the direction of their policies before the debt crisis began. Colonial powers of the First World and transnational corporations (TNCs) dictated the main thrust of economic policy throughout the Third World. The IMF, World Bank and other supranational lending agencies along with the TNCs took over the mechanism of power from the former colonial powers. Moreover, proponents of SAPs/PRSPs point out that once the debt had become a reality, there were no policies other than those given in Box 10.4 which could realistically be imposed to support global capital accumulation.

Box 10.4 Policies and conditions associated with structural adjustment programmes and poverty reduction strategy papers

Increase earnings of foreign capital

- *Boost export production*

  One implication of this is that the production of goods for local and national needs is relegated in importance as a result of the drive for export production. Tourism is seen as an export product, attracting foreign currency into the country. Many Third World countries which used to be self-sufficient in foodstuffs now have to import basic grains and other foodstuffs. In 1996, for instance, Honduras began to import beans for the first time in its history.

- *Devalue the national currency*

  Devaluation makes the country more attractive to foreign investors seeking a cheap place to produce their goods. The cheapness of its currency against other currencies also increases its attraction to foreign tourists.

- *Reduce and/or abolish import tariffs*

  Import tariffs are designed to protect the price of the production of goods for home consumption against the entry of cheaper imported goods. In practice, removing them can have the effect of flooding the country with expensive luxury goods and durables with high value added in production from the First World – for purchase by tourists and the rich elite.

Reduce state involvement in the economy

- *Privatise state-run enterprises*

  Tourism is a flagship of the private sector.

- *Cut public spending on activities which cannot be privatised*

  For example, social services, health provision, education – although increasingly attempts are being made to privatise these services too.

- *Deregulation*

  Reduction of all forms of controls, especially economic regulations, which inhibit industry's ability to maximise profits.

Fiscal measures

- *Reduce inflation*

  This increases security of investment, including investments in tourist developments.

- *Cut interest rates*

  This increases investment, including investment in the tourism industry.

- *Encourage investment rather than saving*

The loans which form part of the SAPs/PRSPs are made only on strict condition that the national economy is restructured according to these policies (see Box 10.4). And this conditionality has meant that the economic and fiscal policies of most African and Latin American countries have been dictated from Washington. In the words of Duncan Green:

> In the aftermath of the debt crisis, the IMF and later World Bank's use of conditionality allowed the powerful industrialised nations to revamp one Third World economy after another along free-market lines. Critics believe that in the process, the IMF has systematically put the powerful nations' self-interest before the welfare of the Third World poor.
>
> (D. Green 1995: 37)

In 2007, an Independent People's Tribunal on the World Bank Group in India also declared that 'the Bank had a disturbingly negative influence in shaping India's national policies . . . and called the Bank "pro-rich, pro-urban and anti-environment"' (World Development Movement 2008: 7).

The initial hope that SAPs/PRSPs may be successful in delivering widespread prosperity through highly dubious mechanisms such as the 'trickle-down' effect and belief in the benefits of economic globalisation is confounded somewhat by the evidence. As Ayesha Imam (1994) states:

> In most countries in Africa, SAPs have not led to increases in production or investment . . . In fact, according to the UN Economic Commission for Africa, production has decreased over the decade since SAPs have been implemented. Investment as a proportion of GDP has fallen too. Budget deficits have grown and a greater proportion of export-earned money is now used solely to service debts . . . The general standard of living has fallen. There have been painfully deep cuts in the provision of health, education and other social services and in the subsidies on basic necessities.
>
> (Imam 1994: 13)

In 2005, the UNDP's Human Development Report showed that 'the gap between the average citizen in the richest and in the poorest countries is wide and getting wider. In 1990 the average American was 38 times richer than the average Tanzanian. Today the average American is 61 times richer' (UNDP 2005: 37). It further noted that the idea of 'championing globalisation while turning a blind eye to global equity concerns is increasingly anachronistic' (UNDP 2005: 38), which appears to imply an acknowledgement that economic growth on its own will not reduce poverty unless it is progressively redirected to the poor.

Even within the World Bank, there is awareness of the effects of these policies. Herman Daly, a renowned and respected economist in the Bank's Environment Department, identified some major conflicts between the ultimate goal of international free trade, at which the Bank's policies aim, and national policies pursued in the national interest (Daly 1993: 124). Joseph Stiglitz, a Nobel laureate in Economics and former chief economist at the World Bank, has suggested that 'the IMF was run by hard-line "market bolsheviks"' (World Development Movement 2000: 6), and that:

> The 'invisible hand' of the market is just nonsense. The liberalisation of capital markets has produced more instability but no more economic growth. We knew this fine, economic science did not recommend such a step, but the IMF has gone

on promoting it . . . the IMF's policies have little or nothing to do with concern
for the poor.

(Quoted in Nicaragua Network Hotline 2002)

Figure 10.2 illustrates the conflicts identified by Daly and Stiglitz and the IMF's
inability to acknowledge, concede to and account for a reality that its executives do not
accept.

In the world of tourism, as in many other industries, this process of structural adjustment effectively delivers control of development to the TNCs and consultancies, most
of which are based in the First World. This relegates the role of the national government
to one of providing the necessary infrastructure. For this, it must seek more loans, leaving
it yet more indebted than before. Some countries, such as Honduras and Panama, are now
even offering free trade zones for tourism companies wishing to build and operate resorts.
Paul Gonsalves (1995: 39) concludes that 'the economies of the so-called Third World
are in effect mere extensions of the economic priorities of the First World'.

The list of development disasters caused by the supranational organisations in funding
such schemes as large-scale dams, roads through sensitive eco-systems, monocultiva-
tion practices, infrastructural developments for the promotion of tourism, and many
others, is long and well documented. And they have not just been confined to the IMF and
World Bank. During the 1980s, the USAID also gained notoriety for its arrogant and
callous disregard for the environments and people affected by its development projects.
Organisations such as World Bank Watch, Survival International, the World Development
Movement, Corporate Watch, FOEI and others have documented these effects, exerted
heavy pressure against them, and publicised the events and examples to such an extent
that the start of the 1990s saw a growing recognition by the supranational organisations
themselves of the effects of their policies. Both the IMF and World Bank increased their
staff who examine the social and environmental impacts of their programmes. The USAID

... SO YOU SEE, THE ENTIRE FUTURE OF THE INTERNATIONAL FINANCIAL SYSTEM HINGES ON
YOUR CAPACITY FOR QUICK RECOVERY AND VAST ECONOMIC GROWTH.

Figure 10.2 Poverty and the 'hard-line market bolsheviks' of the IMF

also made strong efforts to green its image and to promote, for instance, community-based tourism schemes rather than grandiose resort projects. A tourism-related example of the misdirection and failure of their efforts in this regard was given on pp. 240–242.

That these efforts to foster a new environmentally friendly image for themselves have failed should come as no surprise, given that their modus operandi of forcing national governments to pursue a dictated development path has remained unchanged. In the wake of the protests at the trade talks in Seattle in 1999 and elsewhere in subsequent years, arguably the gathering momentum of a global social movement has prompted further attempts at image improvement by the supranational organisations such as the metamorphosis of the structural adjustment programmes into poverty reduction strategy papers. The Doha talks in 2001 became known as the 'development round' of trade talks. The failure to reach agreement between representatives of the rich world and those of the middle and low income countries in the Doha development round necessitated further talks, which finally collapsed in Germany in 2007.

With this increasing awareness of the need to demonstrate that they are changing, even if they are not doing so in reality, the IMF and World Bank have acknowledged the role of NGOs and have made attempts to enlist their advice. Despite this, Pierre Galand, Oxfam-Belgium's general secretary, resigned from the NGO working group on the World Bank and its Steering Committee. In his letter of resignation he said to the World Bank:

> You have stolen speeches from NGOs concerned with development, eco-development, poverty and popular participation . . . The Bank has learned to do excellent analyses and is capable of saying what is important: popular participation – particularly of women – the popular struggle against poverty, and the necessity to protect the environment . . . Why are so many beautiful speeches accompanied by such scandalous practices?
>
> (Galand 1994: 12–13)

More recently, in 2000, Ravi Kanbur, lead author of the World Bank's *World Development Report 2000/2001* (World Bank 2000b), resigned in protest at the Bank's decision to expunge criticism of globalisation from his work (World Development Movement 2000: 6).

In 1995, the Uruguay Round of the GATT gave birth to the World Trade Organisation (WTO), essentially set up to continue and further the process of international trade liberalisation. The organisation establishes international agreements on rules of trade that are binding on parties to the agreements and it adjudicates in trade-related disputes between member states and between a member state and the rules. It has powers to impose sanctions on member states which break its rules.

To support her claim that the WTO is leading us into an era of 'corporate-led, corporate-driven globalisation', Susan George cites David Hartridge, director of WTO Services Division, as stating that: 'without the enormous pressure generated by the American financial services sector . . . there would have been no services agreement and therefore perhaps no Uruguay Round and no WTO' (George in Bircham and Charlton 2001: 14). Brookes and Harrison (2000: 32) also refer to the WTO as 'heavily influenced by the interests of big business' and expose the large number of powerful industry associations which influence the WTO through their positions on consultative committees.

That the WTO is driven by the corporate lobby is rarely denied, but its proponents point to the need for agreed sets of rules to govern international trade, the need for a body to manage and adjudicate on such rules, and to the fact that: 'The WTO's consensus decision-making means that in theory even the smallest country can hold up a trade deal if it doesn't like what is on the table' (Denny 2002: 14).

On the other hand, the perception that the WTO works solely in the interests of corporations and their profits is the driving force behind the rapid growth and conspicuousness of the anti-globalisation movement. Essentially their arguments are that the WTO ignores the needs of the poor, the environment and the rights of workers and consumers, and 'locks countries into a system of rules that means it is effectively impossible for governments to change policy, or for voters to elect a new government that has different policies' (Barry Coates in Bircham and Charlton 2001: 28).

The WTO affects the activities associated with tourism through the General Agreement on Trade in Services (GATS). Service industries, of which tourism is a part, now make up the greater part of the world's economic activity and the GATS negotiations give major new rights to TNCs and reduce the power of national governments to restrict their activities for reasons of environmental protection, labour rights, planning regulations, or similar considerations. It is to the GATS and its effects on tourism that we now turn.

Tourism and the General Agreement on Trade in Services

The GATS is one of several trade agreements administered and controlled by the WTO. As we discussed in Chapter 2, the principal motive for the recent round of economic globalisation, of which GATS is the newest agreement, is arguably to ensure the continued expansion of markets for goods and services produced in and provided from the industrialised nations.

In theory, the WTO works on a consensus principle to create a rules-based system for the conduct of multilateral trade, and GATS works on the same principle for the same aim. Under the agreement, any signatory country has to accord the same treatment to foreign companies as it does to domestic companies in specified sectors of trade to which they have committed themselves. As Kalisch (2001) explains, it

> is designed to ensure that host governments, confronted with powerful transnational corporations who import both their own staff and the majority of goods needed for their tourism operation, cannot compel them to use local materials and products to enhance the 'multiplier effect', or to take special measures to secure a competitive base for their domestic businesses.
>
> (Kalisch 2001: 4)

The GATS defines import tariffs, subsidies and other measures designed to assist domestic companies as 'trade-restrictive', and thereby effectively undermines the power of governments to legislate in the national interest.

In practice, according to a report by the WDM, one of the major UK critics of GATS: 'WTO staff and negotiators openly acknowledge that GATS exists only because of pressure from service multinationals and that this influence has continued since GATS came into effect' (WDM 2002: 5).[2] Moreover, the European Commission's website described GATS as 'not just something that exists between governments. It is first and foremost an instrument for the benefit of business' (European Community 2000).

The mechanism of reaching agreement by consensus appears to be given as a major justification for the GATS. But whether a genuine consensus was achieved at the Doha Ministerial meeting in Qatar in November 2001, called to discuss GATS, or at previous and subsequent similar meetings, is highly debatable: for instance, the 481 delegates from the G7 nations present at the Doha meeting was almost double the 276 delegates from the 39 Least Developed countries, and it is difficult if not impossible to achieve consensus agreement in such unbalanced circumstances. Indeed, the WDM also reports that widespread pressure is exerted by industrialised countries on Third World country negotiators, including suggestions that aid would be withdrawn (WDM 2002: 54). In general, the GATS appears to reflect and reinforce rather than challenge the existing unevenness and inequality in the global economic system. The eventual collapse of the Doha development round of trade talks in 2007 reflected the growing Third World disenchantment with this situation and a resulting willingness within Third World governments to challenge such moves.

In GATS, tourism falls into the sector labelled Tourism and Travel-Related Services which has four categories: hotels and restaurants; travel agencies and tour operator services; tourist guide services; other (unspecified). Kalisch (2001) points out that under GATS,

> it would be impossible for a government to impose on foreign companies practices to limit negative environmental, social and cultural impacts in their country, such as restricting the mushrooming of foreign owned development, (including all-inclusive hotel developments, which are highly controversial among local people because they contribute almost nothing to the local economy), and making the employment of local workers, the use of local products and materials a condition of their investment. It could also render a government powerless to stop tourism developments on indigenous land, including sacred sites, in response to community protests.
>
> (Kalisch 2001: 5)

Kalisch's analysis also suggests that GATS is likely to place fair trade in tourism (see Chapter 7) at a disadvantage since 'the beneficiaries of Fair Trade in Tourism are poor communities, for whom the informal sector (vendors, traders, guides) is an important source of income, as well as small entrepreneurs and workers employed in the tourism industry' (Kalisch 2001: 5). Representing the UK NGO Tourism Concern, Kalisch (2001) identifies the following concerns about the role that GATS plays in the emerging tourism industries in Third World countries: the manner in which GATS negotiations are carried out in the WTO disadvantages and marginalises Third World countries and civil society organisations; GATS fails to integrate sustainable development as identified in major international treaties and agreements, such as the Convention on Biological Diversity which includes decisions on tourism; GATS does not address the specific environmental, social, economic and cultural impacts of tourism in a destination; expanding global tourism increases the threat of biopiracy under the guise of ecotourism, when tour operators and clients collect rare medicinal plants from areas inhabited by tribal peoples but without consulting them; and GATS makes no provision for the monitoring and regulation of TNCs (Kalisch 2001: 5–6).

Of all these concerns of relevance to the activity of tourism, it is perhaps the fear that the GATS will further deepen the increasing poverty and inequities experienced in some Third World countries that prompts the greatest protest. In the 1980s and 1990s, structural adjustment was the major economic policy employed in up to 90 Third World countries under programmes imposed by the IMF and World Bank. During this time, 'the people of sub-Saharan Africa have become 15 per cent poorer, and growth rates in South Asia and Latin America have stagnated while inequality has risen dramatically' (Barry Coates in Bircham and Charlton 2001: 29). The possibility that the GATS will further deepen these effects is feared not just by anti-globalisation protestors but also by the protagonists of fairly traded tourism and pro-poor tourism. The last of these, pro-poor tourism, is a relatively recent extension of governmental activity into the field of tourism development, and we turn to this issue in Chapter 11. Initially however, it is necessary to understand this new direction in the context of donor development activity in the tourism sector more broadly.

Donor development activity

In one of the few analyses of the current approach of development agencies to tourism, the UK DFID Sustainable Tourism and Poverty Elimination Study (Deloitte & Touche et al. 1999) concluded that few agencies regard tourism as a key development sector and yet many agencies, such as the World Bank and European Union, do indeed provide considerable support to tourism-related projects, ranging from economic development and employment generation to sector planning and product development. This study, compiled at the end of the 1990s, noted that no agency could be regarded as developing and implementing a tourism and poverty elimination agenda.

Reviewing the approaches of both supranational (multilateral) and bilateral agencies, the DFID study identified two contrasting approaches of agencies to tourism as a poverty elimination strategy. The first, exemplified at that time by the Danish and Dutch government agencies (Danish Agency for Development Assistance (Danida) and Netherlands Development Assistance (NEDA) respectively) is that tourism cannot and should not compete with traditional development sectors (such as education and primary health care). The second, promoted by some bilaterals and multilaterals such as the United Nations Development Programme and EuropeAid, is that as a result of the considerable impact of tourism on poor peoples' livelihoods, intervention is critical. Such interventions at the local level include the support of small and micro-enterprises, support to rural tourism projects, the development of partnerships between communities and private sector operators, participatory pro-poor planning and a range of measures that have clear benefits for the poor, such as working condition improvements, training and staff development and protected area management (Deloitte & Touche et al. 1999).

An overview carried out at the University of Plymouth during 2007 provides an indication of the current state of play with the multilateral and bilateral donor community.[3] While this overview is not definitive and is not a detailed compilation of past and present initiatives, the review nonetheless provides a good picture of how donors are currently relating to the tourism sector. It suggests that while the volume of activity may have increased substantially, the caricature of tourism within the development sector drawn by the DFID study largely remains. The institutions profiled include the development banks, key institutions within the United Nations system (programmes and commissions), European institutions as well as eleven of the most prominent so-called bilateral development agencies – that is, the development departments and agencies of individual countries which forge bilateral relationships and programmes with Third World governments.

Of the total twenty-four institutions reviewed, ten have officially adopted tourism sector strategies or policies. Beyond UNWTO, this includes UNESCO and the European Commission, USAID, Japan International Cooperation Agency (JICA) and SNV (the former Netherlands development agency). Some, such as the UN Economic and Social Commission for Asia and the Pacific (UNESCAP), have adopted targeted plans of action: in this case the Plan of Action for Sustainable Tourism (PASTA) aimed at enhancing the role of tourism in sustainable socio-economic development and poverty reduction in the region.

As might be expected, it is those agencies that have formally adopted tourism strategies and policies that demonstrate most activity in the sector through project funding and implementation. The UNWTO is the most prominent global agency whose sole mission is to promote tourism and protect and enhance its opportunities for the future. Surprisingly, therefore, UNWTO has been relatively slow to effectively engage in the global struggle against poverty. We will discuss UNWTO and the development of ST–EP (Sustainable Tourism – Eliminating Poverty) further in Chapter 11. There are several bilateral agencies that appear particularly active in the tourism sector. USAID, for example, recognises the role and potential of sustainable tourism in promoting economic growth, conserving natural resources and alleviating poverty, although it is worth adding once again that many critics would point out that the goal of poverty alleviation may, at times, be impeded by the goals of growth and conservation. The organisation has completed 123 projects in 72 countries since the year 2000, within which tourism has either been a primary focus or a significant component in a broader programme. USAID has also developed a Global Sustainable Tourism Alliance as a vehicle for promoting the development of partnerships between international tourism organisations. Similarly, SNV recognises the significance of sustainable tourism and through the deployment of international expertise has supported tourism sector programmes (focused on strategy and business development, community mobilisation and environmental conservation) in countries such as Bhutan, Nepal, Tanzania and Vietnam. It is a major partner in the UNWTO's ST–EP initiative.

The absence of a formal tourism sector strategy does not however signal that agencies are not engaged in tourism sector activities (either directly or indirectly). The World Bank, for example, funds a large number of tourism-related projects through its funding channels, emphasising tourism's significance for economic growth, and its contribution to cultural heritage and sustainable development. Similarly, the Asian Development Bank (2006a) has underlined the importance of tourism in its region and funds a variety of projects such as the Tourism Development Plan of South Asia, including ecotourism-based projects and those emphasising cultural heritage. It is also promoting the significance of tourism through the Greater Mekong Subregion strategies (which will be further discussed in Chapter 11).

The presence of tourism within UN interventions certainly transcends the activities in UNWTO. The UNDP, for example, has supported tourism-centric projects in a number of countries (including East Timor and Jordan) and supported activities through small grant projects in Bhutan, Iran and Mozambique. Other interventions include supporting the introduction of a new legal framework for land management in Panama aimed at pro-hibiting non-environmentally sustainable tourism projects. As Chapter 9 discussed, UNESCO is also a significant player in promoting tourism that helps protect cultural and natural heritage, particularly through its awarding of World Heritage Site status (to some extent 'badges of honour' that crystallise these sites as major tourism destinations). Beyond this, UNESCO's tourism-oriented activities range from promoting poverty eradication through sustainable tourism development in Central Asia and the Himalayas, to promoting community participation in tourism development in Iran. Other UN agencies, such as the UN Economic Commission for Africa (UNECA), is encouraging local level partnerships between the private sector, government and the poor to raise incomes in the informal sector (in this case in Gambia), and encouraging the use of government policy to steer private sector initiatives to adopt poverty reduction practices in the tourism sector. UNECA has also launched a Pro-poor Tourism Tools and Tips Project in support of activities in Southern Africa.

Bilateral agencies too have made varying overtures in the tourism sector; DFID's notable support for pro-poor tourism research, for instance, is further reviewed in Chapter 11. Danida has supported small-scale projects in Cambodia, Mozambique and Vietnam, while the activities of Sweden's development agency, Sida, include supporting employment generation through cultural heritage preservation in Palestine and tourism development in Namibia. Both of Australasia's agencies, AusAid (Australia) and NZAid (New Zealand), have supported nature-based projects in the Asia and Pacific region. For the former this includes support for marine research and coastal regeneration in Thailand following the 2004 Indian Ocean Tsunami (see Chapter 11) and encouraging eco-friendly tourism and the sustainable harvesting of marine resources through establishing protected areas in the Maldives. AusAid is also supporting broad-based economic growth in the Pacific, of which tourism is a significant element. For the latter (NZAid), activities range from supporting ecotourism projects and the improvement of rural livelihoods in Laos and Lombok (Indonesia), to promoting a Nature Tourism Project in Tonga (2001–04) and supporting small-scale tourism operators on Samoa.

What can we conclude from this overview? What is clear is that tourism has steadily crept up the development agenda (noticeably so since the first edition of this book was published). It is slowly being seen as a 'legitimate' sector and activity within a 'development portfolio' – a potential poverty-busting mechanism – and certainly a sector that simply cannot be discounted within development frameworks dominated by economic growth assumptions. Some indications are incremental, such as the increase in projects and programmes with tourism actively referenced. Other signs are far more compelling, ranging from the UN's International Year of Ecotourism (which we discuss later in this chapter) to the formalisation of the World Tourism Organisation as a United Nations agency (an indication of the widespread acceptance of tourism as an agent of development), and the adoption of tourism (as briefly discussed above) by national agency tourism sector strategies. Collectively, and to be anticipated, the approaches promoted by these agencies have a development orientation and most commonly reference the cultural and heritage potential of tourism and ecotourism (ranging from marine conservation to wildlife and area protection) as a specific favoured subsector.

Tourism's contribution to broader livelihood strategies and its poverty-reducing potential are also referenced, and we will return to this critical theme in more detail in Chapter 11. The overview made especially for this book also emphasises the wide range of interventions, some of which are primarily tourism-oriented – that is, attempts to temper the impacts of tourism or to use tourism to promote cultural and natural heritage and conservation – and others where tourism is incidental or of secondary significance (though it is undeniably there). Future research could usefully track the building of this development sector and help encourage the sharing of knowledge and experience in policy and practice among the wide range of agencies involved.

For further discussion on pro-poor tourism we have added Chapter 11 to this, the third edition of the book.

Agenda 21 for the travel and tourism industry

An examination of the global governance of tourism must include the activities and approach of tourism's supranational agencies such as the WTTC and the UNWTO. Together with the Earth Council, these organisations elaborated the role of different tourism interests and their responsibilities following the establishment of Agenda 21 which arose from the 1992 Earth Summit at Rio de Janeiro (see Box 2.1) in a document entitled *Agenda 21 for the Travel and Tourism Industry: Towards Environmentally Sustainable Development* (WTTC 1995). Table 10.2 extracts the duties placed upon central and local governments (and national tourism authorities and representative trade organisations such as the WTTC and UNWTO) in facilitating 'sustainable' tourism development.

There are a number of observations that can be made about the priorities set out in this document. First, it is clearly set within a framework of environmental sustainability. This is indicated in both the report's rubric, *Towards Environmentally Sustainable Development*, and in the respective forewords of the collaborating bodies:

Table 10.2 *Agenda 21: responsibilities of governments*

Priority area	Priorities and objectives	Responses and examples of success
i	Assessing the capacity of existing regulatory, economic and voluntary framework to bring about sustainable tourism, and developing policies to facilitate sustainable tourism	Charge negative externalities to producer; provide cost incentives to companies minimising waste; encourage responsible entrepreneurship through codes of practice. More than 100 codes now exist.
ii	Assessing the economic, energy, social, cultural and environmental implications of an organisation's own operations	Ranges from a more efficient use of water to encouraging staff to use environmentally benign transport and ensuring the destination's character is truly represented in marketing. *Green Globe* was established (see Chapter 7).
iii	To train and educate all stakeholders in travel and tourism about the need to develop more sustainable forms of tourism	Organisations should work with government departments. The UNWTO has established a worldwide Network of Education and Training Centres to promote tourism education and training.
iv	Develop and implement effective land use planning to maximise the environmental and economic benefits of travel and tourism and minimise environmental and cultural damage	Includes working with government planning authorities on sustainable tourism planning, i.e. transport planning and coastal zone management. Bermuda – one of the world's most affluent countries because of planning regulations?
v	Facilitate information exchange between 'developed' and 'developing' countries	Includes developing partnerships with 'less developed' countries, advising them on sustainable tourism.
vi	Provide opportunities for all sectors of society to participate in sustainable tourism	Promote and ensure the participation of women, indigenous people, etc. Develop appropriate training courses. A key example would be the CAMPFIRE project in Zimbabwe.
vii	Design new tourism products that are sustainable: socially, culturally, economically and environmentally	Ranges from defining what makes resorts sustainable to using local materials in construction and ensuring government departments realise the benefits of tourism in conservation. Examples include the game parks of Kenya and South Africa.
viii	Measure progress in sustainable development by setting realistic indicators at national and local level	Ranges from assessing which data are appropriate to exchanging experiences with other organisations.
ix	Develop partnerships to facilitate sustainable tourism development and responsible entrepreneurship	Includes government departments providing a coordinating mechanism for those responsible for sustainable tourism develop-ment and ensuring the necessary infrastruc-ture (e.g. sewage treatment, recycling facilities, etc.) is in place.

Source: World Travel and Tourism Council (1995)

A key conclusion of the Earth Summit was the importance of harnessing the entrepreneurial drive of the private sector in the cause of environmentally compatible development . . . The environment is our core asset, the key component of product quality, and an increasing priority for our consumers.
(Geoffrey Lipman, President, WTTC)

A good deal of our Travel & Tourism activity relies on . . . fragile natural or cultural resources, so it is in our interests to protect them for the future . . . Travel & Tourism will inevitably continue to increase. Meeting this growth in a responsible, sustainable way, that preserves and enhances the beauty of the attraction, is the challenge we all face.
(Antonio Enriquez Savignac, Secretary General, WTO/OMT)

The overriding concern for the industry must be to seek out ways to enhance rather than degrade its core product, the environment.
(Maurice Strong, Chairman, The Earth Council)

Reflecting for a moment on some of the words used here to describe the environment – 'core asset', 'core product', 'product quality', 'resources', 'preserve' – the treatment of the environment as a marketable product is clear. It is also clear that the drive 'towards environmentally sustainable development' is a commercial necessity, in order to maintain the marketability and attractiveness of this commodity. More noticeable is the way in which much, if not all, the discussion bypasses the interests of local people. Where social considerations do appear they are an adjunct. Social development, it is assumed, will be achieved at the same time as improving the environment; the benefits will magically 'trickle down' to local people in the same way as the World Bank and IMF suggest. The ability for local people to choose whether they wish to engage with tourism is not on offer. Rather, it appears to be an agenda for removing the 'growing resentment of residents in some destinations' (WTTC 1995: 3) and allowing for the expansion of the industry.

As the above quotations imply, the report fails to reflect the inherent inequality expressed through global tourism. It is written from a thoroughly First World perspective, where the majority of the tourism industry, tourists and tourism interests are located. The forewords talk of 'our Travel and Tourism activity', 'our consumers', 'our interests'. The supporting statement for priority (WTTC 1995: vi), for example, extols the participation of all sectors of society in sustainable tourism and identifies the promotion of the partici-pation of women and indigenous people in 'appropriate forms of tourism development'. But, of course, these remain the receivers of tourism and the report is silent on the gross inequality of who is, and who is not, able to travel. The Third World, so it would appear (and as argued in Chapter 3), is there for our entertainment. It exists to serve us and see to our needs. As the report states, the 'challenge . . . is to ensure that tourists don't "love nature to death"' (WTTC 1995: 49).

Even 'peace' is invoked as an essential prerequisite to successful sustainable tourism development, further reflecting the overall intention of protecting the tourism product while allowing for overall global expansion. Priority area (vii) supporting statement urges the use of 'the potential of the tourism industry to promote peace between nations and people' (WTTC 1995: 50), acknowledging that 'Peace is a pre-requisite for a sustainable Travel & Tourism industry and, by facilitating contact and understanding between cultures, well-designed tourism products can make a significant contribution to world peace'. While peace is of course a noble goal, once again there is an insipid suggestion that we need to move towards one happy global community, 'In recognition of economic and social cohesion among the peoples of the world as a fundamental principle of sustainable development' with implicit suggestions about the need for global security, a concern that has been greatly enhanced since the turn of the century.

Not only are Third World populations the receivers of visitors, but also, according to this report they are receivers of (First World) information. Table 10.2 reflects this bias. Priority area (v), for example, involving the facilitation of information exchange, more accurately involves advising Third World countries on how to make tourism sustainable. Rather than exchange, this appears to be information delivery and represents a First World perspective on how certain goals are to be achieved. Similarly, priority area (viii) supporting statement encourages the 'exchange' of indicators with other organisations 'especially those in developing countries' (WTTC 1995: 51). Cynically, many of the priorities and objectives advocating 'training' courses are a further indication of the cycles of dependency and development aid-giving – a 'jobs for the boys' syndrome – that such frameworks maintain.

The participatory processes which the report anticipates are passive, and it may be of interest to consider the contents of Table 10.2 in relation to Pretty's typology of participation given in Table 8.1. Participation, so the report suggests, is to be undertaken after key initial decisions have been made. In the wake of the discovery of Mayan ruins near the city of Tekax (in south-east Mexico), and the realisation that this had significant tourism potential, it was agreed that the 'people of the area should participate in the development in order to maximise economic benefits'. The critical point here is that the decision to develop the site and to maximise its economic benefits had already been taken.

Finally, the report offers a Framework for Action, although on the subject of the means of implementation the report is baffling and also reflects the First World perspective. It is difficult to conceive of the processes and procedures set out in the report – that is, securing the commitment of top management to the concept of sustainability, communicating this to all staff, establishing realistic and achievable targets, implementing action programmes, monitoring progress and so on – working in practice. In short, it amounts to top-draw management-speak: many windy words within a highly ambiguous and inappropriate framework. It is difficult to believe that many grossly under-resourced and understaffed government departments would be in a position to undertake such a grand scheme. How the report's recommendations are to be implemented in practice represents a serious failing, and may explain why Agenda 21 has slipped so far down the priorities of development.

The report and its Framework for Action are in effect just one more in a pile of con-ventions, codes of conduct, declarations and the like weighing down on the governments of Third World countries in respect of tourism development (see Figure 10.3). All these pressures are full of fine words and lofty principles, but few if any of them offer the resources required to enable Third World governments to consider sustainability in the development of their tourism industries.

The United Nations International Year of Ecotourism 2002

In November 1998, the United Nations' General Assembly declared the year 2002 as the International Year of Ecotourism (IYE) and invited UN states, members of specialised agencies and pertinent intergovernmental and governmental organisations to participate in the programme, especially regarding ecotourism in developing countries. The UNEP and the UNWTO were mandated by the UN to organise events around the IYE. Along with UNEP and the UNWTO, TIES (see Chapter 5) was given the role of coordinating the organisation of a series of six regional preparatory conferences in partnership with local organisations, 'to focus on fostering a genuine dialogue among all stakeholders involved,

Figure 10.3 Regulating the Third World

especially local communities and ecotourism practitioners' (TIES website 2002). These were held in Belize, India, Peru, Thailand, Kenya and Sweden from November 2001 to April 2002, and each preparatory conference presented reports to the World Ecotourism Summit in Quebec, Canada in May 2002, in which TIES was also an active organisational participant.

The IYE was met with enthusiasm by many tourist boards and tour operators, but with varying degrees of scepticism by many NGOs, especially some in the Third World. Most prominent among these was the Tourism Investigation and Monitoring Team (usually referred to as the TIM-Team) of Bangkok, Thailand, with support from the Third World Network based in Malaysia. Soon after the declaration of the IYE, the TIM-Team called for a fundamental reassessment of ecotourism and for a change of the name to the International Year of Reviewing Ecotourism. In an October 2000 letter to Oliver Hillel, UNEP's tourism programme coordinator, the TIM-Team and a coalition of other groups from Asia and beyond stated:

We, the undersigned NGOs from the South and the North, feel compelled to warn all concerned parties not to skirt the critical issues of ecotourism and the fact that a mountain of money will be spent and a flood of projects initiated around the IYE in order to boost the ecotourism industry. In contrast to advocates who tend to portray ecotourism development as a 'win-win' approach, a means to protect biodiversity and enhance the well-being of local people, we are gravely concerned that this IYE will result in a 'lose-lose' situation for communities and the environment in destination countries.

(TIM-Team 2000)

Citing work conducted on projects including Taman Negara, a national park in western Malaysia, a number of development projects in Thailand and the Greater Mekong Subregion covering six countries, the organisation argued that:

Tourism provides the physical infrastrucure and logistics for freer movement of people and goods within countries and across borders in general . . . Accordingly, ecotourism development has opened opportunities for a whole range of investors to gain access to remote rural, forest, coastal and marine areas . . . more transportation systems are established into remote areas, the more encroachments, illegal logging, mining and plundering of biological resources occur, including biopiracy by unscrupulous individual and corporate collectors.

(TIM-Team 2000)

Three months later the same coalition of groups sent the same letter to Kofi Annan, then UN Secretary General, along with an urgent appeal that the name and focus of the year should be changed. The substance of their appeal and the contrast in the different approaches to the year were reflected in one sentence:

The new name will convey an unmistakeable message to the international community that the year 2002 is not the time for celebrating the ecotourism industry, but is primarily meant as a period of reflection, stock taking, learning and intensive search for solutions to the various problems associated with ecotourism.

(TIM-Team letter to Kofi Annan, 18 January 2001)

The TIM-Team continued to produce regular articles and letters critical of the IYE, culminating in a January 2002 article entitled 'UN "International Year of Ecotourism 2002" in a Deep Muddle – Scrap it!' (TIM-Team 2002). The UK-based Tourism Concern also expressed its misgivings thus:

How can ecotourism to the Ecuadorian rainforest be taken seriously given the resources consumed to take tourists to the product in the first place? The IYE also seems to be failing to address many critical issues: the fact that people are still displaced from their homes because of tourism, including 'ecotourism', denied access to local resources and losing their livelihoods because of so-called ecotourism; that ecotourism rarely takes on board issues of core labour standards, or gender; or that large environmental NGOs managing 'ecotours' too often behave like large corporations, employing 'top down' methods that rarely support local decision making. Perhaps most crucially of all, Tourism Concern believes that promoting ecotourism means promoting a niche market that does nothing to change the industry as a whole.

(Tourism Concern 2001)

A statement presented by a range of indigenous peoples' representatives and NGOs to the Convention on Biological Diversity in Nairobi, Kenya, in May 2000 went further still, naming several of the 'large environmental NGOs managing ecotours':

> In our experience, large nature conservation and development organisations do not respect (local peoples') rights ... several activities undertaken by the [International] Ecotourism Society, Conservation International and IUCN do not respect the rights and interests of Indigenous Peoples and local communities, particularly in regard to the Year of Ecotourism activities, and often threaten cultural and biological diversity.
>
> (Quoted in Pleumarom 2000)

Many of the criticisms and dangers of ecotourism raised in the run-up to the IYE were acknowledged by its organisers. Although the title of the year was not altered, some efforts were made to improve the inclusivity of the events by spreading invitations to participate far and wide. The TIES IYE website was peppered with terms such as 'critical evaluation of ecotourism', 'share the benefits of ecotourism equitably', 'multi-stakeholder consultation', 'foster dialogue', and makes repeated references to 'local communities and indigenous groups'. Nevertheless, as Vivanco (2001) points out, 'Since its inception, the IYE has been aggressively marketed as a celebration of ecotourism's role in facilitating sustainable development' (Vivanco 2001). Despite acknowledgement of the criticisms of ecotourism, the celebratory tone cannot be reconciled with the views of those who consider it inappropriate and who would have preferred a more critical approach.

The UNWTO reported on the outcomes of the IYE to the UN General Assembly.[4] Not surprisingly, the organisation reported positively that 'an extremely wide range of activities [were] stimulated by the IYE at the national and local levels, with the participation of the various sectors involved in ecotourism' (UNWTO/IYE website 2008). The report provides detailed information on follow-up activities and outlines recommendations for further action, although other than the publications produced as a result of the event it has to be asked whether some or all of the listed activities and regulations might have arisen without the IYE. The Global Ecotourism Conference 2007 (GEC07) held in Oslo, Norway, was not a direct outcome or product of the IYE but clearly saw itself as the successor to the IYE. In similar vein, it also declared itself to be 'a great success' (GEC07 website 2008). In particular, the GEC07 addressed the threat that climate change poses to the ecotourism industry but whether its calls for 'stronger leadership and strategies ... in order to substantially decrease ecotourism's carbon footprint' will be either realised or effective or whether these will serve as nothing more than hot air, like so many other conference declarations, will only become clearer as the years roll by.

Sustainable tourism as political discourse

The ecotourism boom in Belize

'Beautiful, multi-faceted jewel' (*Financial Times*, 22 September 1992); 'To sea is to Belize' (*Guardian*, 3 April 1993); 'Mounts, mountains and Mayans' (*Observer*, 27 March 1994); and 'Too good for tourists' (*Independent*, 3 November 1992) – this is how four UK broadsheet newspapers referred to the small Central American country of Belize. Virtually all tourist sources on Belize cite large tracts of 'pristine' rainforest, numerous Mayan archaeological sites, and the second-longest barrier reef (after Australia) in the world. Writers point to a number of successful community-based and conservation projects, including the Community Baboon Sanctuary, Sandy Beach Lodge (Women's Cooperative), the Toledo Ecotourism Association (see Chapter 8), the Hol Chan Marine Reserve, the Cockscomb Basin Wildlife Sanctuary (Jaguar Reserve), and Crooked Tree Wildlife Sanctuary. On paper, much of the surface area of the country enjoys some form of designated protection from development (see Figure 10.4). Growth rates in tourist arrivals and receipts are shown in Figure 10.5.

Up to the late 1980s few would have heard of the former British colony of British Honduras, which gained independence in 1981. In the post-independence period, tourism has become an attractive strategy for diversifying the economy away from the export of agricultural produce. Belize is an excellent example of how a country uses the notion of sustainability to promote the development of new types of tourism and how a government's use of the language of sustainable tourism is often sufficient to win international accolade, even among academics and environmentalists. In the early 1990s it seemed as if Belize had emerged as the ecotourism capital of the world, hosting a number of international conferences. The experience of Belize provides an example of sustainability as a political discourse.

Unlike its Caribbean Community and Common Market (CARICOM) partners, Belize recognised the need to avoid the negative externalities, foreign exchange leakages and the foreign control of the mass tourism industry developed throughout much of the

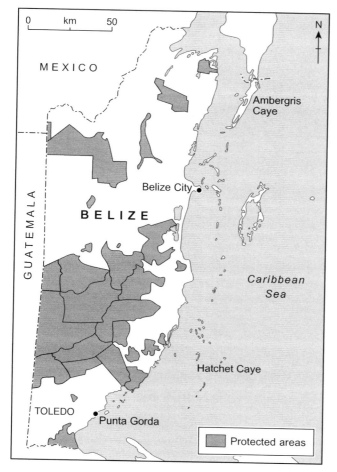

Figure 10.4 Belize

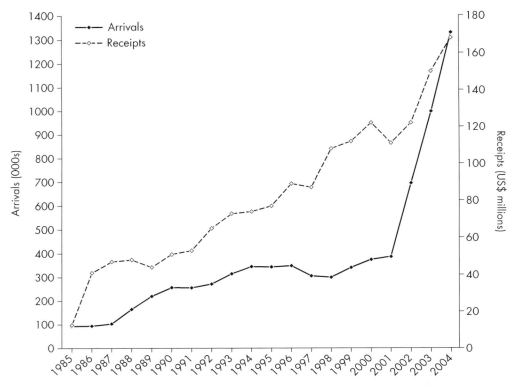

Figure 10.5 Tourist arrivals and receipts, Belize, 1985–2004

Source note: UNWTO (various years) *Compendium of Tourist Statistics*. Please note that in some cases different editions of the *Compendium* give different figures for the same years. Data for some intervening years were taken from the UNWTO's *Tourism Market Trends*, in which sometimes (but not always) conforms to the data given in the Compendiums.

Caribbean. It also acknowledged the potentially disastrous impact that large-scale tourism development would have on the relatively fragile subtropical environment.

In its first term of office, the United Democratic Party (UDP), a right-of-centre and pro-US free market economics party, was the first Belizean administration to focus upon tourism. It approved an Integrated Tourism Policy and Strategy Statement in its last year in office (Belize, Government of 1989), acknowledging the significance of tourism to the government, which had made it the second economic priority next to agriculture. In this way, significant emphasis was placed on the overall potential of tourism, 'its economic stimulus and linkage with other subsectors of the economy' (1989: 1), and the guiding objectives stressed the maximisation of economic benefits. This included a particular emphasis on 'long stay and upper income travellers'.

In the 1989 general election the victorious People's United Party (PUP – a nationalist party and like its increasingly indistinguishable rival, a pro-US and free market-oriented party) had spelled out clearly the advantages of ecotourism in the party's manifesto (Belizeans First 1989–1994) in what it also termed Tourism with Dignity:

> Belize has in abundance what few countries in the world still have left – nature. Ours, for the most part, is still intact. And nature is rapidly becoming tourism's Holy Grail. Belize has so far escaped the most brutal blows of man's destructive

hand . . . But, we must ask, for how long? The answer is, forever if we safeguard them. The protection of our national environment and development of eco-tourism are compatible goals which will be given high priority. The new PUP Government will place the portfolios of Environment and Tourism in one Ministry.

(People's United Party 1989)

In the wake of its election victory, the PUP did indeed combine these portfolios and Belize has subsequently enjoyed popular international support and praise for its stand on the environment and promotion of sustainable tourism development, and has hosted a number of conferences on ecotourism. As we argued earlier, however, these conferences hint at the serious deficiencies and contradictions in ecotourism, not least that the conference fees, half the average monthly Belizean wage, ensured that the only local representation was from professionals representing and paid for by government departments and large non-governmental organisations. Higinio and Munt (1993) describe two particular conferences held in Belize thus:

> Hosted in a new luxury hotel – the construction of which necessitated stripping the protective mangrove cover – on the ecologically fragile outskirts of Belize City, two recent major international conferences on ecotourism highlighted the conflict between environmentally conscious activists and entrepreneurs who have seized upon ecotourism as a convenient marketing tool.
>
> The keynote speaker at the First Caribbean Ecotourism Conference in 1991, Voit Gilmore from the American Society of Travel Agents, promised 'millions of Americans just waiting to come'. Howard Hills, a US investor speaking at the 1992 First World Congress on Tourism and the Environment, promised money. Ecotourism, Hills argued, is just like any other business. He boasted that the Overseas Private Investment Corporation would lend up to $50 million to any 'environmentally sound' tourism development, conditional upon a 25 per cent US stake in the project (see p. 219).
>
> Environmentalists consider such sentiments anathema to the very ethos of ecotourism. They argue that a small-scale, locally controlled and ecologically sensitive industry can neither sustain many visitors, nor be a big money-maker. In contrast to the Belizean government's view of ecotourism as a way to develop economically and to earn foreign exchange, environmental activitists see eco-tourism as justification for ecological and cultural conservation. An incredulous anthropologist from Scotland's Edinburgh University could only express dismay at Hill's remarks: 'The whole idea of ecotourism,' he said, 'is that it is like no other business.'

(Higinio and Munt 1993)

In the wake of these conferences, commentators began to express concern over the nature of the business of ecotourism and what it had to offer Belize. Despite these concerns, the Belizean government has used its commitment to sustainable tourism to full effect. At an ecology and tourism symposium in Costa Rica, for example, the then Belizean Tourism and Environment Minister, Glenn Godfrey, emphasised the relationships between environment, conservation and ecotourism and the fundamental requirement of local control: 'Conservation and therefore ecotourism thrives best where the sunlight penetrates to the lowest levels of autonomous local and community government.' It was a stance he later reiterated at the 1992 Rio Summit, assuring the audience that the Belizean government was committed to 'community-based ecotourism'. But the government's actions have not always matched their statements and a succession of examples from Belize provide a rather less optimistic picture of the use of sustainable tourism.

In 1992 the popular Belizean weekly *Amandala* carried the headline 'Eco-terrorism at Hatchet Caye', following a US resort owner's attempt to blow up part of the ecologically fragile coral reef in order to make his resort more accessible to visiting boats. The developer, Donald McKenzie, was subsequently arrested, fined US$2,000 (for 'dynamiting without a permit') and fled the country (Otis 1992). Of course similar incidents have been reported elsewhere in the world where government rhetoric extolling the virtues of the controlled growth of tourism has been contradicted in practice. In some cases this is a direct result of weak and ineffectual legislation controlling environmental impacts, particularly the inability to enforce legislation. In other cases it reflects the complete absence of necessary controls or the government's willingness to turn a blind eye.

More than environmental damage, however, the Hatchet Caye incident further underlined the unease with the degree and growth of foreign ownership of tourism resources in Belize. *Amandala* reflects these concerns in statements that clearly resonate the political economy of tourism – discussed in Chapter 3.

> As Belize struggles to find the balance between the development and conservation of our natural resources, more and more we will run into the Hatchet Caye syndrome: foreign nationals and corporations who do not believe that they are answerable to us . . . Belize and Belizeans are beginning to be trampled in the rush by those who regard our laws and traditions as inconvenient at best, and our sovereignty as theirs, bought and paid by their money. They wheedle, bribe and connive to extract their fortunes. And when all the above fail, they will resort to terrorism; covert campaigns, destabilisation and destruction, to have their way.
>
> (Amandala 1992: 6)

A more recent example from the same newspaper under the headline 'The Battle of Belize' suggests that this situation has not improved:

> Over the last twenty years, Belize went completely globalist. First we sold our passports. Then we sold our lands. Finally, we sold our public companies. Some of our politicians and lawyers became wealthy beyond their wildest dreams. The Lord of Chichester even decided to become a naturalised Belizean citizen. Ostentatious wealth became the order of the day among the elite.
>
> (Amandala Online 2007a)

The concerns of Belizeans, especially the government's interpretation of what is and what is not ecotourism (see Figure 10.6), were heightened by the announcement of a joint venture with a Mississippi-based company for a US$1 billion development to cover two-thirds of Ambergris Caye, the biggest of Belize's offshore islands. The Ambergris Caye controversy raised serious doubts about the scale and nature of tourist development. Presented as a major contribution to reinvesting in the Belizean people (a reflection of the PUP's 1989 election spin 'Belizeans First'), the government bought back the northern two-thirds of the 20,000-acre Ambergris Caye from its US owner. It was expected that a newly established Ambergris Caye Planning Authority (ACPA) would have its jurisdiction and planning powers extended over this area of land, known as the Pinkerton Estate.

Without the knowledge of ACPA, however, the government set up a semi-autonomous development corporation, appointed by and answerable to the then Tourism and Environment Minister, Glenn Godfrey (who was also the attorney general). It was planned to earmark approximately half of the 20,000 acres for conservation and 2,500 acres for

Figure 10.6 Ecotourism on Ambergris Caye – or not?

Source: Martin Mowforth

Belizeans. But it was on the remaining 7,500 acres that the corporation proposed a US$50 million 'sustainable development'. In the relatively fragile ecology of Ambergris Caye, it is difficult to reconcile Minister Godfrey's statement that 'in Belize we must keep tourist developments small' (address to First World Congress on Tourism and the Environment, April 1992) with the claim that the proposed development represents 'an integrated and ecologically sound resort development' when the development was to comprise at least one international hotel, two 'all-inclusive spa hotels', three to five upscale lodges, two golf-courses, town houses and villas, a thousand luxury homes, polo fields and stables.

Although the scheme did not progress in the 1990s, it is illustrative of the 'seductions of mass tourism' as Godfrey referred to it, and the manner in which local control is compromised, if it is not entirely absent. Just two days before the contract was to be signed between the developer and the Belizean government, Godfrey presented the document to ACPA for the first time in spite of discussions having proceeded over the two previous years. A furious member of the ACPA encapsulated the mood of local people, arguing that if the proposed agreement of North Ambergris Caye was as good as claimed, why had the minister not let the people know:

> When the government acquired the Pinkerton Estate, we were told in no uncertain terms that we were getting back control of the land and meaningful participation. To me, 75 per cent to foreigners and 25 per cent to Belizeans is not fulfilling the promise.

As Otis (1992) reported in *The Nation* newspaper, one Belize-based diplomat commented, 'There is a latent resentment of foreign ownership. But Belizeans made a devil's bargain. They sold the place.'

As noted, the scheme did not progress in the 1990s, but a similar plan emerged in 2007 when it was reported that the Government of Belize was in the process of selling 8,255 acres on North Ambergris Caye for $25 million to an American investor and a further 3,400 acres of land in the same area to another foreign investor (Amandala Online 2007b). The area in question covered the Pinkerton Estate. At the time of writing it is uncertain whether the deal will be completed, partly because of the lack of information about the investors.

A further case in point was the proposed multi-million-dollar Belize City Tourism District Project, which would have involved large-scale foreign capital investment or large-scale multilateral borrowing. New Orleans-based architects completed a USAID-sponsored initial study for a waterfront-style regeneration project. The planners hoped to build an exclusive downtown area complete with new hotels and shopping facilities, a promenade with seafront restaurants and cafés, an ethnic crafts and food market and marina for small yachts cruising the Caribbean. It was, in other words, a mini Miami – the consultants even used Baltimore's Haborplace as an example of good practice and an aim to strive for – designed to transform the seafront district into a vacation pied-à-terre for would-be ecotourists, who would presumably feel protected from the urban problems that have become legendary in guidebooks to Belize. The scheme was also underpinned by the philosophy that the benefits would, somehow, trickle down to the rest of the city's population. This would-be gentrification and a facsimile approach to new developments of certain city quarters globally – aping the style of western waterfront developments – is becoming a particularly notable trend. (See also the discussion of urban tourism in Chapter 9.)

Not only would ordinary Belizeans gain little from this upscale project, but also local representatives were not consulted in the planning process. In a bizarre twist, the city's planning department first learnt of the scheme when a three-dimensional model appeared in the window of the city's supermarket. In a city of 60,000 where the city council was severely strapped for money and poverty was on the increase, the consultants also lauded the scheme for allowing the authorities to 'concentrate its resources' (and by inference cutting back spending in other, poorer, neighbourhoods) in one area. In this way, it was argued, the district would be a showcase, with cleaning, street lighting and road maintenance, for example, maximised.

Much of this expressed concern over the control of tourism development in Belize is also reflected in the ownership of tourism resources and facilities (a point already emphasised by the Amandala quotes above). Some commentators have observed that much of the tourism industry is already in the hands of the country's small, but powerful expatriate community (Cater 1992; Everitt 1987). Cater, a delegate at the ecotourism conference in 1991, found that 25 per cent of the registered delegates were US citizens, and more significantly, that among the 43 per cent registered as Belizeans were many expatriate Americans. This is a phenomenon witnessed in many countries with expatriates running bars, restaurants, tour operations, lodges and hotels. Similarly, at the 1992 First World Congress on Tourism and The Environment in Belize City, more than 40 per cent of the 300 delegates came from the USA and Canada, and again, many of the 29 per cent of delegates from Belize were expatriate Americans. As Linda Baker (2001) has pointed out:

> Two decades after it won independence from Britain, Belize finds itself yoked to another kind of colonial enterprise; the foreign-dominated ecotourism trade . . . although Belize is one of the most environmentally conscious countries in the world, a significant percentage of its ecological wealth is concentrated in the hands of expatriates and foreign investors . . . In 1998, a Belize Department of Environment report found that foreign-owned resorts and hotels were flouting environmental laws, causing damage to coral reefs and fishing grounds.
>
> (Baker 2001)

In Belize the expatriate community is a resolute and well-represented lobby. Expatriate riverside lodge owners from the western district of Cayo, for example, demanded strict zoning laws to protect their interests from further tourist development. The depth of expatriate ownership is further reflected through the USAID-initiated Belize Tourism Industry Association (BTIA), an umbrella group representing private interests. In 1992, BTIA was thrown into turmoil when, during the annual election of officers, a dispute erupted over whether seats on the committee should be restricted to Belizeans, even though 65 per cent of the membership were expatriates.

It appears, therefore, that the Belizean government's stated commitment to community tourism development is becoming more elusive as foreign control of the industry takes over. It is not just the large multinational companies and developers singled out in political economy approaches, but small-scale, alternative tourism businesses as well. As Everitt (1987) comments of the post-independence period in Belize, with foreign (mainly Canadian and US) interests in control of the country's most successful tourist facilities, tourism has emerged as a symptom or contributory factor in the creation and aura of neocolonialism. And it is difficult to escape the impression that Belize is primarily a piece of real estate; an advertisement for Ambergris Caye's Club Caribbean in the US publication *Belize Currents* claims: 'own your own piece of paradise . . . Prices start as low as $9,950 . . . Values are starting to soar.'

With the importance of tourism growing in the national economy, and the lure of perceived gains from the growth and development of the industry increasing as one of the most positively cited laboratories of successful ecotourism, the Belizean experience must be critically assessed. The government's need for foreign exchange is often in direct conflict with the need to control the scale and content of tourism development (Simons 1988). Expensive new infrastructure projects such as airports, roads and ports require a return on investment. Locally based and controlled tourism projects, while worthy, do not provide the return necessary for foreign investors and multilateral lending institutions. Indeed, a government's aspirations and expectations of tourism development often stand in contradiction to the aspirations of local people.

Sloan (1993) concludes that 'Belize is fast becoming one of the most popular ecotourism destinations in the hemisphere. But there's a hitch: All those ecotourists are threatening to trample the place to death – literally.' For others, the problems are considerably more deep-rooted. As Everitt (1987: 51) concludes of tourism, it is 'clear that economic imperialism is rife in Belize despite independence, and ranges throughout the economy which also sells its produce to the United States'. Like citrus fruit or bananas, tourism has emerged as another cash crop in the national economy, and one that is challenging the image of small-scale locally owned projects.

Belize is also a perfect example of the complex interaction between a range of interests and why it is problematic to examine individual interests in isolation. In this case the interests of local people, multinational environmental NGOs, aid agencies and the Belizean government are intimately intertwined.

It is this relationship between government policy and the discourse of sustainable tourism development that we further explore with the case of Costa Rica.

Costa Rica: environmental leader?

Since the 1970s Costa Rica has developed a system of national parks and other protected areas which now cover just over a quarter of the country's land area. Despite severe underfunding for the protection of these areas (Blake and Becher 1991: 82–3), this system has formed the base resource of a tourism industry that has relied heavily on the attraction of the country's natural diversity of flora, fauna and landscape.

Consequently, it has become renowned as a destination for ecotourists. Outside San José, the capital city, its tourism industry has been largely based on small-scale, locally owned lodges and hotels which form an integral part of both the communities and natural environments in which they are located. Even within San José, there is a good supply of small hotels and only a limited supply of large-scale, luxury hotels.

Partly through the efforts of its ministers speaking at international conferences, Costa Rica also gained an international reputation as a leader in environmental conservation. Whether this was deserved or not is rather debatable given its high rate of deforestation, especially during the 1980s, outside its protected areas.

As the 1990s progressed, however, international admiration of the Costa Rican government's shunning of large-scale, mass tourism developments began to fade as numerous contracts were signed with international consortia to build tourist condominia not only offering the 'four Ss' of mass tourism but also claiming to be environmentally sensitive and offering nature as part of the attraction. A number of features of these schemes are briefly outlined in the following paragraphs.

In Tambor on the Pacific coast peninsula of Nicoya, a subsidiary of the Barceló Group, a Spanish holiday firm which owns four-star and five-star hotels in Tunisia, Mexico, Spain, the Dominican Republic, Costa Rica and the USA, built a 400-room hotel complex within 50 metres of the high watermark. In so doing they stripped a hillside, filled in a swamp, extracted sand from a nearby river, destroyed a hillside to quarry its stone and threatened local species, all without the necessary building permits from the ministries of Housing, Health, Public Works and Transport and Natural Resources. Additionally, it was reported that white sand was removed from a nearby beach and used to cover the original black sand beach at the complex. The hotel opened in 1992 and has been accused of depositing its sewage in the Rio Pánica.

In 1996 just to the north of Jacó on the Pacific coast of Costa Rica, construction began on the Los Sueños resort hotel, condominiums and marina at the once tranquil village of Playa Herradura. Apart from its negative aesthetic effect on the village, the project altered the extent and ecosystem of a tidal estuary and a marsh that used to stretch along the entire beach. These wetlands were an essential habitat, food source and nursery for a wide range of wildlife. The new yachting marina at the Los Sueños resort was formally opened by the then President of Costa Rica and the then Minister of Tourism in February 2001.

The massive Papagayo project included the construction of 1,144 homes, 6,270 condo-hotel units, 6,584 hotel rooms, a shopping centre and a golf-course. It formed a large part of the Costa Rican Institute of Tourism's development of Culebra Bay, also on the Pacific coast. The scheme went under the title 'Papagayo Ecodevelopment', but despite the planning which obviously went into it, the 'eco' more appropriately referred to the economic wealth it was intended to generate for its investors rather than to the local ecology it was supposed to save. During the planning of the project, Jeff Marshall, a regular commentator on Costa Rican tourism, said:

> The enormous scope of this project is entirely inconsistent with the concept of sustainable and socially responsible ecotourism, which forms the very foundation of the highly successful tourism industry in this country . . . It is nothing more than a high-profit real estate scheme designed to make a bundle of money for a few Costa Rican insiders and their foreign corporate allies.
>
> (Marshall 1994: 2)

In the mid-1990s, after work on the project had begun, it ground to a halt and people in the region were unsure of its future. Smaller-scale development continued, however, 'with hotels with golf courses being the latest development craze – a seemingly mad idea in this semi-arid landscape' (McNeil 1999). Environmentalists expressed concern that the enormous amount of water needed to maintain the golf-courses would be taken from wetlands, mangroves and other delicate habitats, while the locals were worried that their water supplies may be curtailed. While the original extravagant plans may not be realised in this case, it appears that the scale of development will still be grandiose and may outstrip the area's carrying capacity.

The Arenal Volcano National Park provides a further illustration of the differences between the Costa Rican government's stated policies and its practices. The Arenal Volcano has two craters which erupt 24 hours a day giving a spectacular exhibition of nature which attracts ecotourists and gave rise to the development and expansion of the Arenal Volcano National Park. This process of park expansion led to the forcible eviction of some local communities and the formation of an unholy alliance between the park authorities and hotel and resort owners to market the natural environment in an expensive and socially exclusive manner. Ana Isla (2001) reports on one example of this process in the Arenal Volcano National Park:

> Communities also lost the Tabacon river's hot spring. This spring is highly regarded as medicinal; it has been privatised by Tabacon Resort, which is an international private consortium with the involvement of some 'national' capital. Tabacon Resort has built 11 swimming pools, using the hot spring water, at the very foot of the Arenal Volcano . . . Tourism increased deforestation to build

> cabins and resort centres . . . Forest conversion into resort centres endangers wildlife habitat by provoking mudslides, biotic impoverishment, and species forced migration. A Tabacon Resort worker, who wishes to remain anonymous, says, 'Before building the swimming pools, we carried hundreds of frogs out of the area. After five years of activity, the frogs disappeared and the toucanos do not stand in the trees any longer. Massive use of chemicals to clean the swimming pools, washrooms, etc. left chemical residuals that force the animals to leave the surrounding areas.'
>
> (Isla 2001)

The link between the burden of international debt and government capitulation to the pressure of transnational companies to develop large-scale tourism projects in order to gain much-needed foreign exchange may be denied in some quarters. But while it may be difficult to prove the direct existence of the link, it is also difficult to refute with any degree of credibility. The examples outlined above illustrate the subversive power of the pressure exerted by international consortia – baying at the heels of the international lending agencies which stipulate the need for national governments to raise foreign exchange to service their debt. As Michael Kaye (1994), president of Costa Rica Expeditions tour company, has stated: 'the short-term temptations of the fast and easy money from mass tourism development in the context of an economy the size of Costa Rica's should not be under-estimated'.

Under such pressure, it should be no surprise that the government of Costa Rica has felt obliged to abandon its previously cautious support for a sensitive and responsible type of community-based tourism development from which it, the government, gained relatively little financial benefit as most of this was received directly by Costa Ricans.

In 1992, Deirdre Evans-Pritchard, an English academic who formerly ran the Eco Institute of Costa Rica, pointed out that the Costa Rican Chamber of Tourism (CANATUR) was under the same kind of pressure as the government: 'the 1992 CANATUR congress championed noble causes and concerns in the name of small scale sustainable tourism development and then, in the final ceremony, awarded prizes to the four largest, most resort-like hotels in the country' (Evans-Pritchard 1993: 779).

In 1994, the Costa Rican government suffered from another loss of power over its tourism industry which may account, at least in part, for the 1995 reduction in tourist numbers. The then Minister of Environment, René Castro, increased the entry fees to most national parks for foreigners by a factor of ten, ostensibly to further the government's sustainable development programme

> by limiting the number of tourists using the country's natural resources. In one sense, his policy 'worked' – between '94 and '96, the nation's parks recorded an overall 26.5 per cent decrease in visits. Castro may have been motivated by more than the principle of sustainable development. The majority of foreign visitors to Costa Rica are 'shoestring' travellers, low-budget backpackers from the US, Europe and Australia. Both Castro and Figueres [then president] . . . may have hoped that the tourist industry could recover any losses incurred by the decrease in backpackers through its promotion of Costa Rica as a travel destination for affluent visitors.
>
> (Quoted in Mesoamérica 1996: 9)

In response to the widespread complaints arising from the price increase and the significant decline in tourism receipts, from 1 April 1996, the entry fees were lowered from $15 to $6. Additionally, hotel and car rental companies reduced their daily rates by 15 per cent.

CANATUR president, Mauricio Ventura, asked 'Who is the winner in this?' and answered, 'The tourist' (Mesoamérica 1996).

The Costa Rican Institute of Tourism's (ICT) claim that the country is a model of sustainable development now rings a little hollow. As Jeff Marshall asks:

> Instead of promoting massive centralised tourism projects tied to foreign capital, why doesn't the ICT concentrate on the careful planning and development of a network of quality, locally owned and operated, small scale tourism businesses that are compatible with the natural environment and integrated within the existing structures of local communities and economies?
>
> (Marshall 1994: 36)

And from her 2001 analysis of land management and ecotourism in Costa Rica, Ana Isla concluded that:

> sustainable development suppresses the human rights of local communities and the rights of nature in favour of the rights of corporations, and vulnerable local nature and local people have become connected to the international markets and the world economy on disadvantageous terms. Communities have had to surrender their safe local food system and their role as agricultural producers. Thus, local environments and communities are impoverished. Sustainable development aggravated poverty and environmental destruction for the short term benefit of capital . . . In Costa Rica, these acts of colonisation are called acts of 'sustainable development'.
>
> (Isla 2001)

To set against these frequent criticisms, Costa Rica can cite its Certification for Sustainable Tourism (CST), a programme initiated and run by the Costa Rican Tourism Institute (2005). The CST differentiates between and certifies tourism businesses according to the extent to which they comply with a model of sustainability in terms of four 'fundamental aspects':

- *Physical-biological parameters* – which evaluate the interaction between the company and its surrounding natural habitat;
- *Infrastructure and services* – which evaluate the management policies and the operational systems within the company and its infrastructure;
- *External clients* – which evaluate the interaction of the company with its clients in terms of its involvement of the client as an active contributor in its sustainability policies;
- *Socio-economic environment* – which evaluates the interaction of the company with the local communities.

The CST is widely acknowledged as one of the more convincing attempts to establish visible sustainability credentials within the tourism industry through a system of certification (see Chapter 7). Costa Rica still contains many best case examples of the conduct of nature tourism and ecotourism. But the very dubious examples cited above are not exceptional – the construction of more huge tourist condominia are in plan (Baxter-Neal 2007: 9–10) – and lead us to question its role as 'environmental leader'.

Human rights and tourism in Burma

Aung San Suu Kyi's National League for Democracy (NLD) won 81 per cent of the parliamentary seats in elections in Burma (Myanmar) in 1990. The response by the generals of the SLORC (State Law and Order Council) was to imprison most of the newly elected parliamentarians and to turn the country into what Amnesty International has called 'a prison without walls'.[5]

Since that time, the United Nations Commission on Human Rights (1993) has described the country's violations of human rights as:

> extremely serious, in particular concerning the practice of torture, summary and arbitrary execution, forced labour, including forced portering for the military, abuse of women, politically motivated arrests and detention . . . important restriction on the exercise of fundamental freedoms and the imposition of oppressive measures directed, in particular, at minority groups.
>
> (United Nations Commission on Human Rights 1993)

The military regime's brutal crackdown against pro-democracy uprisings in September 2007, in which thousands were detained and hundreds killed, demonstrates that the political and human rights situation within the country has not changed since its early days.

The issue of tourism in Burma highlights a number of points of interest in this chapter. These concern not only the relationship of the Burmese government to the tourism industry, but also the relationship of First World governments and tourism-related TNCs to the Burmese government and the role and responsibilities of tourism more generally.

Since 1990, 'The SLORC has been wooing new investors – especially in the area of tourism and hotel development – with offers of ten year tax breaks and full repatriation of profits' (Mahr and Sutcliffe 1996: 28). As Mahr and Sutcliffe (1996) point out, such investment from TNCs based in the First World lends the regime political legitimacy, which at first was confirmed by the arrival of First World tourists – international tourist arrivals increased steadily from 26,607 in 1992 to almost 200,000 in 1998 (WTO/OMT, annual). Thereafter the increase stalled with the junta claiming that 192,000 tourists arrived in the year to March 2007 (Goldenberg 2007). Other indications and headlines such as 'Visitor numbers already low due to boycott' (Guardian, 11 October 2007) suggest that since the September 2007 violence numbers have fallen more dramatically.

The SLORC/SPDC declared 1996 the 'Visit Myanmar Year' and increased its efforts to profit from the tourism industry by aiming at a target of half a million visitors for 1996.[6] To prepare for the expected increase in tourists, the SLORC/SPDC put its people to work – forcibly – as described by Mahr and Sutcliffe (1996):

> Whole communities have been forcibly relocated from their homes to make way for foreign hotels and to 'clean up' areas for the eyes of tourists . . . Thousands of people have been forced to work without pay on tourist sites – restoring the moat around Mandalay Palace, for example. Plans have been put into action to relocate the more 'picturesque' ethnic peoples to special villages where they can be visited by tourist groups in what amounts to a human zoo. Model villages for this purpose are under construction near Rangoon and some Padaung people have already been resettled to Inle Lake. There tourists can visit – for a charge – and take photos of 'long-necked' Padaung women.
>
> (Mahr and Sutcliffe 1996: 29)

These human rights violations have been documented in detail throughout the SLORC/ SPDC's regime to the present day by the UK-based Burma Action Group (Sutcliffe 1995) and reported by Tourism Concern and increasingly in the western media.

In 1996 Aung San Suu Kyi specifically asked visitors to stay away from her country until the military government hands over power to the NLD.[7] Lonely Planet Publications, however, has been publishing its Burma guidebook since 1979 and has continued to do so since the first calls for visitors to stay away. In May 2000, the Burma Campaign UK and Tourism Concern launched a boycott campaign against Lonely Planet Publications because of its Burma travel guidebook, written by Joe Cummings and Michael Clark. Under the banner 'There are times when travellers have to take a stand. This is one: BOYCOTT LONELY PLANET', they asked consumers of travel guides to boycott all Lonely Planet publications until the Burma guidebook is withdrawn. Glenys Kinnock, European parliamentarian, in an article in the UK *Guardian* newspaper, summarises the request behind the boycott thus:

> Aung San Suu Kyi asks a simple thing . . . She's asked for sanctions so that the junta will be starved out of existence. We can impose our own sanctions and not go on holiday to Burma. And we should certainly not buy from publishers that suggest that we should.
>
> (Kinnock 2000)

In a letter to concerned members of the UK Parliament, Tony Wheeler, Lonely Planet publisher, retorted by accusing the two groups of censorship. The debate between the two has been joined with some heat ever since (Burma Campaign UK and Tourism Concern 2000a, 2000b; Wheeler 2000, 2001). In response to the boycott campaign's claim that visiting the country is tantamount to approval of the SLORC/SPDC, Wheeler claims that only a tiny proportion of tourist revenues will reach the illegitimate military government's coffers and that the Burmese people want to talk to outsiders. The cartoon (Figure 10.7) is one of many representations of the boycott campaign's view of the Lonely Planet attitude. Wheeler claims that the NLD voices asking for a boycott are those of NLD

Figure 10.7 The Burma boycott debate

Source: Tourism Concern (2000) *In Focus* 37

members living in exile who are not aware of the current reality in Burma. In October 2000, the debate also reached the UK BBC's *Newsnight* programme, which revealed that the Sittwe Hotel, promoted in Lonely Planet's guide along with other government guesthouses in the area, is included in an International Labour Organisation (ILO 1998) report as having been built with forced labour. Critics describe Wheeler's approach as one of arrogant defiance and stubborn self-certainty.

In the UK *Observer* newspaper, Nick Cohen (2000) used the debate to describe the attitudes of both the new tourists who are likely to visit Burma and the alternative guidebook publisher who is likely to benefit from sales to them:

> the Burma spat has been distinguished by the extremism of the invective of bohemian travellers. Dea Birkett, a travel writer, whined rhetorically in the Guardian of all places: 'Aren't holidays supposed to be carefree times for suntans and self-indulgence? Is it really such a crime to seek out somewhere where you can simply enjoy yourself? Tourism Concern and the Burma Campaign's moral outrage is designed to make us feel bad about being good to ourselves. To restrict freedom of movement is the hallmark of totalitarian regimes.'
>
> Forget her wheedling style for a second, and consider the implications of Birkett's words. The opponents of dictatorship, who fight without resources, international support or any military force which might defeat the junta on the field of battle, are totalitarians. Rich western tourists, by contrast, are the true victims even when they stay in Rangoon hotels built on the site of the homes of the Burmese poor – which were bulldozed without compensation – or travel on a moving staircase built by forced labour to catch the marvellous view at sunset from Mandalay Hill. Only a sadist would want to make them 'feel bad about being good to ourselves' . . . Though I'm all for boycotting Lonely Planet, the real significance of the argument is cultural. Discerning liberal consumers are now so self-confident and self-pitying that they pose, without irony, as the victims of Stalin and Hitler when anyone suggests they might make the tiniest moral choice. It says so much about them.
>
> (N. Cohen 2000)

Whatever the pros and cons of the argument, few if any British tour operators can currently be unaware of the debate around the ethical grounds for not visiting Burma. New hotels are being built by TNCs from Hong Kong, Japan, Malaysia, Singapore, South Korea and Thailand, and French and Swiss interests are also involved. Companies such as Voyages Jules Verne, Coromandel and David Sayers Travel all run cultural or trekking tours to Burma. Coromandel describes its relation with Burma thus:

> We are probably Britain's leading specialists for Burma, firmly believing that contact with the free world does more good than isolation. In spite of a number of good new private sector hotels and new airlines, Burma remains a destination for the traveller rather than the tourist, but the rewards are great, with splendid scenery, magnificent monuments, and a warm and smiling welcome from the Burmese themselves.
>
> (Coromandel brochure 2005/6: 7)

Explore, on the other hand, ceased covering Burma on its tours largely because of UK Foreign Office advice regarding the nature of the regime. Bales also ceased to cover Burma in its main brochure in 1999 for the same reason, although Bales is still willing to arrange tailor-made tours upon specific request.

Burma represents an attractive Third World destination for the new middle-class tourists. But more importantly from the standpoints of the First World corporate decision-makers, it represents a tempting investment opportunity for tourism businesses. First World governments can and do assist in these processes of tourism investment and associated human rights violations.

The UK government's stated policy is not to promote investment in Burma because of the SLORC/SPDC's human rights violations. In 1996, however, despite its stated policy, the UK government's Department of Trade and Industry (DTI) was exposed for trying to cover up public knowledge of a trade mission between the two governments (Tourism Concern 1996). Since that time, a number of UK and European companies (such as Rolls-Royce, Adidas, many clothing retailers and many tourism companies) have cut ties with the regime, but tour companies such as Orient Express and Asean Explorer continue 'to make a handsome profit on the suffering of the Burmese people' (Pilger 2007: 33).

The example of Burma is indicative of the 'reality' to which governments and the industry work (see Chapter 7). It is a reality which they use to justify their actions and investments. If ethical considerations contradict this reality, then they must be hidden from view.

Tourism in Burma gives us an illustration of the way in which investment is used to serve specific interests, those of the Burmese government, First World governments, tourists and crucially tourism-related businesses. Other interests, those of the visited, the resettled, the majority of Burmese people, are decidedly not served by these investments and developments. But whereas in many Third World destination countries the governments have to be coerced by First World governments, international lending agencies and TNCs, in the case of Burma the SLORC/SPDC actively enlists the assistance of these bodies. Power is still very much based in the First World, but with a willing partner in Burma's military and business elites.

Despite the clarity and documentation of the link between Burma's human rights violations and its exploitation of nature and cultural tourism, the brochures of some new tourism companies still extol its virtues. In such circumstances, then, it is hardly surprising that for many tourists the relationship between travel, tourism and human rights has remained rather nebulous. Indeed, as was suggested in Chapters 3 and 5, the occurrence of human rights abuses and their coincidence with perceived danger may actually act as a fillip to tourism in certain circumstances and among certain types of tourists.

Conclusion

It has been argued that an understanding of issues of power is essential to an analysis of tourism developments and the role of governments in these. The necessity for such an understanding is as great for analyses of new forms of tourism to Third World destinations as it is for mass tourism developments – perhaps even more so given the relationship of new forms of tourism to 'development'.

It is commonly perceived, especially from the level of the local community, that power rests with the national government. This may indeed be the case in some instances, but we have suggested that the actions and policies pursued by national governments are often circumscribed at best and are sometimes dictated by the influence of external organisations. This includes both the inevitability that tourism must increase, as well as the contest over which form of tourism development should take place. It is especially so for Third World governments weighed down by a burden of debt – a burden that was largely foisted upon them by First World banks and governments as well as their own incautiousness resulting from the unseemly rush for western-style development.

For the most part, the outcome of this process of First World influence on Third World development has been government policies which increase the unevenness and inequality of development within national boundaries and which further widen the gap between First and Third Worlds. The policies pursued for the development of tourism are largely those most suitable to the profits and enjoyment of First World investors and tourists rather than those most beneficial to Third World communities and governments. This is not to say that there are no opportunities for positive and beneficial change. Our discussion of pro-poor tourism initiatives in Chapter 11, for example, suggests that such opportunities may exist. But given the global analysis and the enormity of global poverty, the ability of tourism initiatives to tackle global inequality is slight.

We have tried to show how, even where governments have attempted to follow environmentally benign, socially and economically beneficial and culturally sensitive tourism developments, forces beyond their control have subverted the policies and have served only to emphasise the seemingly ever-widening gaps. In the process, government attempts to satisfy outsiders' views of the content of tourism have turned sustainable tourism into an item of political discourse.

11 New tourism and the poor: making poverty history?

Firmly placing the study of tourism within the field of development has been one of the aims of this book. This is not the same as practising tourism as a form of development, but it is fitting that the start of the twenty-first century has seen efforts to stimulate development through tourism. Increasingly, the perceived need for development has been targeted at the poorer sectors of society and the efforts of most development related organisations and agencies have become explicitly focused on the poor. Poverty alleviation has become the overriding goal for all development activities marshalled under the United Nations Millennium Development Goals (MDGs), and within the tourism industry pro-poor tourism has become the latest in a long line of terms and types to attract attention, funding and energy.

Few if any would argue that poverty alleviation is not a laudable goal, but if pro-poor tourism is not to suffer the same fate as other terms such as 'ecotourism', 'alternative tourism', or 'green tourism', for example, then it must avoid fixing itself to a label that has no substance behind it. It must therefore answer a range of questions about its ideas, processes and goals if it is to gain any success. Who are the poor? What are the mechanisms by which tourism income can reach the poor? What are the tourism activities most suitable for involvement of the poor? How can the tourism industry direct its efforts, policies and energies towards the security of provision of the basic human needs of the poor? This chapter discusses these questions and issues.

Understanding poverty

As Mitchell and Ashley (2007: 10) suggest, 'defining how tourism affects "the poor" and "poverty reduction" is futile without defining who are the "poor" and what is meant by poverty reduction'. Until the 1980s it was common to indicate the wealth of a country, and conversely its poverty, by the measure of its gross domestic product (GDP). The GDP is the total value of the output of goods and services produced by an economy, by both residents and non-residents. It is used to show the relative wealth of different countries and through yearly comparisons to show levels of economic growth. The GDP and its associated measure, gross national product (GNP),[1] are still widely used. Economic growth, however, is not necessarily the same as economic health, as Griffiths (1996) makes clear:

> El Salvador's macroeconomic figures for 1995 looked excellent . . . The country's 6% growth rate in 1995 led Central America and the Caribbean, and was only surpassed by Chile and Peru in Latin America . . . While financial experts lauded the figures as proof of El Salvador's success in implementing neoliberal economic stabilisation, statistics compiled by the ombudsman's office reveal how very little

the macroeconomic surge has so far meant to many citizens left behind. 51% of Salvadorans live in absolute poverty, 40% of the population lacks access to health services, 53% have no potable water and 29% are illiterate, preventing them, tragically, from reading Milton Friedman.

(Griffiths 1996)

Levels of poverty and of economic health can be shown in a number of ways and no single indicator will suffice to demonstrate them accurately on its own. Despite this, the most commonly cited indicator is through reference to the dollar income per day per individual – those living in poverty being defined as having an income of less than US$2 per day, and those living in extreme poverty an income of less than US$1 per day. According to the UNDP 2006 Annual Report, there are over two and a half billion people in the world living on less than US$2 a day. One billion of these live on less than US$1 a day, and the gap between the rich and poor of the world is growing (UNDP 2006). Such figures hide a range of differences within them. For instance, the richest 2 per cent of the world's adult population now own over a half of global household wealth while the bottom half own less than 1 per cent (Korinek et al. 2006; UNDP 2007). So poverty and economic health can be interpreted in both absolute and relative terms, and Vandana Shiva makes clear that there is a distinction between poverty as subsistence and misery through scarcity and want:

It is helpful to distinguish between a cultural concept of a simple and sustainable life, understood as poverty, from the material experience of poverty as a result of dispossession and scarcity. Poverty perceived as such from a cultural perspective is not necessarily real material poverty: subsistence economies which satisfy basic needs by means of self-supply are not poor in the sense that they are wanting. The ideology of development, however, declares them to be poor for not participating significantly in the market economy and for not consuming goods produced in the global economy.

(Shiva 2005)

Moreover, the United Nations High Commissioner for Human Rights (UNHCR) defines poverty as 'a human condition characterized by the sustained or chronic deprivation of the resources, capabilities, choices, security and power necessary for the enjoyment of an adequate standard of living and other civil, cultural, economic, political and social rights'.[2] This again emphasises that, although economic deprivation is an element of any definition of poverty, it is far from being the only element. So the measurement of poverty is not an easy matter, and people's well-being cannot be measured simply by income or value of production. Well-being requires good health, access to satisfaction of the basic human needs, a safe environment and a strong communal life. Taken together, some of these measures are often used as indicators of a nation's level of development, and in 1990 the United Nations Development Programme published the first Human Development Report (UNDP, annual) which included data on the human development index (HDI) as a composite measure of human development. By combining both social and economic indicators, the HDI attempts to reflect a general sense of well-being that people may or may not feel in their lives. The specific measures they combine are life expectancy, literacy and per capita income. The mathematical basis for, rational justification for, drawbacks of and results of the measure are given in much greater detail in various yearly editions of the Human Development Report of the UNDP.

Since 1990, other supplementary composite measures of development have been produced by the UNDP, one of these being the human poverty index (HPI). The HPI reflects levels of deprivation in a country through the same variables used in the calculation of the HDI. The HPI-1 measures poverty in developing countries through longevity (as the probability at birth of not surviving to age 40), knowledge (as the adult illiteracy rate), and overall economic provision (as the percentage of people not using improved water sources and the percentage of children under 5 who are underweight).

As the UNDP states, '*all* definitions of poverty involve social judgements' (UNDP 1996: 67, emphasis in original). The UNDP defines absolute poverty by comparing personal or household income with the cost of buying a given quantity of goods or services, relative poverty by comparing that income with the incomes of others, and subjective poverty by comparing actual income against the income earner's expectations and perceptions. And as they add, 'There is no scientific, unequivocal definition of who is and is not poor' (UNDP 1996: 67). Commenting on papers focused on poverty commissioned for conferences and roundtables, Chambers (1995: 179) pointedly asks: 'One may speculate on what topics the poor and powerless would commission . . . if they could convene conferences and summits: perhaps on greed, hypocrisy and exploitation.'

Within tourism studies the paucity of definition is especially acute and is 'generally attempted in only a rather crude way' (Mitchell and Ashley 2007). Mitchell and Ashley continue their commentary by noting that the documents reviewed for evidence and methods in pro-poor tourism for the World Bank,

> often do not address the issue directly and assume implicitly that, for instance, tourism growth inevitably benefits the poor. Alternatively, studies focus on specific groups of people who may act as proxies for the poor (for instance 'rural' residents or 'crafters').
>
> (Mitchell and Ashley 2007: 10)

As will be reviewed below this has resulted in the somewhat bizarre categorisation of the 'not-quite-so-poor' (Deloitte & Touche et al. 1999: 55) in some of the pro-poor tourism literature.

Despite this definitional problem, measuring poverty and development is important and the use of composite indices, such as the HDI and HPI, is now widely acknowledged to be more appropriate than single measures of income or value of productivity, such as GDP. But the HDI and HPI also fail to tell the whole story. Also important, especially in the field of tourism, is income inequality within countries and regions. Economic growth on its own will not reduce poverty unless governments redirect it to the poor.

Inequality exacerbates the effects of market and policy failures on growth and thus on progress against poverty. The poor, for example, generally find it more difficult to borrow as they have no collateral, and this impedes their ability to set up small businesses, as pertinent in the tourism industry as in others. Concentration of income at the top also undermines public policies, on education and health, for instance, which will advance human development and the individual's potential to better themselves. Moreover, inequality erodes social capital and the ability of society to provide reliable institutions and services, and thereby renders the individual's trust in society less worthwhile, so that their participation loses significance (UNDP 2003: 17).

That inequality is important in Third World countries is demonstrated by Oliver Marshall, co-author of the *Rough Guide to Brazil*: 'everywhere you go, extreme social and economic disparities are striking, nowhere more so than in the cities, where conspicuous wealth is displayed side by side with extreme poverty' (Marshall 2004). The lesson from this is starkly put by José Antonio Ocampo, Executive Secretary of the Economic Commission for Latin America and the Caribbean (ECLAC), stating that:

depending solely on economic growth to deal with the problem of poverty in Latin America will make it hard to meet the goal set for 2015. It is becoming ever more necessary to resort to economic policies that, aside from seeking to expand the productive base and increase national output, include the progressive redistribution of income as a viable alternative for meeting the Millennium targets.

(Ocampo 2002)

A country's levels of poverty, inequality and development matter primarily for the sake of that country's population. But they also matter to potential investors in the country, to potential visitors to the country (some attracted, others 'put-off') and increasingly of late to the promoters of neoliberal economic development (often referred to as the Washington Consensus). To the supranational institutions (the World Bank, IMF, WTO and the regional development banks) high levels of poverty and low levels of development are a mark of failure of their model of economic development, one of whose principal tenets is that economic well-being will trickle down to the poor thereby spreading, eventually and theoretically, the benefits of this form of development to the whole population. If poverty is seen to increase and development is seen to stagnate or worse, then clearly the model is failing, even by its own criteria.

The need to examine the modus operandi of the prevailing economic model in order to reduce poverty and inequality significantly has recently been shown in the acknowledgement by the supranational organisations and international financial institutions of the importance of promoting poverty reduction. As Bridget Wooding and Richard Moseley-Williams point out: 'recognition of the links between wealth creation and poverty reduction in highly unequal societies has now led to national anti-poverty programmes in developing countries, backed by international financial institutions' (Wooding and Moseley-Williams 2004). The MDGs are the most notable global manifestation of this recent recognition, and we introduce these goals and their application to tourism in the next section. A number of developments in bilateral donor policies have also given cause for some optimism in this regard, among them the development of pro-poor tourism initiatives, and this is the focus of the subsequent section.

For our purposes, the point about poverty is that its incidence and depth are not always easy to discern, and the tourism industry is renowned for its ability to conceal the unpleasant realities of life from people whose primary purpose is to enjoy their time away from home. So can tourism with a focus on alleviating the poverty of local people really reveal poverty without offending the visitor? And can it ensure that the economic benefits of tourism are directed effectively to the poor?

The Millennium Development Goals and tourism

As Chapter 2 discussed, the evolution of development theory and practice has witnessed a range of approaches to addressing 'under-development' and poverty. But the MDGs stand out for the manner in which they have galvanised almost universal acceptance that poverty is unacceptable and that global action is required to tackle it. In September 2000, the UN General Assembly's Millennium Summit 'Millennium Declaration' heralded the agreement of a set of time-bound and measurable targets that became known as the MDGs. Signed by nearly 190 UN member states, the MDGs are the single most significant global development framework for combating poverty, hunger, disease, illiteracy, environmental degradation and discrimination against women. The goals and their targets are shown in Table 11.1. For each target a number of indicators are used to measure the success or failure of their achievement. The table also indicates how tourism can be applied to the MDGs, and initiatives that might be considered as supporting these goals (Green Hotelier 2006). For the arguments developed throughout this book, however, the effectiveness of these initiatives can be disputed, not least because they are highly generalised and do not address the issue of enforcement mechanisms.

As important as the MDGs are in pinning the 'international community's' colours to the development flag and providing a single and unifying vision of some basic human needs and values – they also suffer some fundamental drawbacks. In Chapter 2 we referred tentatively to an age of 'alternative modernisation' signalling a return to a top-down and technically driven approach to development. The MDGs are the centrepiece of this turn – they are the rallying point for multilateral and bilateral donor agencies – and there are now few, if any, donor materials that do not use the MDGs as a significant if not central focus of attention. Indeed, even the tourism industry is becoming attuned to the global MDG challenge (see, for example, the discussion on the Greater Mekong Tourism Strategy later in this chapter, and the Green Hotelier 2006).

While the goal of eliminating global poverty is laudable, it is important to ask if the MDG approach is the right one in pitch and process. The core drawbacks are threefold. First, and reflecting back on the economic primacy and dominance of development agendas reviewed in Chapter 2, the MDGs remain situated in a global mode of development that remains oriented to economic growth as the engine for development, as pursued by one of the most important Bretton Woods institutions, the World Bank. The Independent People's Tribunal on the World Bank Group in India has been just one of many voices that have pointed out that the World Bank must be made 'accountable for policies and projects that in practice directly contradict its mandate of alleviating poverty for the poorest' (Independent People's Tribunal website 2007).

Charged with supporting member states in monitoring and achieving the MDG targets, the UN agencies have begun in earnest their exhortations to national governments of the industrialised, technocratic countries and international financial institutions to adapt policies in order to meet the MDGs – to follow, in other words, what might be called a reformist agenda; a fine-tuning of the existing and dominant model of development. Tourism is being increasingly invoked as an agent of development, a means through which development can be achieved. This is witnessed by the recent efforts of First World government overseas development agencies and the supranational organisations such as the UNDP and the UNWTO to promote pro-poor tourism initiatives. The UNWTO, for example, established in 2003 a special initiative on poverty called 'Sustainable Tourism – Eliminating Poverty' or ST-EP. Box 11.1 gives a brief introduction to the ST–EP in the UNWTO's own words.

The MDGs themselves make it clear that they are to be achieved, or not, in collaboration with the private sector, and former UN Secretary General Kofi Annan talks of a 'successful, development-oriented result [which] could boost investment flows and help revive the global economy' (Annan 2003). Indeed, it is interesting to note that all the rhetoric about poverty reduction and elimination that has emanated from the supranational institutions in recent years has assumed that there will be and can be no change in the prevailing model of development. The point is not that the intentions to adjust policy are ill-inspired, but that they are contingent upon a system which has manifestly failed to date to deliver development to a majority of the world's population.

Second, and beyond the global macro-economic framework within which the MDGs work, there are also the practicalities of applying the ostensibly international targets expressed in the MDGs to the national and local level. The goals and targets may represent the hard-won output of diplomacy and the international civil service, but they remain somewhat perplexing for those at the receiving end – the global poor. For example, the signature dollar-a-day benchmark for measuring poverty (MDG 1 – Eradicate extreme

Table 11.1 *MDGs and the travel and tourism industry*

	Goal		Target between 1990 and 2015	How travel and tourism can engage
1	Eradicate extreme poverty and hunger	1	Reduce by half the proportion of people living on less than one dollar a day	Recruit and train local people. Practise sustainable supply chain management (SSCM), i.e. local sourcing of produce and services, build pro-poor partnerships and linkages and provide training and support to small independent enterprises. Help generate opportunities for local enterprise and ownership. Pay a fair or above-average wage. Set up profit-sharing schemes
		2	Reduce by half the proportion of people who suffer from hunger	
2	Achieve universal primary education	3	Ensure that all boys and girls complete a full course of primary schooling	Never use child labour and ensure your suppliers are not using child labour. Support/develop local education programmes
3	Promote gender equality and empower women	4	Eliminate gender disparity in primary and secondary education preferably by 2005 and at all levels no later than 2015	Employ more women and help build their potential or promotion through training. Develop strong policies and take action against sexual harassment. Buy from women's cooperatives and support women's issues directly or through NGOs
4	Reduce child mortality	5	Reduce by two-thirds the mortality rate in children under five	Support foundations which enable the provision of improved nutrition, health, access to water and sanitation such as the World Health Organisation, UNICEF, etc.
5	Improve maternal health	6	Reduce by three-quarters the maternal mortality rate	Support women's NGOs and community health services. Combat stigmatisation in countries where cultural and religious factors hinder open discussion and action on reproductive health issues. Raise staff awareness of contraception, sexually transmitted disease and nutritional issues
6	Combat HIV/AIDS, malaria and other diseases	7	Halt and begin to reverse the spread of HIV/AIDS	Raise staff and general awareness of HIV/AIDS, malaria and other major diseases where these are key health issues in your country. Raise funds for, support or develop health and immunisation programmes
		8	Halt and begin to reverse the incidence of malaria and other major diseases	
7	Ensure environmental sustainability	9	Integrate the principles of sustainable development into country policies and programmes and reverse loss of environmental resources	Practise resource efficiency by using less energy, water and creating less waste. Ensure that your operations do not impact negatively upon the environment and biodiversity

Table 11.1 *Continued*

Goal	Target between 1990 and 2015	How travel and tourism can engage
	10 Halve the proportion of people without sustainable access to safe drinking water	Support climate change initiatives such as carbon offsetting and initiate local 'clean-ups'. Enable community access to resort drinking supplies.
	11 Achieve a significant improvement in the lives of at least 100 million slum dwellers by 2020	Support initiatives such as 'Just a Drop'. Donate items such as bathroom amenities, old linens, uneaten food items to groups working with the disadvantaged
8 Develop a global partnership for development	12 Develop an open, rule-based, predictable, non-discriminatory trading and financial system	Be transparent in your business operations, report publicly on your environmental and socio-economic initiative. Apply the principles of the Global Compact
	13 Address the least developed countries' special needs. Includes tariff and quota-free access for their exports; enhanced debt relief for heavily indebted poor countries; cancellation of official bilateral debt; and more generous official development assistance for countries committed to poverty reduction	Promote business investment in least developed countries Argue for improved market access for least developed countries through NGOs and the UN
	14 Address the special needs of landlocked and Small Island Developing States (SIDS)	Promote youth employment and entrepreneurship. Consult the UN's Youth Employment Network, hire young people, run apprenticeship programmes and build linkages with enterprises run by young people
	15 Deal with developing countries' debt problems through national and international measures to make debt sustainable in the long term	Support NGO or community-led projects by donating resources, expertise and technology
	16 In cooperation with the developing countries, develop decent and productive work for youth	Increase the flow of communication with the local community Educate tourists about how they can contribute through purchasing locally
	17 In cooperation with pharmaceutical companies, provide access to affordable essential drugs in developing countries	
	18 In cooperation with the private sector, make available the benefits of new technologies – especially information and communications technologies	

Source: *Green Hotelier* 39 (April 2006), a publication of the International Tourism Partnership

Box 11.1 Sustainable Tourism – Eliminating Poverty (ST–EP)

The ST–EP project seeks to refocus and generate incentives for sustainable tourism – social, economic and ecological – to make it a primary tool for eliminating poverty in the world's poorest countries (particularly the lesser developing countries): bringing development and jobs to people who are living on less than a dollar a day. The core of ST–EP is a tripartite institutional framework, which raises substantial funds, targets best practice research and creates an operating system that specifically encourages sustainable tourism geared to the elimination of poverty.

The steps are:

1 Creation of an International Foundation, whose purpose is to secure a sustained revenue source to advance ST–EP goals in the research, operational and promotional fields.
2 Creation of a Research Base, where a small institute will organise the worldwide networks of academic communities, to focus research on the linkages between sustainable tourism and eliminating poverty, and identify practical approaches capable of replication.
3 Development of 'Sustainable Operations'. This program will seed small- and medium-sized projects to benefit the world's poorest communities by enabling them to secure sustainable livelihoods through engaging in tourism.

The World Tourism Organisation (UNWTO) and UNCTAD are implementing this concept throughout the world by engaging stakeholders – government, the private sector and civil society.

Source: UNWTO website

poverty and hunger, target 1 – Reduce by half the proportion of people living on less than one dollar a day), may make for clear and receptive communication but is close to meaningless in local contexts that are so many and varied.

As the discussion at the outset of this chapter emphasised, poverty is not experienced as a set of neatly defined and quantified goals and supporting targets, but as complex and multifaceted systems of deprivation. More importantly though, development is something that is most likely to happen in communities, villages, towns and cities, and through the organisational structures (the local committees and non-governmental organisations, councils and local governments, for example) that are closest to those in need. And yet the very notion of decentralisation in decision-making and crafting locally fashioned visions of the future is eerily silent in the MDGs. As Satterthwaite (2005: 4) argues, 'while successful development is intensely local, most development actions and investments are planned, implemented and evaluated centrally – by national governments and international agencies' (see also Mitlin and Satterthwaite 2007). Although the UN continues to work with central governments to translate these global MDGs into meaningful national targets, the same cannot be said of 'localising' the MDGs – in development-speak, providing ownership to local communities in setting and meeting their own goals and targets. Local governments, for example, struggle to see the relevance of the MDGs in their day-to-day business. For the most part, this is not because they care little about education, health or the provision of adequate water and sanitation, but that the power, authority and finance to act is not vested locally in these institutions. This is a serious, potentially crippling, problem with MDG implementation, as development requires a rebalancing, if not wholesale shift, in the distribution of power and authority. As Sen (1999: 53) remarks, people need to be 'actively involved – given the opportunity – in shaping their own destiny, and not just as passive recipients of the fruits of cunning development programs'. It is also symptomatic of arguably the most potent and deep-seated problems with the so-called 'architecture of international development aid', the 'obsession with always working through the government' (Easterly 2006: 156), the institutional entity furthest from those in poverty and whom aid is supposedly there to help.

Finally, as with all numbers games, there is now a tendency with the MDGs for the distribution of effort to be allocated to proving that targets have been met. Quantitative monitoring is not so much a problem in contexts where travel companies are proving to their shareholders that the number of holidays sold or profit margins are on target, but is far less adequate for demonstrating that real and lasting progress has been made in social and economic (human) development. Careers are made and broken on hitting targets – and the development industry is no different. The MDGs are in danger of triggering a development melee in the headlong rush to 2015 in which the major effort is in proving that targets have been met rather than responding to the needs of the poor and providing the means of meeting their targets and broader human development.

Of course this begs the question of how else a form of global consensus might be formed around such a complex process as development and the eradication of extreme poverty: how should it be expressed and measured? And this is a valid counterpoint. Poverty (extreme or relative) is indeed an unacceptable scourge and immediate action is required. It is difficult not to be moved by the passionate advocacy to 'get the job done' through massive increases in international aid, debt relief and fairer international trade policies espoused by Special Advisor to the UN Secretary General, Jeffrey Sachs (2005). We agree too that it is not acceptable to hide behind counter-criticisms of corruption and poor governance in Third World countries as the major brake on development and reason for withholding aid. As Sachs (2005: 312) rightly argues, governance and poverty go hand in hand; 'Africa's governance is poor because Africa is poor'. Assistance to the Third World is pitiful and international targets and funds are both fundamental. And yet the lure and notion of a Rostow-style stages of growth 'big push' is equally problematic. As Easterly (2006: 51) latterly argues, it is a fallacy within which aid agencies have remained fixated with the 'fixed objective of stimulating higher growth, although evidence does not support an effect of aid on growth'.

Whether the MDGs serve as a yoke on Third World countries – as shown in Figure 11.1 – will be easier to judge when we reach the target date of 2015, although midway through the allotted time period the signs of promise are relatively few. The drawbacks sketched above, therefore, are not levelled at the ethics and need, but at the process and delivery and at considering how the goal of development can be sustainably achieved. In other words, to what extent are the MDGs tackling the symptoms of (under)development rather than the causes (the structures of power and influence)? And to what extent are they targeted at the right level? Endemic, structural and systematic poverty and disadvantage will need a wholesale shift in the balance of power and opportunity (or freedom from a Senist perspective). It is in this vein that we must consider the efficacy of pro-poor tourism. Critically, to what extent does pro-poor tourism help fulfil the development goals of eradicating absolute poverty, and to what extent does it challenge and transform unequal systems of power? It is to these questions that we turn next.

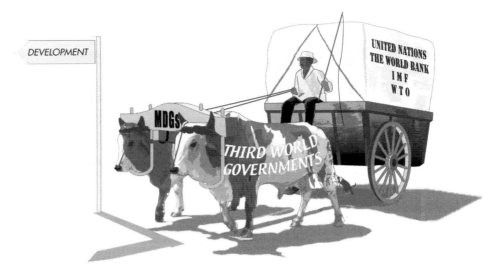

Figure 11.1 The drive towards the MDGs

Source: Brian Rogers, Cartographic Resources Unit, University of Plymouth

Pro-poor tourism

With the emergence of development approaches focused on the poor, by the end of the 1990s development practitioners had begun to think about the possibility of applying poverty elimination goals to tourism. As reviewed in the first part of this book, tourism had traditionally been viewed as a commercial sector, and considered a 'frivolous or elitist industry' (Deloitte & Touche et al. 1999: 11) and not a 'serious' development-related activity. The UK's DFID, however, has supported a number of studies (including case study research) that seek to elaborate a pro-poor tourism approach and promote tourism as a 'legitimate' development activity.

The emergence and development of pro-poor tourism must be understood and analysed within the global development orthodoxy that economic growth is fundamental to pro-poor development. Ultimately, it is an effort to see if tourism can become more pro-poor and if, in the much repeated sound-bite of its proponents, it can meet the need to 'tilt the tourism cake rather than expanding it'.

Pro-poor tourism is a direct intervention in the debate between the advocates and detractors of tourism. Advocates point to four potential advantages of tourism as an economic sector capable of facilitating pro-poor growth: the high potential of linkage, labour intensity, tourism's potential in poor countries and the ability to build tourism on natural and cultural assets. Detractors are critical of the level of leakage, negative impacts on the poor, displacement and socio-cultural disruption (Deloitte & Touche et al. 1999).

Pro-poor tourism is defined by its proponents as tourism

> that generates net benefits for the poor . . . [it] is not a specific product or sector of tourism, but an overall approach. Rather than aiming to expand the size of the sector, pro-poor tourism strategies aim to unlock opportunities – for economic gain, other livelihood benefits, or engagement in decision-making – for the poor.
>
> (Ashley et al. 2001)

Table 11.2 lists the main components of pro-poor tourism. Pro-poor tourism, it is argued, differs from other tourism types that claim to have some developmental value (principally ecotourism, sustainable tourism, community-based tourism, fair trade tourism and other ethically based tourisms) in that it focuses directly on the needs of the poor. As such it is premised upon the sustainable livelihoods approach, now a mainstay in discourses of so-called 'people-centred' approaches to development that seek to build upon the capabilities, assets and activities of the poor (Carney 1998). However, it is also recognised that general support for and integration of pro-poor strategies with the mainstream tourism sector (for example, developing tourist boards or the provision of key infrastructure such as roads or water supplies) is essential to complement pro-poor tourism, and that 'interventions do not always need to be poverty-focused to benefit the poor' (Deloitte & Touche et al. 1999: 56). In this sense there are clear similarities with other forms of new tourism, in that the success of pro-poor tourism relies on an appropriate policy environment that allows such forms of tourism to grow; and thus, pro-poor tourism initiatives may also include so-called 'top-down' government-promoted approaches (Mahony and Van Zyl 2001; Renard et al. 2001). The issues and potential strategies that have been identified in the initial work are shown in Table 11.3.

Like other forms of tourism, as an economic activity pro-poor tourism must compete effectively in the tourism marketplace, and access to the market and ability to compete will depend on a number of fundamental factors common to all forms of tourism: geography (the desirability of the destination and its accessibility), the level of understanding and knowledge of the tourism sector and level of disposable resources, and access to decision-making (much of which is done internationally). Beyond this however, and in order to fulfil the stated goal of pro-poor tourism, two key obstacles are detectable in overcoming existing problems and encouraging new approaches to development through tourism. The first relates to an understanding of power relationships in tourism; the second to the capacity of pro-poor tourism to reduce poverty. As these are fundamental issues we discuss each at some length, with extended discussion of the latter point.

Table 11.2 *Pro-poor tourism*

Participation	Poor people must participate in tourism decisions if their livelihood priorities are to be reflected in the way tourism is developed
Holistic livelihoods	The range of livelihood concerns of the poor – economic, social, environmental, short term and long term – need to be recognised. Focusing simply on cash or jobs is inadequate
Distribution	Promoting pro-poor tourism requires some analysis of the distribution of benefits and costs – and how to influence them
Flexibility	Blueprint approaches are unlikely to maximise benefits to the poor. The pace or scale of development may need to be adapted; appropriate strategies and positive impacts will take time to develop; situations are widely divergent
Commercial realism	Ways to enhance impacts on the poor within the constraints of commercial viability need to be sought
Learning	As much is untested, learning from experience is essential. Pro-poor tourism also needs to draw on lessons from poverty analysis, environmental management, good governance and small enterprise development

Source: Ashley et al. (2000)

Table 11.3 *Pro-poor tourism strategies*

Issues	Strategies
Barriers to participation in the industry for small and micro-enterprises	• Credit training • Policy reform and planning mechanisms • Partnerships with the private sector
Apparent potential for linkages with other sectors rarely realised	• Improved quality, reliability and competitiveness of local products • Change incentives and attitudes in the business sector towards local products • Facilitate linkage process, reduce transaction costs
The *number* of tourism jobs available to the poor limited by their lack of skills, and *quality of employment* in the industry can be low	• Training in hospitality skills, targeted at poor people • Development and implementation of labour standards • Initiatives aiming to reverse the growth of sex tourism and child labour
Tourism can result in lost or reduced access to land and natural resources, or their degradation. Environmental mitigation measures sometimes conflict with, rather than enhance, livelihoods	• Development/reform of planning frameworks • Devolve tenure rights over land or wildlife to local people • Employ appropriate mitigation measures
Tourism can undermine basic services of poor communities by overburdening water and sewage systems; can result in cultural intrusion and loss of privacy; and has led to a growth of the sex industry	• Combine investment in basic services for tourism with provision for local needs • Local codes of conduct and local tourism initiatives • Raise awareness of cultural issues among tourists
Institutions and decision-making processes often structured in such a way that interests of the poor are not taken into account	• Participatory planning processes • Local institutional capacity-building • Cross-sectoral coordination • Incentives and capacity for pro-poor tourism
Many governments see tourism as a means to generate foreign exchange rather than address poverty. Where commitment exists, it can be difficult to identify and implement the wide range of policy reforms needed	• Combine investment in basic services for tourism with provision for local needs • Local codes of conduct and local tourism initiatives • Raise awareness of cultural issues among tourists
The reasons for business to promote the pro-poor agenda are not clear	• Business partnerships promote pro-poor tourism on the business agenda
Pro-poor tourism not currently on the international agenda, yet many international activities have potential to influence pro-poor tourism	• International voluntary codes • Consumer education • Assessment of the impact of European Union tourism regulations on poor producers • 'Good practice' guidelines in pro-poor tourism • Promote pro-poor tourism on international agenda

Source: Deloitte & Touche et al. (1999)

Pro-poor tourism and power

First, the DFID report is critical of the misrepresentation of tourism as an 'industry where foreign interests dominate' (Deloitte & Touche et al. 1999: 7), and it directly challenges critiques of tourism that have emphasised the significance of power relationships. However, the authors of the report themselves point out that 'Developing countries have only a minority share of the international tourism market', refer to the major players in tourism and their 'low level of commitment to any one destination' (Deloitte & Touche 1999: 8, 22) and point to the significance of 'customer requirements' – the fundamental needs and desires of tourists that must be met for tourism to be a going concern. And as subsequent pro-poor tourism analysis suggests, 'Most of the critical decisions that affect the sector are made outside of the country or by a few powerful local interests' (Ashley et al. 2001: 28).

All such elements seem to confirm, rather than question, the significance of the spatially uneven and unequal nature of tourism production, consumption and development. Indeed, this appears to be reflected in claims that many of the disadvantages that have been attributed to tourism are in fact 'characteristics of growth and globalisation' and that tourism is little different from other economic sectors in that 'negative impacts that arise as a result of tourism development would also occur with development in other sectors' (Deloitte & Touche et al. 1999: 11). While, for the arguments advanced in Chapter 3 we would stop short of claiming that tourism is no different from other economic sectors, some of the oft-quoted problems of tourism such as leakage or economic volatility or the structure of social relations and power that involve local elites as well as foreign interests, are indeed characteristic of other economic sectors too and that the 'difference between tourism and other sectors might be perceived rather than real' (Deloitte & Touche et al. 1999: 11).

For these reasons we have argued that tourism must be understood primarily within an analysis of relationships of power (foreign or otherwise) and that tourism must be considered within the same frame of analysis as other economic sectors especially when assessed within a development context. In this sense, it is a 'prism' for understanding broader global issues and relations. Our challenge is to try and understand tourism within the context of power and how this affects local livelihoods, and to measure the veracity of the claims that have been made about the ability of new forms of tourism to assist development. A key question is, given the size and strength of growth in tourism, is new tourism capable of facilitating development?

Can pro-poor tourism reduce poverty?

Given the global 'consensus' on the need to dramatically reduce absolute poverty, the second fundamental question, and acid test, centres on the ability of pro-poor tourism to deliver poverty reduction. As might be expected at the initial stage in the development of a prospective approach that is 'relatively untried and untested' and where a blueprint for implementation does not exist (Ashley et al. 2001: viii), accounts of pro-poor tourism are replete with imponderables (the 'may', 'should', 'likely to') and with acknowledgement that impacts are not easy to measure objectively (Deloitte & Touche et al. 1999). Equally, there has been justifiable caution in extracting too many conclusions from initial case analysis. While this is certainly no good reason for not embarking on pro-poor tourism initiatives, it does perhaps call into question the confidence invested in pro-poor tourism approaches and the manner in which it has been preferenced and mainstreamed into bilateral development assistance.

Like any other form of activity or approach, pro-poor tourism has needed to capture its unique selling point and appeal to the emergent development policy agendas. In this case the UN-inspired MDGs reviewed earlier in this chapter and the consensus surrounding the needs of the poor as the focus for development activities has meant that pro-poor tourism advocates have been careful to define their approach in contrast to other types of tourism, even where the needs of the poor are already being addressed. Inevitably, this has meant trying to rationalise away potential inconsistencies. So while the UN targets are unequivocally focused on the reduction of the poor living in 'absolute poverty', on less than a dollar a day, pro-poor tourism advocates point to the problems of defining poverty and the poor. In terms of an intervention in poverty, the DFID report's initial conclusions were that 'tourism probably does not compare with more direct tools, such as investment in health, education and agriculture. But as a strategy for promoting broad-based growth (also essential for achieving poverty elimination), pro-poor tourism has good potential' and it is the so-called 'not-quite-so-poor' (Deloitte & Touche et al. 1999: ii, 55) that are in a better position to be involved in tourism.

On the basis of a review of supporting pro-poor tourism case studies from Ecuador, Namibia, Nepal, St Lucia, South Africa and Uganda, Ashley et al. (2001: 11) conclude that a 'focus on "the poor" often translates in practice to a focus on local residents or the "community"'. Researchers have highlighted both positive and negative impacts on the livelihoods of the poor through their initial case studies which at the time of analysis were less than ten years old, but show uncertainty about the strength of positive impacts because it is 'difficult to tell how much this is because pro-poor initiatives have fewer negative consequences, or how much it reflects researchers' and respondents' quest for positive results' (Ashley et al. 2001: 23).

More recent research conducted jointly by the World Bank and the UK-based Overseas Development Institute (ODI) held as categorical proof of the success of pro-poor tourism initiatives also needs to be read with a note of caution. As Mitchell and Ashley (2007) comment on research on the tourism value chain in Luang Prabang (Laos):

> benefits earned by the poor are equivalent to approximately 27% of total tourism expenditure within the destination. The poor are defined here as the informal sector and unskilled and semi-skilled workers . . . This result probably lies at one end of the spectrum, as the definition of the poor is broad (including local wholesalers and retailers of food, not just farmers, and micro entrepreneurs), while the tourism expenditure is only the net amount that reaches the destination (measured net of booking fees).
>
> (Mitchell and Ashley 2007: 21)

Demonstrable evidence is therefore expressed as a likelihood that the 'share of the benefits of tourism to the poor at the destination rarely fall below the 10% level' and is based on a 'narrow empirical foundation' (Mitchell and Ashley 2007: 22). In other words, the effect is to be welcomed, but is likely to be small, and may indeed help temper the some-times wildly exaggerated claims that tourism is the answer to addressing development and tackling Third World poverty. Table 11.4 offers a crude representation – it includes aggregate national figures of all tourism receipts only and does not therefore indicate how income is distributed at the local destination level – of what a 10 per cent baseline of tourism receipts that may reach the poor amounts to.

This information impasse continues to undercut a clearer understanding of the pro-poor credentials of new forms of tourism. As the World Bank/ODI review by Mitchell and Ashley concludes:

Table 11.4 *What 10 per cent of a country's tourism receipts might mean to the poorest households*

Country	International tourism receipts (US$ millions) (2005)	Population (millions) (2004)	International tourism receipts per head of population per year (US$)	10% of previous column per year (US$)
Honduras	429	7.0	61.28	6.1
Grenada	92	0.1	1033.71	103.4
Tanzania	621	36.1	17.20	1.7
Malawi	24	12.4	1.94	0.2
South Africa	6282	44.4	141.49	14.1
Laos	119	6.1	19.51	2.0
Samoa	71	0.2	398.88	39.9
Malaysia	8198	23.5	348.85	34.9
India	6121	1065.0	5.75	0.6
Peru	1078	27.5	39.20	3.9

Source: UNWTO (2007) *Compendium of Tourism Statistics: Dated 2001–2005*, Madrid: UNWTO

It is surprising that, despite a huge literature related to tourism and poverty reduction, very few studies can answer the seemingly simple question which lies at the heart of the pro-poor tourism debate 'What share of the financial benefits of tourism are enjoyed by poor people?'

(Mitchell and Ashley 2007: 21)

Neither it seems is there agreement on the non-financial benefits of tourism (such as the accumulation of new skills and capacity, the establishment of community-based organisations and the empowerment and organisational skills that are accumulated by members through this) and the contribution of these to reducing poverty.

As advocates of pro-poor tourism concur, it is certainly not a panacea: 'it is probably true that the small size of . . . initiatives has meant that tourism provides a *minor dent in the national poverty* even when multiplier effects are taken into account' (Ashley et al. 2001: 28, emphasis in original). Pro-poor tourism is not, therefore, a tool for eliminating nor necessarily alleviating absolute poverty, but rather is principally a measure for making some sections of poorer communities 'better-off' and of reducing the vulnerability of poorer groups to shocks (such as hunger). As the DIFD commissioned research suggests:

Economic benefits generated by pro-poor tourism may not reach the poorest – workers and entrepreneurs are unlikely to be from the poorest quintile. Nevertheless, those with sufficient assets to make crafts, sell tea/food, service the accommodation sector, or work in infrastructure may still be poor by either the international definition or by national poverty lines (particularly where tourism occurs in places where the majority are 'poor').

(Deloitte & Touche et al. 1999: 36)

Additionally, it is difficult to draw conclusions from national aggregate data and analysis that 'shows that in most countries with high levels of poverty, tourism is significant (contributing over 2 per cent of GDP or 5 per cent of exports) or growing (aggregate growth of over 50 per cent between 1990 and 1997)' (Deloitte & Touche et al. 1999: 9). While this growth may well present opportunities for the poor, it must also recognise the geographically uneven and unequal nature of tourism development and activity. Tourism, and its potential benefits, will not be distributed evenly and may well display sharp regional and social disparities. The Maoist insurrections in areas of Nepal, for example, emphasise the potential volatility of uneven regional growth. As a prerequisite therefore, there needs to be a 'large number of poor people, in areas with tourism assets' (Deloitte & Touche et al. 1999: 26) and the commitment of government to pro-poor approaches to stand a chance of meeting the goal of pro-poor tourism: poverty reduction.

In contrast, it could be argued that in absolute terms the main effort and resourcing should be deployed in optimising tourism revenues, reining back the negative impacts of the mainstream (mass) tourism industry where it is operating and seeking to develop mainstream poor-focused initiatives (in health, education, shelter and so on). As tentative analysis suggests (Cox 1999), and as conjectured in pro-poor tourism analysis, 'Even if tourism does not directly involve poor people, it may have pro-poor impact, if it improves government revenue and if that revenue is used in pro-poor ways' (Ashley et al. 2001: 39). Indeed, the thesis of this book is to critically question the assertion that new tourism was indisputably superior to mass tourism in its outlook, 'culture' and impact (the last of which was seen as anathema to development). As Mitchell and Faal (2006) suggest, 'out-of-pocket spend per tourist differs little between tourists in 2, 3, 4 and 5 star accommodation . . . up-market tourists are not necessarily better for the local economy' (cited in Mitchell and Ashley 2007: 71). Similarly in the analysis of Luang Prabang referenced above, Ashley (2006) found that while the expenditure of a trip per day incurred by different types of tourists (up-market, mid-range and budget) differed substantially, there was less difference when comparing the direct and indirect income flowing to the poor from the various tourist categories. Of Gambia, the authors conclude:

> This study finds little evidence to support the view that small, up-country tourist product will necessarily be more pro-poor than the current product. The key mechanisms for poverty reduction through tourism are in activities like staffing and supplying large hotels and restaurants and the operation of craft markets, excursions and local transport facilities. If product development focuses upon dispersing the tourism product across the country in a shotgun approach there is a danger that the mechanisms which link the poor to the benefits of tourism will not reach a critical mass for viability and, therefore, will not take place. In this sense, mass tourism could be more accessible to the poor than niche ecotourism operations.
>
> (Quoted in Mitchell and Ashley 2007: 72)

Mitchell and Ashley's (2007) report for the ODI is a full and important analysis of the effectiveness of the mechanisms by which tourism can be used to alleviate poverty. Although the coverage of its analysis is much greater than we can summarise here, Box 11.2 gives a brief synthesis of some of the major findings with regard to these mechanisms and their effects on macro and local economies.

Given that the implementation of most pro-poor tourism initiatives requires subsidies (and this would especially be the case if such initiatives were scaled up and replicated on a broader basis), and in light of the need to allocate funds where the impact on poverty reduction will be optimal, Ashley et al. (2001) conclude that donors, NGOs and indeed governments must ultimately consider how cost effective pro-poor tourism is in reducing poverty in relation to other approaches.

Box 11.2 The story so far: dynamic effects on macro and local economies

There has been a lot of analysis but so far with few clear results. A key finding is that the sustained long-run effects of tourism will depend on a host of other factors.

- Most of the empirical regression analyses conclude that tourism boosts national economic growth, although it can clearly make economic growth more volatile . . . There is sufficient evidence, however, to question an assumption that tourism development necessarily equates to national economic development;
- several analyses suggest that tourism will have a positive impact in the short-run, but that this may not be sustainable into the long-run, principally if there is low productivity growth in tourism;
- tourist development is more appropriate for some countries than others. For instance, small tourist-dependent island states or very poor countries may have fewer options to encouraging tourism . . . compared with a large and diversified economy. On the other hand, inter-sector linkages between the tourism and non-tourism economy should be deeper in more sophisticated economies;
- several studies emphasise the role of tourism as a source of taxation for government, and thus transfers to the poor, where government has a redistributive policy programme. The efficiency, magnitude and pro-poor potential of tourism taxes are a feature of several analyses; and
- the few analyses with a distributional focus suggest that a policy of simply growing tourism as quickly as possible is insufficient to guarantee pro-poor growth. The poor can benefit from tourism at a national level, but this requires thoughtful and deliberate complementary policies – often outside the tourist sector.

Source: Mitchell and Ashley (2007: 67)

Of course, there are advantages to the growth of new tourism (and pro-poor tourism) and there are a number of case studies that appear to demonstrate the real advantages and benefits to the poor that arise from tourism. As advocates of pro-poor tourism understandably maintain, while in absolute terms the scale of benefits may appear small, they can be relatively very significant when viewed from the perspective of the beneficiary groups. Our tentative conclusions, however, remain the same: given the growth in global inequality and poverty, the benefits to the poor of poverty elimination (however real they may be where they occur) will be marginal within the overall context of tourism.

Finally, it is not unreasonable to consider the degree to which pro-poor tourism analysis and promotion is a repackaging of existing initiatives so that they fit within the prevailing development paradigm with an emphasis on poverty reduction, sustainable livelihoods and a focus on the poor and pro-poor growth. The readjustment and representation of existing projects and the structure of new initiatives is a key feature of the development merry-go-round in the competition for limited development funds. In addition, it will prove necessary in the long run to consider the cumulative effect of supporting (through multilateral and bilateral aid programmes focused on economic growth) the expansion of capitalist relations and the manner in which this may undercut 'sustainable livelihoods' and exacerbate, rather than alleviate, poverty.

Beyond the local: pro-poor tourism as a strategic objective

One of the most enduring criticisms of pro-poor tourism is that it is too small scale and localised to make any significant impact (given the size of the tourism industry) and that it is insufficiently (if at all) 'mainstreamed' into local and national development strategies and policies. Trans-boundary approaches are therefore of potential significance in breaking from the mould of 'village-based' approaches and promoting a strategic approach to poverty reduction through tourism; the approach adopted in South East Asia illustrates such an approach.

In late 2005, six countries sharing the Mekong River (Cambodia, China, Laos, Burma (Myanmar), Thailand and Vietnam) unveiled a Joint Tourism Strategy billed as promoting an integrated, prosperous, and equitable vision of the subregion for five to ten years. Through funding and technical assistance from the Asian Development Bank (ADB), the Greater Mekong Subregion (GMS) countries are undertaking a Tourism Project to develop a long-term Tourism Strategy and Five-Year Action Plan. The stated goal and objectives of the Strategy are:

> To develop and promote the Mekong as a single destination, offering a diversity of good quality and high-yielding subregional products that help to distribute the benefits of tourism more widely; add to the tourism development efforts of each GMS country; and contribute primarily to poverty reduction, gender equality and empowerment of women, and sustainable development, while minimizing any adverse impacts.
>
> (ADB 2005: 1)

The strategy is a major component of the overall approach to the region and starts from the premise that tourism is a 'major growth engine for socio-economic development and poverty alleviation, as a promoter of the conservation of natural and cultural heritage, and as a harbinger of peace'. With a population of over 257 million people in the subregion the need to eradicate poverty is paramount, though the distribution of poverty (and therefore the need to spread tourism in the region) varies between and within countries and is generally highest in remote areas. The strategy is clear that tourism development in practice has failed to target these areas of extreme poverty:

> although international tourism to the subregion in 2004 probably generated over $22.2 billion in output, $18.6 billion in income, $2.3 billion in government revenue, and sustained 3.8 million jobs, the bulk of this is concentrated in one country and the distribution of these benefits to those in greatest need – disadvantaged groups such as the poor, women and ethnic communities – minimal.
>
> (ADB 2005: 1)

Figure 11.2 illustrates the causal links identified in the strategy as having the most bearing on the under-performance of tourism as a potential poverty alleviator. According to consultants for the ADB, among other key shortcomings resulting in this imbalanced sector are high leakage levels of tourism expenditures (30 to 40 per cent), weak linkages between tourism and non-tourism economic sectors, and the absence of appropriate so-called enabling policies. The thrust of the strategy is therefore developed on the realisation that a 'more socially responsible, culturally and environmentally sustainable, and economically viable tourism sector' (ADB 2005: 1) requires investing in a range of components including the provision of infrastructure that spreads the benefits of tourism and has greater pro-poor impacts. It also requires control of the negative effects of tourism

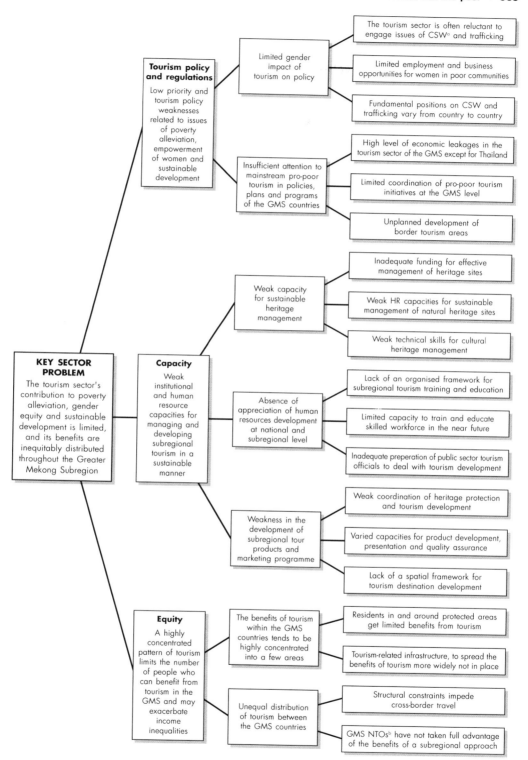

Figure 11.2 Greater Mekong Subregion tourism sector problem tree

Notes: [a] Child sex workers; [b] National Tourism Organisation

Source: Asian Development Bank (2005)

on women, children and ethnic communities (especially through the spread of HIV-AIDS, commercial sex work, child prostitution, and human trafficking) and its negative impact on the image of the subregion.

A core problem noted by the strategy concerns pro-poor tourism development focused largely on preparing single local communities to engage in tourism (through, for example, day trips or overnight stays). It is an approach that tends to downplay the development of sustainable linkages in the local economy. The report concludes that while a 'single community-based approach is certainly necessary, it is not sufficient if tourism is to play anything other than a marginal role in contributing towards the goals of the GMS Cooperation Framework and its roots in the MDGs' (ADB 2005: 65) and overcome the spatial bias between and within countries (for example, urban areas are reported to fare much better than rural areas). As a result the strategy notes:

> much of the current benefits of tourism in the subregion by-pass the vast majority of persons living below the poverty line, especially in the high-poverty incidence provinces in border areas of the subregion (estimated at 38 million people living in 66 provinces excluding Yunnan Province, PRC).
>
> (ADB 2005: 65)

The strategy defines pro-poor tourism as characterised by economic and employment gain (including small/medium-sized enterprise (SME) opportunities) together with other livelihood benefits (such as improved water, sanitation, health and education services), better communications and roads (providing access to markets) and improved protection of natural and cultural resources. It also includes improved opportunity and capacity to participate in decision-making. The key issues for developing subregional pro-poor tourism are moving beyond localised approaches to pro-poor tourism to mainstreaming poverty eradication in the tourism policies, plans and programmes of the participating countries, while attempting to apply the principles and practices of effective pro-poor tourism including the assurance of broad-based participation in the development process (see Box 11.3). (We refer the reader back to our discussion of participation in Chapter 8.)

Of course the degree to which this strategy merely incorporates the current discourse on pro-poor tourism development, rather than addressing the root causes of uneven and unequal development in the subregion is clearly a matter for critical discussion; there is only so much that a sector strategy can do to address imbalances in the distribution of power. If, the strategy's potential was to be initially (and somewhat crudely) measured on the level of resources calculated for the development of such initiatives and the delivery of the pro-poor component (see Box 11.4), the signs are not encouraging. The total allocation amounts to just $13.5 million over five years (with funding sourced from public sector consolidated revenue, special levies, loans, multilateral and bilateral sources).

Migration and remittances

The previous section has looked at pro-poor tourism and strategies in some detail. But do we need to set a discussion of pro-poor tourism in a broader reading of pro-poor 'migrations' more generally, particularly in a context where the benefits of pro-poor tourism are likely to be limited and locally specific (especially so because of the areas deemed desirable to new tourists)? The prevailing economic 'consensus' which favours First World countries and their populations, allowing so many of them the wealth, freedom and 'right' to travel globally, is the same model that can (at least in part) be indicted for the structural poverty of Third World countries. As with the central argument developed

Box 11.3 Pro-poor and equitable tourism development

To ensure that tourism development in the subregion is more pro-poor orientated, the strategy seeks to:

- broaden the pro-poor tourism development approach by continuing to create opportunities to bring tourists to the villages and towns in high poverty incidence provinces; widen the tourism livelihood opportunities available; provide the framework for the local population to engage in businesses and in direct employment opportunities in the tourism area, and create indirect employment opportunities from the production of tourism-related inputs. The program will be implemented using six pilot pro-poor tourism development projects; and
- mainstream all relevant aspects of poverty reduction strategies and activities with tourism strategies and activities based on a cross-sectoral approach at the national, provincial and local levels, coordination of legislative policy and planning, establishment of National Poverty Reduction and Tourism Ministerial Task Force or Implementation Committees, effective coordination with donors and NGOs, targeted training programs for policy makers and planners on tourism and poverty alleviation, and relevant successful tourism case studies shared through site visits, workshops and websites.

Source: Asian Development Bank (2005)

Box 11.4 Principles and practice: the case of the Greater Mekong Subregion

Principles

- A diversity of actions from micro to macro level, including product development and marketing planning, policy and investment.
- Government policies, good governance, and interdepartmental/ministry co-operation is crucial.
- Pro-poor tourism action works well where the wider tourism destination is working well.
- The poverty impact can be greater in remote areas, although tourism may be on a small scale, e.g. Nam Ha in Lao PDR.
- Pro-poor tourism strategies often involve using current markets to develop new products, e.g. rafting in Burma (Myanmar), hot-air ballooning in Cambodia.
- Ensuring products and services are commercially viable, e.g. the successful handicraft development of Artisans d'Angkor.
- Non-financial benefits can reduce vulnerability and bring produce/products to markets, e.g. roads and bridges.
- Both parts of the business equation need addressing – the demand for the product/service from the market and the quality/quantity and diversity of the product/services.

- Poverty alleviation is a long-term investment and expectations must be realistic.
- External funding and assistance may be required.
- The main stakeholders – governments, the private sector, civil society (NGOs), the poor and donors all have essential roles.

Proposals (national pro-poor tourism pilots)

- **Cambodia** (Stung Treng and Rattanakiri provinces in north east, population 175,300): nature-based destination.
- **Lao PDR** (Phongsaly, Houaphan and Xienghuang, population 1,198,000): important ecological karst landscapes with excellent tourism potential.
- **Burma (Myanmar)** (Mount Popa area, population 900,000): important national park that is an old volcano, heavily forested and has significant cultural value.
- **Thailand** (N.E. in the Emerald Triangle, population 3,170,000): rural border area and poorest in Thailand, situated on a high plateau that butts into the Mekong and is in the 'Emerald Triangle'.
- **Vietnam** (N.W. provinces of Son La and Dien Bien, population 1,086,315): highland, forested, ethnically diverse area shares a common border with Lao PDR pro-poor tourism priority area, and in the GMS North East Lao PDR and North West Vietnam Highlands Zone.
- **Yunnan** (Xishuangbanna area of the Golden Quadrangle, population 900,000): main tropical area for China PRC, ethnically diverse and rich in wildlife resources.
- **Guangxi** (Southwest region of the Autonomous region): borders Vietnam, with large concentrations of minority nationalities living in poverty. Pro-poor development may be identified within several counties, all with high-quality karst-based scenic views, adventure-oriented tourist activities, and rich cultural traditions and historic resources.

Source: Asian Development Bank (2005)

in the first part of the book, it is necessary to understand tourism and economic migration within a context, and as counterveiling forces, of a broader reading of globalisation and of uneven and unequal development. We have argued, for example, that the right to travel is unevenly shared among the First and Third Worlds and between different social groups within these worlds. In Figure 11.3 we have represented the statistics of tourist arrivals to a number of Third World regions along with data on international migration from those countries; it is clear that the growth rates of both are not dissimilar. We would not argue that there is a correlation nor a causal relationship between these two sets of data, but it is clear that the two phenomena are not completely unrelated.

Tourism is a form of migration, normally short term in duration, temporary and often seasonal. Internationally, the direction of movement has been predominantly from rich countries to poor countries – see Table 2.3 – though as Chapter 4 argued domestic tourism and so-called 'south–south' tourism is on the increase as the middle classes in Third World countries begin to expand. The contra-flow to those following this migration pattern is the flow of economic migrants from Third to First World countries. Of course, describing this migration flow as in the 'opposite' direction is too simple. The flows of economic migrants

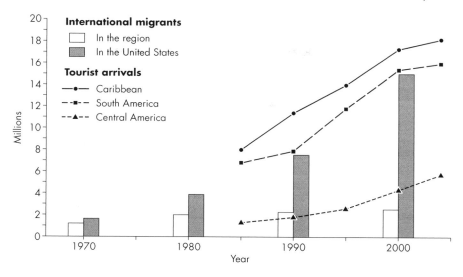

Figure 11.3 Tourist arrivals to and migrants from Latin American and Caribbean countries

Sources: Tourist arrivals: WTO (2005) *Tourism Market Trends, 2005* (data as collected by WTO November 2005). Migration data: Economic Commission on Latin America and the Caribbean (ECLAC) notes, May 2002 (no. 22).

are complex and are not entirely from the Third to First Worlds. Just as with the growth of south–south tourism, the World Bank (2006) estimates that somewhere between 30 and 45 per cent of total remittances received by developing countries is a product of 'south–south' migration. There are clearly many factors which have driven both the rise in international tourism and the increasing incidence of emigration of all types from Third World countries. There is no straightforward balance between 'tourists in' and 'locals out', and their durations of stay are normally very different: tourists in general are short-term migrants; many of the emigrants from Third World countries, on the other hand, are long term. These factors do not mean, however, that there is no association between the two flows.

It is clear that both forms of migration offer sources of income for poorer countries. The earnings of many migrants 'leak out' of the First World as savings and are sent home to those countries from which they have migrated. It is not surprising therefore that supranational organisations, donor agencies and commentators on international development have become increasingly drawn to the comparisons between international development assistance (aid) and remittances. Through formal banking channels alone (and thus not counting the return of money that is not formally banked) the World Bank calculates that remittances amounted to a staggering US$167 billion in 2005; and in the World Bank's view informal remittances might amount to an additional 50 per cent (or more) of that total. By way of comparison, the OECD estimated that official development assistance (crudely the total global aid budget) amounted to below US$107 billion (from OECD fact sheets) in 2005; and UNWTO data for 2004 give total global tourism receipts to all countries in Africa, Latin America, the Caribbean, and most of South and South East Asia (not including Australia, New Zealand, Japan, Singapore, Taiwan and South Korea) as US$135 billion.

As the World Bank argues, the 'growing importance of remittances as a source of foreign exchange is reflected in the fact that remittance growth has outpaced private capital flows and official development assistance (ODA) over the last decade', and in countries such as Morocco remittances 'are larger than tourism receipts' (World Bank 2006: 88). Similarly, in some countries of Latin America and the Caribbean, remittances and the tourism industry rival each other as major sources of inward investment, and Tremlett (2006) argues that in most Latin American countries the money sent home in tiny individual sums adds up to more than both foreign aid and foreign direct investment put together. A United Nations Economic Commission for Latin America and the Caribbean (2003) report has noted that an 'explosive rise in remittances' from Latin American and Caribbean emigrants amounts to the region's second largest source of external financing after foreign direct investment. Despite the estimated funds sent home by the almost 20 million Latin Americans and Caribbean individuals living overseas barely exceeding US$200 per month, the report notes that about US$25 billion in remittances entered the region in 2002. Funds that emigrants send back home have risen by an annual average of 12.4 per cent since the early 1980s, the highest growth rate for the different regions around the world; the region received 31.3 per cent of all remittances flowing into developing countries.

Similarly, a United Nations Economic and Social Commission for Asia and the Pacific (2006) survey places great emphasis on the role of migrants' remittances as a major contribution to development. 'Over the last two decades or more, remittances by migrants or temporary workers in foreign countries to their countries of origin have made a major socio-economic contribution in many countries,' the survey states. 'At the social level, remittances have added to family incomes and boosted consumption. At the national level, remittances have reduced, in some cases substantially, the current account deficit of many developing countries' (UNESCAP 2006).

There are both macro- and micro-economic advantages of these targeted flows. The World Bank suggests that remittances are generally stable and in macro-economic terms may be counter-cyclical, help improve a country's creditworthiness and help raise external financing, though there is inconclusive evidence on the effect of remittances on long-term economic growth. Notwithstanding the welfare of individual migrants themselves, and reducing this discussion to one merely of 'income' and flows of money (an important consideration and one that reflects Sen's advocacy of broader and non-traditional interpretations of what development means in practice, as discussed in Chapter 2), remittances also have important local (and micro) economic benefits. The World Bank summarises the effect of remittances on households and poverty: remittances can

> reduce poverty . . . help smooth household consumption by responding positively to adverse shocks (for example, crop failure, job loss, or a health crisis); ease working capital constraints on farms and small-scale entrepreneurs; lead to increased household expenditure in areas considered important for development, particularly education, entrepreneurship, and health.
>
> (World Bank 2006: 117)

Unsurprisingly, donor aid agencies have also recognised the potential benefit of remittances in 'targeted' local developments which are controlled, for the most part, by the migrants themselves (and not therefore requiring monolithic development bureaucracies to hand out and administer aid). The UK's DFID (2006), for example, notes: 'money sent back by migrants plays an important part in sustaining the local economy', and that migration is a legitimate component of international development policy. But remittances do not find favour with all. Jeffrey Sachs, for example, launches an acerbic assault on remittances. The 'U.S. government also tried to argue incredibly,' Sachs asserts, 'that remittances of foreign workers in the United States back to their home country should somehow count as a form of aid.' He rounds on such an assertion as 'ridiculous' and declares that the 'remittances are returns for work. They are no more a form of aid than are remittances of U.S. profits from Mexico a form of aid from Mexico to the United States' (J. Sachs 2005: 303). In our view, while remittances should not be discounted against aid budgets, they are of undoubted importance in helping to redistribute economic benefits. Moreover, remittances from migration call into question the ethical, social and economic justification for increasingly tight barriers to Third World migration to the First World, while simultaneously seeking to roll back the restrictions on movement in the opposite direction.

For our purposes within a critical discussion of pro-poor tourism, juxtaposing migration and tourism helps emphasise the framework of global unevenness and inequality in which all forms of people movement must be contextualised and are conditioned. Equally, and arguably, it is a reminder that pro-poor tourism needs to be read as a component of pro-poor movements more generally, and its relative merits and effectiveness measured against and compared to other forms of movement. At the same time that the barriers to the movement of tourists are increasingly forced down and the development potential of tourism talked up, the barriers to emigration (temporary or not) are made increasingly tougher. In June 2006, for example, the European Union announced a 35 per cent increase in its aid budget (a total of 22 billion euros between 2008 and 2013) for African, Caribbean and Pacific countries as part of efforts to stem migration into the 'wealthy bloc'. 'Managing migration for the benefit of development is a new priority of EU development assistance', the EU's Executive Commission commented (report by Reuters).

Of course in practice global development assistance is likely to be most effective through a diversified set of initiatives (including increases in and the more intelligent use of aid, debt cancellation, fairer trade, migration and so on). However, on the basis of a somewhat crude comparison of the most significant and asymmetrical flow of people (tourists and migrants) from First to Third World and vice versa, one might conclude that the most efficient and targeted form of 'pro-poor tourism' (and one that is far less cyclical and vulnerable to disasters and fashion) is extended stays of Third World migrants overseas. Given their relative effectiveness, in time it may be judged that there is a strange irony of encouraging new, pro-poor, forms of tourism in Third World countries through aid budgets, while simultaneously blocking labour migration to First World economies.

Human security

This section critically assesses the potential of pro-poor tourism from a rather different angle, that of human security. As Chapter 2 discussed, human security has emerged as an additional facet of development (complementing approaches that focus on basic human rights and human development). The degree to which this is just further evidence of the growing gulf between reality (the challenges faced by those in poverty) and the intellectualisation of development (characterised by overlapping, sometimes contradictory or barely distinguishable theories and concepts) is a matter of debate. What human security does provide, however, is a definite spotlight on the individual and acknowledges that development is impossible within a context of violence.

There are two immediate reflections when holding up the human security concept to the practice of tourism. First, the development of pro-poor tourism is much more difficult, and virtually impossible, in countries where security is low (including those under violent conflict and very poor Third World countries). Equally the human security approach is a timely reminder for the tourism sector that just as sector growth can provide benefits, downturns in the words of Amartya Sen are also felt very unequally (see Box 11.5). Second, there are issues that sit beyond the control of tourism – those so-called 'acts of

---

Box 11.5 Human development, human rights, human security: what's the difference?

The human development approach . . . has helped to shift the focus of development attention away from an overarching concentration on the growth of inanimate objects of convenience, such as commodities . . . to the quality and richness of human lives . . . There is of course no basic contradiction between the focus of human security and the subject matter of the human development approach. . . . But the emphasis and priorities are quite different in the cautious perspective of human security from those typically found in the relatively sanguine and upward oriented literature of the human focus of development approaches . . . which tend to concentrate on 'growth and equity' . . . In contrast, human security requires that serious attention be paid to 'downturns with security' since downturns may inescapably occur from time to time . . . There is much economic evidence that even if people rise together as the process of economic expansion proceeds, when they fall, they tend to fall very divided . . . The economic case merely illustrates a general contrast between the two perspectives of *expansion with equity* and *downturn with security*.

There is a similar complementarity between the concepts of human rights and human security . . . The basically normative nature of the concept of human rights leaves open to question which particular freedoms are crucial enough to count as human rights that society should acknowledge, safeguard and promote. This is where human security can make a significant contribution by identifying the importance of freedom from basic insecurities.

Source: Commission on Human Security (2003: 8–9)

God' – that are manifested in natural disasters including earthquakes, fires, hurricanes and typhoons, mudslides and tidal waves. Of course, the impacts of such events stretch well beyond tourism, but it is nonetheless instructive to look at the insecurities arising from the co-presence of new tourism and such potential events and most significantly what happens in the aftermath of such events. In this section we present a number of cases that reflect directly on the role of tourism and the nature of development within a framework of insecurity and how these insecurities can be addressed.

Surviving the second tsunami

On 26 December 2004 an Indian Ocean undersea earthquake resulted in a tsunami unparalleled in recorded history. Around 230,000 people were tragically killed with five countries most affected: India, Indonesia, Maldives, Sri Lanka and Thailand. In some important respects, this cataclysmic event and its aftermath embody many of the relatively 'silent' battles that occur globally through the expansion and spread of the tourism industry, and undercut the security of those in the wake of such expansion.

One of the most fundamental problems for the survivors of the tsunami was returning to, and reclaiming, what they considered rightly theirs: the land upon which their homes had formerly stood. While land is a critical and finite resource, upon which tourism in particular heavily relies, land issues, including access to it and its ownership, are relatively undiscussed in the tourism literature. We have introduced this theme elsewhere as land and property often lie at the heart of conflicts, but it is of such fundamental significance to the well-being and livelihoods of communities that it is worth spending a few moments re-emphasising. The key question often revolves around how the right to ownership of, or access to, housing, land and property is expressed and shared – a point that was introduced through the discussion of de Soto's work in Chapter 2. In many Third World countries there is so-called 'legal pluralism' characterised by the coexistence of a mix of state, customary or traditional and religious law, reflected in different forms of and approaches to tenure (a general term describing the access to and ownership of land, property and housing). In other words, not all tenure rights are written and lodged in formal land administration institutions and agencies; they may be oral rights that are passed from one generation to the next. Moreover, the body of law and land tenure arrangements may differ significantly from place to place within a country and are dynamic (changing over time) (Food and Agriculture Organisation (FAO) 2002).

Dramatic events and disasters exacerbate already relatively insecure contexts and dramatically increase insecurity. In the context of the tsunami, the Asian Coalition of Housing Rights (ACHR) argues that the result was 'to tear open and aggravate all these already difficult issues of land: who determines how it's supposed to be used and who has the right to use it' (ACHR 2006b). Box 11.6 briefly outlines the problems of loss of land and livelihood of fishermen and their families in Sri Lanka as a result of the tsunami, due largely to the pressures exerted by the tourism industry.

Box 11.6 A tsunami exclusion zone for some, but not for others

On the day after Christmas 2004 . . . an outpouring of charitable giving from individuals and aid from governments around the world followed the news that the tsunami had claimed more than 200,000 lives; one of history's deadliest disasters inspired historic sympathy and generosity.

Three years later, it seems worth noting how some of that international aid is being put to use in Sri Lanka, one of the nations hardest hit in 2004: to facilitate Sir Lanka's plans to turn itself into a 'world-class tourist destination,' in the Sri Lankan government's words.

Shortly after the tsunami, that government announced an up-to-200-meter exclusion zone along the Sri Lankan coast. Ostensibly, this zone was meant as a buffer against future tsunamis. In practice, it meant the displacement of and loss of livelihood for many thousands of fishermen and their families who have lived for generations beside the ocean where they plied their trade. They were among those most devastated by the tsunami – in addition to the many killed, most who survived lost homes, fishing vessels and equipment.

And not long after the Sri Lankan government created the exclusion zone, it began granting exceptions to the tourism industry.

Now, with fishermen relocated inland, Sri Lanka is teaming with private capital to abet a hotel and resort building boom along the coast (and within the 'exclusion zone'). International aid is helping to make this possible – both directly, through grants to the Sri Lankan government, and indirectly, through the charitable work of NGOs that have stepped in to provide the basic services and survivor aid that the government has neglected.

Source: Rather (2007)

In situations such as the tsunami, and areas in conflict, the significance of land becomes especially charged, and land and property grabbing and the destruction and falsification of land titles are common. Put simply, such events provide unparalleled opportunities to secure land and in the context of tourism provide for the expansion of tourist facilities. Unsurprisingly, land rapidly emerged as one of the most pressing and complicated issues in tsunami reconstruction and rehabilitation. Beyond upholding human rights there are more broad-ranging reasons for effectively tackling secure access to land and property. Access to land and property is critical to freedom from basic insecurity, though it is often overlooked in both post-conflict and disaster situations.[3]

Equally, such events also provide an important opportunity for local people to resist expulsion and address broader forces of economic, social and cultural change that would otherwise compromise their human security, and to address historical inequalities and reassert the right to use and own land and cultural identity. As the ACHR observes of Thailand's southern fishing communities, the 'waves were not the only thing that threatened the ways of life of Asia's coastal fishing communities. Tourism, development, market forces and globalisation have all taken their toll on the delicate and richly varied indigenous cultures' (ACHR 2006a). For the ACHR, it is a case of communities 'surviving the second tsunami'.

As the ACHR reported, in the six tsunami-hit provinces of Thailand it was recorded that of the 428 coastal villages affected, 89 had 'extremely shaky land status' and over a third of these had become 'embroiled in very nasty land conflicts soon after the tsunami' (ACHR 2006a: 2). These represent an intensification of the quiet battles over Thailand's coastline that have bubbled away for years, in which local communities have been dispossessed of their land by tourist resorts, shrimp farms and tin mines (ACHR 2006a). As ACHR summarises: 'Despite having lived there for decades or even centuries, many of these villages remained messy patchworks of uncertain land status, overlapping ownership claims and tenurial vulnerability: evictions just waiting to happen' (ACHR 2006a: 2).

However, there are some positive outcomes from this tragic event. As Indian activist Kirtee Shah suggests, disasters are an opportunity to rebuild communities, to strengthen people, to take a step forward (ACHR 2006c). It is yet one more reminder that given the right conditions poor communities are able to address their own needs and to reassert their rights to land and property (ACHR 2006c). They do not conform to the stereotype concretised by the world's media in the immediate aftermath of disasters: helpless individuals reliant on the mercy of the West. It is also a reminder that the universal solutions so often advocated and imposed on poor communities regarding land and property (centred on the need for formal legal titles) are not necessarily the best, nor the only way forward, as some have advocated. The United States Agency for International Development (USAID) concludes that the 'tendency over the years has been to treat land as a technical issue that requires a technical solution or simple legal solution. Land related initiatives have therefore been characterised by rigidity and a general lack of flexibility.' Rather the 'complex and politically sensitive nature of land conflict requires a strategic, creative and flexible approach' (USAID 2005).

In Thailand, for example, some communities have pioneered their own solutions by returning to their land and fighting for new collective land tenure which makes the community the 'unit of security' and collective strength through 'collective leases, collective title and collective user rights, which cannot be sold off individually, bit by bit' (ACHR 2006a: 3). Alternatively, some communities, such as with the Tung Wah sea gypsy fisher folk village, have negotiated a land-sharing deal whereby the community returns to part of its land safeguarded through a long-term communal lease, with a remaining portion returned to the local administration (ACHR 2006a). Finally, as Box 11.7 suggests, such

Box 11.7 Using the crisis of the tsunami to spearhead a whole island's cultural revival

Over the past two decades, Koh Lanta island, which is just two short ferry rides from the mainland, has been developing into another tourist destination for the beach-seeking set, with resorts, guesthouses, restaurants, and boutiques springing up along the island's sandy western beaches. The eastern side of the island, however, retains the sleepy, local ambience of the island's long history before the tourist boom.

When the tsunami struck, the waves left the touristy side untouched, and struck parts of this quieter side of the island . . . the tsunami was the catalyst that opened up issues which had been brewing since long before the tsunami: the cultural marginalisation and precarious land tenure of Lanta's original communities, environmental degradation, rampant commercialisation and increasing control of the island's development by outside commercial interests and government tourism development schemes.

The post-tsunami program covers issues of land, housing, development, tourism, livelihood, natural resource management and environment, and gives a great deal of freedom to local groups to develop their own projects. Environmental management on a small island like Lanta is a delicate issue, so some projects are focusing on solid waste management, creating garbage-free communities and organic farming. But the overwhelming majority of projects focus on reviving cultural practices of Koh Lanta's indigenous Muslim, sea Gypsy, Buddhist and Chinese communities, and involve master craftspeople, artists, builders, dancers, musicians, fisherman, boat-builders . . . and many others. All these activities are creating new strength among the communities and boosting their confidence in dealing with local authorities, which earlier were reluctant to go along.

Source: Asian Coalition of Housing Rights (2006a: 6)

disasters can also initiate a wholesale review of the development trajectory of entire communities; an opportunity to take stock and negotiate a different future. The 'unusual kind of participatory, post-disaster planning process' described in Box 11.7 offers a glimpse of what pro-people and pro-poor action might achieve when there is an 'opportunity to plunge right into some deeper development issues' (ACHR 2006a: 6).

Hurricane Mitch and disaster tourism

While the 2004 Boxing Day tsunami illustrates the land disputes and dispossessions that have arisen in the aftermath of the disaster, the effect of Hurricane Mitch on Central America in 1998 highlights the strong relationship between poverty and the vulnerability of the lives of rural populations in particular to extreme natural events. The relationship between these high levels of vulnerability and the neoliberal economic policies forced upon the governments of the region by the international financial institutions was also exposed (Mowforth 2001). As we shall see, however, land disputes were also not uncommon.

Hurricane Mitch hit Central America with devastating effect in late October and early November 1998. In just five days it deposited more than the region's annual average rainfall on Nicaragua, Honduras, El Salvador and Guatemala. The flooding and landslides caused death, injury and illness and the destruction of houses, crops, roads, bridges and other infrastructure on a large scale. Mitch exposed the vulnerability of a large proportion of the population in terms of its food security, flimsy housing, family cohesion and poor health provision.

The event gained much international publicity with images of the floods, landslides, destruction and general chaos flashed across television screens around the world for several weeks. The tourist money on which many Central Americans depended, especially on the Caribbean coast of Honduras where tourism had previously provided a substantial proportion of the population with their livelihoods, dried up. Six months later the tourists were still absent and the industry was still failing to recover. In response to the dire situation, the Honduran Institute of Tourism (IHT by its Spanish initials) teamed up with US and Honduran tour operators to promote a kind of disaster tourism under the banner of 'Help the Environment Recover from Hurricane Mitch' (Honduran Institute of Tourism 1999). Details of the scheme (named the Trujillo Project) are given in Box 11.8.

Slowly the northern coast of Honduras managed to recover, and by 2007 tourism was rivalling the production of fruit crops and palm oil in economic importance. That is not to say that everyone involved in the industry has recovered – many individuals were prompted to migrate in search of better opportunities, and some households depend very heavily on the remittances sent back by their family members.

Land disputes and claims have also become more evident since Hurricane Mitch, partly as a result of the hurricane and the 'opportunities' it created for land grabs, but also because of legislation passed shortly before Mitch. The issue of land titles and ownership became

---

Box 11.8 The Trujillo Project

Spend your vacation on a Tropical Beach – And Help the Environment Recover from Hurricane Mitch

Imagine taking a vacation on a remote tropical beach, and returning with the feeling that you had helped endangered species and a Honduran National Park recover from the ravages of Hurricane Mitch.

Well you can. The Honduras Ministry of Transport has partnered with the conservation organisation FUCAGUA to sponsor a six day working vacation program that mixes tropical relaxation and exploration with the restoration and recovery of key natural resources. Participants will help replant the forest of Capiro Calentura National Park, which suffered extensive wind damage as Hurricane Mitch (the strongest Atlantic hurricane in over 200 years) barrelled on shore. They will also help clear the famous beaches around Trujillo of debris (mostly driftwood) driven ashore by the storm's giant waves and restore nesting beaches for turtles . . .

Participants in the project will receive discounted airfare from Houston and Miami to San Pedro Sula (Honduras) where they will be met and helped through customs . . . Each day's activities will include a mix of recovery work (tree planting, beach clean up, sea turtle nest monitoring) and play (beach time, walking tours, museums, boat tours, shopping).

Source: Honduran Institute of Tourism (1999)

front-page news in Honduras in 1998 when the National Congress proposed reforming the constitution to permit foreigners to buy land for tourism development projects. Article 107 of the constitution previously prohibited foreigners from acquiring land less than 40 km from the coast or on any of the Honduran islands. Since that time, the Honduran government has made every effort to attract foreign investment in the tourism industry.

The indigenous Garífuna, who are scattered along the most popular tourist areas on the Caribbean coast and the Bay Islands, however, have ancestral titles to their land, some of the documents for which are not recognised by the Honduran government (Mills 1998). Commercial interests of many industries have previously threatened the Garífuna land rights, but they have fought hard to gain various degrees of recognition for their communal ownership of land, which is, at least theoretically, free of market forces. As Thorne explains, 'These titles grant the community rights to a given area in perpetuity. They may not sell the land or transfer its ownership outside the community' (Thorne 2004: 24). These Garífuna victories with respect to land rights 'cannot be understood without taking into account the ways in which this community has successfully politicised and linked identity and land' (Thorne 2004: 23). Despite these measures of success in defence of their land, the tourism boom in recent years 'has greatly amplified the intensity and dimensions of the threat' (Thorne 2004: 24), and some lands have been sold on through fear of loss of land if they refuse to sell and/or through bribery. The threats, however, have given rise to a large-scale grassroots movement of protection, although the struggle is certainly not finished – the area is subject to powerful commercial, political and military influences which are not generally concerned about the tactics they have to employ to achieve their ends.

Terrorism

Thus far we have examined the issue of security in tourism from the perspective of local people affected by the industry, a perspective we consider to be perfectly justified in any analysis of the alleviation of poverty. It is not the tourists who are poor. But the link between human security and the tourist is of course a clear and immediate one, as illustrated by the Luxor massacre in 1997 (refer back to Figure 10.1).

In November 1997, Islamic terrorists massacred 57 mostly North American tourists at the Temple of Hatshepsut at Luxor in Egypt. The effect on international tourism to Egypt was instantaneous. Governments issued no-go warnings and tour operators pulled tour groups out and switched destinations for tours and tour groups thereafter. Hotel occupancy rates all over Egypt plummeted in the November–May winter peak season that followed. For 1998, international visitor arrivals decreased by 12.8 per cent over 1997 and international tourism receipts decreased by 31.2 per cent over 1997.

Figure 10.1 shows the immediacy of the effect in the downturn in both arrivals and receipts in 1998. What is also clear, however, is the rapid return to normality. For 1999, the number of international tourists arriving appears as if nothing had happened at the end of 1997 – the graph of growth could have been continued from earlier years to give the 1999 figure that actually occurred. Receipts also recovered rapidly, but not quite to the same level as arrivals. In the absence of appropriate data with greater explanatory power, it might be hypothesised that this slightly slower recovery of receipts (compared with arrivals) might have been due to visitors spending a shorter time in the country, this perhaps reflecting a still evident lack of confidence.

Despite the fact that the immediate effect was disastrous for many local people in Luxor whose incomes had depended on catering for tourists, it is clear that the effect was not much greater than a short-term 'blip' in the inevitable growth of international tourism to

Egypt. Perhaps it would not be wise to underestimate the impacts on local lives of such an event, but Richardson (1998) pointed out one beneficial result of this attack:

> An incidental but pleasing fall-out from the massacre is that prices for flights and hotels have been slashed and Egyptians – for the first time in many cases – can afford to go on holiday in their own country, acquaint themselves with the astonishing civilisation of their ancestors.
>
> (Richardson 1998)

This may suggest a form of tourism with less malign effects on climate change than one that is dependent on international air travel, but the tourist perspective also suggests a blindness to local hardship.

It is worth pointing out that the rapid recovery in this case was a function not only of the short-term memories of tourists, but also of aggressive security operations by the Egyptian authorities and aggressive marketing by the tour industry, local, national and international. And perhaps it is a lesson to us all that rapid recovery in such instances is unlikely without such policies, practices and efforts. Memory loss, however, certainly seems to have had some influence on the recovery as Jean-Claude Baumgarten (President of the World Travel & Tourism Council) remarked: 'Look at Egypt and the Luxor incident, which had a terrible impact on Egypt. Two years later, the market forgot about it. We are getting used to living in an uncertain environment' (Travel Impact Newswire 2003).

It might be hypothesised that the effect of the Luxor massacre would be generalised across the Middle East region. Conversely, it might be considered that any generalised effect of the Luxor massacre on the whole Middle East region would be counterbalanced (at least in part) by a substitution effect which suggests that many of the visitors who would have gone to Egypt but who changed their minds as a result of the massacre went elsewhere in the Middle East region instead.

Although the global tourism industry did not exactly collapse after the terrorist attacks in New York and Washington in September 2001, it certainly suffered major setbacks for a short time. But there is a crucial difference between the insecurity of violent events (natural or man-made) and the insecurity of poverty and vulnerability. The former tend to have only a short-term effect on tourism (Mowforth 2003) while the latter serve as either a long-term deterrent to tourism or a short-term attractor to the modern adventurous tourist who wants to see poverty.

As described, these cases show the effects of insecurity on tourism. At its most fundamental, the aim of pro-poor tourism must be to counter the insecurity, whether it comes from extreme natural events or human violence. And we have already alluded to pro-poor tourism's mechanism of 'tilting the cake' or spreading wealth rather than creating extra wealth or new economic growth. Nevertheless, pro-poor tourism does not aim to alleviate poverty by promoting alternatives to the prevailing model of economic growth and free trade. Yet to the critics of this prevailing neoliberal dogma it is precisely this model of free trade and economic growth which creates and/or exacerbates poverty and vulnerability to the shocks caused by extreme natural or human events. The Centro de Intercambio y Solidaridad (CIS 2001) (Exchange and Solidarity Centre in San Salvador) refers to this as 'the permanent emergency generated by the neoliberal economic model'; Helen Collinson (1999) points out the hypocrisy of promoting 'respect for civil and political rights on the one hand while imposing market-oriented policies which undermine the poor majority's socio-economic rights on the other', and Mowforth (2001) considers that 'while meteorological phenomena are unable to distinguish between rich and poor, neoliberal economic policies reduce the ability of the poor to protect themselves against these phenomena'. Many other critics question the contradiction between support for the prevailing economic model, that 'series of gyrations and chaotic experiments' (Harvey 2005: 13), and attempts to alleviate poverty and vulnerability.

Conclusion

This chapter has tried to set a discussion of new forms of tourism within broader considerations of the poverty-reducing potential of tourism, an issue that we have tackled at various points throughout this book. Initially we discussed the meaning of poverty and the poor, for it is futile to discuss tourism as a poverty-busting mechanism unless it is clear who it is benefiting. In order to situate the discussion within current global frameworks of poverty reduction, the centrepiece UN MDGs were presented and their significance discussed.

The main part of the chapter discussed the current thinking and practice on pro-poor tourism itself. The conclusion of pro-poor analysis is that the opportunities will vary from place to place (country to country) and between social groups (of which the poor are one), and will be fashioned by the varied contingent factors in each place. Opportunities will come in one or more of three forms or pathways (direct and indirect benefits of tourism to the poor, and as the result of dynamic effects on the economy more broadly). Current analysis concludes that given the right circumstances, tourism 'can demonstrably benefit poor people' (though the ODI report on *Pathways to Prosperity* (Mitchell and Ashley 2007) is less explicit if this equates to reducing 'poverty'). This is good news for the poor, but the evidence remains extraordinarily patchy despite the existence of research facilities focused on pro-poor tourism that have been operational since the year 2000.

In part this is a reflection of the historic scepticism with which overseas development agencies and donors regard tourism as a 'legitimate' development sector (and a general unwillingness to fund tourism-related initiatives, including research and analysis), and in part a reflection of how notoriously difficult it is to accurately measure financial and more importantly broader livelihood benefits that flow from tourism. Even from observation it is obvious that tourism has a propulsive effect on local economies, but it is much more problematic understanding how this then translates and filters through to the poor. In the absence of hard data, the inevitable temptation has been to fall back on reasoned conjecture regarding linkage and leakage.

The weight of evidence for the 'pro-poor' tourism potential is arguably no greater than, say, the pro-poor potential of the manufacture of sports equipment for leading global brands, the production of electrical and computing equipment, or the fairly traded bananas, cocoa or coffee. Why is it then that the hope invested in tourism's ability to be pro-poor and a weapon in the arsenal of poverty reduction is so much greater than other sectors? These also have value chains, provide local employment and have a variety of linkages, and some would argue have powered the development of the South East Asian tiger economies. As we suggested earlier, part of the answer is simply that tourism in a Third World context is unique in bringing together consumers and producers, and in some circumstances exposes tourists to the 'poor'. Of course the veracity of tourism's pro-poor credentials depends much on the broader public policy arena (the ability to use economic sectors to support poverty reduction) rather than tourism per se. It might be argued therefore that critiques of mass tourism should be re-evaluated as the failure of public policy rather than as a failure of tourism.

The chapter also extended the argument of uneven and unequal development to a comparison of tourism and migration. It was argued that this was of considerable importance in that while the barriers to tourist movements have generally gone down, the restrictions to economic migration from Third to First World countries have tightened considerably. Yet it is the latter, in terms of revenue generation and the return of remittances, that significantly outstrips the former. It would therefore be misguided to assume that there is no association, however weak, between these contrasting flows. The growing numbers of visitors to Third World countries are a barometer of increasing levels of wealth in the rich First World, while the flows of emigrants from Third to First worlds reflect the increasing levels of poverty and inequality in those countries, and globally.

The final section of the chapter assessed tourism (and its pro-poor qualities) from a rather different angle, one of human security, and we make the link between poverty, insecurity and vulnerability. This perspective shifts attention from the singular spotlight of economic poverty and focuses it instead on the basic human right to live in a secure environment, free from the extremes of human violence and with a degree of resilience to the effects of extreme natural events. The creation of the conditions for such security is as much an aim of poverty alleviation as are the efforts to lift people above given financial thresholds such as a dollar a day. Again, however, we find that the prevailing global pressures of economic growth and free trade do little to assist in this direction and the new forms of tourism can also play only a subsidiary role in this respect.

Tourism undoubtedly has a role to play in the global fight against poverty and the promotion of justice, but as this chapter suggests, this will vary in its extent and configuration from place to place. But casting tourism as a Cinderella sector, as some commentators have been tempted to do, is as misplaced as it is dangerous. Equally, equating new tourism as the sole source of such benefits and denying any transformative value to mass tourism would in our view be a grave misjudgement. As this chapter emphasises, much depends on the commitment and ability of Third World governments to harness tourism and use broader public policy to redistribute its benefits.

12 Conclusion

This book has sought to provide a broad debate as a means of critically assessing the important developments in contemporary tourism. It has attempted to explore the ways in which new forms of tourism are interrelated with notions of sustainability and development and are reflected in the Third World. This approach has enabled us to view and analyse tourism in a broad context and to stress that tourism is not the sole focus of our discussion. Instead, we set out to explore the way in which socio-cultural, economic and political processes operate on and through tourism; in other words, tourism is a mirror of these wider processes and a means of critically engaging in debates centred upon notions of 'development'. In our view, it was inappropriate to provide further analysis and evidence of the now well-documented environmental, economic and socio-cultural impacts of tourism.

Key themes and key words

In order to achieve this broader analysis, a number of key themes (uneven and unequal development, relationships of power and globalisation) and key words (new tourism, the Third World and sustainability) were identified. Table 12.1 summarises these positions and seeks to illustrate the relationships between the key words and key themes by drawing some examples from and questions raised by the book. Clearly the relationships summarised in the cells of the table are not exclusive and there is considerable overlap between both the key words and key themes.

Through this intersection, or dialectic, it was argued, globalisation, sustainability and development have emerged as the most significant contemporary processes and ideas. It has also been argued throughout the book that globalisation, sustainability and development are rife with competing meanings and interpretations: to a backpacker, the Third World may represent exotic cultures and environments and an endless world of adventure; to mass tourists in the USA, Europe or the Pacific Rim, the notion of the Third World may represent a threat, and a place either to avoid when holidaying or where a holiday needs to be carefully controlled, through all-inclusive enclave tourism, for example. Arguably therefore, the most important purpose for this book was to establish that these ideas are hotly contested and that relationships of power underlie the construction, representation and implementation of the processes of globalisation, sustainability and development, which are manifest unevenly and unequally. In other words, power and uneven development are critical to understanding contemporary tourism and must lie at the centre of tourism analysis.

This necessitates a more nuanced and extended discussion of inequality and shows that power is transmitted in and through tourism. Such a discussion must extend from the power wielded by private business consortia such as the WTTC, to the priorities of environmental sustainability, to the way in which new tourism draws in and controls new Third World

Table 12.1 Key themes and key words

Key words \ Key themes	Uneven and unequal development	Relationships of power	Globalisation	Tourism and geographical imagination
New tourism	Majority of tourists from the First World, especially new tourists. Do new forms of tourism adequately address the shortcomings of conventional tourism? Can new tourism reduce poverty? Can it address structural inequality?	First World owns majority of tourism resources	Places drawn into global tourism, and new tourisms spread	'Benidorm' vs. 'Himalayas' Independent vs. mass tourism
Third World	Third World is structurally disadvantaged – global inequality. Who sets the development agenda? Is it achievable? Is it working?	Third World required to adjust structurally for First World institutions	Third World increasingly interdependent with the First World. How far is globalisation an even and harmonious process? And to what degree does it exclude and marginalise places?	Third World as poverty stricken Third World as environmental paradise
Sustainability	Is the First World prescribing solutions: 'Think globally, act locally'? Are new forms of tourism sustainable?	Third World governments and NGOs clash with the First World over environmental responsibility	'Think globally, act locally' Planet Earth – global environmental interdependency	Third World environments as sizeable tourism assets Third World environments need saving How far has the environmental agenda seeped into the production and consumption of tourist places?

destinations. As Hutnyk (1996: 216) observes: 'social sciences have sometimes been complicit in maintaining distinctions and privileges among different classes of people, and between First and Third World, by choosing to investigate the conditions of "the poor" and "disempowered" rather than pursuing research into power'. We have traced these relationships of power and variable development through the Third World, though we have recognised that in itself the Third World is a socially constructed and contested entity that is inexorably related to development.

In Table 12.1 we have added 'tourism and geographical imagination' to the themes. The notion of tourism and geographical imagination seeks to emphasise that we are dealing with the nature of representation, how the context and meanings of words such as sustainability and development are socially constructed by host communities, governments, TNCs, smaller operators, environmental organisations and tourists. They are words whose meaning is open to interpretation and are continually contested between different interest groups – a point that is reflected in the examples provided in the final column of Table 12.1.

New forms of Third World tourism

The new forms of tourism are summarised in Chapter 4. As the early chapters of the book argue and as Figure 12.1 and Table 12.1 portray, they have arisen as a result of a wide variety of economic, socio-cultural and environmental factors. Factors such as a structural economic shift from a Fordist to a post-Fordist mode of production, accompanied by cultural shifts characterised as moving from modernism to postmodernism, and a growing environmentalism, help explain the increase in the number of new forms of tourism. Independent travel has sought to distance itself from mass tourism, and a variety of benevolent terms (appropriate, alternative, acceptable, pro-poor, responsible, sustainable, and so on) have been employed in an attempt to assert that it is these forms of tourism that

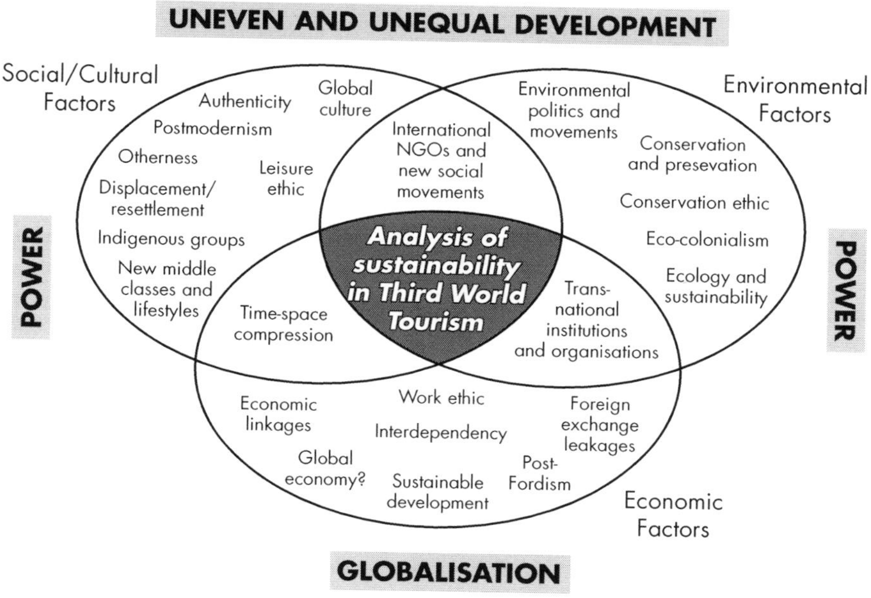

Figure 12.1 Tourism, sustainability and globalisation

provide an ethical and practically acceptable response to 'development' and to the structure of disadvantage of the Third World.

We have attempted to demonstrate a tentative link between the economic and cultural shifts in contemporary capitalism and the growth of the new middle classes. In Chapter 5 we suggested that it is these class fractions in particular (though not exclusively) that can be linked to the growth of new tourism. Furthermore, these new groups and their motives are more formally represented by the international NGOs which campaign around issues of conservation, ecology, pollution, poverty and human rights and cultural identity to which many of the new middle classes subscribe. These organisations (discussed in Chapter 6) represent the vanguard of this movement to gain a moral base and a justification for their own definition of sustainability, which is often constrained by a restricted association with ecology. The subliminal promotion of a conservation ethic and the assured certainty of one form of sustainability are responsible for the emergence of attitudes which in some cases have been characterised as eco-colonialist.

Chapter 7 examined the ways in which the new tour operators, as well as the more established and conventional ones which also lay claim to sustainability, are defining the notion to fit conveniently with their own practices and purposes. This can be achieved through a range of techniques. The chapter also discussed how supposed 'business realities' can subvert all these efforts and turn them into relatively ineffective palliatives and marketing ploys. Thus the claims to sustainability may be seen as having little or no effect on business as usual. Despite this, the new forms of tour operation have brought into existence a variety of new posts and new techniques within the industry to promote its claims to sustainability.

The new forms of tourism have changed the relationships of power and exchange for many destination communities and these were discussed in Chapter 8. The roles of destination communities in Third World countries may not have changed so drastically as a result of the emergence of new forms of tourism, but local participation in the decision-making and running of tourism schemes is highly variable. For destination communities, particular problems have arisen as a result of the conservation movement and its penchant for promoting area protection measures (such as the designation of national parks) which have often excluded local populations and prevented them from continuing 'life as normal'. This has been particularly so for indigenous groups who may find themselves fulfilling the role of 'zoofied' objects for the entertainment of new tourists. Host populations have found themselves drawn into the new tourism activities as local entrepreneurs, as local elites, as applicants for government funds, as local service providers and local operators, rather than simply as the objects to be viewed. These changing relationships have occasionally had positive effects, but in other cases have led to the perpetuation of unequal development.

In an attempt to rebalance the focus of the new tourism literature that is almost exclusively on rural areas, Chapter 9 looks more closely at why cities have fared so badly in 'development' and in new tourism more specifically. The chapter discussed two notable forms of new urban tourism. The first centred on the presence of superb cultural heritage in some Third World cities which presents a considerable draw card for new tourists. However, as with naturally and ecologically diverse rural environments, the conservation of urban areas can initiate a gentrification process with detrimental impacts on local residents and the urban poor. As might be predicted of new tourism's pursuit of different, edgy and other experiences, the second hones in on the presence of urban slums and poverty. The degree to which this is locally rooted and responsive pro-poor tourism (as discussed in Chapter 11), or further evidence of the aestheticisation and trivialisation of poverty, needs further careful examination.

The governance of new forms of tourism was demonstrated in Chapter 10. At this level the power relationships are often dominated by external influences upon nation states and the whole process of 'development' is often determined by the requirements of inter-national finance and considerations of ideology and associated international political strategies.

Our final substantive chapter (11) addressed what might be considered as the logical orientation of new tourism's supposed development credentials: pro-poor tourism. We set our discussion within the broader global development framework – the Millennium Development Goals – and discussed the prospects of tourism in serving as an agent of development. Our conclusions were guarded. While in some cases tourism has acted as an important local development mechanism, its ability to interact with the 'poor' is limited, and the privileged claims of new tourism over conventional mass tourism as a developmental vehicle may be considerably overdrawn.

Globalisation, sustainability and development

Figure 2.3 presented a graphic illustration of the interplay of the key factors involved in an analysis of sustainability in Third World tourism. In the chapters that followed we attempted to broaden the notion of sustainability and to analyse the new forms of tourism from different perspectives using an interdisciplinary approach. Figure 2.3 was a much-simplified version, presented to show the framework for the analysis in subsequent chapters. This is fleshed out in Figure 12.1 with some of the detail of the analyses of the key relationships, using examples from throughout the book. The figure links the key themes and key words discussed above and expands the notion of sustainability in Third World tourism by presenting relationships between the key themes and the factors which feed into an analysis of tourism.

The core of the analysis has been an examination of the notion of sustainability in Third World tourism and how this is intimately related to the processes of globalisation and development. This analysis is affected by socio-cultural, environmental and economic factors. Figure 12.1 provides examples of the forms that these factors take, which were discussed at appropriate places throughout the book. The economic factors, for example, range from the influences of post-Fordist modes of production giving more flexible forms of tourism (discussed in Chapter 2) to the global economic linkages exploited by transnational companies (Chapters 7 and 10). It is not just the effects of tourism on the economies of Third World destinations that should concern us, but the manner in which an understanding of global and economic forces bears upon an understanding of contemporary tourism.

Similarly, the discussion of socio-cultural factors also seeks to set the debate within a broad analysis. Factors range from postmodernism and the leisure ethic to the problems of displacement and resettlement caused by these holiday pursuits for those at the receiving end (Chapter 8). Linking the socio-cultural and environmental factors are the new international socio-environmental movements (Chapter 6) whose interests and concern for environmental issues show a considerable overlap with those of the new middle classes whose tourism practices were discussed in Chapter 5. Again, an understanding of both environmental politics at a global level and the effects and impacts of environmentalism at a local level are required in an analysis of tourism.

The examples drawn into Figure 12.1 are by no means exclusive and you may wish to consider other examples which illustrate these factors and their interrelationships. Overall, the figure serves as a reminder that a deeper understanding of sustainability and Third World tourism may be gained from a critical assessment of an array of global and local factors within an interdisciplinary framework. Arguably of most significance is the way in which power and uneven and unequal development are manifested through these processes and reflected through tourism. The core of the diagram in Figure 12.1, where the three sets of factors and the key themes overlap, provides the rationale for Chapters 5–11 of the book, in which different aspects of sustainability in new forms of tourism were examined.

Sustainability and power

A key characteristic of our argument, as Figure 12.1 emphasises, is the need to understand sustainability within a broad context. We have argued that such a context provides a richer basis for understanding and analysing the emergence of these new forms of tourism; in a nutshell, sustainability and new tourism are intimately related. It is also suggested throughout the book that sustainability as a key word is open to a wide variety of competing meanings. At its most basic, the perception and understanding of sustainability for the residents of the leafy suburbs of Vancouver, London or Melbourne is considerably different from that of local tourist destination communities in Bali, Bolivia or Uganda. Moreover, as has been suggested, sustainability is a conduit, as it were, of power; and power is reflected in and transmitted through notions of sustainability, as both Figure 12.1 and Table 12.1 seek to illustrate through examples.

Chapter 2 set the scene by discussing the power of globalisation, sustainability and development as ideas that for the most part have been generated and driven from the First World. This does not imply that the Third World is simply a passive recipient of global trends but that its ability to withstand the force of 'global consensus' is circumscribed. We took the discussion of power further in Chapter 3 by assessing the various conceptual bases for analysing power. It was suggested that sustainability is ideological in the sense that it is from the First World that the notion has emerged and, in the main, it is First World interests that are served through the promotion of sustainability. But it is also ideological in the way in which notions of sustainability are forced upon the Third World and enforced by First World interests and institutions. While this is a generalisation, it helps capture the inequality felt by Third World countries at the Rio and Rio + 10 summits and at numerous global trade talks. It is also a point underlined by the power and pervasiveness of global environmentalism.

Sustainability was also referred to as a discourse, a question, as Eagleton (1991: 9) suggests, of asking 'who is saying what to whom and for what purposes'. Sustainability, as the latter chapters of the book stress, is used by a variety of interests in a variety of ways as a means of supporting and enhancing their bases of power. This ranges from the activities of the tourism industry to the interpretation adopted by or foisted upon Third World governments or the discourse adopted by the myriad environmental organisations. In short, sustainability and sustainable tourism reflect a discourse that is contested and through which power circulates.

The final concept of power discussed was hegemony, and again (as with discourse) this relates usefully to the contested meaning of sustainability, and the manner in which it must be continually renewed, redefined, defended and so on. This is demonstrably the case with the range of new tourisms that have emerged and the protracted debates over the most 'appropriate' way to holiday (as Figure 3.3 suggests). For example, Chapter 5 sought to demonstrate the practical strategies adopted by new tourists and the way they both seek to sustain lifestyles and cut out a tourism niche for themselves; a similar analysis was applied to the new tourism operators. Another way of illustrating the hegemony of sustainability and its relation to tourism is through a continuum of meanings adopted by the socio-environmental movements (Table 6.1). Overall we have tried to demonstrate why notions of power and contest are fundamental to an understanding of sustainability and especially the emergence of new forms of tourism.

In Chapter 4 we made reference to what some see as a guiding spiritual or philosophical maxim which is often used to explain dominant behaviour patterns and lifestyles: the ethics. The abstract notions of the work ethic and the leisure ethic – not mutually exclusive – were briefly discussed with respect to the prevailing patterns of holidaymaking. We also introduced the idea of the emergence of what might be called a conservation ethic as a limited descriptor of the holidaymaking behaviour of some sectors of the new tourists. All of the 'ethics' were considered to be rather weak, and it may be more appropriate to see them merely as descriptors of lifestyles rather than to use them as explanatory variables.

New tourisms, new critiques

The emergence of new forms of tourism begs many questions. We have attempted to identify some of these and answer others. Tourism is a vast and complex industry and field of study, and we have provided a partial analysis of some of its facets.

We have adopted a thematic approach in order to re-contextualise the analysis of Third World tourism. In part this seeks to avoid a case study approach, analysing examples of new tourism development and extrapolating lessons from these. There are many excellent and an increasing number of accounts of individual cases that have come to the fore since the late 1990s – of Annapurna in Nepal, CAMPFIRE in Zimbabwe or the Toledo village guesthouse scheme in Belize – to name three of the most oft-quoted examples. This is not to say that a thematic approach is superior to case study analyses – both have an important part to play. But we have been conscious of the need to place the analysis of local tourism development within a global context; in our opinion, as suggested above, an understanding of new tourism must be grounded in an assessment of the broader forces at work.

The analysis of new tourism should also embrace a much fuller understanding of the tourism process itself and the theoretical frameworks that can help us make better sense of new developments. As suggested at the outset, the study of tourism tends to be descriptive and theoretically light. We have therefore attempted to develop a critique of new tourism by extending and applying a political economy approach and engaging in a multidisciplinary discussion. Far from heralding the death of dominance and control of Third World destinations by foreign-owned mass tourism interests, the central argument is that the emergence of new forms of tourism has the potential to supplement and reinforce, if in a considerably more nuanced way, the relationships of power already transmitted and circulated.

As the above discussion implies, there is a need for more analysis of new tourism to supplement the existing studies. Arguably, new tourism is currently a western phenomenon. Indeed, those Third World countries experiencing major increases in numbers of visitors from Chinese, Indian and South East Asian countries, for example, are affected far more by the impact of conventional mass tourism. It would nevertheless be interesting to consider whether the South East Asian middle classes will resemble the new middle classes elsewhere in their cultural lifestyles and the subsequent production of new forms of tourism. If this were to be the case, it would dramatically increase the number of people who desire a holiday with a difference.

Whither new forms of tourism?

At the end of Chapter 4 we briefly speculated on the question of 'Whither sustainability in tourism?' The need for a political analysis of the tourism industry in order to reveal the broader context in which it is set was emphasised. Without such analysis it is unlikely that the currently prevailing power structures which hold up their own definitions of sustainability in tourism will be seriously challenged. It is to this consideration that we now return, speculating on the likely future development of new forms of tourism. It is important to stress here that this brief section covers only what we consider to be likely and does not examine what might be considered desirable.

Despite the immediate effects of the 11 September terrorist attacks on the USA and the resulting 'war on terror', the growth in the demand for new forms of tourism to Third World destinations is likely to continue. In the First World, the appeal of an experience that is adventurous, authentic and alternative is unlikely to decline. The recent rapid growth of volunteer tourism (or voluntourism – see Chapter 5) offers us a preview of the likely increase in adventurous holiday experiences, for today's volunteers in the Third World are tomorrow's middle classes who will seek ever more adventurous and interesting holidays in the future. Temporal fluctuations arising out of First World recessions, terrorist attacks or spatial fluctuations due to local circumstances such as war will serve only as local and short-term 'blips' in what seems to be the inevitable growth of new forms of tourism. Moreover, it is likely that the diversity of forms which are demanded will continue to grow – there are still more hidden corners of the planet and underexposed cultures for the new middle classes to consume and 'tourisms' with which the new middle classes are anxious to differentiate themselves from others. Indeed, as a group, the new middle classes of the First World look likely to grow in size and proportion; and differentiation of fashions and tastes within this group is likely to increase, leading to ever greater demand for holidays with a difference. This may be less likely in the case of the tourist patterns of the Asian new middle classes, who, as has already been noted, appear to date to have pursued, in large part, conventional mass tourism activities rather than the new forms.

The differentiation of fashion in holidays is no less likely to be associated with sustainability and the conservation ethic in the future than it is now. And this association is likely to be fostered, successfully, by the international environmental and conservation NGOs. Some of these organisations, such as Conservation International and TIES, have produced manuals on 'how to do ecotourism right' as Whelan (1991: 4) suggests. Such assistance effectively offers local communities and others guidelines on how to make the prevailing system work for them, in particular how to work within the system of current power relationships to promote their own project or development. Some critics will argue that it is possible that this will only strengthen and perpetuate existing power relationships rather than bring about change; others will argue that it is only in this way that the system will change, even if only incrementally.

The industry is unlikely to change its modus operandi. There seems little prospect of changes to the dominant imperatives of capitalism and capital growth, and the growth in 'corporate consciousness' and corporate social responsibility is unlikely to have more than a minimal impact on global justice and equity in tourism. It is likely, however, that the industry will become the major proponent of 'sustainability'. The industry's definition of sustainability is unlikely to become much clearer than it is at present, but the techniques of conveying it, especially to the tourists, will become more refined, more complex and more sophisticated. This applies to operators and service providers in both the mainstream industry and the new forms of the industry and at all scales and sizes. As the large operators diversify their activities, they are likely to associate their new products increasingly with the notion of sustainability; and as new operators emerge to cater for new, alternative forms of tourism, so they too will increasingly deploy links with conservation, ecology and matters ethical, to their own ends.

We envisage that the industry will find ways around the dilemma of flying as a major element of global tourism. These ways will involve the rebranding of the travel element rather than its elimination and a growth in the use of carbon-offsetting techniques despite increasing awareness that such techniques only allow polluters to continue polluting.

Local communities will continue to adapt themselves to offer the type of holiday that the new tourists demand. This is a major source of optimism, and despite the structurally unequal context within which smaller enterprises exist, there are opportunities for 'progress' here. This is unlikely, however, to alter drastically the relationships of power that operate between the relevant interest groups as smaller projects are reliant on the support of and their adaptation to external factors bearing on tourism activities, these ranging from supportive trade associations to marketing overseas. These are after all largely commercial businesses and must make a profit. There exists the possibility that some Third World communities will take a degree of control over their own exploitation of tourism, and particularly new forms of tourism, which will represent, at least for them, a rebalancing of power. In this regard, the pro-poor tourism initiatives reviewed in Chapter 11 provide some small cause for optimism. Although the number of communities in such a position is likely to increase, it is likely to be limited (albeit significant for the communities concerned). Some examples of this possibility were discussed in Chapter 8. Overall, however, change is unlikely to be significant or substantial, and given the enormity of matching the development targets and the scale of global poverty, tourism might be expected to play a relatively minor role, particularly when contextualised by the size and financial power of the sector. For effective change to happen and the ability of small community projects to succeed, national government policy will need to change to reflect the needs of the local sector.

It is our assessment, however, that national governments are unlikely to promote change that will alter the balance of power between all the players in the field of tourism. We have tried to show throughout this book that even countries which are held up as models of sustainable tourism development, such as Costa Rica and to a lesser extent Belize, are just as likely to end up promoting the development of mass tourism enclaves. More significantly, where they do genuinely promote small-scale, sensitive and 'low-impact' tourism, they are no more likely to break out of the existing power structures. Of course this is not always a result of a desire to stay within the confines of 'acceptable' economic development processes; but the external pressures of the transnational companies, First World governments and supranational lending agencies are generally too great for Third World countries to withstand. In particular, there are real pressures placed on governments to maximise revenues from tourism leading to understandable preferences for large-scale mega projects over small-scale community-based projects. It has also been suggested (Chapters 7 and 10) that where supranational organisations advocate green or ethical tourism, the body of policy often remains largely intact, unchanged and insensitive to alternative approaches.

If this conclusion appears rather pessimistic, short on hope and delivering the message that we can expect 'more of the same', then we should also point out that the viewpoint was a global one: in other words, while we have used local case studies to illustrate the book, we have not organised the arguments around them. As Chapter 2 stressed, too local an analysis would tend to ignore the wider global forces and too global an analysis would tend to ignore local differences and examples. There is scope for communities and others to take control of the development of the industry around them and a few examples given in Chapter 8 illustrated this, even if they all had their imperfections. Notwithstanding imperfections, there are successes, and the obstacles to change are not insurmountable. The possibilities for change, however, are unlikely to come from the top or from the middle, where the power of vested interest is too great; it is more likely to come from below, where the need for change is the greatest. At this level, the resources of nature, power and finance and the control over them are small in dimension and change is unlikely to be significant on anything but a local scale.

Notes

1 Introduction

1 Equally, this critique of First World package tourism and its effects on the Third World has not accounted for the phenomenal growth of package tourism demanded by the burgeoning middle classes in South Asia (including Indonesia and Malaysia: see Economist 1995; Ghimire 2001; Hitchcock et al., 1993). However, this is not the focus of debate in this book.

2 Where categorisations and tables of data are used in this book we adopt the standard definitions given by the World Tourism Organisation (*Tourism Market Trends* (annual), Madrid: UNWTO) as follows: *Visitor*: Any person who travels to a country other than that in which s/he has his/her usual residence but outside his/her usual environment for a period not exceeding 12 months and whose main purpose of visit is other than the exercise of an activity remunerated from within the country visited. *Tourist:* A visitor who stays at least one night in a collective or private accommodation in the country visited. *Same-day visitor:* A visitor who does not spend the night in a collective or private accommodation in the country visited. *Arrivals:* All data refer to arrivals and not to actual number of people travelling. One person visiting the same country several times during the year is counted each time as a new arrival. Likewise, the same person visiting several countries during the same trip is counted each time as a new arrival. See also list of abbreviations, UNWTO.

3 It should be noted that the growth of the middle classes in some countries, especially those in South Asia and the Pacific Rim and increasingly in Latin America, is also now of considerable significance (Economist 1995). Most of the patterns of tourism arising from these countries resemble conventional mass forms of tourism rather than the new forms discussed in this book which are characteristic largely of the new middle classes of North America, Europe and Australasia.

2 Globalisation, sustainability, development

1 We urge all users of WTO data to treat its statistics with considerable caution and to apply a high degree of scepticism to its interpretation. For this third edition of the book we have again dropped from Tables 2.1 and 2.2 the average mean annual changes in global tourist arrivals and receipts, as the WTO's calculations for the data given in the first edition were inaccurate. Neither of the two accepted methods of calculating average mean annual changes would yield the results published by the WTO in its two major yearly publications, *Tourism Market Trends* and *Compendium of Tourism Statistics*; moreover, the discrepancies between the WTO figures and the correct figures are undoubtedly significant. Our correspondence with the WTO regarding this matter, conducted throughout 1997 and 1998, did not resolve the problem, and their failure to reply to our final letter hints at their inability to find an answer. It would appear from the following that our experience with WTO statistics is not exceptional:

> The agency [the UNWTO] also exercises its power through its pervasive control over data collection, economic impact studies and market research in the field of tourism. 'Member governments, private tourism companies, consulting firms, universities and the media all

recognise [UN]WTO as the world's most complete and reliable source of global tourism statistics and forecasts,' it claims. In contrast, independent statistics experts and economists have argued that [UN]WTO statistics, forecasts and technical analysis are highly suspect and directed to hoodwink everyone about the supposed benefits of tourism.

(Pleumarom 2001: 7)

The caution and scepticism that we urge towards use of the UNWTO's average mean annual changes might well be advisable for other statistics collected, collated, calculated and disseminated by the UNWTO.

2   In 1983 the World Commission on Environment and Development was set up, with Gro Harlem Brundtland as its chair, in response to a United Nations General Assembly resolution. The Commission's report (the 'Brundtland Report') was submitted to the United Nations in 1987. Its often quoted definition of sustainable development is 'development which meets the needs of the present without compromising the ability of future generations to meet their own needs'.

3   Some authors argue that the new middle classes are inseparable from postmodernism. Lash (1991: 252) argues that 'economic growth and cultural change (post-Fordism and postmodernism) constitute . . . the two sides of these new post-industrial urban middle classes'. Savage et al. (1992: 1) refer to a growing convergence that postmodernism is 'some kind of middle class phenomenon'.

4   Easterly (2006) appears to have subsequently toned down his presentation in a more recent economic discussion, *The White Man's Burden: Why the West's Efforts to Aid the Rest Have Done So Much Ill and So Little Good*. The central argument here is that it is locally fashioned responses to the challenge of poverty that have the most to offer, contrasted to the failed attempts of 'planned' development interventions and the 'big push' methods adopted by the multilateral and bilateral development agencies. He reflects:

> I am among the many who have tried hard to find the answer to the question of what the end of poverty requires of foreign aid. I realized only belatedly that I was asking the question backward; I was captive to a planning mentality. Searchers ask the question the right way around: What can foreign aid do for poor people? Setting a prefixed (and grandiose) goal is irrational because there is no reason to assume that the goal is attainable at a reasonable cost with the available means.

(Easterly 2006: 11)

5   This is a criticism that even Easterly (2001: 64) concedes in commenting upon the use of economic data from rich countries simply because 'those were the countries that had the good quality data'.

3   Power and tourism

1   Sex tourism has emerged as a major activity in a number of countries, especially in South East Asia. Edward Said (1991: 25) depressingly catalogues the 'formidable structure of cultural domination' in his study of *Orientalism*. It represents such a 'system of truths', he argues, that it has 'rarely offered the individual anything but imperialism, racism, and ethnocentrism for dealing with "other" cultures' (Said 1991: 204). The most persistent value of western cultural life, and one which is remarkably persistent, is sex: sexual experimentation and fantasy, promise, desire, delight and unlimited sensuality. It is this key feature of Orientalism (carefully nurtured by the nineteenth century's colonial wayfarers) that has been so cleverly refined in the world of mass communications, but remains a standardised cultural stereotyping of 'the mysterious Orient' – smiling, servile and sexy.

2   Susan Sontag (1979) makes this point about the 'aesthetising tendency of photography', that photography 'develops in tandem' with tourism and that, ultimately, the 'medium which conveys distress ends up by neutralising it'. Or in other words, photographs preserve and consecrate the status quo, and 'aesthetise the injuries of class, race and sex' (Sontag 1979: 9, 109, 178).

4 Tourism and sustainability

1 The motor car, however, has an extremely powerful lobby and many supporters who would pit the arguments of personal freedom and inevitability against environmental sustainability.

2 Details of other techniques in this category and their application are not given here, although a number of examples are discussed in Chapters 5–11. Instead, you are referred to Briassoulis (1992), Fletcher (1989), H. Green and Hunter (1992), Lee and George (2000) and Witt and Moutinho (1994).

3 The G8 countries are: USA, UK, Japan, France, Germany, Italy, Canada and Russia. Of these, the five chosen in Hunter and Shaw's work were USA, UK, Japan, France and Germany.

5 A new class of tourist

1 For this chapter we have drawn heavily on our field notes and on an analysis of tour brochures offering holidays to a new middle-class clientele.

2 Reference is sometimes made to the *lifestyle* movement. This was especially associated with yuppies and was encapsulated in the idea that your 'style of consumption could both indicate and influence [your] position in a new social and cultural order' (Shurmer-Smith and Hannam 1994).

3 Baudrillard is poignant here in his exposition of the 'into', being the 'key to everything': 'Into your sexuality, into your own desire . . . The hedonism of the "into"' (Baudrillard 1988: 35). It is within this context that control and self are most eloquently expressed. Not only are the new middle classes able to map their futures in time, but also they are able to extend this control spatially. 'Done it', 'Chill out', 'The world's a breeze!'

4 The representations of tourists and tourism in a number of tourist brochures and an analysis of travel reviews appearing in UK broadsheet newspapers are of particular significance in reflecting the characteristics of contemporary travel. As a number of authors (Massey 1995b; Munt 1994b) have argued, these are a significant medium through which 'geographical imaginations' are formed and reflected (Walter 1982). In particular, the manner in which the *Independent* newspaper mirrors the cultural and 'political' lifestyles of the new middle class is of special interest. It is characterised by distaste for 'yuppie' culture, a veneer of classlessness and political independence and an adoption of fractured non-party political issues (new socio-environmental movement): the campaigns on Bosnia, dog mess and café street life are illustrative of this. Similarly, tourism has become a focal point, with the appearance of the *Independent Traveller* section and its campaigning stance on more sustainable forms of travel. In this respect, the *Independent* is a mouthpiece of new middle-class travel and frequently has contributions from organisations such as Tourism Concern.

5 Interestingly, the brochure says:

> Essentially our tours are holidays; we have fun and although we focus on the crafts and textiles of a country, we do not miss out on any of the general sightseeing. So in India, we will still visit the Taj Mahal, in Peru we will explore the ruins of Machu Picchu and so on . . . We fly with comfortable scheduled airlines . . . Our accommodation is always of a good standard and is chosen for its ambience and cleanliness whether in colonial house, castles, hotels, pensions or even tents.

6 This point is strongly reflected in Sontag's (1979) arguments regarding photography, another principal means of recording and translating the experience of tourism:

> To take a picture is to have an interest in things as they are, in the status quo remaining unchanged . . . to be in complicity with whatever makes a subject interesting, worth photographing – including, when that is the interest, another person's pain or misfortune.
>
> (Sontag 1979: 12)

In so doing this process aestheticises reality, a tendency in photography that Sontag (1979: 109) observes as 'the medium which conveys distress ends by neutralising it'.

6 Socio-environmental organisations

1 'Largely' is an important qualification here, for clearly it would be to overstate the case considerably to suggest that environmentalism and the middle classes are inseparable and mutually inclusive: not all middle classes are supportive of environmentalism.

2 This latter position is reflective of Dobson's (1995) attention to ecologism. It is interesting to note Dobson's call for a separation of environmentalism – which is easily integrated into other ideologies, such as socialism or feminism – from ecologism which he argues is an ideology in its own right. Much of the debate over tourism is about the absorption or integration of environmental concerns into existing approaches, whether they be political ideology (eco-socialism) or hegemonic strategies (the adoption of green concerns by major transnational tourism companies, for example).

7 The industry

1 FTSE: Financial Times Stock Exchange. The FTSE is a trade mark of the UK *Financial Times* and the London Stock Exchange.

2 www.defra.gov.uk/news/2006/060719b.htm – and as reported in Friends of the Earth press release, 21 July 2006.

3 As in the *Guardian* (2006) 'How to be a responsible tourist', London, 17 July 2006 – associated with Esther Addley's article 'Boom in green holidays as ethical travel takes off'.

4 The exercise of consultation is closely intertwined with the issue of participation, which is examined in some detail in Chapter 8. In particular, the types of consultation shown in Figure 7.2 and discussed here should be referred to Jules Pretty's typology of forms of participation shown in Table 8.1.

5 European Commission Directive of 13 June 1990 (990/314/EEC), *Official Journal of the European Communities*, no. L158/59, 23 June 1990.

6 In 1998, this issue of land titles and ownership became front-page news in Honduras when the National Congress proposed reforming the constitution to permit foreigners to buy land for tourism development projects. Article 107 of the constitution previously prohibited foreigners from acquiring land less than 40 km from the coast or land on any of the Honduran islands. The reform was expected to foment investment in tourism. The indigenous Garífuna, who are scattered along the most popular tourist areas on the Atlantic coast and the Bay Islands, however, have ancestral titles to their land, some of the documents for which are not recognised by the Honduran government. The Garífuna fear that big hotel investors will push them out of the tourism industry and out of their homes.

8 'Hosts' and destinations

1 There is also a rich heritage to the theory and method of participation in the field of urban enquiry and action which underlines the significance of development as a global issue (for example, refer to Abbott 1996; Arnstein 1969; Gibson 1996; Hamdi and Goethert 1997; Wates 2000).

2 SELVA: Somos Ecologistas en Lucha por la Vida y el Ambiente (We are ecologists struggling for life and the environment), a local organisation formed in 1991 by individuals of like mind to face some of the environmental and social problems confronting their region in a political context where central government had withdrawn its interest in and resources from the area.

3 David Western's succession of Dr Leakey as director of the KWS brought about a change of attitude to the wildlife/human conflict in Kenya since 1994. Most Maasai, however, still await a change in the balance of power that will favour them.

4 Rigoberta Menchú is a Guatemalan Quiché Indian and Nobel Peace Prize winner.

9 Cities and tourism

1 Slum is used as a catch-all phrase here to refer to lower-quality or informal housing areas that exist across the Third World (see UN-HABITAT 2003).
2 A *favelado* is an inhabitant of a *favela*.

10 Governance, governments and tourism

1 See list of abbreviations.
2 Among other things, the WDM cites a 1997 speech made by the WTO's Director of Trade in Services Division at a conference entitled Opening Markets for Banking Worldwide: The WTO General Agreement on Trade in Services.
3 We are extremely grateful to Elizabeth Elliott and Matthew Lane of the Department of Geography, University Plymouth, for their research into the tourism policies and implications of the development strategies pursued by a wide range of governmental and supranational organisations and agencies. They undertook this work from May to September 2007. The agencies included in the research were:

World Bank
Asian Development Bank
African Development Bank
Inter-American Development Bank
European Bank of Reconstruction and Development
European Commission
United Nations Development Programme
United Nations Educational, Scientific and Cultural Organisation
International Labour Organisation
United Nations Regional Commissions:
Economic Commission for Africa (ECA)
Economic and Social Commission for Asia and the Pacific (ESCAP)
Economic Commission for Europe (ECE)
Economic Commission for Latin America and the Caribbean (ECLAC)
Economic and Social Commission for Western Asia (ESCWA)
United Nations World Tourism Organisation (UNWTO)
Department for International Development (DFID) (UK)
Danida (Denmark)
Sida (Sweden)
United States Agency for International Development (USAID)
AusAid (Australia)
NZAID (New Zealand)
French Development Agency (AFD)
Spanish Agency for International Cooperation (AECI)
Italian Development Cooperation
Japan International Cooperation Agency (JICA)
SNV (Netherlands)

4 UN General Assembly resolution A/RES/53/200.
5 In 1997 the Orwellian-sounding SLORC renamed itself the State Peace and Development Council (SPDC).
6 SLORC renamed the country 'Myanmar' in 1989.
7 Early in 2002, Aung San Suu Kyi was released from house arrest by the SLORC/SPDC, but she was rearrested in May 2003. The NLD's request for visitors to stay away from the country remained in force and has since been strengthened by the general revulsion at the regime's crackdown against pro-democracy protests in September 2007. In May 2008, Cyclone Nardis caused catastrophic devastation in Burma, with over 100,000 fatalities. The same month, Aung

San Suu Kyi, whose roof was blown off in the cyclone, was detained for a further year (to May 2009): www.cnn.com/2008/WORLD/asiapcf/05/08/aung.myanmar.ap/

11 New tourism and the poor

1 GDP is the total value of the output of goods and services produced by an economy, by both residents and non-residents, regardless of the allocation to domestic and foreign claims. The gross national product (GNP) is the total domestic and foreign value added claimed by residents and therefore equals the GDP + net income from abroad (which is the income residents receive from abroad for services (labour and capital) less similar payments made to non-residents who contribute to the domestic economy).

2 The Office of the UN High Commissioner for Human Rights: www.unhchr.ch/development/poverty-01.html

3 The situation faced by women is especially precarious. Comparative research shows that women have experienced inferior access to property and secure tenure both through legislation (or policy) and/or customary and traditional laws and practice (UN-HABITAT/ISDR 2005). Situations where relative stability and security are severely disrupted by disaster or conflict therefore exacerbate already discriminatory environments and increase the vulnerability of women (Zevenbergen and Van Der Molen 2004).

⬤ Appendix 1

Travel and tourism-related websites

www.acs-aeg.org	Association of Caribbean States (ACS)
www.aito.co.uk	Association of Independent Tour Operators (AITO)
www.iccwbo.org/basd/	Business Action for Sustainable Development
www.c-e-r-t.org	Campaign for Environmentally Responsible Tourism (CERT)
www.realholiday.co.uk	Campaign for Real Travel Agents (CARTA)
www.ecotour.org	Conservation International
www.ihei.org/csr	Corporate Social Responsibility Forum
www.corporatewatch.org	Corporate Watch
www.earthsummit.biz	Corporate Watch's site about the Rio + 10 greenwash
www.turismo-sostenible.co.cr	Costa Rican Institute of Tourism Certification for Sustainable Tourism
www.culturalsurvival.org	Cultural Survival
www.dfid.gov.uk	Department for International Development (DFID)
www.challengefunds.org	DFID and Deloitte & Touche Tourism Challenge Fund
www.ecpat.org	End Child Prostitution, Pornography and Trafficking (ECPAT)
www.enn.com	Environmental News Network
www.ethicalvolunteering.org	Ethical Volunteering
www.egroups.com/group/fairtradetourism	Fair Trade in Tourism Network
www.footprintbooks.com	Footprint guidebooks
www.forumforthefuture.org.uk	Forum for the Future
www.foe.co.uk/campaigns/corporates	Friends of the Earth (FOE) Corporate Accountability Campaign
www.foei.org	Friends of the Earth International
www.ecotourismglobalconference.org	Global Ecotourism Conference 2007 (GEC07)
www.greenglobe21.com	Green Globe
www.greenpeace.org.uk	Greenpeace (UK)
www.green-travel.com	Green Travel
www.hotelbenchmark.com	Hotel Benchmark Team at Deloitte & Touche

www.iblf.org	International Business Leaders Forum (IBLF) Corporate Social Responsibility Forum
www.cfrt.org.uk	International Centre for Responsible Tourism
www.iied.org	International Institute for Environment and Development (IIED) – publications on pro-poor tourism and tourism and the environment
www.ippg.net	International Porter Protection Group (IPPG)
www.tourismpartnership.org	International Tourism Partnership (formerly the International Hotels Environment Initiative – IHEI)
wcpa.iucn.org	International Union for Conservation of Nature and Natural Resources (IUCN), known as the World Conservation Union
www.world-tourism.org/sustainable/ 2002ecotourism	International Year of Ecotourism
www.odi.org.uk	Overseas Development Institute (ODI)
www.panos.org.uk	Panos Institute
www.planeta.com	Planeta Platica (Latin America and the Caribbean ecotourism website)
www.propoortourism.org.uk	Pro-poor tourism briefings (IIED and ODI)
www.rethinkingtourism.org	Rethinking Tourism Project (USA)
www.roughguides.com	Rough Guides
www.stakeholderforum.org	Stakeholder Forum for Our Common Future (formerly UNED Forum)
www.survival-international.org	Survival International
www.rainforest-alliance.org/tourism	Sustainable Tourism Stewardship Council
www.tearfund.org	Tearfund
www.ecotourism.org	The International Ecotourism Society (TIES)
www.ecotourism.org/iye/	TIES website for the International Year of Ecotourism
www.toinitiative.org	Tour Operators Initiative for Sustainable Tourism Development
www.tourismconcern.org.uk	Tourism Concern UK
www.twnside.org.sg/tour.htm	Tourism Investigation and Monitoring Team (TIM-Team)
www.undp.org	United Nations Development Programme (UNDP)
www.unep.org	United Nations Environment Programme (UNEP)
www.uneptie.org	UNEP Division of Technology, Industry and Economics
www.unescap.org	United Nations Economic and Social Commission for Asia and the Pacific
www.unesco.org	United Nations Educational, Scientific and Cultural Organisation (UNESCO)

www.unhabitat.org	United Nations Human Settlements Programme (UN Habitat)
www.vso.org.uk	Voluntary Service Overseas (VSO)
www.wcmc.org.uk	World Conservation Monitoring Centre (WCMC)
www.wdm.org.uk	World Development Movement (WDM)
www.worldlandtrust.org	World Land Trust
www.earthsummit2002.org	World Summit on Sustainable Development
www.world-tourism.org	World Tourism Organisation (UNWTO)
www.wto.org	World Trade Organisation (WTO)
www.wttc.org	World Travel and Tourism Council (WTTC)
www.worldwatch.org	Worldwatch Institute
www.wwf.org.uk	World Wide Fund for Nature (WWF)
www.benchmarkhotel.com	WWF/IHEI benchmarking tool

 # Appendix 2

Websites relating to carbon budgets and carbon offsetting

www.cheap-parking.net/flight-carbon-emissions.php	Advance Parking (uses Google maps flight emission calculator)
www.atmosfair.de	Atmosfair
www.bestfootforward.com/footprintlife.htm	Best Foot Forward – footprint calculator
www.thetravelfoundation.org.uk/reducing_our_ carbon_footprint.asp	Carbon Footprint Calculation Guide (available through the Travel Foundation)
www.carbonfootprint.com	Carbon Footprint Ltd
www.carbonindependent.org/index.htm	Carbon Independent
www.chooseclimate.org/flying	Choose Climate website (created by Dr Ben Matthews)
www.climatecare.org/living/calculator_info/index.cfm	Climate Care
www.co2balance.uk.com/co2calculators	co2balance.com
www.ecofoot.org	Earth Day Network
www.nef.org.uk/energyadvice/co2calculator.htm	National Energy Foundation
www.resurgence.org/carboncalculator/index.htm	Resurgence Carbon Dioxide Calculator
www.thecarbonaccount.com	The Carbon Account
www.wwf.org.uk/climatechange/climate_main.asp	World Wide Fund for Nature (WWF) – footprint calculator

References

Abbott, J. (1996) *Sharing the City: Community Participation in Urban Management*, London: Earthscan.

Adams, K. (1984) 'Come to Tana Toraja, "Land of the Heavenly Kings": travel agents as brokers of ethnicity', *Annals of Tourism Research* 11 (3): 469–85.

Adams, K. (1991) 'Distant encounters: travel literature and shifting images of the Toraja of Sulawesi, Indonesia', *Terrae Incognitae* 16: 84–92.

Adams, W. (1990) *Green Development: Environment and Sustainability in the Third World*, London: Routledge.

Adams, W. (2001) *Green Development: Environment and Sustainability in the Third World*, 2nd edn, London: Routledge.

Adler, J. (1989) 'Origins of sightseeing', *Annals of Tourism Research* 16: 7–29.

Aguilar, M. and Garcia de Fuentes, A. (2007) 'Barriers to achieving the water and sanitation-related Millennium Goals in Cancun, Mexico at the beginning of the twenty-first century', *Environment and Urbanization* 19 (1): 243–60.

Albers, P. and James, W. (1988) 'Travel photography: a methodological approach', *Annals of Tourism Research* 15: 134–58.

Allan, N. (1988) 'Highways to the sky: the impact of tourism on South Asian mountain culture', *Tourism Recreation Research* 13: 11–16.

Allen, J. (1992) 'Post-industrialisation and post-Fordism', in S. Hall and T. McGrew (eds) *Modernity and its Future*, Milton Keynes: Open University Press.

Allen, J. and Hamnett, C. (eds) (1995) *A Shrinking World? Global Unevenness and Inequality*, Milton Keynes: Open University Press.

Allen, J. and Massey, D. (eds) (1995) *Geographical Worlds*, Milton Keynes: Open University Press.

Amandala (1992) 'Ecology, economics, tourism and terrorism', editorial, *Amandala* 14 August: 6.

Amandala Online (2007a) 'The Battle of Belize', www.amandala.com.bz/newsadmin (accessed 16 November 2007).

Amandala Online (2007b) 'Foreign interests offer $70 million for nearly 12,000 acres of Ambergris Caye', www.amandala.com.bz/ (accessed 16 November 2007).

Angotti, T. (1995) 'The Latin American metropolis and the growth of inequality', *NACLA Report on the Americas* 28 (4): 13–18.

Annan, K. (2003) 'Trade', *Guardian Special on Trade*, 8 September: 9.

Arden-Clarke, C. (1992) 'Presentation to trade and environment policies after UNCED seminar', in A. Taylor and J. Gordon (eds) *Trade and Environment Policies after UNCED: Reconciling the Irreconcilable?*, London: South–North Centre for Environmental Policy and Global Environment Research Centre.

Arif, H. (2003) 'Why do we need more slums in Asia?, Development Reform, Asian Coalition for Housing Rights: http://www.achr.net.

Arif, H. (2007) 'Global capital and the cities of the South', Development Reform, Asian Coalition for Housing Rights: http://www.achr.net.

Arlt, W.G. (2006) 'Will China change the face of global tourism?', Keynote speech presented at the Beijing International Travel & Tourism Market Seminar, Beijing, 4 April.

Arnstein, S. (1969) 'A ladder of citizen participation', *Journal of the American Institute of Planners* 4: 216–24.

Ashley, C. (2006) *Participation by the Poor in Luang Prabang Tourism Economy: Current Earning and Opportunities for Expansion*, ODI-SNV Working Paper, London: Overseas Development Institute.

Ashley, C., Boyd, C. and Goodwin, H. (2000) *Pro-poor Tourism: Putting Poverty at the Heart of the Tourism Agenda*, Natural Resource Perspectives 51, London: Overseas Development Institute.

Ashley, C., Roe, D. and Goodwin, H. (2001) *Pro-poor Tourism Strategies: Making Tourism Work for the Poor: A Review of Experience*, London: Overseas Development Institute.

Asian Coalition of Housing Rights (2006a) *Tsunami Update*, Bangkok: ACHR.

Asian Coalition of Housing Rights (2006b) 'Land Issue in Thailand' (accessed through www.achr.net).

Asian Coalition of Housing Rights (2006c) *Community Driven Tsunami Rehabilitation*, Bangkok: ACHR.

Asian Development Bank (2005) *Greater Mekong Subregion: Tourism Sector Strategy*, Final Report Volume 1, TA Number 6179, prepared by Asia Pacific Projects Incorporated, Manila, Philippines.

Asian Development Bank (2006a) 'Urban poverty. Tourism: more than sight-seeing', www.adb.org (accessed July 2006).

Asian Development Bank (2006b) *Urbanization and Sustainability: Case Studies of Good Practice*, Manila: Asian Development Bank.

Association of Independent Tour Operators (1996a) *An Introduction to the Association of Independent Tour Operators*, London: AITO.

Association of Independent Tour Operators (1996b) *The AITO Directory of Real Holidays*, London: AITO.

Association of Independent Tour Operators (2002) 'The independent holiday website', www.aito.co.uk/corporate_Responsible-Tourism.asp (accessed 8 January 2008).

Atkinson, R. and Bridge, G. (eds) (2005) *Gentrification in a Global Context: The New Urban Colonialism*, London: Routledge.

Attenborough, D. (1986) *State of the Ark*, London: Routledge.

Baker, L. (2001) 'Enterprise at the expense of the environment?', Environmental News Network www.enn.com, 7 March.

Bali Tourism Authority (2004) 'The 2002 Bali bombings – the effects upon the industry', www.balitourismauthority.net/ (accessed 19 October 2006).

Banana Link (2004) 'Editorial: WTO – Wiser Tariffs Opportunity?', *Banana Trade News Bulletin* 30, April: 1.

Barrett, F. (1989) *The Independent Guide to Real Holidays Abroad*, London: The Independent.

Barrett, F. (1990) *The Independent Guide to Real Holidays Abroad*, London: The Independent.

Barrett, F. (1994) 'Slow boats to China replace lager louts', *Independent* 10 January: 10.

Barthes, R. (1981) *Camera Lucida*, London: Fontana.

Baudrillard, J. (1988) *America*, London: Verso.

Bauman, Z. (1998) *Globalisation: The Human Consequences*, Cambridge: Polity Press.

Baxter-Neal, L. (2007) 'Big spending in Costa Rica', *Mesoamerica, Journal of the Institute of Central America Studies*, August.

Beall, J. (2002) 'Living in the present, investing in the future – household security among the poor', in C. Rakodi and T. Lloyd-Jones (eds) *Urban Livelihoods: A People-Centred Approach to Reducing Poverty*, London: Earthscan.

Beall, J. and Fox, S. (2007) *Urban Poverty and Development in the 21st Century: Towards an Inclusive and Sustainable World*, Oxfam Research Report, Oxford: Oxfam UK.

Becken, S. (2004) 'Climate change and tourism in Fiji: vulnerability, adaptation and mitigation', University of the South Pacific, August.

Beder, S. (1997) *Global Spin: The Corporate Assault on Environmentalism*, Totnes, Devon: Green Books.

Belize Centre for Environmental Studies (1994) 'The complexity of carrying capacity', *The Centre Forum*, January–February: 1.

Belize, Government of (1989) *Integrated Tourism Policy and Strategy Statement*, Belmopan: Government of Belize.

Belize Tourism Industry Association (BTIA) (1992) *Tourism Link*, April, Belize City: BTIA.

Bennett, J. (1997) 'San Blas: the role of control and community participation in sustainable tourism development', unpublished MA dissertation, University of North London, UK.

Berwari, N. and Mutter, M. (2005) 'Introduction', in J. Samuels (ed.) *Removing Unfreedoms: Citizens as Agents of Change in Urban Development*, Rugby: ITDG Publishing.

Betz, H. (1992) 'Postmodernism and the new middle class', *Theory, Culture and Society* 9: 93–114.

Bhutan Tourism Corporation Limited (2007) www.kingdomofbhutan.com/visitor/visitor_.html (accessed 14 December 2007).

Bianchi, R. (2002a) 'Conceptualising the relations of place and power in tourism development', *Tourism, Culture and Communication*, New York: Cognizant Communication Corporation.

Bianchi, R. (2002b) 'Towards a new political economy of global tourism', in R. Sharpley and D. Telfer (eds) *Tourism and Development: Concepts and Issues*, Clevedon: Channel View Publications.

Biles, A. (2001) 'A day-trip to the urban jungle', *Regeneration and Renewal* 17 August: 15.

Bircham, B. and Charlton, J. (eds) (2001) *Anti-Capitalism: A Guide to the Movement*, London: Bookmarks.

Bird, C. (1995) 'Communal lands, communal problems', *In Focus* 16: 7–8, 15.

Bishop, M. (1983) *Selected Speeches 1979–1981*, Casa de las Americas.

Black, M. (1994) 'Mega-slums: the coming sanitary crisis', WaterAid, London, March.

Blake, B. and Becher, A. (1991) *The New Key to Costa Rica*, San José: Publications in English.

Blum, W. (2002) *Rogue State: A Guide to the World's Only Superpower*, London: Zed Books.

Blunt, A. (1994) *Travel, Gender, and Imperialism*, London: Guilford Press.

Boggan, S. and Williams, F. (1991) 'WWF bankrolled rhino mercenaries', *Independent on Sunday* 7 November: 6.

Boo, E. (1990) *Ecotourism: The Potentials and Pitfalls*, vols 1 and 2, Baltimore, MD: World Wildlife Fund.

Boorstin, D. (1961) *The Image: A Guide to Pseudo Events in America*, New York: Harper & Row.

Borzello, A. (1991) 'Postcard from a truck', *In Focus* 2: 20.

Borzello, A. (1994) 'The myth of the traveller', *In Focus* 14: 7.

Bourdieu, P. (1984) *Distinction: A Critique of the Judgement of Taste*, London: Routledge & Kegan Paul.

Bourdieu, P. (1986) 'From rules to strategies', *Cultural Anthropology* 1: 110–20.

Bourdieu, P. (1987) 'What makes a social class? On the theoretical and practical existence of groups', *Berkeley Journal of Sociology* 22: 1–17.

Bourdieu, P. and Eagleton, T. (1994) 'Doxa and common life: an interview', in S. Zizek (ed.) *Mapping Ideology*, London: Verso.

Bradshaw, S., Linneker, B. and Zúñiga, R. (2002) 'Social roles and spatial relations of NGOs and civil society: recent participation and effectiveness in the development of Central America and the middle America region', in C. McIlwaine and K. Willis (eds) *Challenges and Change in Middle America: Perspectives on Development in Mexico, Central America and the Caribbean*, London: Pearson Education.

Bradt, H. (1994) *Backpacker's Africa: East Guide*, Chalfont St Peter, UK: Bradt.

Brandon, K. (1993) 'Basic steps toward encouraging local participation in nature tourism projects', in K. Lindberg and D. Hawkins (eds) *Ecotourism: A Guide for Planners and Managers*, North Bennington, VT: Ecotourism Society.

Breslin, P. (2007) 'What Sachs Lacks', *Grassroots Development, Journal of the Inter-American Foundation*, Arlington, VA: Inter-American Foundation.

Briassoulis, H. (1992) 'Environmental impacts of tourism: a framework for analysis and evaluation', in H. Briassoulis and J. van der Straaten (eds) *Tourism and the Environment*, London: Kluwer Academic.

British Airways (2001) *From the Ground Up: Social and Environmental Report 2001*, Hounslow: British Airways.

British Airways (2007) 'Air transport and climate change', www.britishairways.com/travel/cr globalwarm/public/en_gb (accessed 12 February 2007).

British Airways (2008) www.ba.com/responsibility (accessed 8 January 2008).

Britton, S. (1981a) *Tourism and Economic Vulnerability in Small Pacific States: The Case of Fiji*, Monograph 23, Canberra: Development Studies Centre, Australian National University.

Britton, S. (1981b) *Tourism, Dependency and Development: A Mode of Analysis*, Occasional Paper 23, Canberra: Development Studies Centre, Australian National University.

Britton, S. (1981c) 'The spatial organisation of tourism in a neo-colonial economy: a Fiji case study', *Pacific Viewpoint* 21 (2): 144–65.

Britton, S. (1982) 'The political economy of tourism in the Third World', *Annals of Tourism Research* 9: 331–58.

Britton, S. (1991) 'Tourism, capital and place: towards a critical geography of tourism', *Environment & Planning D: Society & Space* 9 (4): 451–78.

Britton, S. and Clarke, W. (1987) *Ambiguous Alternative: Tourism in Small Developing Countries*, Fiji: University of the South Pacific.

Brockelman, W. and Dearden, P. (1990) 'The role of nature trekking in conservation: a case study in Thailand', *Environmental Conservation* 17 (2): 141–8.

Brookes, B. and Harrison, R. (2000) 'The WTO and the corporate lobby', *Ethical Consumer* 62: 32–3.

Brooks, E. (1990) 'The hidden Kingdom of Mustang', *Geographical Magazine* September: 15.

Bruner, E. (1989) 'Of cannibals, tourists, and ethnographers', *Cultural Anthropology* 4 (4): 438–45.

Bryden, J. (1973) *Tourism and Development: A Case Study of the Commonwealth Caribbean*, London: Cambridge University Press.

Budowski, G. (1995) 'Responsible tourism: new trends', address to Conference on Sustainable Tourism, San José, Costa Rica.

Budowski, G. (1996) 'Ecotourism and conservation: avoiding conflicts and building a mutually profitable relationship', paper presented to the International Meeting on Ecotourism, Manaus, Brazil, January.

Bugnicourt, J. (1977) *Tourism with No Return, Development Forum*, Geneva: United Nations.

Burgess, R., Carmona, M. and Kolstee, T. (1997) 'Contemporary policies for enablement and participation: a critical review', in R. Burgess, M. Carmona and T. Kolstee (eds) *The Challenge of Sustainable Cities: Neoliberalism and Urban Strategies in Developing Countries*, London: Zed Books.

Burma Campaign UK and Tourism Concern (2000a) 'Lonely Planet boycott update', autumn, Burma Campaign UK and Tourism Concern.

Burma Campaign UK and Tourism Concern (2000b) 'Lonely Planet replies and some suggested responses', June, Burma Campaign UK and Tourism Concern.

Burns, P. and Holden, A. (1995) *Tourism: A New Perspective*, London: Prentice-Hall.

Business Action for Sustainable Development (BASD) (2001) www.iccwbo.org/basd

Butcher, J. (2003) *The Moralisation of Tourism: Sun, Sand . . . and Saving the World?*, London: Routledge.

Butler, R. (1975) *Tourism as an Agent of Social Change*, Occasional Paper 4, Peterborough, Ont: Department of Geography, Trent University, Ontario.

Butler, R. (1980) 'The concept of a tourist area cycle of evolution: implications for management of resources', *Canadian Geographer* 24 (11): 5–12.

Butler, R. (1991) 'Tourism, environment, and sustainable development', *Environmental Conservation* 18 (3): 201–9.

Bygrave, M. (2002) 'Where have all the protesters gone?', *Guardian Weekly* 1–7 August: 21–2.

Calder, S. (1994a) 'The biodegradable tourist', *Independent*.

Calder, S. (1994b) 'The myth makers' guide to the world', *In Focus* 14: 6.

CarbonNeutral Company (2007) www.carbonneutral.com/pages/whatwedo.asp (accessed 12 February 2007).

Cardenal, L. (1991) 'Río San Juan: El repoblamiento amenaza la vida del bosque', *Barricada* 17 January: 4.

Carney, D. (ed.) (1998) *Sustainable Rural Livelihoods: What Contribution Can We Make?* London: Department for International Development.

Carney, D., Drinkwater, M., Rusinow, T., Neefjes, K., Wanmali, S. and Singh, N. (1999) *Livelihood Approaches Compared*, London: Department for International Development.

Carothers, A. (1993) 'The green machine', *New Internationalist* 246: 14–16.

Carriere, J. (1991) 'The crisis in Costa Rica: an ecological perspective', in D. Goodman and M. Redclift (eds) *Environment and Development in Latin America: The Politics of Sustainability*, Manchester: Manchester University Press.

Carroll, R. (2007) 'Fearful rich keep poor at bay with gated homes and razor wire', *Guardian* 25 April: 17.

Cater, E. (1992) 'Profits from paradise', *Geographical Magazine* March: 17–20.

Ceballos-Lascurain, H. (2001) 'Land use and site planning', lecture presented to the University for Peace, San José, Costa Rica.

Centre On Housing Rights and Evictions (COHRE) (2006) *Global Survey on Forced Evictions: Violations of Human Rights*, Geneva: COHRE.

Centro de Intercambio y Solidaridad (CIS) (2001) 'El Salvador and the Earthquakes', *CIS Bulletin*, 13 February to 9 March, San Salvador.

Chambers, R. (1994a) 'The origins and practice of participatory rural appraisal', *World Development* 22 (7): 953–69.

Chambers, R. (1994b) 'Participatory rural appraisal (PRA): analysis of experience', *World Development* 22 (9): 1253–68.

Chambers, R. (1994c) 'Participatory rural appraisal (PRA): challenges, potentials and paradigm', *World Development* 22 (10): 1437–54.

Chambers, R. (1995) 'Poverty and livelihoods: whose reality counts?' *Environment & Urbanization* 7 (1): 173–204.

Chatterjee, P. (1995) 'Mexico: World Bank to bail out banks by cutting environment, other loans', Inter-Press Third World News Agency, Washington, DC.

Chatterjee, P. and Finger, M. (1994) *The Earth Brokers: Power, Politics and World Development*, London: Routledge.

Child, B. (1996) 'The practice and principles of community-based wildlife management in Zimbabwe: the CAMPFIRE programme', *Biodiversity and Conservation* 5 (3): 369–98.

Child, G. (1996) 'The role of community-based wild resources management in Zimbabwe', *Biodiversity and Conservation* 5 (3): 355–68.

Chomsky, N. (1985) *Turning the Tide: US Intervention in Central America and the Struggle for Peace*, Boston, MA: Pluto Press.

Chomsky, N. (1989) *The Culture of Terrorism*, London: Pluto Press.

Chomsky, N. (2000) *Rogue States: The Rule of Force in World Affairs*, London: Pluto Press.

Chung, H. K. (1994) 'People's spirituality and tourism', *Contours* 6 (7–8): 19–24.

Clark, J. (1990) 'Carrying capacity: the limits to tourism', paper presented to the Congress on Marine Tourism, Hawaii.

Clark, S. (1983) 'Introduction', in B. Marcus and M. Taber (eds) *Maurice Bishop Speaks: The Grenada Revolution 1979–1983*, New York: Pathfinder.

Cleaver, F. (1999) 'Paradoxes of participation: questioning participatory approaches to development', *Journal of International Development* 11: 597–612.

Cleaver, F. (2001) 'Institutions, agency and the limitations of participatory approaches to development', in B. Cooke and U. Kothari (eds) *Participation: The New Tyranny*, London: Zed Books.

Cleverdon, R. and Kalisch, A. (2000) 'Fair trade in tourism', *International Journal of Tourism Research* 2: 171–87.

Climate Care (2007) www.climatecare.org (accessed 12 February 2007).

Cohen, E. (1972) 'Toward a sociology of international tourism', *Social Research* 39 (1): 164–82.

Cohen, E. (1974) 'Who is a tourist? A conceptual clarification', *Sociological Review* 22 (4): 527–55.

Cohen, E. (1979a) 'A phenomenology of tourist experiences', *Sociology* 13: 179–201.

Cohen, E. (1979b) 'The impact of tourism on the hill tribes of Northern Thailand', *Internationales Asienforum* 10 (1–2): 5–38.

Cohen, E. (1979c) 'Rethinking the sociology of tourism', *Annals of Tourism Research* 6 (1): 18–35.

Cohen, E. (1985) 'Tourism as play', *Religion* 15: 291–304.

Cohen, E. (1989) '"Primitive and remote" hill tribe trekking in Thailand', *Annals of Tourism Research* 16 (1): 30–61.

Cohen, M. (2004) *Reframing Urban Assistance: Scale, Ambition, and Possibility*, Urban Update 5, Washington, DC: Woodrow Wilson International Centre for Scholars.

Cohen, N. (2000) 'Burma's shame', *Observer* 4 June: 6.

Colantonio, A. and Potter, R. (2006) 'The rise of urban tourism in Havana since 1989', *Geography* 91 (1): 23–33.

Colchester, M. (1994) *Salvaging Nature: Indigenous Peoples, Protected Areas and Biodiversity Conservation*, Discussion Paper, Geneva: United Nations Research Institute for Social Development.

Collinson, H. (1999) 'Rights talk', in *CIIR News*, March, London: Catholic Institute of International Relations (now Progressio).

Commission on Human Security (2003) *Human Security Now*, New York: Commission on Human Security.

Cook, I. (1993) 'Constructing the exotic: the case of tropical fruit', paper presented at the annual conference of the Institute of British Geographers, Royal Holloway and Bedford New College, Egham, Surrey, 5–8 January.

Cooke, B. (2001) 'The social psychological limits of participation', in B. Cooke and U. Kothari (eds) *Participation: The New Tyranny*, London: Zed Books.

Cooke, B. and Kothari, U. (eds) (2001) *Participation: The New Tyranny*, London: Zed Books.

Corporate Social Responsibility Forum (2002) 'International Business Leader's Forum', www.ihei.org/csr/csrwebassist.nsf.

Corporate Watch (2000) 'Can't we just talk it out . . . NGOs engaging with business', Oxford, UK: *Corporate Watch*: 12.

Costa Rican Tourism Institute (2005) 'Certification for Sustainable Tourism', www.turismo-sostenible.co.cr.

Coward, R. (1996) 'Sun, sand and encounters with otherness', *Guardian* 27 May: 11.

Cox, A. (1999) *DAC Scoping Study of Donor Poverty Reduction Policies and Practices*, Synthesis Report, ODI and ARID for the DAC Informal Network on Poverty Reduction.

Crick, M. (1989) 'Representations of international tourism in the social sciences: sun, sex, sights, savings, and servility', *Annual Review of Anthropology* 18: 307–44.

Croall, J. (1996) 'On the road to disaster', *Guardian* 14 August: 5.

Crompton, R. (1993) *Class and Stratification*, Oxford: Polity Press.

Culler, J. (1988) 'The semiotics of tourism', in J. Culler (ed.) *Framing the Sign: Criticisms and its Institutions*, Oxford: Basil Blackwell.

Cultural Survival Quarterly (1982) 'The tourist trap: who's getting caught?', *Cultural Survival Quarterly* 6 (3).

Cultural Survival Quarterly (1990a) 'Breaking out of the tourist trap: part one', *Cultural Survival Quarterly* 14 (1).

Cultural Survival Quarterly (1990b) 'Breaking out of the tourist trap: part two', *Cultural Survival Quarterly* 14 (2).

Curran, J., Ecclestone, J., Oakley, G. and Richardson, A. (eds) (1986) *Bending Reality: The State of the Media*, London: Pluto Press.

Dag Hammaskjöld Foundation (1975) *What Now: Another Development, Development Dialogue*, 1 (2), Uppsala: Dag Hammarskjöld Foundation.

Daltabuit, M. and Pi-Sunyer, O. (1990) 'Tourism development in Quintana Roo, Mexico', *Cultural Survival Quarterly* 14 (1): 9–13.

Daly, H. (1993) 'From adjustment to sustainable development: the obstacle of free trade', in Earth Island Press (ed.) *The Case Against Free Trade: GATT, NAFTA and the Globalisation of Corporate Power*, San Francisco, CA: Earth Island Press.

Daniels, P., Bradshaw, M., Shaw, D. and Sidaway, J. (2001) *Human Geography: Issues for the 21st Century*, Harlow, UK: Prentice-Hall.

Davies, P. (2000) 'Calcutta: the gifted city', in *A Meeting by the River*, a report of the Conference on Calcutta and Howrah Waterfronts, London Rivers Association, London, 9–11 February.

Davis, M. (2002) *Dead Cities and Other Tales*, New York: The New Press.

Davis, M. (2006) *Planet of Slums*, London: Verso.

Dearden, P. and Harron, S. (1991) 'Tourism and the hilltribes of Thailand', in B. Weiler and M. Hall (eds) *Special Interest Tourism*, Chichester: Wiley.

Dearden, P. and Harron, S. (1993) 'Alternative tourism and adaptive change', *Annals of Tourism Research* 21: 81–119.

Deihl, C. (1985) 'Wildlife and the Maasai', *Cultural Survival Quarterly* 9 (1): 37–40.

de Kadt, E. (ed.) (1979) *Tourism: Passport to Development?*, Oxford: Oxford University Press.

Deloitte & Touche, International Institute for Environment and Development (IIED) and Overseas Development Institute (ODI) (1999) 'Sustainable tourism and poverty elimination study: a report to the Department for International Development', unpublished report, London: Department for International Development.

Denny, C. (2001) 'Profit motive', *Guardian Weekly* 15–21 November: 27.

Denny, C. (2002) 'WTO to benefit from a monk's patience', *Guardian Weekly* 6–12 June: 14.

Department for Environment, Food and Rural Affairs (DEFRA) (2005) 'Emissions Trading Schemes', www.defra.gov.uk/environment/climatechange/trading/index.htm (accessed 11 July 2005).

Department for Environment, Food and Rural Affairs (2006) www.defra.gov.uk/news/2006/060719b.htm (accessed November 2007).

Department for International Development (1997) *Eliminating World Poverty: A Challenge for the 21st Century*, London: The Stationery Office.

Department for International Development (1999) *Tourism and Poverty Elimination: Untapped Potential*, London: Department for International Development.

Department for International Development (2000a) *Eliminating World Poverty: Making Globalisation Work for the Poor*, London: The Stationery Office.

Department for International Development (2000b) *Halving World Poverty by 2015: Economic Growth, Equity and Security*, London: Department for International Development.

Department for International Development (2006) *Eliminating World Poverty: Making Governance Work for the Poor*, London: The Stationery Office.

Department for International Development (undated) *Changing the Nature of Tourism: Developing an Agenda for Action*, London: Department for International Development.

de Rivero, O. (2001) *The Myth of Development: The Non-viable Economies of the 21st Century*, London: Zed Books.

Desai, V. (1995) *Community Participation and Slum Housing: A Study of Bombay*, London: Sage.

de Soto, H. (2001) *The Mystery of Capital: Why Capitalism Triumphs in the West and Fails Everywhere Else*, London: Black Swan.

Devas, N. (2002) 'Urban livelihoods: issues for urban governance and management', in C. Rakodi and T. Lloyd-Jones (eds) *Urban Livelihoods: A People-Centred Approach to Reducing Poverty*, London: Earthscan.

Dobson, A. (1995) *Green Political Thought*, London: Routledge.

Dodds, K. (2002) 'The Third World, developing countries, the South, poor countries', in V. Desai and R. Potter (eds) *The Companion to Development Studies*, London: Arnold.

Doxey, G. (1975) 'A causation theory of visitor–resident irritants: methodology and research inferences', in *Proceedings of the Travel Research Association Sixth Annual Conference*, San Diego, CA.

Doxey, G. (1976) 'When enough's enough: the natives are restless in old Niagara', *Heritage Canada* 2: 26–7.

Drake, S. (1991) 'Local participation in ecotourism projects', in T. Whelan (ed.) *Nature Tourism*, Washington, DC: Island Press.

Drukier, C. (2001) 'Hollow eco', *Bangkok Post* 14 January: 10.

Duncan, I. and Woolcock, M. (2003) 'Arrested development: the political origins and socio-economic foundations of common violence in Jamaica', mimeo.

Dunkley, G. (1997) *The Free Trade Adventure, the Uruguay Round and Globalism: A Critique*, Melbourne: Melbourne University Press.

Dwek, D. (2004) 'Favela tourism: innocent fascination or inevitable exploitation?', MA dissertation, Institute of Latin American Studies, London.

Eade, D. (2002) 'Preface', in D. Westendorff and D. Eade (eds) *Development and Cities*, Oxford: Oxfam GB.

Eagleton, T. (1991) *Ideology*, London: Verso.

Eames, A. (1994) 'Train cruises', *High Life* September: 86–96.

Easterly, W. (2001) *The Elusive Quest for Growth: Economists' Adventures and Misadventures in the Tropics*, Cambridge, MA: MIT Press.

Easterly, W. (2006) *The White Man's Burden: Why the West's Efforts to Aid the Rest Have Done So Much Ill and So Little Good*, Oxford: Oxford University Press

Eckersley, R. (1986) 'The environment movement as middle-class elitism: a critical analysis', *Regional Journal of Social Issues* 18: 24–36.

Eckersley, R. (1989) 'Green politics and the new class: selfishness or virtue?', *Political Studies* 37: 205–23.

Eckersley, R. (1992) *Environmentalism and Political Theory: Towards an Ecocentric Approach*, London: UCL Press.

Economist, The (1995) 'Asia goes on Holiday', *The Economist* 20 May: 57–8.

Economist, The (2006a) 'China and tourism: the golden years', *The Economist* 13 May.

Economist, The (2006b) 'Chinese tourism – outward bound: the Chinese are starting to travel abroad. But getting them to spend is difficult', *The Economist* 22 June.

Economist, The (2006c) 'Of property and poverty', *The Economist* 26 August: 11–12.

Economist, The (2006d) 'The mystery of capital deepens', *The Economist* 26 August: 66.

Economist Intelligence Unit (EIU) (1995) 'The Gambia', in *EIU Country Report*, Third Quarter, London: EIU.

Edwards, R. (1992) 'Rape of the Himalayas', *Guardian Weekly* 19 and 21 June.

Elkington, J. (2001) 'The "triple bottom line for 21st-century business"', in R. Starkey and R. Welford (eds) *The Earthscan Reader in Business and Sustainable Development*, London: Earthscan.

Elkington, J. and Hailes, J. (1992) *Holidays that Don't Cost the Earth*, London: Gollancz.

Elliott, J. (2002) 'Development as improving human welfare and human rights', in V. Desai and R. Potter (eds) *The Companion to Development Studies*, London: Arnold.

Environment & Urbanization (2004) 'Urban violence and insecurity', *Environment & Urbanization* 16 (2).

Epler Wood, M. (1991) 'Global solutions: an ecotourism society', in T. Whelan (ed.) *Nature Tourism*, Washington, DC: Island Press.

Epler Wood, M. (1994) 'Membership Directory background', *Ecotourism Society 1994 International Membership Directory*, North Bennington, VT: Ecotourism Society.

Errington, F. and Gewertz, D. (1989) 'Tourism and anthropology in a post-modern world', *Oceania* 60: 37–54.

Escobar, A. (1995) *Encountering Development*, Princeton, NJ: Princeton University Press.

Esteva, G. (1992) 'Development', in W. Sachs (ed.) *The Development Dictionary: A Guide to Knowledge as Power*, London: Zed Books.

Ethical Consumer (1994) 'The EC Ecolabel', *Ethical Consumer* 32, November.

Ethical Trading Initiative (2001) 'Introducing the Ethical Trading Initiative', www.ethicaltrade.org.

European Community (2000) 'Opening world markets for services: towards GATS 2000', http://gats-info.eu.

Evans, K., O'Hare, G. and Thompson, C. (1994) 'Soft tourism facing hard choices in The Gambia', *In Focus* 13: 10 and 17.

Evans-Pritchard, D. (1993) 'Mobilisation of tourism in Costa Rica', *Annals of Tourism Research* 20 (4): 778–9.

Everitt, J. (1987) 'The torch is passed: neocolonialism in Belize', *Caribbean Quarterly* 33 (3–4): 42–59.

Fair Trade in Tourism Network (2002) *Corporate Futures: Social Responsibility in the Tourism Industry*, London: Fair Trade in Tourism Network.

Fanon, F. (1967) *The Wretched of the Earth*, London: Penguin.

Featherstone, M. (1987) 'Lifestyle and consumer culture', *Theory, Culture and Society* 4 (1): 55–70.

Featherstone, M. (1988) 'In pursuit of the postmodernism: an introduction', *Theory, Culture and Society* 5 (2–3): 195–217.

Featherstone, M. (1991) *Consumer Culture and Postmodernism*, London: Sage.

Fennell, D. (2006) *Tourism Ethics*, Clevedon, UK: Channel View.

Ferguson, J. (1990) *Grenada: Revolution in Reverse*, London: Latin America Bureau.

Fernandes, D. (1994) 'The shaky ground of sustainable tourism', *TEI Quarterly Environment Journal* 2 (4): 4–35.

Finca Sonador (2000) 'Volunteer work in Longo Maï and UNAPROA, Costa Rica', Finca Sonador Information Sheet, San Isidro de El General, Costa Rica: Finca Sonador.

Fletcher, J. (1989) 'Input–output analysis and tourism impact studies', *Annals of Tourism Research* 16 (4): 514–29.

Flynn, M. (1996) 'Report on Guatemala', *Mesoamerica* 15 (8): 3–4.

Food and Agriculture Organisation (FAO) (2002) *Gender and Access to Land: Land Tenure Studies*, Rome: FAO.

Forsyth, T. (1996) *Sustainable Tourism: Moving from Theory to Practice*, London: Tourism Concern.

Forum for the Future (2001) *Annual Report 2001*, London: Forum for the Future.

Forum for the Future (2002) www.forumforthefuture.org.uk (accessed 28 March 2002).

Foucault, M. (1972) *The Archaeology of Knowledge*, New York: Harper Colophon.

Foucault, M. (1980) *Power and Knowledge*, Hemel Hempstead: Harvester Wheatsheaf.

Frank, A. G. (1966) 'The development of underdevelopment', *Monthly Review* 18 (4): 17–31.

Frank, A. G. (1967) *Capitalism and Underdevelopment in Latin America: Historical Studies of Chile and Brazil*, New York: Monthly Review Press.

Frank, A. G. (1969) *Latin America: Underdevelopment or Revolution*, New York: Monthly Review Press.

Frank, A. G. (1970) *Lumpen-Bourgeoisie: Lumpen-Development, Dependency, Class, and Politics in Latin America*, New York: Monthly Review Press.

Frank, A. W. (1991) 'For a sociology of the body: an analytical review', in M. Featherstone, M. Hepworth and B. Turner (eds) *The Body: Social Process and Cultural Theory*, London: Sage.

Friends of the Earth (1992) 'The eleven days that tried to change the world', letter to members and supporters, London: Friends of the Earth.

Friends of the Earth International (FOEI) (2001) 'FOEI Statement on the World Summit on Sustainable Development', London: September.

Fukuyama, F. (1989) 'The end of history?', *The National Interest* summer: 1–18.

Fukuyama, F. (1992) *The End of History and the Last Man*, London: Penguin.

Fussell, P. (1980) *Abroad: British Literary Travelling between the Wars*, Oxford: Oxford University Press.

Galand, P. (1994) 'I don't want to be an accomplice', *Envío* 12 (153): 12–13.

Gellhorn, M. (1990) 'Too good for tourists', *Independent Magazine* 3 November: 70–4.

Gentleman, A. (2006) 'Slum tours: a day trip too far?', *Observer* 7 May.

George, S. (1988) *A Fate Worse than Debt*, London: Penguin.

Ghimire, K. (2001) *The Native Tourist: Mass Tourism within Developing Countries*, London: Earthscan.

Gibson, T. (1996) *The Power in our Hands: Neighbourhood Based-World Shaking*, Charlbury, UK: Jon Carpenter.

Giddens, A. (1989) *Sociology*, Cambridge: Polity Press.

Ginzburg, O. (2006) *There You Go!*, London: Survival International.

Global Ecotourism Conference 2007 (GEC07) (2007) www.ecotourismglobalconference.org (accessed 1 February 2008).

Goffman, E. (1997) 'The presentation of self in everyday life', in C. Lemert and A. Branaman (eds) *The Goffman Reader*, Oxford: Blackwell.

Goldenberg, S. (with Reuters in Rangoon) (2007) 'Burmese hotels hit as crackdown prompts drop in tourism', *Guardian* 11 October

Gonsalves, P. (1993) 'Divergent views: convergent paths. Towards a Third World critique of tourism', *Contours* 6 (3–4): 8–14.

Gonsalves, P. (1995) 'Structural adjustment and the political economy of the Third World', *Contours* 7 (1): 33–9.

Goodall, B. (1992) 'Environmental auditing for tourism', in C. Cooper (ed.) *Progress in Tourism, Recreation and Hospitality Management*, vol. 4, Environmental Issues, London: Belhaven.

Gordon, R. (1990) 'The prospects for anthropological tourism in Bushmanland', *Cultural Survival Quarterly* 14 (1): 6–8.

Gordon-Walker, R. (1993) 'A real adventure (and so cheap!)', *Independent* 13 August: 19.

Gouldner, A. (1979) *The Future of Intellectuals and the Rise of the New Class*, London: Macmillan.

Gramsci, A. (1973) *Selections from Prison Notebooks*, edited by Q. Hoare and G. Nowell Smith, London: Lawrence & Wishart.

Green, D. (1995) *Silent Revolution: The Rise of Market Economics in Latin America*, London: Cassell.

Green, H. and Hunter, C. (1992) 'The environmental impact assessment of tourism development', in P. Johnson and B. Thomas (eds) *Perspectives on Tourism Policy*, London: Mansell.

Green Horizons Travel (1995) 'Holidays that don't cost the earth', information pack, Green Horizons Travel.

Green Hotelier (2006) 'Tourism and the Millennium Development Goals', *Green Hotelier* 39.

Gregory, D. (1994) *Geographical Imaginations*, Oxford: Basil Blackwell.

Griffiths, T. (1996) 'El Salvador: economic round-up', *Mesoamerica* 15 (2).

Guardian (1993) 'Pass notes: package holidays', *Guardian* 20 August: 3.

Guardian (2006) 'How to be a responsible tourist', *Guardian* 17 July.

Guerrero, R. (1999) *Violence Prevention – Technical Note 8: Violence Control at the Municipal Level*, Sustainable Development Department, Washington, DC: Inter-American Development Bank.

Guerrin, M. (1991) 'For Cartier-Bresson the focus is now firmly on drawing', *Guardian Weekly* 8 December: 14.

Gujit, I. and Shah, M. (1998) *The Myth of Community: Gender Issues in Participatory Development*, London: IT Publications.

Gunn, C. (1994) *Tourism Planning*, London: Taylor & Francis.

Gunson, P. (1996) 'Marketing men put curse of tourism industry on Mayas', *Guardian* 28 September: 14.

Gutiérrez, R. (2001) *The Urban Architectural Heritage of Latin America*, available from http://www.icomos.org/studies/latin-towns.htm (accessed July 2006).

Habermas, J. (1981) 'New social movements', *Telos* 49: 33–7.

Hague, C., Wakely, P., Crespin, J. and Jasko, C. (2006) *Making Planning Work: A Guide to Approaches and Skills*, Rugby: Intermediate Technology Publications.

Hailey, J. (2001) 'Beyond the formulaic: process and practice in South Asian NGOs', in B. Cooke and U. Kothari (eds) *Participation: The New Tyranny*, London: Zed Books.

Hall, C. and Jenkins, J. (1995) *Tourism and Public Policy*, London: Routledge.

Hall, D. and Kinnaird, V. (1994) 'Ecotourism in Eastern Europe', in E. Cater and G. Lowman (eds) *Ecotourism: A Sustainable Option?*, Chichester: Wiley.

Hall, S. (1992a) 'The question of cultural identity', in S. Hall, D. Held and T. McGrew (eds) *Modernity and its Future*, Oxford: Polity Press.

Hall, S. (1992b) 'The West and Rest: discourse and power', in S. Hall and B. Gieben (eds) *Formations of Modernity*, Oxford: Polity Press.

Hall, S. (1995) 'New cultures for old', in D. Massey and P. Jess (eds) *A Place in the World? Places, Cultures and Globalisation*, Milton Keynes: Open University Press.

Ham, M. (1995) 'Cashing in on the silver-backed gorilla', *Features: Ecotourism Special*, London: Panos Institute.

Hamdi, N. and Goethert, R. (1997) *Action Planning for Cities*, Chichester: Wiley.

Hamilton, A. (1995) 'Ecocentrics', *Guardian* 17 June: 54–6, 59.

Harrison, P. (1979) *Inside the Third World*, London: Penguin.

Harvey, D. (1973) *Social Justice and the City*, London: Arnold.

Harvey, D. (1989a) *The Urban Experience*, Oxford: Blackwell.

Harvey, D. (1989b) *The Condition of Postmodernity*, Oxford: Blackwell.

Harvey, D. (2005) *A Brief History of Neoliberalism*, Oxford: Oxford University Press.

Harvey, D. (2006) *Spaces of Global Capitalism: Towards a Theory of Uneven Geographical Development*, London: Polity Press.

Hawkins, D., Epler Wood, M. and Bittman, S. (eds) (1995) *The Ecolodge Sourcebook for Planners and Developers*, North Bennington, VT: International Ecotourism Society.

Hawkins, R. and Middleton, V. (1993) 'The environmental practices and programme of travel and tourism companies', in J. Brent Richie and D. Hawkins (eds) *World Travel and Tourism Review: Indicators, Trends and Issues*, Wallingford, UK: CAB International.

Hawkins, R. and Middleton, V. (1994) 'International environmental regulation and control', in S. Witt and L. Moutinho (eds) *Tourism Marketing and Management Handbook*, Hemel Hempstead: Prentice-Hall.

Hayward, T. (1994) *Ecological Thought: An Introduction*, Oxford: Polity Press.

Heeks, R. (1999) *The Tyranny of Participation in Information Systems: Learning from Development Projects*, Development Informatics Working Paper 4, Institute for Development and Management Policy, University of Manchester.

Hemer, O. (2005) 'Aches to assets', in Sida (ed.) *Urban Assets: Cultural Heritage as a Tool for Development*, Stockholm: Sida.

Henkel, H. and Stirrat, R. (2001) 'Participation as spiritual duty: empowerment as secular subjection', in B. Cooke and U. Kothari (eds) *Participation: The New Tyranny*, London: Zed Books.

Herman, E. (1992) *Beyond Hypocrisy: Decoding the News in an Age of Propaganda*, Boston, MA: South End Press.

Hettne, B. (1995) *Development Theory and the Three Worlds: Towards an International Political Economy of Development*, Harlow, UK: Longman.

Hettne, B. (2002) 'Current trends and future options in development studies', in V. Desai and R. Potter (eds) *The Companion to Development Studies*, London: Arnold.

Hickman, L. (2007) 'Sun, sand and slavery', *Guardian* 22 March.

Higinio, E. and Munt, I. (1993) 'Eco-tourism gone awry', *NACLA Report on the Americas* 46 (4): 8–10.

Hills, T. and Lundgren, T. (1977) 'The impact of tourism in the Caribbean', *Annals of Tourism Research* 4: 248–57.

Hines, C. (2000) *Localization: A Global Manifesto*, London: Earthscan.

Hirsch, F. (1976) *The Social Limits to Growth*, Cambridge, MA: Harvard University Press.

Hitchcock, M., King, V. and Parnwell, M. (1993) *Tourism in South-East Asia*, London: Routledge.

Holder, J. (1990) 'The Caribbean: far greater dependency on tourism likely', *Courier* 122 July–August: 74–9.

Holmberg, J., Thomson, K. and Timberlake, L. (eds) (1993) *Facing the Future: Beyond the Earth Summit*, London: International Institute of Environment and Development and Earthscan.

Honduran Institute of Tourism (1999) 'The Trujillo Project', email communication, Tegucigalpa.

Hong, E. (1985) *See the Third World While it Lasts: The Social and Environmental Impact of Tourism with Special Reference to Malaysia*. Penang: Consumers' Association of Penang.

Hughes, L. (2007) 'Land rights: the unfinished business of colonialism', *The Independent* 6 March.

Hunning, S. and Novy, J. (2006) *Tourism as an Engine of Neighbourhood Regeneration? Some Remarks towards a Better Understanding of Urban Tourism beyond the 'Beaten Path'*, Center for Metropolitan Studies Working Paper 006-2006, Technical University of Berlin.

Hunter, C. and Shaw, J. (2006) 'Applying the ecological footprint to ecotourism scenarios', *Environmental Conservation* 32 (4): 1–11.

Hutnyk, J. (1996) *The Rumour of Calcutta: Tourism, Charity and the Poverty of Representation*, London: Zed Books.

Huyssen, A. (1984) 'Mapping the postmodern', *New German Critique* 33: 5–52.

Imam, A. (1994) 'SAP is really sapping us', *New Internationalist* 257: 12–13.

Independent People's Tribunal on the World Bank Group in India website, www.worldbank tribunal.org (accessed 5 February 2008).

Inglehart, R. (1977) *The Silent Revolution: Changing Values and Political Styles among Western Publics*, Princeton, NJ: Princeton University Press.

Inglehart, R. (1981) 'Post-materialism in an environment of insecurity', *American Political Science Review* 75: 880–900.

Inskeep, E. (1987) 'Environmental planning for tourism', *Annals of Tourism Research* 14: 118–35.

Intergovernmental Panel on Climate Change (IPCC) (2007) *Climate Change 2007: The Physical Science Basis*, Geneva: IPCC

International Ecotourism Society (2002) www.ecotourism.org/iye/.

International Hotels Environment Initiative (IHEI) (1991) *Charter for Environmental Action in the International Hotel and Catering Industry*, London: IHEI.

International Institute for Environment and Development (IIED) (1994) *Whose Eden? An Overview of Community Approaches to Wildlife Management*, London: IIED.

International Labour Organisation (1998) *Forced Labour in Myanmar (Burma)*, Report of the Commission of Enquiry, Geneva: International Labour Organisation.

International Tourism Partnership (2007) www.tourismpartnership.org (accessed 10 May 2007).

International Union for the Conservation of Nature and Natural Resources (IUCN) (1985) *Guidelines for Protected Area Management Categories*, Gland, Switzerland: IUCN.

International Union for the Conservation of Nature and Natural Resources (1989) *Recursos: Annual Report IUCN Programme for Central America*, San José: IUCN.

International Union for the Conservation of Nature and Natural Resources (1994) *Guidelines for Protected Area Management Categories*, Cambridge: IUCN.

Isla, A. (2001) 'Land management and ecotourism: a flawed approach to conservation in Costa Rica', paper presented at the Natural Capital, Poverty and Development Conference, University of Toronto, September.

Iyer, P. (1989) *Video Night in Kathmandu and Other Reports from the Not-So-Far-East*, London: Black Swan.

Jackson, P. (1992) *Maps of Meaning*, London: Routledge.

Jacobs, M. and Stott, M. (1992) 'Sustainable development and the local economy', *Local Economy* 7, 3.

Jacques, M. (2005) 'The end of the world as we know it?', *Guardian Review* 23 July.

Jager, M. (1986) 'Class definition and the esthetics of gentrification: Victoriana in Melbourne', in N. Smith and P. Williams (eds) *Gentrification of the City*, London: Allen & Unwin.

James, C. (2001) 'Tourism and remittances are essential', *Financial Times* 19 March: V.

Jameson, F. (1984) 'Postmodernism, or the cultural logic of late capitalism', *New Left Review* 146: 53–92.

Jameson, F. (1991) *Postmodernism, or the Cultural Logic of Late Capitalism*, London: Verso.

Jamieson, W. (2002) *Workshop Report: Regional Workshop – Urban Tourism and Poverty Reduction*, Colombo, Sri Lanka: ESCAP/CITYNET, 20–22 November.

Jardine, C. (1994) 'Beware: rough road ahead', *Daily Telegraph* 2 November: A.

Jaworski, A and Pritchard, A. (eds) (2005) *Discourse, Communication and Tourism*, Clevedon, UK: Channel View.

Johnston, B. (1990) 'Introduction: breaking out of the tourist trap', *Cultural Survival Quarterly* 14 (1): 2–5.

Jonas, S. (1991) *The Battle for Guatemala: Rebels, Death Squads, and US Power*, Boulder, CO: Westview Press.

Jones, A. (2004) *Review of Gap Year Provision*, Research Report 555, University of London.

Josephides, N. (1994) 'Tour operators and the myth of self-regulation, *In Focus* 14: 10–11.

Kalisch, A. (2001) *Tourism as Fair Trade; NGO Perspectives*, London: Tourism Concern.

Kay, C. (1989) *Latin America Theories of Development and Underdevelopment*, London: Routledge.

Kaye, M. (1994) 'Costa Rica at the crossroads: mass development or nature based tourism?', in 'A call for an international dialogue on the future of Costa Rica tourism', San José: mailing list on Internet.

Kincaid, J. (1988) *A Small Place*, London: Virago.

King, A. (1995) 'Migrations, globalisation and place', in D. Massey and P. Jess (eds) *A Place in the World?*, Milton Keynes: Open University.

Kinnock, G. (2000) 'Burma deserves a boycott', *Guardian* 28 June: 8.

Klak, T. (2002) 'World-systems theory: centres, peripheries and semi-peripheries', in V. Desai and R. Potter (eds) *The Companion to Development Studies*, London: Arnold.

Klatchko, J. (1991) 'Sherpa trek', *New Internationalist* 222: 27.

Klein, N. (2001) *No Logo*, London: Flamingo.

Kohl, J. (1996) 'Recruiting a PRA team for Ecuador', email communication, 21 March.

Koonings, K. and Kruijt, D. (eds) (2006) *Fractured Cities: Social Exclusion, Urban Violence and Contested Spaces in Latin America*, London: Zed Books.

Koptiuch, K. (1991) 'Third-Worlding at home transforming new frontiers in the urban U.S.', *Cultural Anthropology* 37 (5): 737–62.

Korinek, A., Mistiaen, J. and Ravallion, M. (2006) 'Survey nonresponse and the distribution of income', *Journal of Economic Inequality* 4 (2): 33–55.

Korten, D. (1996) 'Development is a sham', *New Internationalist* 278: 12–13.

Kothari, U. (2001) 'Power, knowledge and social control in participatory development', in B. Cooke and U. Kothari (eds) *Participation: The New Tyranny*, London: Zed Books.

Krippendorf, J. (1987) *The Holidaymakers: Understanding the Impact of Leisure and Travel*, London: Heinemann.

Kuhn, T. (1962) *The Structure of Scientific Revolutions*, Chicago, IL: University of Chicago Press.

Kutay, K. (1989) 'The new ethic in adventure travel', *Buzzworm: The Environmental Journal* 1 (4): 31–6.

Kvalöy, S. (1993) 'A tale of two countries', *Resurgence* 159: 14–17.

Landry, C. (2000) *The Creative City: A Toolkit for Urban Innovators*, London: Comedia and Earthscan.

Landry, C. (2006) *The Art of City Making*, London: Earthscan.

Larrain, J. (1989) *Theories of Development: Capitalism, Colonialism and Dependency*, Oxford: Polity Press.

Lascelles, D. (1992) 'Conflict and dilemmas', in J. Quarrie (ed.) *Earth Summit '92*, London: Regency Press.

Lash, S. (1991) *Sociology of Postmodernism*, London: Routledge.

Lash, S. and Urry, J. (1987) *The End of Organised Capitalism*, Cambridge: Polity Press.

Lash, S. and Urry, J. (1994) *Economies of Signs and Space*, London: Sage.

Lavery, P. (1971) *Recreational Geography*, London: David & Charles.

Lea, J. (1988) *Tourism and Development in the Third World*, London: Routledge.

Lea, J. (1993) 'Tourism development ethics in the Third World', *Annals of Tourism Research* 20: 701–15.

Lee, N. and George, C. (eds) (2000) *Environmental Assessment in Developing and Transitional Countries*, Chichester: Wiley.

Leiss, W. (1983) 'The icons of the marketplace', *Theory, Culture and Society* 1 (3): 10–21.

Lewis, D. (1990) 'Conflict of interest', *Geographical Magazine* December: 18–22.

Lindberg, K. and Hawkins, D. (eds) (1993) *Ecotourism: A Guide for Planners and Managers*, North Bennington, VT: Ecotourism Society.

Lindberg, K. and Huber, R. (1993) 'Economic issues in ecotourism management', in K. Lindberg and D. Hawkins (eds) *Ecotourism: A Guide for Planners and Managers*, North Bennington, VT: Ecotourism Society.

Linden, I. (1993) 'Free lunches and free markets', *CIIR News* December: 3.

Lipietz, A. (1995) *Green Hopes: The Future of Political Ecology*, Oxford: Polity Press.

Lipman, G. (1992) 'The role of travel and tourism industry: promoting sustainable ecotourism', paper presented at the Royal Geographical Society Conference, London.

Lohmann, L. (1991) 'Who defends biological diversity?', in V. Shiva, P. Anderson, H. Schuching, A. Gray, L. Lohmann and D. Cooper (eds) *Biodiversity: Social and Ecological Perspectives*, Penang: World Rainforest Movement.

McAfee, K. (1991) *Storm Signals: Structural Adjustment and Development Alternatives in the Caribbean*, London: Zed Books.

MacCannell, D. (1973) 'Staged authenticity: arrangements of social space in tourist settings', *American Journal of Sociology* 79 (3): 589–603.

MacCannell, D. (1976) *The Tourist: A New Theory of the Leisure Class*, New York: Sulouken.

MacCannell, D. (1992) *Empty Meeting Grounds: The Tourist Papers*, London: Routledge.

McCellan, D. (1986) *Ideology*, Milton Keynes: Open University Press.

McClarence, S. (1995) 'A discrete charm', *Independent on Sunday* 3 September: 55.

McClintock, A. (1994) 'Soft-soaping empire: commodity racism and imperial advertising', in G. Robertson, M. Mash, L.Tickner, J. Bird, B. Curtis and T. Putnam (eds) *Travellers' Tales*, London: Routledge.

MacEwan, A. (2001) 'The neoliberal disorder: the inconsistencies of trade policy', *NACLA Report on the Americas* 35, 3: 27–33. New York: NACLA.

McGrew, A. (1992) 'The state in advanced capitalist countries', in J. Allen, P. Braham and P. Lewis (eds) *Political and Economic Forms of Modernity*, Oxford: Polity Press.

McGrew, A. (1995) 'World order and political space', in J. Anderson, C. Brook and A. Cochrane (eds) *A Global World? Re-ordering Political Space*, Milton Keynes: Open University Press.

Machlis, G. and Tichnell, D. (1985) *The State of the World's Parks: An International Assessment for Resource Management, Policy and Research*, Boulder, CO: Westview Press.

McIvor, C. (1994) *Management of Wildlife, Tourism and Local Communities in Zimbabwe*, Discussion Paper, Geneva: United Nations Research Institute for Social Development.

McIvor, C. (1995) 'Clash on environment', *New African* December: 35.

McKercher, B. (1993) 'The unrecognized threat to tourism: can tourism survive sustainability?', *Tourism Management* 14 (2): 131–6.

McLean, G. (2006) 'Where we're headed', *Guardian Weekend* 1 April: 38–43.

McNeil, J. (1999) *Costa Rica: The Rough Guide*, London: Rough Guides.

Madeley, J. (1996) *Foreign Exploits: Transnationals and Tourism*, Briefing Paper, London: Catholic Institute for International Relations.

Mader, R. (1996) 'Honduras notes', email communication, 8 October.

Mahony, K. and Van Zyl, J. (2001) *Practical Strategies for Pro-poor Tourism. Case Studies of Makuleke and Manyeleti Tourism Initiatives*, Working Paper 2, South Africa: Pro-Poor Tourism.

Mahr, J. and Sutcliffe, S. (1996) 'Come to Burma', *New Internationalist* 280: 28–30.

Maldonado, T., Hurtado De Mendoza, L. and Saborio, O. (1992) *Análisis de Capacidad de Carga para Visitación en las Áreas Silvestres de Costa Rica*, San José: Fundación Neotrópica.

Mann, M. (2000) *The Community Tourism Handbook: Exciting Holidays for Responsible Travellers*, London: Earthscan.

Marcus, B. and Taber, M. (eds) (1983) *Maurice Bishop Speaks: The Grenada Revolution 1979–1983*, New York: Pathfinder Press.

Marshall, J. (1994) 'Papagayo isn't "eco-tourism"', *Tico Times* 8 April: 2 and 36.

Marshall, O. (2004) 'Introducing South America', *Rough News* 23, summer.

Marx, K. (1965) *Capital*, London: Lawrence & Wishart.

Mason, P. and Mowforth, M. (1995) *Codes of Conduct in Tourism*, Occasional Paper 1, University of Plymouth.

Mason, P. and Mowforth, M. (1996) 'Codes of conduct in tourism', *Progress in Tourism and Hospitality Research*, 2: 151–67.

Massey, D. (1991) 'A global sense of place', *Marxism Today* 24–9 June.

Massey, D. (1993) 'Questions of locality', *Geography* 78 (2): 142–9.

Massey, D. (1995a) 'The conceptualization of place', in D. Massey and P. Jess (eds) *A Place in the World?*, Oxford: Oxford University Press.

Massey, D. (1995b) 'Imaging the world', in J. Allen and D. Massey (eds) *Geographical Worlds*, Milton Keynes: Open University Press.

Massey, D. (1995c) *Spatial Divisions of Labour*, 2nd edn, London: Macmillan.

Mathieson, A. and Wall, G. (1982) *Tourism: Economic, Physical and Social Impacts*, London: Longman.

Meadows, D. (1997) 'Putting a price tag on mother nature', *Valley News*, New Hampshire, USA, 24 May.

Mesoamerica (1996) 'Costa Rica', *Mesoamerica* 15 (4): 9.

Miller, D. (1995) *Acknowledging Consumption*, London: Routledge.

Miller, G. (2001) 'The development of indicators for sustainable tourism: results of a Delphi survey of tourism researchers', *Tourism Management* 22: 351–62.

Mills, H. (1998) 'Garífuna Denounce Land Sales to Foreigners', *Mesoamerica* 17 (7).

Mitchell, J. and Ashley, C. (2007) *Pathways to Prosperity: How Can Tourism Reduce Poverty?* Draft Report, London: Overseas Development Institute and World Bank.

Mitchell, J. and Faal, J. (2006) 'Package holiday tourism in The Gambia', *Development Southern Africa*, Special Edition on Tourism.

Mitlin, D. (2002) 'Sustaining markets or sustaining poverty reduction?', *Environment & Urbanisation* 14 (1): 173–7.

Mitlin, D. and Satterthwaite, D. (2007) 'Strategies for grassroots control of international aid', *Environment & Urbanisation* 19 (2): 483–500.

Monbiot, G. (1994) *No Man's Land*, London: Macmillan.

Monbiot, G. (1995) 'No man's land . . .', *In Focus* 15: 10.

Montero Mejía, A. (2001) 'La última etapa de le globalización', *Tiempos Del Mundo*, Comentarios, Costa Rica, 29 November.

Morrison, P. (1995) 'No more heroes', *Wanderlust* January: 68.

Morrison, P. (1996) 'Tales of the unexpected', *Wanderlust* February–March: 84.

Morrow, L. (1995) 'I came, I saw, I spoiled everything', *Time* 10 July.

Mortlock, E. (1988) *Postguide to Indonesia*, Ashbourne, UK: Moorland.

Moser, C. (2004) 'Urban violence and insecurity: an introductory roadmap', *Environment & Urbanisation* 16 (2): 3–16.

Mosse, D. (2001) 'People's knowledge, participation and patronage: operations and representations in rural development', in B. Cooke and U. Kothari (eds) *Participation: The New Tyranny*, London: Zed Books.

Mowforth, M. (1996) 'Co-operativo Longo Maï, Costa Rica', *Newsletter of the Environmental Network for Central America* 19: 6–7.

Mowforth, M. (1998) *Feasibility Study of SELVA's Ecotourism Project on the Cosigüina Peninsula of Nicaragua*, London: Methodist Relief and Development Fund.

Mowforth, M. (2001) *Storm Warnings: Hurricanes Georges and Mitch and the Lessons for Development*, London: Catholic Institute for International Relations

Mowforth, M. (2003) 'Tourism, terrorism and climate change', paper prepared for NATO Advanced Research Workshop on 'Climate Change and Tourism: Assessment and Coping Strategies', Warsaw, November.

Mowforth, M., Charlton, C. and Munt, I. (2008) *Tourism and Responsibility: Perspectives from Latin America and the Caribbean*, London: Routledge.

Mulberg, J. (1993) 'Economics and the impossibility of environmental evaluation', paper presented to conference on Values and the Environment, University of Surrey, Guildford.

Mumtaz, B. (2001) 'Slums are good for you', *Habitat Debate* 17 (3) September.

Munt, I. (1992) 'A great escape?', *Town and Country Planning* 61 (7–8): 212–14.

Munt, I. (1994a) 'Eco-tourism or ego-tourism?', *Race and Class* 36 (1): 49–60.

Munt, I. (1994b) 'The Other postmodern tourism: travel, culture and the new middle class', *Theory, Culture and Society* 11 (3): 101–24.

Munt, I. (1995) 'The travel virtuosos', *Contours* 7 (2): 29–34.

Murphree, M. (1996) *Wildlife in Sustainable Development: Approaches to Community Participation*, African Wildlife Policy Consultation, London: Overseas Development Administration.

Murphy, P. (1983) 'Perceptions and attitudes of decision-making groups in tourism centres', *Journal of Travel Research* 21 (3): 8–12.

Murphy, P. (1985) *Tourism: The Community Approach*, London: Routledge.

Mutal, S. (2005) 'Heritage and urban development – considerations on new approaches to heritage', in Sida (ed.) *Urban Assets: Cultural Heritage as a Tool for Development*, Stockholm: Sida.

Naipaul, V.S. (1962) *The Middle Passage*, London: Picador.

Nash, D. (1989) 'Tourism as a form of imperialism', in V. Smith (ed.) *Hosts and Guests: The Anthropology of Tourism*, Oxford: Blackwell.

Nederveen Pieterse, J. (2000) 'After post-development', *Third World Quarterly* 21 (2): 175–91.

Nederveen Pieterse, J. (2001) *Development Theory: Deconstructions/Reconstructions*, London: Sage.

Nepal, S.K. (2000) 'Tourism in protected areas: the Nepalese Himalaya', *Annals of Tourism Research* 27 (3): 661–81

Neuwirth, R. (2006) *Shadow Cities: A Billion Squatters, a New Urban World*, London: Routledge.

New Economics Foundation (2000) *Corporate Spin*, London: New Economics Foundation.

New Internationalist (1984) 'Visions of poverty, visions of wealth: tourism in the Third World', *New Internationalist* 142.

New Internationalist (1994) 'Simply . . . how Bretton Woods re-ordered the world', *New Internationalist* 257, July.

New Internationalist (2005) 'BINGOs! The big charity bonanza', *New Internationalist* 383, October.

New Internationalist (2006) 'Conned: carbon offsets stripped bare', *New Internationalist* 391, July.

New Internationalist (2007) 'Corporate responsibility unmasked', *New Internationalist* 407, December.

Nicaragua Network Hotline (2002) 'Former IMF Chief Economist bashes neoliberalism', Nicaragua Network, Managua, 8 July.

Noi, Chang (2001) 'Welcome to Thighlandia', *New Frontiers* May–June: 6.

Noronha, R. (1979) 'Paradise reviewed: tourism in Bali', in E. de Kadt (ed.) *Tourism: Passport to Development?*, Oxford: Oxford University Press.

Norris, S. (1994) 'Credit on a high road with no bank', *Independent* 23 July: 38.

Ocampo, J.A. (2002) *Report of the Economic Commission on Latin America and the Caribbean.*

Offe, C. (1985) 'New social movements: challenging the boundaries of institutional politics', *Political Science Review* 6 (4): 483–99.

Olerokonga, T. (1992) 'What about the Maasai?', *In Focus* 4: 6–7.

Olindo, P. (1991) 'The old man of nature: Kenya', in T. Whelan (ed.) *Nature Tourism*, Washington, DC: Island Press.

Open University (1995) *The Shape of the World: Study Guide 3*, Milton Keynes: Open University.

O'Reilly, C. (2005) 'Tourist or traveller? Narrating backpacker identity', in A. Jaworski and A. Pritchard (eds) *Discourse, Communication and Tourism*, Clevedon, UK: Channel View.

O'Riordan, T. (1978) 'Participation through objection: some thoughts on the UK experience', in B. Sadler (ed.) *Involvement and Environment: Proceedings of the Canadian Conference on Public Participation*, Edmonton, Alta: Environment Council of Alberta.

Otis, J. (1992) 'Belize counting the cost of pact with tourism developers', *The Nation* 7 September: 33.

Ó Tuathail, G. (1994) '(Dis)placing geopolitics: writing on the maps of global politics', *Environment and Planning D: Society and Space* 12: 525–46.

Overseas Private Investment Corporation (OPIC) (1995) *Program Handbook*, Washington, DC: OPIC.

Overseas Private Investment Corporation (2004) *Environmental Handbook*, Washington, DC: OPIC.

Oxfam (1995) *The Oxfam Poverty Report: A Summary*, Oxford: Oxfam.

Pacific Asia Travel Association (PATA) (2007) 'Total Tourism India: a strategic research report', email communication from PATA Strategic Intelligence Centre, Bangkok, 31 October.

Painter, J. (1989) *Guatemala: False Hope, False Freedom*, London: Latin America Bureau.

Painter, J. (2000) 'Pierre Bourdieu', in P. Crang and N. Thrift (eds) *Thinking Space*, London: Routledge.

Panos Institute (1995) *Ecotourism: Paradise Gained or Paradise Lost?*, Panos Media Briefing 14, London: Panos Institute.

Parnwell, M. (2002) 'Agropolitan and bottom-up development', in V. Desai and R. Potter (eds) *The Companion to Development Studies*, London: Arnold.

Patterson, K. (1992) 'Aloha for sale', *In Focus* 4.

Pattullo, P. (2005) *Last Resorts: The Cost of Tourism in the Caribbean*, 2nd edn, London: Latin America Bureau.

Payer, C. (1991) *Lent and Lost: Foreign Credit and Third World Development*, London: Zed Books.

Pearce, D. G. (1989) *Tourist Development*, London: Longman.

Pearce, D. G. (1995) *Tourism Today: A Geographical Analysis*, London: Longman.

Pearce, D. W. (1993) *Economic Values and the Natural World*, London: Earthscan.

Pearce, D. W. and Moran, D. (1994) *The Economic Value of Biodiversity*, London: Earthscan.

Pearce, F. (1990) 'Exchanging chances', *Guardian* 13 April: 10.

Pearce, J. (1982) *Under the Eagle: US Intervention in Central America and the Caribbean*, Boston, MA: South End Press.

Pearce, P. and Moscardo, G. (1986) 'The concept of authenticity in tourist experiences', *Australia and New Zealand Journal of Sociology* 22 (1): 121–32.

Pels, D. and Crebas, A. (1991) 'Carmen or the invention of a new feminine myth', in M. Featherstone, M. Hepworth and B. Turner (eds) *The Body: Social Process and Cultural Theory*, London: Sage.

People's United Party (PUP) (1989) *Belizeans First*, Belize: PUP.

Pérez, J. and Richmond, L. (2003) 'Ecotourism keeps coffee farmers afloat in Costa Rica', *Newsletter of the Environmental Network for Central America* 34: 9–10

Perez, L. (1974) 'Aspects of underdevelopment in the West Indies', *Science and Society* 37: 473–80.

Perez, L. (1975) 'Tourism in the West Indies', *Journal of Communications* 25: 136–43.

Phillips, T. (2005) "Blood, sweat and fears in favelas of Rio', *Guardian* 29 October.

Picard, M. (1991) 'Cultural tourism in Bali', in M. Hitchcock, V. King and M. Parnwell (eds) *Tourism in South East-Asia*, London: Routledge.

Pilger, J. (2007) 'The politics of hypocrisy', *Guardian* 27 October.

Pleumarom, A. (1990) 'Alternative tourism: a viable solution?', *Contours* 4 (8): 12–15.

Pleumarom, A. (1994) 'The political economy of tourism', *Ecologist* 24 (4): 142–7.

Pleumarom, A. (2000) *Do we need the International Year of Ecotourism?*, Bangkok: TIM-Team.

Pleumarom, A. (2001) *Campaign on Corporate Power in Tourism (COCPIT)*, Clearinghouse for Reviewing Ecotourism, Bangkok: TIM-Team.

Plog, S. (1972) 'Why destination areas rise and fall in popularity', *Cornell HRA Quarterly* November: 13–16.

Poon, A. (1989a) 'Competitive strategies for Caribbean tourism: the new versus the old', *Caribbean Affairs* 2 (2): 74–91.

Poon, A. (1989b) 'Competitive strategies for a "New Tourism"', in C. Cooper (ed.) *Progress in Tourism, Recreation and Hospitality Management*, vol. 1, London: Belhaven.

Poon, A. (1993) *Tourism, Technology and Competitive Strategies*, Wallingford, UK: CAB International.

Porritt, J. (1984) *Seeing Green*, Oxford: Blackwell.

Postman, N. (1985) *Amusing Ourselves to Death: Public Discourse in the Age of Show Business*, London: Methuen.

Potter, R. (2002) 'Theories, strategies and ideologies of development', in V. Desai and R. Potter (eds) *The Companion to Development Studies*, London: Arnold.

Power, S. (2007) *Gaps in Development: An Analysis of the UK International Volunteering Sector*, London: Tourism Concern.

Pratt, M. (1992) *Imperial Eyes: Travel Writing and Transculturation*, London: Routledge.

Pretty, J. (1995) 'The many interpretations of participation', *In Focus* 16: 4–5.

Pretty, J. and Hine, R. (1999) *Participatory Appraisal for Community Assessment: Principles and Methods*, Colchester, UK: Centre for Environment and Society, University of Essex.

Price, C. (1993) 'The irrelevance of discounted cash flow to environmental values', in *Proceedings of the Conference on Values and the Environment*, Guildford, UK: Faculty of Human Studies, University of Surrey.

Programme for Belize (1989) *Goal and Objectives of Programme*, Belize City: Programme for Belize.

Prosser, R. (1994) 'Societal change and the growth in alternative tourism', in E. Cater and G. Lowman (eds) *Ecotourism: A Sustainable Option?*, Chichester: Wiley.

Quintero, G.D. and Vega, A.N. (2006) *Manual of Procedures for Trainers in Sustainable Tourism*, Association of Caribbean States.

Rahnema, M. (1992) 'Participation', in W. Sachs (ed.) *The Development Dictionary: A Guide to Knowledge as Power*, London: Zed Books.

Rahnema, M. and Bawtree, V. (1997) *The Post-development Reader*, London: Zed Books.

Rao, N. (1991) 'Tourist as a pilgrim: a critique', Conference on Postmodernism and the Search for the Other, Delhi University.

Rather, D. (2007) 'Christmas, Sri Lanka and New Orleans', *Mexia Daily News*, www.mexia dailynews.com.

Reddy, T. (2006) 'Blinded by the light', *New Internationalist* 391 July.

Reid, D. (2003) *Tourism, Globalisation and Development: Responsible Tourism Planning*, London: Pluto Press.

Renard, Y., Darcheville, A. and Krishnarayan, V. (2001) *Practical Strategies for Pro-poor Tourism: A Case Study of the St. Lucia Heritage Tourism Programme*, Pro-Poor Tourism Working Paper 7.

Rice, X. (2008) 'Sun, sea and miles of empty beach: the paradise that faces disaster', *Guardian* 8 February.

Richards, G. and Wilson, J. (eds) (2004) *The Global Nomad: Backpacker Travel in Theory and Practice*, Clevedon, UK: Channel View.

Richardson, N. (1998) 'Blood and sand', (London) *Daily Telegraph* 28 February.

Richter, L. (1994) 'The political fragility of tourism in developing nations', *Contours* 6 (7–8): 32–8.

Richter, L. and Richter, W. (1985) 'Policy choices in South Asian tourism development', *Annals of Tourism Research* 12: 201–17.

Rist, G. (1997) *History of Development*, London: Zed Books.

Ritzer, G. (1993) *The McDonaldization of Society*, Newbury Park, CA: Pine Forge Press.

Robbins, N. (2002) 'Loot: in search of the East India Company, the world's first transnational corporation', *Environment & Urbanization* 14 (1): 79–88.

Robins, K. (1991) 'Tradition and translation: national culture in its global context', in J. Corner and S. Harvey (eds) *Enterprise and Heritage: Crosscurrents of National Culture*, London: Routledge.

Rodney, W. (1988) *How Europe Underdeveloped Africa*, London: Bogle L'Ouverture.

Rojas, E. (1999) *Old Cities, New Assets: Preserving Latin America's Urban Heritage*, Washington, DC: Inter-American Development Bank.

Rojas, E. (2002) *Urban Heritage Conservation in Latin America and the Caribbean: A Task for All Actors*, Washington, DC: Inter-American Development Bank.

Rojek, C. (1993) *Ways of Escape: Modern Transformations in Leisure and Travel*, London: Macmillan.

Rosenthal, R. (1991) 'Sustainable tourism: optimism v. pessimism', *In Focus* 1: 2–3.

Rosselson, R. (2001) 'Ethical tourism', *Ethical Consumer* 69: 28–9.

Rostow, W. (1960) *The Stages of Economic Growth: A Non-communist Manifesto*, Cambridge: Cambridge University Press.

Rughani, P. (1993) 'From tourist to target', *New Internationalist* 245: 7–12.

Ryel, R. and Grasse, T. (1991) 'Marketing ecotourism: attracting the elusive ecotourist', in T. Whelan (ed.) *Nature Tourism: Managing for the Environment*, Washington, DC: Island Press.

Sachs, J. (2005) *The End of Poverty: How We Can Make It Happen in our Lifetime*, London: Penguin.

Sachs, W. (1992a) 'Whose environment?', *New Internationalist* 232 June.

Sachs, W. (ed.) (1992b) *The Development Dictionary: A Guide to Knowledge as Power*, London: Zed Books.

Sachs, W. (ed.) (1993) *Global Ecology*, London: Zed Books.

Sachs, W. (1999) *Planet Dialectics: Explorations in Environment and Development*, London: Zed Books.

Said, E. (1991) *Orientalism*, London: Penguin.

Samuels, J. (ed.) (2005) *Removing Unfreedoms: Citizens as Agents of Change in Urban Development*, Rugby: ITDG Publishing.

San José Audubon Society (1992) 'An invitation to ecotourism', San José, Costa Rica.

Sansom, A. (2001) 'The gap widens', *Sunday Telegraph Review* 2 September: 17.

Sarre, P. (1995) 'Paradise lost, or the conquest of the wilderness', in P. Sarre and J. Blunden (eds) *An Overcrowded World? Population, Resources and the Environment*, Oxford: Oxford University Press.

Satterthwaite, D. (2001) 'Reducing urban poverty: constraints on the effectiveness of aid agencies and development banks and some suggestions for change', *Environment & Urbanization* 13 (1): 137–57.

Satterthwaite, D. (2005) 'Introduction: why local organizations are central to meeting the MDGs', in T. Bigg and D. Satterthwaite (eds) *How to Make Poverty History: The Central Role of Local Organizations in Meeting the MDGs*, London: International Institute for Environment and Development.

Savage, M., Barlow, J., Dickens, P. and Fielding, T. (1992) *Property, Bureaucracy and Culture: Middle-class Formation in Contemporary Britain*, London: Routledge.

Saville, N. (2001) *Practical Strategies for Pro-poor Tourism. Case Study of Pro-poor Tourism and SNV in Humla District, West Nepal*, Pro-Poor Tourism Working Paper 3.

Scarpaci, J. (2005) *Plazas and Barrios: Heritage Tourism and Globalization in the Latin American Centro Historico*, Tucson, AZ: University of Arizona Press.

Schumacher, E. (1974) *Small is Beautiful: A Study of Economics as if People Mattered*, London: Abacus.

Schuurman, J. (2002) 'The impasse in development studies', in V. Desai and R. Potter (eds) *The Companion to Development Studies*, London: Arnold.

Scott, G. (2001) *Ecotourism Discovers the Last Frontier*, Clearinghouse for Reviewing Ecotourism 8, Bangkok: TIM-Team.

Seabrook, J. (1995) 'Far horizons', *New Statesman and Society* 11 August: 22–3.

Selwyn, T. (1993) 'Peter Pan in South East Asia: the brochures', in M. Hitchcock, V. King and J. Parnwell (eds) *Tourism in South East Asia*, London: Routledge.

Selwyn, T. (1994) 'Tourism and myth', *In Focus* 14: 4–5.

Selwyn, T. (1995) *The Tourist Image: Myth and Myth Making in Tourism*, Chichester: Wiley.

Sen, A. (1999) *Development as Freedom*, Oxford: Oxford University Press.

Sharpley, R. and Telfer, P. (eds) (2002) *Tourism and Development: Concepts and Issues*, New York: Channel View.

Shaw, B. (2006) 'Urban heritage conservation and tourism development in Southeast Asia', in T. Wong, B. Shaw and K. Goh (eds) *Challenging Sustainability: Urban Development and Change in Southeast Asia*, Singapore: Marshall Cavendish.

Shaw, G. and Williams, A. (1994) *Critical Issues in Tourism: A Geographical Perspective*, Oxford: Blackwell.

Shepherd, J. (2007) 'The price of a year out', *Education Guardian* 9 January.

Sherman, P. and Dixon, J. (1991) 'The economics of nature tourism: determining if it pays', in T. Whelan (ed.) *Nature Tourism*, Washington, DC: Island Press.

Shiva, V. (1988) *Staying Alive: Women, Ecology and Development*, London: Zed Books.

Shiva, V. (1993) 'The greening of the global reach', in W. Sachs (ed.) *Global Ecology*, London: Zed Books.

Shiva, V. (1999) 'The round to the Citizens', *Guardian* 8 December: 8.

Shiva, V. (2005) 'Hacer que la pobreza sea historia, y la historia de la pobreza', Znet on www.zmag.org.

Shivji, I. (1973) *Tourism and Socialist Development*, Dar-as-Salaam: Tanzania Publishing.

Short, L. (1991) 'Crafts: bridge or melting pot?', *Orbit* 42: 6.

Shurmer-Smith, P. and Hannam, K. (1994) *Worlds of Desire, Realms of Power*, London: Arnold.

Sida (2004) *A Future for the Past: Historic Cities in Development*, Stockholm: Sida.

Sida (2005a) *Urban Assets: Cultural Heritage as a Tool for Development*, Stockholm: Sida.

Sida (2005b) *Caring for the Historic Environment*, Stockholm: Sida.

Sidaway, J. (2002) 'Post-development', in V. Desai and R. Potter (eds) *The Companion to Development Studies*, London: Arnold.

Sidaway, R. (1994) *The Limits of Acceptable Change*, report prepared for the Countryside Commission, Edinburgh.

Simon, J. (1997) *Endangered Mexico: An Environment on the Edge*, London: Latin America Bureau.

Simons, P. (1988) 'Belize at the crossroads', *New Scientist* 29 October: 61.

Simpson, M. (2007) 'The impacts of community benefit tourism on rural livelihoods and poverty reduction', PhD thesis, University of Oxford.

Simpson, M. (2008) 'Progress in tourism management: community benefit tourism initiatives – a conceptual oxymoron?', *Tourism Management* 29 (1).

Sloan, E. (1993) 'Eco-tourists threatening to trample tiny Belize to death', *The Nation* 28 January: 32.

Smith, A. (1961) *The Wealth of Nations*, London: Methuen.

Smith, N. (1984) *Uneven Development: Nature, Capital and the Production of Space*, New York: Blackwell.

Smith, N. (1987) 'Of yuppies and housing: gentrification, social restructuring, and the urban dream', *Environment and Planning D: Society and Space* 5: 151–72.

Smith, N. (2002) 'New globalism, new urbanism: gentrification as global urban strategy', *Antipode* 34 (3): 427–50.

Smith, V. (ed.) (1989) *Hosts and Guests: The Anthropology of Tourism*, Philadelphia, PA: University of Pennsylvania Press.

Smith, V. and Brent, M. (eds) (2001) *Hosts and Guests Revisited: Tourism Issues in the 21st century*, New York: Cognizant.

Sniffen, J. (1995) 'UNEP impact assessment meetings', email communication, 20 June.

Soares, M. (1992) *Debt Swaps, Development and Environment*, Rio de Janeiro: Brazilian Institute for Economic and Social Analysis.

Sontag, S. (1979) *On Photography*, London: Penguin.

Srisang, K. (1992) 'Third World tourism: the new colonialism', *In Focus* 4: 2–3.

Stancliffe, A. (1995) 'Agenda 21 and tourism: an introductory guide', paper available from Tourism Concern, London.

Starkey, R. and Welford, R. (eds) (2001) *The Earthscan Reader in Business and Sustainable Development*, London: Earthscan.

Stauth, G. and Turner, B. (1988) 'Nostalgia, postmodernism and the critique of mass culture', *Theory, Culture and Society* 5: 509–26.

Stewart, J. and Hams, T. (1991) *Local Government for Sustainability*, Luton: Local Government Management Board.

Stiglitz, J. E. (2002) *Globalization and its Discontents*, London: Penguin.

Stonich, S., Sorenson, J. and Hundt, A. (1995) 'Ethnicity, class and gender in tourism developments: the case of the Bay Islands, Honduras', *Journal of Sustainable Tourism* 3 (1): 1–28.

Survival International (1991) 'Tourism: special issue', *Survival* 28.

Survival International (1995) 'Tourism and tribal peoples', background sheet, London: Survival International.

Survival International (1996) 'Parks or people?', *Survival* 35.

Survival International (undated) *Survival: A Unique Organisation for Tribal Peoples*, booklet, London: Survival International.

Sutcliffe, S. (1995) *Burma: The Alternative Guide*, London: Burma Action Group.

Tannerfeldt, G. and Ljung, P. (2006) *More Urban Less Poor: An Introduction to Urban Development and Management*, London: Earthscan.

Taylor, H. (2001) 'Insights into participation from critical management and labour process perspectives', in B. Cooke and U. Kothari (eds) *Participation: The New Tyranny*, London: Zed Books.

Tearfund (2001) *Tourism: Putting Ethics into Practice*, Teddington, UK: Tearfund.

Tearfund (2002) *Worlds Apart: A Call to Responsible Global Tourism*, Teddington, UK: Tearfund.

Telfer, D. (2002) 'The evolution of tourism and development theory', in R. Sharpley and D. Telfer (eds) *Tourism and Development: Concepts and Issues*, Clevedon, UK: Channel View.

Teye, V. (1986) 'Liberation wars and tourism development in Africa: the case of Zambia', *Annals of Tourism Research* 13: 589–608.

Teye, V. (1988) 'Coups d'état and African tourism: a study of Ghana', *Annals of Tourism Research* 15: 329–56.

Theroux, P. (1979) *The Old Patagonian Express: By Train through the Americas*, London: Penguin.

Thomas, C. (1988) *The Poor and the Powerless: Economic Policy and Change in the Caribbean*, London: Latin America Bureau.

Thomas, N. (1994) *Colonialism's Culture: Anthropology, Travel and Government*, Oxford: Polity Press.

Thompson, C., O'Hare, G. and Evans, K. (1995) 'Tourism in The Gambia: problems and proposals', *Tourism Management* 16 (8): 571–81.

Thompson, J. (1991) 'Editor's introduction', in P. Bourdieu, *Language and Symbolic Power*, Cambridge: Polity Press.

Thorne, E.T. (2004) 'Land Rights and Garífuna Identity', *NACLA Report on the Americas* 38 (2), September–October, New York: North American Congress on Latin America

Thrupp, L. (1990) 'Environmental initiatives in Costa Rica: a political ecology perspective', *Society and Natural Resources* 3: 243–56.

Thurot, J. and Thurot, G. (1983) 'The ideology of class and tourism: confronting the discourse of advertising', *Annals of Tourism Research* 10: 173–89.

Tickell, O. (1992) 'After the Summit', *Green Line* July, 98: 3.

Tinker, J. (1992) 'Endpiece', United Nations Conference on Environment and Development: A User's Guide, special bulletin 1, United Nations Environment Programme–UK and International Institute for Environment and Development, London.

Tourism Concern (1992) *Beyond the Green Horizon: Principles for Sustainable Tourism*, Godalming, UK: World Wide Fund for Nature.

Tourism Concern (1994a) 'Greenwash contradictions', *In Focus* 14: 13.

Tourism Concern (1994b) 'Rainforest SOS', *In Focus* 13: 8.

Tourism Concern (1996) 'Our holidays, their homes', *Campaign Update*, spring, London: Tourism Concern.

Tourism Concern (2001) 'UN declares 2002 International Year of Ecotourism', *In Focus* 38: 16.

Tourism Concern (2002a) 'Trekking wrongs: porters' rights', *In Focus* 41.

Tourism Concern (2002b) 'Ecotourism evictions', *In Focus* 42.

Tourism Concern (2007) 'Gaps in development', *In Focus* May.

Tourism Concern (undated) 'Trekking in the Himalayas', information sheet, London: Tourism Concern.

Tourism Investigation and Monitoring Team (2000) 'Call for a fundamental reassessment of the UN International Year of Ecotourism 2002: Letter to Oliver Hillel, UNEP Tourism Programme Coordinator', TIM-Team, Bangkok, Thailand.

Tourism Investigation and Monitoring Team (2001) 'Mekong Region: tourism and indigenous peoples', Clearinghouse for Reviewing Ecotourism 12, TIM-Team, Bangkok, Thailand.

Tourism Investigation and Monitoring Team (2002) 'UN "International Year of Ecotourism 2002" in a deep muddle – scrap it!', Clearinghouse for Reviewing Ecotourism 19, TIM-Team, Bangkok, Thailand.

Traidcraft (2001a) 'Fair trade? Ethical trade? What is the difference?', www.traidcraft.co.uk.

Traidcraft (2001b) 'What is fair trade, and why is it important?', www.traidcraft.co.uk.

Travel Impact Newswire (2003) 'WTTC focus more on cure, less on prevention. Thailand', *Travel Impact Newswire* March.

Tremlett, G. (2006) 'Latin American migrants send home £27bn', *Guardian* 15 November.

Truck Africa (2007) www.truckafrica.com (accessed 20 November 2007).

Truong, D. (1991) *Sex, Money and Morality: Prostitution and Tourism in South East Asia*, London: Zed Books.

Turner, C. and Manning, P. (1988) 'Placing authenticity – on being a tourist: a reply to Pearce and Moscardo', *Australia and New Zealand Journal of Sociology* 24 (1): 136–9.

Turner, L. (1974) *Multinational Companies and the Third World*, London: Allen Lane.

Turner, L. (1976) 'The international division of leisure: tourism and the Third World', *World Development* 4 (3): 253–60.

Turner, L. and Ash, J. (1975) *The Golden Hordes: International Tourism and the Pleasure Periphery*, London: Constable.

United Nations Centre for Human Settlements (UNCHS–Habitat) (1996) *An Urbanizing World: Global Report on Human Settlements 1996*, Oxford: Oxford University Press.

United Nations Centre for Human Settlements (UNCHS–Habitat) (2001) *Cities in a Globalizing World: Global Report on Human Settlements 2001*, London: Earthscan.

United Nations Commission on Human Rights (1993) *Report on the Situation of Human Rights in Myanmar*, New York: UNCHR.

United Nations Conference on Environment and Development (1992) *Resolution 44/228*, New York: UN.

United Nations Department of Economic and Social Affairs (2006) *World Urbanization Prospects: The 2005 Revision*, Working Paper ESA/P/WP/200, UNDESA Population Division, New York.

United Nations Development Programme (1996) *Human Development Report*, Oxford: Oxford University Press.

United Nations Development Programme (2001) *Human Development Report*, New York and Oxford: UNDP and Oxford University Press.

United Nations Development Programme (2003) *Human Development Report*, Oxford: Oxford University Press.

United Nations Development Programme (2005) *Human Development Report 2005: International Cooperation at a Crossroads: Aid, Trade and Security in an Unequal World*, New York: UNDP.

United Nations Development Programme (2006) *UNDP Annual Report: Global Partnership for Development*, New York: UNDP.

United Nations Development Programme (2007) *UNDP Annual Report 2007: Making Globalisation Work for All*, New York: UNDP.

United Nations Economic Commission for Latin America and the Caribbean (2003) 'Foreign Investment in Latin America and the Caribbean', Report presented by ECLAC on 17 May 2004.

United Nations Economic and Social Commission for Asia and the Pacific (2006) *Economic and Social Survey of Asia and the Pacific 2006*, Bangkok: UNESCAP.

United Nations Environment and Development UK (1993) 'Convention update: Biodiversity Convention, 30 ratify', *Connections* autumn: 9.

United Nations Environment and Development UK (1994) 'Travel and tourism: the environment is our business', *Connections* autumn: 7.

United Nations Environment Programme (UNEP) (1998) *Ecolabels in the Tourism Industry*, New York: UNEP.

UNFPA (2007) *UNFPA State of World Population 2007: Unleashing the Potential of Urban Growth*, New York: United Nations Population Fund.

UN-HABITAT (2003) *The Challenge of Slums: Global Report on Human Settlements 2003*, London: Earthscan.

UN-HABITAT (2005) *Themes for the Twenty-first and other Future Sessions of the Governing Council of the United Nations Settlements Programme*, HSP/GC/20/13.

UN-HABITAT (2007) *Enhancing Urban Safety and Security: Global Report on Human Settlements 2007*, Nairobi: UN-HABITAT.

UN-HABITAT/ISDR (2005) *Gender and Disaster Management in Africa: Policy and Practice – Initial Draft*, Nairobi: UN-HABITAT and ISDR.

USAID (2005) *Land & Conflict: A Toolkit for Intervention*, Washington, DC: USAID Office of Conflict Management and Mitigation.

Urry, J. (1990) 'The consumption of "tourism"', *Sociology* 24: 23–35.

Urry, J. (2001) *The Tourist Gaze*, 2nd edn, London: Sage.

van den Abbeele, G. (1980) 'Sightseers: the tourist as theorist', *Diacritics* 10: 3–14.

Veblen, T. (1925) *The Theory of the Leisure Class*, London: Allen & Unwin.

Vick, K. (2002) 'Few tourists now go to see Rwanda's apes', *Washington Post* 7 April.

Vidal, J. (2001) 'A great white lie', *Guardian Review* 1 December.

Vidal, J. (2004) 'Fairtrade sales hit £100 m. a year', *Guardian* 28 February.

Vidal, J. (2007) 'Saving St Lucia: UK supermarket sweeps up 100 m. bananas', *Guardian* 26 February.

Vidal, J. (2008) 'The great green land grab', *Guardian G2* 13 February.

Vivanco, L. (2001) 'The International Year of Ecotourism in an age of uncertainty', paper presented at the University of Vermont Environmental Symposium, Vermont, USA, October.

Wall, G. (1997) 'Sustainable tourism – unsustainable development', in S. Wahab and J. Pilgrim (eds) *Tourism Development and Growth: The Challenge of Sustainability*, London: Routledge.

Wallerstein, I. (1979) *The Capitalist World – Economy*, Cambridge: Cambridge University Press.

Wallerstein, I. (2000) 'Globalization or the age of transition? A long-term view of the trajectory of the world system', *International Sociology* 15: 249–65.

Walter, J. (1982) 'Social limits to tourism', *Leisure Studies* 1: 295–304.

Warde, M. (1992) 'Where thieves often prosper', *Guardian Weekend* 14 November: 51–2.

Wates, N. (2000) *The Community Planning Handbook*, London: Earthscan.

Watson, G. and Kopachevsky, J. (1996) 'Tourist carrying capacity: a critical look at the discursive dimension', in C. Cooper and A. Lockwood (eds) *Progress in Tourism and Hospitality Research* vol. 2, Chichester: Wiley.

Way Ahead, The (1996) 'Forum for the short term?', *The Way Ahead* 26: 4–5.

Weber, W. (1993) 'Private conservation and ecotourism in Africa', in C. Potter, J. Cohen and D. Janczewski (eds) *Perspectives on Biodiversity: Case Studies of Genetic Resource*, Washington, DC: American Association for the Advancement of Science.

Weinberg, B. (1991) *War on the Land: Ecology and Politics in Central America*, London: Zed Books.

Welk, P. (2004) 'The beaten track: anti-tourism as an element of backpacker identity construction', in G. Richards and J. Wilson (eds) *The Global Nomad: Backpacker Travel in Theory and Practice*, Clevedon, UK: Channel View.

Wells, M. and Brandon, K. (1992) *People and Parks: Linking Protected Area Management with Local Communities*, Washington, DC: The World Bank, World Wildlife Fund and US Agency for International Development.

West, P. and Brechin, S. (eds) (1991) *Resident Peoples and National Parks*, Tucson, AZ: University of Arizona Press.

Wheat, S. (1994) 'Tourism Concern interview', *In Focus* 14: 8–9.

Wheeler, T. (2000) 'Lonely Planet Myanmar (Burma) Guide', *Planet Talk*, correspondence, June, London: Lonely Planet Publications.

Wheeler, T. (2001) 'Burma/Myanmar: the great divide', *Planet Talk: 2*, London: Lonely Planet Publications.

Wheeler, T. and Lyon, J. (1994) *Guide to Bali and Lombok*, Hawthorn, Victoria: Lonely Planet Publications.

Wheeller, B. (1992) 'Alternative tourism: a deceptive ploy', in C. Cooper and A. Lockwood (eds) *Progress in Tourism, Recreation and Hospitality Management*, London: Belhaven.

Wheeller, B. (1993a) 'Sustaining the ego', *Journal of Sustainable Tourism* 1 (2): 129–39.

Wheeller, B. (1993b) 'Willing victims of the ego trap', *In Focus* 9: 14.

Wheeller, B. (1996) 'In whose interest?', *In Focus* 19: 14–15.

Whelan, T. (ed.) (1991) *Nature Tourism: Managing for the Environment*, Washington, DC: Island Press.

Whitehorn, K. (1963) Quoted in TIPA (Truth and Integrity in Public Affairs), occasional paper, 7: 9.

Wight, P. (1994) 'Environmentally responsible marketing of tourism', in E. Cater and G. Lowman (eds) *Ecotourism: A Sustainable Option?*, Chichester: Wiley.

Williams, R. (1988) *Keywords: A Vocabulary of Culture and Society*, London: Fontana.

Wilson, J. and Richards, G. (2004) 'Backpacker icons: influential literary "nomads" in the formation of backpacker identities', in G. Richards and J. Wilson (eds) *The Global Nomad: Backpacker Travel in Theory and Practice*, Clevedon, UK: Channel View.

Winton, A. (2004) 'Urban violence: a guide to the literature', *Environment & Urbanisation* 16 (2): 165–84.

Witt, S. and Moutinho, L. (eds) (1994) *Tourism Marketing and Management Handbook*, Hemel Hempstead: Prentice-Hall.

Wolff, I. (1991) 'Weekending in El Salvador', *Independent on Sunday* 27 October: 59–61.

Wood, K. (1991) 'Belize cleans up in eco-tourism stakes', *Observer* 6 October: 65.

Wood, K. and House, S. (1991) *The Good Tourist*, London: Mandarin.

Wooding, B. and Moseley-Williams, R. (2004) 'Worlds apart', *Interact*, spring, London: Catholic Institute for International Relations.

World Bank (1996) *The World Bank Participation Sourcebook: Environmentally Sustainable Development*, Washington, DC: World Bank.

World Bank (2000a) *Urban and Local Government Strategy*, Washington, DC: World Bank.

World Bank (2000b) *World Development Report 2000/2001: Attacking Poverty*, Washington, DC: World Bank.

World Bank (2003) *A Resource Guide for Municipalities: Community Based Crime and Violence Prevention in Urban Latin America*, Washington, DC: World Bank.

World Bank (2006) *Global Economic Prospects 2006: Economic Implications of Remittances and Migration*, Washington, DC: World Bank.

World Commission on Environment and Development (1987) *Our Common Future*, 'Brundtland Report', New York: United Nations.

World Council of Churches (2002) 'Statement of indigenous peoples interfaith dialogue on globalisation and tourism', Chiang Rai, Thailand, 14–18 January.

World Development Movement (WDM) (2000) 'New resignation exposes World Bank dogma on globalisation', *WDM In Action*, autumn, London: WDM.

World Development Movement (2002) *GATS: A Disservice to the Poor*, London: WDM.

World Development Movement (2008) 'World Bank indicted', *WDM In Action*, winter 2007–2008, London: WDM.

World Tourism Organisation (1990) *Regional Economic Statistics*, Madrid: World Tourism Organisation.

World Tourism Organisation (1991) *What is WTO?*, Madrid: World Tourism Organisation.

World Tourism Organisation (1994) *National and Regional Tourism Planning: Methodologies and Case Studies*, London: Routledge.

World Tourism Organisation (1995) *Global Tourism Forecasts in the Year 2000 and Beyond*, vol. 1, Madrid: World Tourism Organisation.

World Tourism Organisation (1999) *Tourism: 2020 Vision*, Executive Summary, Madrid: World Tourism Organisation.

World Tourism Organisation (2000) *Compendium of Tourism Statistics*, Madrid: World Tourism Organisation.

World Tourism Organisation (2001) *Tourism Market Trends: Americas*, Madrid: World Tourism Organisation.

World Tourism Organisation (2002a) *Tourism and Poverty Alleviation*, Madrid: World Tourism Organisation.

World Tourism Organisation (2002b) *Enhancing the Economic Benefits of Tourism for Local Communities and Poverty Alleviation*, Madrid: World Tourism Organisation.

World Tourism Organisation (2004) *Tourism and Poverty Alleviation: Recommendations for Action*, Madrid: UNWTO.

World Tourism Organisation (2005a) *Tourism, Micro-finance and Poverty Alleviation: Recommendations for Action*, Madrid: UNWTO.

World Tourism Organisation (2005b) *Cultural Tourism and Poverty Alleviation – The Asia-Pacific Perspective*, Madrid: UNWTO.

World Tourism Organisation (2006a) *Poverty Alleviation through Tourism: A Compilation of Good Practices*, Madrid: UNWTO.

World Tourism Organisation (2006b) *Tourism Market Trends 2005 Edition: World Overview and Tourism Topics*, Madrid: UNWTO.

World Tourism Organisation (2007) *Compendium of Tourism Statistics: Data 2001–2005*, Madrid: UNWTO.

World Travel and Tourism Council (WTTC) (1991) *Travel and Tourism in the World Economy*, Brussels: WTTC.

World Travel and Tourism Council (1995) *Agenda 21 for the Travel and Tourism Industry: Towards Environmentally Sustainable Development*, London: WTTC.

World Travel and Tourism Council (undated) 'Background Information Sheet', London: WTTC.

World Travel and Tourism Council website, www.wttc.org (accessed 30 November 2007).

Worth, J. (2007) 'Companies who care?', *New Internationalist* 407 (December).

Wright, E. (1985) *Classes*, London: Verso.

Wright, M. (1996) 'Have mouse, will travel', *Independent on Sunday* 10 March: 56.

WWF–UK (1991) 'Reserved for the future', *WWF–UK Review 1991*, Godalming, UK: WWF–UK.

WWF–UK (2000a) *Tourism Certification: An Analysis of Green Globe 21 and Other Certification Programmes*, Godalming, UK: WWF–UK.

WWF–UK (2000b) 'Tourism certification schemes still leave much to be desired', http://www.wwf-uk.org/news.

WWF–UK (2002) *Holidays Abroad Needn't Cost the Earth*, Godalming, UK: WWF–UK.

Yearley, S. (1995) 'The transnational politics of the environment', in J. Anderson, C. Brook and A. Cochrane (eds) *A Global World? Reordering Political Space*, Milton Keynes: Open University Press.

Young, H. (1994) 'Community-based tourism development in Belize: government policies and plans in support of community initiatives', in *Guide to Community-based Ecotourism in Belize*, Belize: Ministry of Tourism and Environment and Belize Enterprise for Sustained Technology.

Zaba, B. and Scoones, I. (1994) 'Is carrying capacity a useful concept to apply to human populations?', in B. Zaba and J. Clarke (eds) *Environment and Population Change*, Liege: Ordina Editions.

Zevenbergen, J. and Van Der Molen, P. (2004) *Legal Aspects Of Land Administration In Post Conflict Areas*, Symposium on Land Administration in Post Conflict Areas, Geneva, 29–30 April.

Zukin, S. (1982) *Loft Living: Culture and Capital in Urban Change*, Baltimore, MD: Johns Hopkins University Press.

Zukin, S. (1987) 'Gentrification: culture and capital in the urban core', *Annual Review of Sociology* 3: 129–47.

Index

Printed in the USA/Agawam, MA
April 6, 2010

540610.082